I0484224

DARWIN AND MODERN SCIENCE

By A.C. Seward and Others

"My success as a man of science, whatever this may have amounted to, has been determined, as far as I can judge, by complex and diversified mental qualities and conditions. Of these, the most important have been—the love of science—unbounded patience in long reflecting over any subject—industry in observing and collecting facts—and a fair share of invention as well as of common sense. With such moderate abilities as I possess, it is truly surprising that I should have influenced to a considerable extent the belief of scientific men on some important points."

Autobiography (1881); "The Life and Letters of Charles Darwin", Vol. 1. page 107.

PREFACE

At the suggestion of the Cambridge Philosophical Society, the Syndics of the University Press decided in March, 1908, to arrange for the publication of a series of Essays in commemoration of the Centenary of the birth of Charles Darwin and of the Fiftieth anniversary of the publication of "The Origin of Species". The preliminary arrangements were made by a committee consisting of the following representatives of the Council of the Philosophical Society and of the Press Syndicate: Dr H.K. Anderson, Prof. Bateson, Mr Francis Darwin, Dr Hobson, Dr Marr, Prof. Sedgwick, Mr David Sharp, Mr Shipley, Prof. Sorley, Prof. Seward. In the course of the preparation of the volume, the original scheme and list of authors have been modified: a few of those invited to contribute essays were, for various reasons, unable to do so, and some alterations have been made in the titles of articles. For the selection of authors and for the choice of subjects, the committee are mainly responsible, but for such share of the work in the preparation of the volume as usually falls to the lot of an editor I accept full responsibility.

Authors were asked to address themselves primarily to the educated layman rather than to the expert. It was hoped that the publication of the essays would serve the double purpose of illustrating the far-reaching influence of Darwin's work on the progress of knowledge and the present attitude of original investigators and thinkers towards the views embodied in Darwin's works.

In regard to the interpretation of a passage in "The Origin of Species" quoted by Hugo de Vries, it seemed advisable to add an editorial footnote; but, with this exception, I have not felt it necessary to record any opinion on views stated in the essays.

In reading the essays in proof I have availed myself freely of the willing assistance of several Cambridge friends, among whom I wish more especially to thank Mr Francis Darwin for the active interest he has taken in the preparation of the volume. Mrs J.A. Thomson kindly undertook the translation of the essays by Prof. Weismann and Prof. Schwalbe; Mrs James Ward was good enough to assist me by translating Prof. Bougle's article on Sociology, and to Mr McCabe I am indebted for the translation of the essay by Prof. Haeckel. For the translation of the botanical articles by Prof. Goebel, Prof. Klebs and Prof. Strasburger, I am responsible; in the revision of the translation of Prof. Strasburger's essay Madame Errera of Brussels rendered valuable help. Mr Wright, the Secretary of the Press Syndicate, and Mr Waller, the Assistant Secretary, have cordially cooperated with me in my editorial work; nor can I omit to thank the readers of the University Press for keeping watchful eyes on my shortcomings in the correction of proofs.

The two portraits of Darwin are reproduced by permission of Messrs Maull and Fox and Messrs Elliott and Fry. The photogravure of the study at Down is reproduced from an etching by Mr Axel Haig, lent by Mr Francis Darwin; the coloured plate illustrating Prof. Weismann's essay was originally published by him in his "Vortrage uber Descendenztheorie" which afterwards appeared (1904) in English under the title "The Evolution Theory". Copies of this plate were supplied by Messrs Fischer of Jena.

The Syndics of the University Press have agreed, in the event of this volume being a financial success, to hand over the profits to a University fund for the endowment of biological research.

It is clearly impossible to express adequately in a single volume of Essays the influence of Darwin's contributions to knowledge on the subsequent progress of scientific inquiry. As Huxley said in 1885: "Whatever be the ultimate verdict of posterity upon this or that opinion which Mr Darwin has propounded; whatever adumbrations or anticipations of his doctrines may be found in the writings of his predecessors; the broad fact remains that, since the publication and by reason of the publication of "The Origin of Species" the fundamental conceptions and the aims of the students of living Nature have been completely changed... But the impulse thus given to scientific thought rapidly spread beyond the ordinarily recognised limits of Biology. Psychology, Ethics, Cosmology were stirred to their foundations, and 'The Origin of Species' proved itself to be the fixed point which the general doctrine needed in order to move the world."

In the contributions to this Memorial Volume, some of the authors have more especially concerned themselves with the results achieved by Darwin's own work, while others pass in review the progress of research on lines which, though unknown or but little followed in his day, are the direct outcome of his work.

The divergence of views among biologists in regard to the origin of species and as to the most promising directions in which to seek for truth is illustrated by the different opinions of contributors. Whether Darwin's views on the modus operandi of evolutionary forces receive further confirmation in the future, or whether they are materially modified, in no way affects the truth of the statement that, by employing his life "in adding a little to Natural Science," he revolutionised the world of thought. Darwin wrote in 1872 to Alfred Russel Wallace: "How grand is the onward rush of science: it is enough to console us for the many errors which we have committed, and for our efforts being overlaid and forgotten in the mass of new facts and new views which are daily turning up." In the onward rush, it is easy for students convinced of the correctness of their own views and equally convinced of the falsity of those of their fellow-workers to forget the lessons of Darwin's life. In his autobiographical sketch, he tells us, "I have steadily endeavoured to keep my mind free so as to give up any hypothesis, however much beloved...as soon as facts are shown to be opposed to it." Writing to Mr J. Scott, he says, "It is a golden rule, which I try to follow, to put every fact which is opposed to one's preconceived opinion in the strongest light. Absolute accuracy is the hardest merit to attain, and the highest merit. Any deviation is ruin."

He acted strictly in accordance with his determination expressed in a letter to Lyell in 1844, "I shall keep out of controversy, and just give my own facts." As was said of another son of Cambridge, Sir George Stokes, "He would no more have thought of disputing about priority, or the authorship of an idea, than of writing a report for a company promoter." Darwin's life affords a striking confirmation of the truth of Hazlitt's aphorism, "Where the pursuit of truth has been the habitual study of any man's life, the love of truth will be his ruling passion." Great as was the intellect of Darwin, his character, as Huxley wrote, was even nobler than his intellect.

A.C. SEWARD.

Botany School, Cambridge, March 20, 1909.

DATES OF THE PUBLICATION Of
CHARLES DARWIN'S BOOKS AND OF
THE PRINCIPAL EVENTS IN HIS
LIFE

1809:

Charles Darwin born at Shrewsbury, February 12.

1817:

"At 8 1/2 years old I went to Mr Case's school." (A day-school at Shrewsbury kept by the Rev G. Case, Minister of the Unitarian Chapel.)

1818:

"I was at school at Shrewsbury under a great scholar, Dr Butler; I learnt absolutely nothing, except by amusing myself by reading and experimenting in Chemistry."

1825:

"As I was doing no good at school, my father wisely took me away at a rather earlier age than usual, and sent me (Oct. 1825) to Edinburgh University with my brother, where I stayed for two years."

1828:

Began residence at Christ's College, Cambridge.

"I went to Cambridge early in the year 1828, and soon became acquainted with Professor Henslow...Nothing could be more simple, cordial and unpretending than the encouragement which he afforded to all young naturalists."

"During the three years which I spent at Cambridge my time was wasted, as far as the academical studies were concerned, as completely as at Edinburgh and at school."

"In order to pass the B.A. Examination, it was...necessary to get up Paley's 'Evidences of Christianity,' and his 'Moral Philosophy'... The careful study of these works, without attempting to learn any part by rote, was the only part of the academical course which...was of the least use to me in the education of my mind."

1831:

Passed the examination for the B.A. degree in January and kept the following terms.

"I gained a good place among the oi polloi or crowd of men who do not go in for honours."

"I am very busy,...and see a great deal of Henslow, whom I do not know whether I love or respect most."

Dec. 27. "Sailed from England on our circumnavigation," in H.M.S. "Beagle", a barque of 235 tons carrying 6 guns, under Capt. FitzRoy.

"There is indeed a tide in the affairs of men."

1836:

Oct. 4. "Reached Shrewsbury after absence of 5 years and 2 days."

"You cannot imagine how gloriously delightful my first visit was at home; it was worth the banishment."

Dec. 13. Went to live at Cambridge (Fitzwilliam Street).

"The only evil I found in Cambridge was its being too pleasant."

1837:

"On my return home (in the 'Beagle') in the autumn of 1836 I immediately began to prepare my journal for publication, and then saw how many facts indicated the common descent of species... In July (1837) I opened my first note-book for facts in relation to the Origin of Species, about which I had long reflected, and never ceased working for the next twenty years... Had been greatly struck from about the month of previous March on character of South American fossils, and species on Galapagos Archipelago. These facts (especially latter), origin of all my views."

"On March 7, 1837 I took lodgings in (36) Great Marlborough Street in London, and remained there for nearly two years, until I was married."

1838:

"In October, that is fifteen months after I had begun my systematic enquiry, I happened to read for amusement 'Malthus on Population,' and being well prepared to appreciate the struggle for existence which everywhere goes on from long-continued observation of the habits of animals and plants, it at once struck me that under these circumstances favourable variations would tend to be preserved, and unfavourable ones to be destroyed. The result of this would be the formation of new species. Here then I had at last got a theory by which to work; but I was so anxious to avoid prejudice, that I determined not for some time to write even the briefest sketch of it."

1839:

Married at Maer (Staffordshire) to his first cousin Emma Wedgwood, daughter of Josiah Wedgwood.

"I marvel at my good fortune that she, so infinitely my superior in every single moral quality, consented to be my wife. She has been my wise adviser and cheerful comforter throughout life, which without her would have been during a very long period a miserable one from ill-health. She has earned the love of every soul near her" (Autobiography).

Dec. 31. "Entered 12 Upper Gower street" (now 110 Gower street, London). "There never was so good a house for me, and I devoutly trust you (his future wife) will approve of it equally. The little garden is worth its weight in gold."

Published "Journal and Researches", being Vol. III. of the "Narrative of the Surveying Voyage of H.M.S. 'Adventure' and 'Beagle'"...

Publication of the "Zoology of the Voyage of H.M.S. 'Beagle'", Part II., "Mammalia", by G.R. Waterhouse, with a "Notice of their habits and ranges", by Charles Darwin.

1840:

Contributed Geological Introduction to Part I. ("Fossil Mammalia") of the "Zoology of the Voyage of H.M.S. 'Beagle'" by Richard Owen.

1842:

"In June 1842 I first allowed myself the satisfaction of writing a very brief abstract of my (species) theory in pencil in 35 pages; and this was enlarged during the summer of 1844 into one of 230 pages, which I had fairly copied out and still (1876) possess." (The first draft of "The Origin of Species", edited by Mr Francis Darwin, will be published this year (1909) by the Syndics of the Cambridge University Press.)

Sept. 14. Settled at the village of Down in Kent.

"I think I was never in a more perfectly quiet country."

Publication of "The Structure and Distribution of Coral Reefs"; being Part I. of the "Geology of the Voyage of the Beagle".

1844:

Publication of "Geological Observations on the Volcanic Islands visited during the Voyage of H.M.S. 'Beagle'"; being Part II. of the "Geology of the Voyage of the 'Beagle'".

"I think much more highly of my book on Volcanic Islands since Mr Judd, by far the best judge on the subject in England, has, as I hear, learnt much from it." (Autobiography, 1876.)

1845:

Publication of the "Journal of Researches" as a separate book.

1846:

Publication of "Geological Observations on South America"; being Part III. of the "Geology of the Voyage of the 'Beagle'".

1851:

Publication of a "Monograph of the Fossil Lepadidae" and of a "Monograph of the subclass Cirripedia".

"I fear the study of the Cirripedia will ever remain 'wholly unapplied,' and yet I feel that such study is better than castle-building."

1854:

Publication of Monographs of the Balanidae and Verrucidae.

"I worked steadily on this subject for...eight years, and ultimately published two thick volumes, describing all the known living species, and two thin quartos on the extinct species... My work was of considerable use to me, when I had to discuss in the "Origin of Species" the principles of a natural classification. Nevertheless, I doubt whether the work was worth the consumption of so much time."

"From September 1854 I devoted my whole time to arranging my huge pile of notes, to observing, and to experimenting in relation to the transmutation of species."

1856:

"Early in 1856 Lyell advised me to write out my views pretty fully, and I began at once to do so on a scale three or four times as extensive as that which was afterwards followed in my 'Origin of Species'."

1858:

Joint paper by Charles Darwin and Alfred Russel Wallace "On the Tendency of Species to form Varieties; and on the perpetuation of Varieties and Species by Natural Means of Selection," communicated to the Linnean Society by Sir Charles Lyell and Sir Joseph Hooker.

"I was at first very unwilling to consent (to the communication of his MS. to the Society) as I thought Mr Wallace might consider my doing so unjustifiable, for I did not then know how generous and noble was his disposition."

"July 20 to Aug. 12 at Sandown (Isle of Wight) began abstract of Species book."

1859:

Nov. 24. Publication of "The Origin of Species" (1250 copies).

"Oh, good heavens, the relief to my head and body to banish the whole subject from my mind!... But, alas, how frequent, how almost universal it is in an author to persuade himself of the truth of his own dogmas. My only hope is that I certainly see many difficulties of gigantic stature."

1860:

Publication of the second edition of the "Origin" (3000 copies).

Publication of a "Naturalist's Voyage".

1861:

Publication of the third edition of the "Origin" (2000 copies).

"I am going to write a little book... on Orchids, and to-day I hate them worse than everything."

1862:

Publication of the book "On the various contrivances by which Orchids are fertilised by Insects".

1865:

Read paper before the Linnean Society "On the Movements and Habits of Climbing plants". (Published as a book in 1875.)

1866:

Publication of the fourth edition of the "Origin" (1250 copies).

1868:

"I have sent the MS. of my big book, and horridly, disgustingly big it will be, to the printers."

Publication of the "Variation of Animals and Plants under Domestication".

"About my book, I will give you (Sir Joseph Hooker) a bit of advice. Skip the whole of Vol. I, except the last chapter, (and that need only be skimmed), and skip largely in the 2nd volume; and then you will say it is a very good book."

"Towards the end of the work I give my well-abused hypothesis of Pangenesis. An unverified hypothesis is of little or no value; but if anyone should hereafter be led to make observations by which some such hypothesis could be established, I shall have done good service, as an astonishing number of isolated facts can be thus connected together and rendered intelligible."

1869:

Publication of the fifth edition of the "Origin".

1871:

Publication of "The Descent of Man".

"Although in the 'Origin of Species' the derivation of any particular species is never discussed, yet I thought it best, in order that no honourable man should accuse me of concealing my views, to add that by the work 'light would be thrown on the origin of man and his history'."

1872:

Publication of the sixth edition of the "Origin".

Publication of "The Expression of the Emotions in Man and Animals".

1874:

Publication of the second edition of "The Descent of Man".

"The new edition of the "Descent" has turned out an awful job. It took me ten days merely to glance over letters and reviews with criticisms and new facts. It is a devil of a job."

Publication of the second edition of "The Structure and Distribution of Coral Reefs".

1875:

Publication of "Insectivorous Plants".

"I begin to think that every one who publishes a book is a fool."

Publication of the second edition of "Variation in Animals and Plants".

Publication of "The Movements and Habits of Climbing Plants" as a separate book.

1876:

Wrote Autobiographical Sketch ("Life and Letters", Vol. I., Chap II.).

Publication of "The Effects of Cross and Self fertilisation".

"I now (1881) believe, however,...that I ought to have insisted more strongly than I did on the many adaptations for self-fertilisation."

Publication of the second edition of "Observations on Volcanic Islands".

1877:

Publication of "The Different Forms of Flowers on Plants of the same species".

"I do not suppose that I shall publish any more books... I cannot endure being idle, but heaven knows whether I am capable of any more good work."

Publication of the second edition of the Orchid book.

1878:

Publication of the second edition of "The Effects of Cross and Self fertilisation".

1879:

Publication of an English translation of Ernst Krause's "Erasmus Darwin", with a notice by Charles Darwin. "I am EXTREMELY glad that you approve of the little 'Life' of our Grandfather, for I have been repenting that I ever undertook it, as the work was quite beyond my tether." (To Mr Francis Galton, Nov. 14, 1879.)

1880:

Publication of "The Power of Movement in Plants".

"It has always pleased me to exalt plants in the scale of organised beings."

Publication of the second edition of "The Different Forms of Flowers".

1881:

Wrote a continuation of the Autobiography.

Publication of "The Formation of Vegetable Mould, through the Action of Worms".

"It is the completion of a short paper read before the Geological Society more than forty years ago, and has revived old geological thoughts... As far as I can judge it will be a curious little book."

1882:

Charles Darwin died at Down, April 19, and was buried in Westminster Abbey, April 26, in the north aisle of the Nave a few feet from the grave of Sir Isaac Newton.

"As for myself, I believe that I have acted rightly in steadily following and devoting my life to Science. I feel no remorse from having committed any great sin, but have often and often regretted that I have not done more direct good to my fellow creatures."

The quotations in the above Epitome are taken from the Autobiography and published Letters:—

"The Life and Letters of Charles Darwin", including an Autobiographical Chapter. Edited by his son, Francis Darwin, 3 Vols., London, 1887.

"Charles Darwin": His life told in an Autobiographical Chapter, and in a selected series of his published Letters. Edited by his son, Francis Darwin, London, 1902.

"More Letters of Charles Darwin". A record of his work in a series of hitherto unpublished Letters. Edited by Francis Darwin and A.C. Seward, 2 Vols., London, 1903.

I. INTRODUCTORY LETTER From Sir Joseph Dalton Hooker, O.M., G.C.S.I., C.B., M.D., D.C.L., LL.D., F.R.S., ETC.

The Camp,

near Sunningdale,

January 15, 1909.

Dear Professor Seward,

The publication of a Series of Essays in Commemoration of the century of the birth of Charles Darwin and of the fiftieth anniversary of the publication of "The Origin of Species" is assuredly welcome and is a subject of congratulation to all students of Science.

These Essays on the progress of Science and Philosophy as affected by Darwin's labours have been written by men known for their ability to discuss the problems which he so successfully worked to solve. They cannot but prove to be of enduring value, whether for the information of the general reader or as guides to investigators occupied with problems similar to those which engaged the attention of Darwin.

The essayists have been fortunate in having for reference the five published volumes of Charles Darwin's Life and Correspondence. For there is set forth in his own words the inception in his mind of the problems, geological, zoological and botanical, hypothetical and theoretical, which he set himself to solve and the steps by which he proceeded to investigate them with the view of correlating the phenomena of life with the evolution of living things. In his letters he expressed himself in language so lucid and so little burthened with technical terms that they may be regarded as models for those who were asked to address themselves primarily to the educated reader rather than to the expert.

I may add that by no one can the perusal of the Essays be more vividly appreciated than by the writer of these lines. It was my privilege for forty years to possess the intimate friendship of Charles Darwin and to be his companion during many of his working hours in Study, Laboratory, and Garden. I was the recipient of letters from him, relating mainly to the progress of his researches, the copies of which (the originals are now in the possession of his family) cover upwards of a thousand pages of foolscap, each page containing, on an average, three hundred words.

That the editorship of these Essays has been entrusted to a Cambridge Professor of Botany must be gratifying to all concerned in their production and in their perusal, recalling as it does the fact that Charles Darwin's instructor in scientific methods was his lifelong friend the late Rev. J.S. Henslow at that time Professor of Botany in the University. It was owing to his recommendation that his pupil was appointed Naturalist to H.M.S. "Beagle", a service which Darwin himself regarded as marking the dawn of his scientific career.

Very sincerely yours,

J.D. HOOKER.

II. DARWIN'S PREDECESSORS. By J. Arthur Thomson.

Professor of Natural History in the University of Aberdeen.

In seeking to discover Darwin's relation to his predecessors it is useful to distinguish the various services which he rendered to the theory of organic evolution.

(I) As everyone knows, the general idea of the Doctrine of Descent is that the plants and animals of the present-day are the lineal descendants of ancestors on the whole somewhat simpler, that these again are descended from yet simpler forms, and so on backwards towards the literal "Protozoa" and "Protophyta" about which we unfortunately know nothing. Now no one supposes that Darwin originated this idea, which in rudiment at least is as old as Aristotle. What Darwin did was to make it current intellectual coin. He gave it a form that commended itself to the scientific and public intelligence of the day, and he won wide-spread conviction by showing with consummate skill that it was an effective formula to work with, a key which no lock refused. In a scholarly, critical, and pre-eminently fair-minded way, admitting difficulties and removing them, foreseeing objections and forestalling them, he showed that the doctrine of descent supplied a modal interpretation of how our present-day fauna and flora have come to be.

(II) In the second place, Darwin applied the evolution-idea to particular problems, such as the descent of man, and showed what a powerful organon it is, introducing order into masses of uncorrelated facts, interpreting enigmas both of structure and function, both bodily and mental, and, best of all, stimulating and guiding further investigation. But here again it cannot be claimed that Darwin was original. The problem of the descent or ascent of man, and other particular cases of evolution, had attracted not a few naturalists before Darwin's day, though no one (except Herbert Spencer in the psychological domain (1855)) had come near him in precision and thoroughness of inquiry.

(III) In the third place, Darwin contributed largely to a knowledge of the factors in the evolution-process, especially by his analysis of what occurs in the case of domestic animals and cultivated plants, and by his elaboration of the theory of Natural Selection, which Alfred Russel Wallace independently stated at the same time, and of which there had been a few previous suggestions of a more or less vague description. It was here that Darwin's originality was greatest, for he revealed to naturalists the many different forms—often very subtle—which natural selection takes, and with the insight of a disciplined scientific imagination he realised what a mighty engine of progress it has been and is.

(IV) As an epoch-marking contribution, not only to Aetiology but to Natural History in the widest sense, we rank the picture which Darwin gave to the world of the web of life, that is to say, of the inter-relations and linkages in Nature. For the Biology of the individual—if that be not a contradiction in terms—no idea is more fundamental than that of the correlation of organs, but Darwin's most characteristic contribution was not less fundamental,—it was the idea of the correlation of organisms. This, again, was not novel; we find it in the works of naturalist like Christian Conrad Sprengel, Gilbert White, and Alexander von Humboldt, but the realisation of its full import was distinctively Darwinian.

AS REGARDS THE GENERAL IDEA OF ORGANIC EVOLUTION.

While it is true, as Prof. H.F. Osborn puts it, that "'Before and after Darwin' will always be the ante et post urbem conditam of biological history," it is also true that the general idea of organic evolution is very ancient. In his admirable sketch "From the Greeks to Darwin" ("Columbia University Biological Series", Vol. I. New York and London, 1894. We must acknowledge our great indebtness to this fine piece of work.), Prof. Osborn has shown that several of the ancient philosophers looked upon Nature as a gradual development and as still in process of change. In the suggestions of Empedocles, to take the best instance, there were "four sparks of truth,—first, that the development of life was a gradual process; second, that plants were evolved before animals; third, that imperfect forms were gradually replaced (not succeeded) by perfect forms; fourth, that the natural cause of the production of perfect forms was the extinction of the imperfect." (Op. cit. page 41.) But the fundamental idea of one stage giving origin to another was absent. As the blue Aegean teemed with treasures of beauty and threw many upon its shores, so did Nature produce like a fertile artist what had to be rejected as well as what was able to survive, but the idea of one species emerging out of another was not yet conceived.

Aristotle's views of Nature (See G.J. Romanes, "Aristotle as a Naturalist", "Contemporary Review", Vol. LIX. page 275, 1891; G. Pouchet "La Biologie Aristotelique", Paris, 1885; E. Zeller, "A History of Greek Philosophy", London, 1881, and "Ueber die griechischen Vorganger Darwin's", "Abhandl. Berlin Akad." 1878, pages 111-124.) seem to have been more definitely evolutionist than those of his predecessors, in this sense, at least, that he recognised not only an ascending scale, but a genetic series from polyp to man and an age-long movement towards perfection. "It is due to the resistance of matter to form that Nature can only rise by degrees from lower to higher types." "Nature produces those things which, being continually moved by a certain principle contained in themselves, arrive at a certain end."

To discern the outcrop of evolution-doctrine in the long interval between Aristotle and Bacon seems to be very difficult, and some of the instances that have been cited strike one as forced. Epicurus and Lucretius, often called poets of evolution, both pictured animals as arising directly out of the earth, very much as Milton's lion long afterwards pawed its way out. Even when we come to Bruno who wrote that "to the sound of the harp of the Universal Apollo (the World Spirit), the lower organisms are called by stages to higher, and the lower stages are connected by intermediate forms with the higher," there is great room, as Prof. Osborn points out (op. cit. page 81.), for difference of opinion as to how far he was an evolutionist in our sense of the term.

The awakening of natural science in the sixteenth century brought the possibility of a concrete evolution theory nearer, and in the early seventeenth century we find evidences of a new spirit—in the embryology of Harvey and the classifications of Ray. Besides sober naturalists there were speculative dreamers in the sixteenth and seventeenth centuries who had at least got beyond static formulae, but, as Professor Osborn points out (op. cit. page 87.), "it is a very striking fact, that the basis of our modern methods of studying the Evolution problem was established not by the early naturalists nor by the speculative writers, but by the Philosophers." He refers to Bacon, Descartes, Leibnitz, Hume, Kant, Lessing, Herder, and Schelling. "They alone were upon the main track of modern thought. It is evident that they were groping in the dark for a working theory of the Evolution of life, and it is remarkable that they clearly perceived from the outset that the point to which observation should be directed was not the past but the present mutability of species, and further, that this mutability was simply the variation of individuals on an extended scale."

Bacon seems to have been one of the first to think definitely about the mutability of species, and he was far ahead of his age in his suggestion of what we now call a Station of Experimental Evolution. Leibnitz discusses in so many words how the species of animals may be changed and how intermediate species may once have linked those that now seem discontinuous. "All natural orders of beings present but a single chain"... "All advances by degrees in Nature, and nothing by leaps." Similar evolutionist statements are to be found in the works of the other "philosophers," to whom Prof. Osborn refers, who were, indeed, more scientific than the naturalists of their day. It must be borne in mind that the general idea of organic evolution—that the present is the child of the past—is in great part just the idea of human history projected upon the natural world, differentiated by the qualification that the continuous "Becoming" has been wrought out by forces inherent in the organisms themselves and in their environment.

A reference to Kant (See Brock, "Die Stellung Kant's zur Deszendenztheorie," "Biol. Centralbl." VIII. 1889, pages 641-648. Fritz Schultze, "Kant und Darwin", Jena, 1875.) should come in historical order after Buffon, with whose writings he was acquainted, but he seems, along with Herder and Schelling, to be best regarded as the culmination of the evolutionist philosophers—of those at least who interested themselves in scientific problems. In a famous passage he speaks of "the agreement of so many kinds of animals in a certain common plan of structure"... an "analogy of forms" which "strengthens the supposition that they have an actual blood-relationship, due to derivation from a common parent." He speaks of "the great Family of creatures, for as a Family we must conceive it, if the above-mentioned continuous and connected relationship has a real foundation." Prof.

Osborn alludes to the scientific caution which led Kant, biology being what it was, to refuse to entertain the hope "that a Newton may one day arise even to make the production of a blade of grass comprehensible, according to natural laws ordained by no intention." As Prof. Haeckel finely observes, Darwin rose up as Kant's Newton. (Mr Alfred Russel Wallace writes: "We claim for Darwin that he is the Newton of natural history, and that, just so surely as that the discovery and demonstration by Newton of the law of gravitation established order in place of chaos and laid a sure foundation for all future study of the starry heavens, so surely has Darwin, by his discovery of the law of natural selection and his demonstration of the great principle of the preservation of useful variations in the struggle for life, not only thrown a flood of light on the process of development of the whole organic world, but also established a firm foundation for all future study of nature." ("Darwinism", London, 1889, page 9). See also Prof. Karl Pearson's "Grammar of Science" (2nd edition), London, 1900, page 32. See Osborn, op. cit. Page 100.))

The scientific renaissance brought a wealth of fresh impressions and some freedom from the tyranny of tradition, and the twofold stimulus stirred the speculative activity of a great variety of men from old Claude Duret of Moulins, of whose weird transformism (1609) Dr Henry de Varigny ("Experimental Evolution". London, 1892. Chap. 1. page 14.) gives us a glimpse, to Lorenz Oken (1799-1851) whose writings are such mixtures of sense and nonsense that some regard him as a far-seeing prophet and others as a fatuous follower of intellectual will-o'-the-wisps. Similarly, for De Maillet, Maupertuis, Diderot, Bonnet, and others, we must agree with Professor Osborn that they were not actually in the main Evolution movement. Some have been included in the roll of honour on very slender evidence, Robinet for instance, whose evolutionism seems to us extremely dubious. (See J. Arthur Thomson, "The Science of Life". London, 1899. Chap. XVI. "Evolution of Evolution Theory".)

The first naturalist to give a broad and concrete expression to the evolutionist doctrine of descent was Buffon (1707-1788), but it is interesting to recall the fact that his contemporary Linnaeus (1707-1778), protagonist of the counter-doctrine of the fixity of species (See Carus Sterne (Ernest Krause), "Die allgemeine Weltanschauung in ihrer historischen Entwickelung". Stuttgart, 1889. Chapter entitled "Bestandigkeit oder Veranderlichkeit der Naturwesen".), went the length of admitting (in 1762) that new species might arise by intercrossing. Buffon's position among the pioneers of the evolution-doctrine is weakened by his habit of vacillating between his own conclusions and the orthodoxy of the Sorbonne, but there is no doubt that he had a firm grasp of the general idea of "l'enchainement des etres."

Erasmus Darwin (1731-1802), probably influenced by Buffon, was another firm evolutionist, and the outline of his argument in the "Zoonomia" ("Zoonomia, or the Laws of Organic Life", 2 vols. London, 1794; Osborn op. cit. page 145.) might serve in part at least to-day. "When we revolve in our minds the metamorphoses of animals, as from the tadpole to the frog; secondly, the changes produced by artificial cultivation, as in the breeds of horses, dogs, and sheep; thirdly, the changes produced by conditions of climate and of season, as in the sheep of warm climates being covered with hair instead of wool, and the hares and partridges of northern climates becoming white in winter: when, further, we observe the changes of structure produced by habit, as seen especially in men of different occupations; or the changes produced by artificial mutilation and prenatal influences, as in the crossing of species and production of monsters; fourth, when we observe the essential unity of plan in all warm-blooded animals,—we are led to conclude that they have been alike produced from a similar living filament"... "From thus meditating upon the minute portion of time in which many of the above changes have been produced, would it be too bold to imagine, in the great length of time since the earth began to exist, perhaps millions of years before the commencement of the history of mankind, that all warm-blooded animals have arisen from one living filament?"... "This idea of the gradual generation of all things seems to have been as familiar to the ancient philosophers as to the modern ones, and to have given rise to the beautiful hieroglyphic figure of the proton oon, or first great egg, produced by night, that is, whose origin is involved in obscurity, and animated by Eros, that is, by Divine Love; from whence proceeded all things which exist."

Lamarck (1744-1829) seems to have become an evolutionist independently of Erasmus Darwin's influence, though the parallelism between them is striking. He probably owed something to Buffon, but he developed his theory along a different line. Whatever view be held in regard to that theory there is no doubt that Lamarck was a thorough-going evolutionist. Professor Haeckel speaks of the "Philosophie Zoologique" as "the first connected and thoroughly logical exposition of the theory of descent." (See Alpheus S. Packard, "Lamarck, the Founder of Evolution, His Life and Work, with Translations of his writings on Organic Evolution". London, 1901.)

Besides the three old masters, as we may call them, Buffon, Erasmus Darwin, and Lamarck, there were other quite convinced pre-Darwinian evolutionists. The historian of the theory of descent must take account of Treviranus whose "Biology or Philosophy of Animate Nature" is full of evolutionary suggestions; of Etienne Geoffroy St Hilaire, who in 1830, before the French Academy of Sciences, fought with Cuvier, the fellow-worker of his youth, an intellectual duel on the question of descent; of Goethe, one of the founders of morphology and the greatest poet of Evolution—who, in his eighty-first year, heard the tidings of Geoffroy St Hilaire's defeat with an interest which transcended the political anxieties of the time; and of many others who had gained with more or less confidence and clearness a new outlook on Nature. It will be remembered that Darwin refers to thirty-four more or less evolutionist authors in his Historical Sketch, and the list might be added to. Especially when we come near to 1858 do the numbers increase, and one of the most remarkable, as also most independent champions of the evolution-idea before that date was Herbert Spencer, who not only marshalled the arguments in a very forcible way in 1852, but applied the formula in detail in his "Principles of Psychology" in 1855. (See Edward Clodd, "Pioneers of Evolution", London, page 161, 1897.)

It is right and proper that we should shake ourselves free from all creationist appreciations of Darwin, and that we should recognise the services of pre-Darwinian evolutionists who helped to make the time ripe, yet one cannot help feeling that the citation of them is apt to suggest two fallacies. It may suggest that Darwin simply entered into the labours of his predecessors, whereas, as a matter of fact, he knew very little about them till after he had been for years at work. To write, as Samuel Butler did, "Buffon planted, Erasmus Darwin and Lamarck watered, but it was Mr Darwin who said 'That fruit is ripe,' and shook it into his lap"... seems to us a quite misleading version of the facts of the case. The second fallacy which the historical citation is a little apt to suggest is that the filiation of ideas is a simple problem. On the contrary, the history of an idea, like the pedigree of an organism, is often very intricate, and the evolution of the evolution-idea is bound up with the whole progress of the world. Thus in order to interpret Darwin's clear formulation of the idea of organic evolution and his convincing presentation of it, we have to do more than go back to his immediate predecessors, such as Buffon, Erasmus Darwin, and Lamarck; we have to inquire into the acceptance of evolutionary conceptions in regard to other orders of facts, such as the earth and the solar system (See Chapter IX. "The Genetic View of Nature" in J.T. Merz's "History of European Thought in the Nineteenth Century", Vol. 2, Edinburgh and London, 1903.); we have to realise how the growing success of scientific interpretation along other lines gave confidence to those who refused to admit that there was any domain from which science could be excluded as a trespasser; we have to take account of the development of philosophical thought, and even of theological and religious movements; we should also, if we are wise enough, consider social changes. In short, we must abandon the idea that we can understand the history of any science as such, without reference to contemporary evolution in other departments of activity.

While there were many evolutionists before Darwin, few of them were expert naturalists and few were known outside a small circle; what was of much more importance was that the genetic view of nature was insinuating itself in regard to other than biological orders of facts, here a little and there a little, and that the scientific spirit had ripened since the days when

Cuvier laughed Lamarck out of court. How was it that Darwin succeeded where others had failed? Because, in the first place, he had clear visions—"pensees de la jeunesse, executees par l'age mur"—which a University curriculum had not made impossible, which the "Beagle" voyage made vivid, which an unrivalled British doggedness made real—visions of the web of life, of the fountain of change within the organism, of the struggle for existence and its winnowing, and of the spreading genealogical tree. Because, in the second place, he put so much grit into the verification of his visions, putting them to the proof in an argument which is of its kind—direct demonstration being out of the question—quite unequalled. Because, in the third place, he broke down the opposition which the most scientific had felt to the seductive modal formula of evolution by bringing forward a more plausible theory of the process than had been previously suggested. Nor can one forget, since questions of this magnitude are human and not merely academic, that he wrote so that all men could understand.

AS REGARDS THE FACTORS OF EVOLUTION.

It is admitted by all who are acquainted with the history of biology that the general idea of organic evolution as expressed in the Doctrine of Descent was quite familiar to Darwin's grandfather, and to others before and after him, as we have briefly indicated. It must also be admitted that some of these pioneers of evolutionism did more than apply the evolution-idea as a modal formula of becoming, they began to inquire into the factors in the process. Thus there were pre-Darwinian theories of evolution, and to these we must now briefly refer. (See Prof. W.A. Locy's "Biology and its Makers". New York, 1908. Part II. "The Doctrine of Organic Evolution".)

In all biological thinking we have to work with the categories Organism—Function—Environment, and theories of evolution may be classified in relation to these. To some it has always seemed that the fundamental fact is the living organism,—a creative agent, a striving will, a changeful Proteus, selecting its environment, adjusting itself to it, self-differentiating and self-adaptive. The necessity of recognising the importance of the organism is admitted by all Darwinians who start with inborn variations, but it is open to question whether the whole truth of what we might call the Goethian position is exhausted in the postulate of inherent variability.

To others it has always seemed that the emphasis should be laid on Function,—on use and disuse, on doing and not doing. Practice makes perfect; c'est a force de forger qu'on devient forgeron. This is one of the fundamental ideas of Lamarckism; to some extent it met with Darwin's approval; and it finds many supporters to-day. One of the ablest of these—Mr Francis Darwin—has recently given strong reasons for combining a modernised Lamarckism with what we usually regard as sound Darwinism. (Presidential Address to the British Association meeting at Dublin in 1908.)

To others it has always seemed that the emphasis should be laid on the Environment, which wakes the organism to action, prompts it to change, makes dints upon it, moulds it, prunes it, and finally, perhaps, kills it. It is again impossible to doubt that there is truth in this view, for even if environmentally induced "modifications" be not transmissible, environmentally induced "variations" are; and even if the direct influence of the environment be less important than many enthusiastic supporters of this view—may we call them Buffonians—think, there remains the indirect influence which Darwinians in part rely on,— the eliminative process. Even if the extreme view be held that the only form of discriminate elimination that counts is inter-organismal competition, this might be included under the rubric of the animate environment.

In many passages Buffon (See in particular Samuel Butler, "Evolution Old and New", London, 1879; J.L. de Lanessan, "Buffon et Darwin", "Revue Scientifique", XLIII. pages 385-391, 425-432, 1889.) definitely suggested that environmental influences—especially of climate and food—were directly productive of changes in organisms, but he did not discuss the question of the transmissibility of the modifications so induced, and it is difficult to gather from his inconsistent writings what extent of transformation he really believed in. Prof. Osborn says of Buffon: "The struggle for existence, the elimination of the least-

13

perfected species, the contest between the fecundity of certain species and their constant destruction, are all clearly expressed in various passages." He quotes two of these (op. cit. page 136.):

"Le cours ordinaire de la nature vivante, est en general toujours constant, toujours le meme; son mouvement, toujours regulier, roule sur deux points inebranlables: l'un, la fecondite sans bornes donnee a toutes les especes; l'autre, les obstacles sans nombre qui reduisent cette fecondite a une mesure determinee et ne laissent en tout temps qu'a peu pres la meme quantite d'individus de chaque espece"... "Les especes les moins parfaites, les plus delicates, les plus pesantes, les moins agissantes, les moins armees, etc., ont deja disparu ou disparaitront."

Erasmus Darwin (See Ernst Krause and Charles Darwin, "Erasmus Darwin", London, 1879.) had a firm grip of the "idea of the gradual formation and improvement of the Animal world," and he had his theory of the process. No sentence is more characteristic than this: "All animals undergo transformations which are in part produced by their own exertions, in response to pleasures and pains, and many of these acquired forms or propensities are transmitted to their posterity." This is Lamarckism before Lamarck, as his grandson pointed out. His central idea is that wants stimulate efforts and that these result in improvements, which subsequent generations make better still. He realised something of the struggle for existence and even pointed out that this advantageously checks the rapid multiplication. "As Dr Krause points out, Darwin just misses the connection between this struggle and the Survival of the Fittest." (Osborn op. cit. page 142.)

Lamarck (1744-1829) (See E. Perrier "La Philosophie Zoologique avant Darwin", Paris, 1884; A. de Quatrefages, "Darwin et ses Precurseurs Francais", Paris, 1870; Packard op. cit.; also Claus, "Lamarck als Begrunder der Descendenzlehre", Wien, 1888; Haeckel, "Natural History of Creation", English translation London, 1879; Lang "Zur Charakteristik der Forschungswege von Lamarck und Darwin", Jena, 1889.) seems to have thought out his theory of evolution without any knowledge of Erasmus Darwin's which it closely resembled. The central idea of his theory was the cumulative inheritance of functional modifications. "Changes in environment bring about changes in the habits of animals. Changes in their wants necessarily bring about parallel changes in their habits. If new wants become constant or very lasting, they form new habits, the new habits involve the use of new parts, or a different use of old parts, which results finally in the production of new organs and the modification of old ones." He differed from Buffon in not attaching importance, as far as animals are concerned, to the direct influence of the environment, "for environment can effect no direct change whatever upon the organisation of animals," but in regard to plants he agreed with Buffon that external conditions directly moulded them.

Treviranus (1776-1837) (See Huxley's article "Evolution in Biology", "Encyclopaedia Britannica" (9th edit.), 1878, pages 744-751, and Sully's article, "Evolution in Philosophy", ibid. pages 751-772.), whom Huxley ranked beside Lamarck, was on the whole Buffonian, attaching chief importance to the influence of a changeful environment both in modifying and in eliminating, but he was also Goethian, for instance in his idea that species like individuals pass through periods of growth, full bloom, and decline. "Thus, it is not only the great catastrophes of Nature which have caused extinction, but the completion of cycles of existence, out of which new cycles have begun." A characteristic sentence is quoted by Prof. Osborn: "In every living being there exists a capability of an endless variety of form-assumption; each possesses the power to adapt its organisation to the changes of the outer world, and it is this power, put into action by the change of the universe, that has raised the simple zoophytes of the primitive world to continually higher stages of organisation, and has introduced a countless variety of species into animate Nature."

Goethe (1749-1832) (See Haeckel, "Die Naturanschauung von Darwin, Goethe und Lamarck", Jena, 1882.), who knew Buffon's work but not Lamarck's, is peculiarly interesting as one of the first to use the evolution-idea as a guiding hypothesis, e.g. in the interpretation of vestigial structures in man, and to realise that organisms express an attempt to make a compromise between specific inertia and individual change. He gave the finest expression that science has yet known—if it has known it—of the kernel-idea of what is called "bathmism," the idea of an "inherent growth-force"—and at the same time he held that "the way of life powerfully reacts upon all form" and that the orderly growth of form "yields to change from externally acting causes."

Besides Buffon, Erasmus Darwin, Lamarck, Treviranus, and Goethe, there were other "pioneers of evolution," whose views have been often discussed and appraised. Etienne Geoffroy Saint-Hilaire (1772-1844), whose work Goethe so much admired, was on the whole Buffonian, emphasising the direct action of the changeful milieu. "Species vary with their environment, and existing species have descended by modification from earlier and somewhat simpler species." He had a glimpse of the selection idea, and believed in mutations or sudden leaps—induced in the embryonic condition by external influences. The complete history of evolution-theories will include many instances of guesses at truth which were afterwards substantiated, thus the geographer von Buch (1773-1853) detected the importance of the Isolation factor on which Wagner, Romanes, Gulick and others have laid great stress, but we must content ourselves with recalling one other pioneer, the author of the "Vestiges of Creation" (1844), a work which passed through ten editions in nine years and certainly helped to harrow the soil for Darwin's sowing. As Darwin said, "it did excellent service in this country in calling attention to the subject, in removing prejudice, and in thus preparing the ground for the reception of analogous views." ("Origin of Species" (6th edition), page xvii.) Its author, Robert Chambers (1802-1871) was in part a Buffonian—maintaining that environment moulded organisms adaptively, and in part a Goethian—believing in an inherent progressive impulse which lifted organisms from one grade of organisation to another.

AS REGARDS NATURAL SELECTION.

The only thinker to whom Darwin was directly indebted, so far as the theory of Natural Selection is concerned, was Malthus, and we may once more quote the well-known passage in the Autobiography: "In October, 1838, that is, fifteen months after I had begun my systematic enquiry, I happened to read for amusement 'Malthus on Population', and being well prepared to appreciate the struggle for existence which everywhere goes on from long-continued observation of the habits of animals and plants, it at once struck me that under these circumstances favourable variations would tend to be preserved, and unfavourable ones to be destroyed. The result of this would be the formation of new species." ("The Life and Letters of Charles Darwin", Vol. 1. page 83. London, 1887.)

Although Malthus gives no adumbration of the idea of Natural Selection in his exposition of the eliminative processes which go on in mankind, the suggestive value of his essay is undeniable, as is strikingly borne out by the fact that it gave to Alfred Russel Wallace also "the long-sought clue to the effective agent in the evolution of organic species." (A.R. Wallace, "My Life, A Record of Events and Opinions", London, 1905, Vol. 1. page 232.) One day in Ternate when he was resting between fits of fever, something brought to his recollection the work of Malthus which he had read twelve years before. "I thought of his clear exposition of 'the positive checks to increase'—disease, accidents, war, and famine—which keep down the population of savage races to so much lower an average than that of more civilized peoples. It then occurred to me that these causes or their equivalents are continually acting in the case of animals also; and as animals usually breed much more rapidly than does mankind, the destruction every year from these causes must be enormous in order to keep down the numbers of each species, since they evidently do not increase regularly from year to year, as otherwise the world would long ago have been densely crowded with those that breed most quickly. Vaguely thinking over the enormous and constant destruction which this implied, it occurred to me to ask the question, Why do some die and some live? And the answer was clearly, that on the whole the best fitted live. From the effects of disease the most healthy escaped; from enemies the strongest, the swiftest, or the most cunning; from famine the best hunters or those with the best digestion; and so on. Then it suddenly flashed upon me that this self-acting process would necessarily IMPROVE THE RACE, because in every generation the inferior would inevitably be killed off and the

15

superior would remain—that is, THE FITTEST WOULD SURVIVE." (Ibid. Vol. 1. page 361.) We need not apologise for this long quotation, it is a tribute to Darwin's magnanimous colleague, the Nestor of the evolutionist camp,—and it probably indicates the line of thought which Darwin himself followed. It is interesting also to recall the fact that in 1852, when Herbert Spencer wrote his famous "Leader" article on "The Development Hypothesis" in which he argued powerfully for the thesis that the whole animate world is the result of an age-long process of natural transformation, he wrote for "The Westminster Review" another important essay, "A Theory of Population deduced from the General Law of Animal Fertility", towards the close of which he came within an ace of recognising that the struggle for existence was a factor in organic evolution. At a time when pressure of population was practically interesting men's minds, Darwin, Wallace, and Spencer were being independently led from a social problem to a biological theory. There could be no better illustration, as Prof. Patrick Geddes has pointed out, of the Comtian thesis that science is a "social phenomenon."

Therefore, as far more important than any further ferreting out of vague hints of Natural Selection in books which Darwin never read, we would indicate by a quotation the view that the central idea in Darwinism is correlated with contemporary social evolution. "The substitution of Darwin for Paley as the chief interpreter of the order of nature is currently regarded as the displacement of an anthropomorphic view by a purely scientific one: a little reflection, however, will show that what has actually happened has been merely the replacement of the anthropomorphism of the eighteenth century by that of the nineteenth. For the place vacated by Paley's theological and metaphysical explanation has simply been occupied by that suggested to Darwin and Wallace by Malthus in terms of the prevalent severity of industrial competition, and those phenomena of the struggle for existence which the light of contemporary economic theory has enabled us to discern, have thus come to be temporarily exalted into a complete explanation of organic progress." (P. Geddes, article "Biology", "Chambers's Encyclopaedia".) It goes without saying that the idea suggested by Malthus was developed by Darwin into a biological theory which was then painstakingly verified by being used as an interpretative formula, and that the validity of a theory so established is not affected by what suggested it, but the practical question which this line of thought raises in the mind is this: if Biology did thus borrow with such splendid results from social theory, why should we not more deliberately repeat the experiment?

Darwin was characteristically frank and generous in admitting that the principle of Natural Selection had been independently recognised by Dr W.C. Wells in 1813 and by Mr Patrick Matthew in 1831, but he had no knowledge of these anticipations when he published the first edition of "The Origin of Species". Wells, whose "Essay on Dew" is still remembered, read in 1813 before the Royal Society a short paper entitled "An account of a White Female, part of whose skin resembles that of a Negro" (published in 1818). In this communication, as Darwin said, "he observes, firstly, that all animals tend to vary in some degree, and, secondly, that agriculturists improve their domesticated animals by selection; and then, he adds, but what is done in this latter case 'by art, seems to be done with equal efficacy, though more slowly, by nature, in the formation of varieties of mankind, fitted for the country which they inhabit.'" ("Origin of Species" (6th edition) page xv.) Thus Wells had the clear idea of survival dependent upon a favourable variation, but he makes no more use of the idea and applies it only to man. There is not in the paper the least hint that the author ever thought of generalising the remarkable sentence quoted above.

Of Mr Patrick Matthew, who buried his treasure in an appendix to a work on "Naval Timber and Arboriculture", Darwin said that "he clearly saw the full force of the principle of natural selection." In 1860 Darwin wrote—very characteristically—about this to Lyell: "Mr Patrick Matthew publishes a long extract from his work on "Naval Timber and Arboriculture", published in 1831, in which he briefly but completely anticipates the theory of Natural Selection. I have ordered the book, as some passages are rather obscure, but it is certainly, I think, a complete but not developed anticipation. Erasmus always said that surely this would be shown to be the case some day. Anyhow, one may be excused in not having discovered the fact in a work on Naval Timber." ("Life and Letters" II. page 301.)

De Quatrefages and De Varigny have maintained that the botanist Naudin stated the theory of evolution by natural selection in 1852. He explains very clearly the process of artificial selection, and says that in the garden we are following Nature's method. "We do not think that Nature has made her species in a different fashion from that in which we proceed ourselves in order to make our variations." But, as Darwin said, "he does not show how selection acts under nature." Similarly it must be noted in regard to several pre-Darwinian pictures of the struggle for existence (such as Herder's, who wrote in 1790 "All is in struggle... each one for himself" and so on), that a recognition of this is only the first step in Darwinism.

Profs. E. Perrier and H.F. Osborn have called attention to a remarkable anticipation of the selection-idea which is to be found in the speculations of Etienne Geoffroy St Hilaire (1825-1828) on the evolution of modern Crocodilians from the ancient Teleosaurs. Changing environment induced changes in the respiratory system and far-reaching consequences followed. The atmosphere, acting upon the pulmonary cells, brings about "modifications which are favourable or destructive ('funestes'); these are inherited, and they influence all the rest of the organisation of the animal because if these modifications lead to injurious effects, the animals which exhibit them perish and are replaced by others of a somewhat different form, a form changed so as to be adapted to (a la convenance) the new environment."

Prof. E.B. Poulton ("Science Progress", New Series, Vol. I. 1897. "A Remarkable Anticipation of Modern Views on Evolution". See also Chap. VI. in "Essays on Evolution", Oxford, 1908.) has shown that the anthropologist James Cowles Prichard (1786-1848) must be included, even in spite of himself, among the precursors of Darwin. In some passages of the second edition of his "Researches into the Physical History of Mankind" (1826), he certainly talks evolution and anticipates Prof. Weismann in denying the transmission of acquired characters. He is, however, sadly self-contradictory and his evolutionism weakens in subsequent editions—the only ones that Darwin saw. Prof. Poulton finds in Prichard's work a recognition of the operation of Natural Selection. "After enquiring how it is that 'these varieties are developed and preserved in connection with particular climates and differences of local situation,' he gives the following very significant answer: 'One cause which tends to maintain this relation is obvious. Individuals and families, and even whole colonies, perish and disappear in climates for which they are, by peculiarity of constitution, not adapted. Of this fact proofs have been already mentioned.'" Mr Francis Darwin and Prof. A.C. Seward discuss Prichard's "anticipations" in "More Letters of Charles Darwin", Vol. I. page 43, and come to the conclusion that the evolutionary passages are entirely neutralised by others of an opposite trend. There is the same difficulty with Buffon.

Hints of the idea of Natural Selection have been detected elsewhere. James Watt (See Prof. Patrick Geddes's article "Variation and Selection", "Encyclopaedia Britannica (9th edition) 1888.), for instance, has been reported as one of the anticipators (1851). But we need not prolong the inquiry further, since Darwin did not know of any anticipations until after he had published the immortal work of 1859, and since none of those who got hold of the idea made any use of it. What Darwin did was to follow the clue which Malthus gave him, to realise, first by genius and afterwards by patience, how the complex and subtle struggle for existence works out a natural selection of those organisms which vary in the direction of fitter adaptation to the conditions of their life. So much success attended his application of the Selection-formula that for a time he regarded Natural Selection as almost the sole factor in evolution, variations being pre-supposed; gradually, however, he came to recognise that there was some validity in the factors which had been emphasized by Lamarck and by Buffon, and in his well-known summing up in the sixth edition of the "Origin" he says of the transformation of species: "This has been effected chiefly through the natural selection of numerous successive, slight, favourable variations; aided in an important manner by the inherited effects of the use and disuse of parts; and in an unimportant manner, that is, in relation to adaptive structures, whether past or present, by the direct action of external conditions, and by variations which seem to us in our ignorance to arise spontaneously."

To sum up: the idea of organic evolution, older than Aristotle, slowly developed from the stage of suggestion to the stage of verification, and the first convincing verification was Darwin's; from being an a priori anticipation it has become an interpretation of nature, and Darwin is still the chief interpreter; from being a modal interpretation it has advanced to the rank of a causal theory, the most convincing part of which men will never cease to call Darwinism.

III. THE SELECTION THEORY, By
August Weismann.

Professor of Zoology in the University of
Freiburg (Baden).

I. THE IDEA OF SELECTION.

Many and diverse were the discoveries made by Charles Darwin in the course of a long and strenuous life, but none of them has had so far-reaching an influence on the science and thought of his time as the theory of selection. I do not believe that the theory of evolution would have made its way so easily and so quickly after Darwin took up the cudgels in favour of it, if he had not been able to support it by a principle which was capable of solving, in a simple manner, the greatest riddle that living nature presents to us,—I mean the purposiveness of every living form relative to the conditions of its life and its marvellously exact adaptation to these.

Everyone knows that Darwin was not alone in discovering the principle of selection, and that the same idea occurred simultaneously and independently to Alfred Russel Wallace. At the memorable meeting of the Linnean Society on 1st July, 1858, two papers were read (communicated by Lyell and Hooker) both setting forth the same idea of selection. One was written by Charles Darwin in Kent, the other by Alfred Wallace in Ternate, in the Malay Archipelago. It was a splendid proof of the magnanimity of these two investigators, that they thus, in all friendliness and without envy, united in laying their ideas before a scientific tribunal: their names will always shine side by side as two of the brightest stars in the scientific sky.

But it is with Charles Darwin that I am here chiefly concerned, since this paper is intended to aid in the commemoration of the hundredth anniversary of his birth.

The idea of selection set forth by the two naturalists was at the time absolutely new, but it was also so simple that Huxley could say of it later, "How extremely stupid not to have thought of that." As Darwin was led to the general doctrine of descent, not through the labours of his predecessors in the early years of the century, but by his own observations, so it was in regard to the principle of selection. He was struck by the innumerable cases of adaptation, as, for instance, that of the woodpeckers and tree-frogs to climbing, or the hooks and feather-like appendages of seeds, which aid in the distribution of plants, and he said to himself that an explanation of adaptations was the first thing to be sought for in attempting to formulate a theory of evolution.

But since adaptations point to CHANGES which have been undergone by the ancestral forms of existing species, it is necessary, first of all, to inquire how far species in general are VARIABLE. Thus Darwin's attention was directed in the first place to the phenomenon of variability, and the use man has made of this, from very early times, in the breeding of his domesticated animals and cultivated plants. He inquired carefully how breeders set to work, when they wished to modify the structure and appearance of a species to their own ends,

18

and it was soon clear to him that SELECTION FOR BREEDING PURPOSES played the chief part.

But how was it possible that such processes should occur in free nature? Who is here the breeder, making the selection, choosing out one individual to bring forth offspring and rejecting others? That was the problem that for a long time remained a riddle to him.

Darwin himself relates how illumination suddenly came to him. He had been reading, for his own pleasure, Malthus' book on Population, and, as he had long known from numerous observations, that every species gives rise to many more descendants than ever attain to maturity, and that, therefore, the greater number of the descendants of a species perish without reproducing, the idea came to him that the decision as to which member of a species was to perish, and which was to attain to maturity and reproduction might not be a matter of chance, but might be determined by the constitution of the individuals themselves, according as they were more or less fitted for survival. With this idea the foundation of the theory of selection was laid.

In ARTIFICIAL SELECTION the breeder chooses out for pairing only such individuals as possess the character desired by him in a somewhat higher degree than the rest of the race. Some of the descendants inherit this character, often in a still higher degree, and if this method be pursued throughout several generations, the race is transformed in respect of that particular character.

NATURAL SELECTION depends on the same three factors as ARTIFICIAL SELECTION: on VARIABILITY, INHERITANCE, and SELECTION FOR BREEDING, but this last is here carried out not by a breeder but by what Darwin called the "struggle for existence." This last factor is one of the special features of the Darwinian conception of nature. That there are carnivorous animals which take heavy toll in every generation of the progeny of the animals on which they prey, and that there are herbivores which decimate the plants in every generation had long been known, but it is only since Darwin's time that sufficient attention has been paid to the facts that, in addition to this regular destruction, there exists between the members of a species a keen competition for space and food, which limits multiplication, and that numerous individuals of each species perish because of unfavourable climatic conditions. The "struggle for existence," which Darwin regarded as taking the place of the human breeder in free nature, is not a direct struggle between carnivores and their prey, but is the assumed competition for survival between individuals OF THE SAME species, of which, on an average, only those survive to reproduce which have the greatest power of resistance, while the others, less favourably constituted, perish early. This struggle is so keen, that, within a limited area, where the conditions of life have long remained unchanged, of every species, whatever be the degree of fertility, only two, ON AN AVERAGE, of the descendants of each pair survive; the others succumb either to enemies, or to disadvantages of climate, or to accident. A high degree of fertility is thus not an indication of the special success of a species, but of the numerous dangers that have attended its evolution. Of the six young brought forth by a pair of elephants in the course of their lives only two survive in a given area; similarly, of the millions of eggs which two thread-worms leave behind them only two survive. It is thus possible to estimate the dangers which threaten a species by its ratio of elimination, or, since this cannot be done directly, by its fertility.

Although a great number of the descendants of each generation fall victims to accident, among those that remain it is still the greater or lesser fitness of the organism that determines the "selection for breeding purposes," and it would be incomprehensible if, in this competition, it were not ultimately, that is, on an average, the best equipped which survive, in the sense of living long enough to reproduce.

Thus the principle of natural selection is THE SELECTION OF THE BEST FOR REPRODUCTION, whether the "best" refers to the whole constitution, to one or more parts of the organism, or to one or more stages of development. Every organ, every part, every character of an animal, fertility and intelligence included, must be improved in this manner, and be gradually brought up in the course of generations to its highest attainable state of perfection. And not only may improvement of parts be brought about in this way, but new parts and organs may arise, since, through the slow and minute steps of individual or "fluctuating" variations, a part may be added here or dropped out there, and thus something new is produced.

The principle of selection solved the riddle as to how what was purposive could conceivably be brought about without the intervention of a directing power, the riddle which animate nature presents to our intelligence at every turn, and in face of which the mind of a Kant could find no way out, for he regarded a solution of it as not to be hoped for. For, even if we were to assume an evolutionary force that is continually transforming the most primitive and the simplest forms of life into ever higher forms, and the homogeneity of primitive times into the infinite variety of the present, we should still be unable to infer from this alone how each of the numberless forms adapted to particular conditions of life should have appeared PRECISELY AT THE RIGHT MOMENT IN THE HISTORY OF THE EARTH to which their adaptations were appropriate, and precisely at the proper place in which all the conditions of life to which they were adapted occurred: the humming-birds at the same time as the flowers; the trichina at the same time as the pig; the bark-coloured moth at the same time as the oak, and the wasp-like moth at the same time as the wasp which protects it. Without processes of selection we should be obliged to assume a "pre-established harmony" after the famous Leibnitzian model, by means of which the clock of the evolution of organisms is so regulated as to strike in exact synchronism with that of the history of the earth! All forms of life are strictly adapted to the conditions of their life, and can persist under these conditions alone.

There must therefore be an intrinsic connection between the conditions and the structural adaptations of the organism, and, SINCE THE CONDITIONS OF LIFE CANNOT BE DETERMINED BY THE ANIMAL ITSELF, THE ADAPTATIONS MUST BE CALLED FORTH BY THE CONDITIONS.

The selection theory teaches us how this is conceivable, since it enables us to understand that there is a continual production of what is non-purposive as well as of what is purposive, but the purposive alone survives, while the non-purposive perishes in the very act of arising. This is the old wisdom taught long ago by Empedocles.

II. THE LAMARCKIAN PRINCIPLE.

Lamarck, as is well known, formulated a definite theory of evolution at the beginning of the nineteenth century, exactly fifty years before the Darwin-Wallace principle of selection was given to the world. This brilliant investigator also endeavoured to support his theory by demonstrating forces which might have brought about the transformations of the organic world in the course of the ages. In addition to other factors, he laid special emphasis on the increased or diminished use of the parts of the body, assuming that the strengthening or weakening which takes place from this cause during the individual life, could be handed on to the offspring, and thus intensified and raised to the rank of a specific character. Darwin also regarded this LAMARCKIAN PRINCIPLE, as it is now generally called, as a factor in evolution, but he was not fully convinced of the transmissibility of acquired characters.

As I have here to deal only with the theory of selection, I need not discuss the Lamarckian hypothesis, but I must express my opinion that there is room for much doubt as to the cooperation of this principle in evolution. Not only is it difficult to imagine how the transmission of functional modifications could take place, but, up to the present time, notwithstanding the endeavours of many excellent investigators, not a single actual proof of such inheritance has been brought forward. Semon's experiments on plants are, according to the botanist Pfeffer, not to be relied on, and even the recent, beautiful experiments made by Dr Kammerer on salamanders, cannot, as I hope to show elsewhere, be regarded as proof, if only because they do not deal at all with functional modifications, that is, with modifications brought about by use, and it is to these ALONE that the Lamarckian principle refers.

III. OBJECTIONS TO THE THEORY OF SELECTION.

(a) Saltatory evolution.

The Darwinian doctrine of evolution depends essentially on THE CUMULATIVE AUGMENTATION of minute variations in the direction of utility. But can such minute variations, which are undoubtedly continually appearing among the individuals of the same species, possess any selection-value; can they determine which individuals are to survive, and which are to succumb; can they be increased by natural selection till they attain to the highest development of a purposive variation?

To many this seems so improbable that they have urged a theory of evolution by leaps from species to species. Kolliker, in 1872, compared the evolution of species with the processes which we can observe in the individual life in cases of alternation of generations. But a polyp only gives rise to a medusa because it has itself arisen from one, and there can be no question of a medusa ever having arisen suddenly and de novo from a polyp-bud, if only because both forms are adapted in their structure as a whole, and in every detail to the conditions of their life. A sudden origin, in a natural way, of numerous adaptations is inconceivable. Even the degeneration of a medusoid from a free-swimming animal to a mere brood-sac (gonophore) is not sudden and saltatory, but occurs by imperceptible modifications throughout hundreds of years, as we can learn from the numerous stages of the process of degeneration persisting at the same time in different species.

If, then, the degeneration to a simple brood-sac takes place only by very slow transitions, each stage of which may last for centuries, how could the much more complex ASCENDING evolution possibly have taken place by sudden leaps? I regard this argument as capable of further extension, for wherever in nature we come upon degeneration, it is taking place by minute steps and with a slowness that makes it not directly perceptible, and I believe that this in itself justifies us in concluding that THE SAME MUST BE TRUE OF ASCENDING evolution. But in the latter case the goal can seldom be distinctly recognised while in cases of degeneration the starting-point of the process can often be inferred, because several nearly related species may represent different stages.

In recent years Bateson in particular has championed the idea of saltatory, or so-called discontinuous evolution, and has collected a number of cases in which more or less marked variations have suddenly appeared. These are taken for the most part from among domesticated animals which have been bred and crossed for a long time, and it is hardly to be wondered at that their much mixed and much influenced germ-plasm should, under certain conditions, give rise to remarkable phenomena, often indeed producing forms which are strongly suggestive of monstrosities, and which would undoubtedly not survive in free nature, unprotected by man. I should regard such cases as due to an intensified germinal selection—though this is to anticipate a little—and from this point of view it cannot be denied that they have a special interest. But they seem to me to have no significance as far as the transformation of species is concerned, if only because of the extreme rarity of their occurrence.

There are, however, many variations which have appeared in a sudden and saltatory manner, and some of these Darwin pointed out and discussed in detail: the copper beech, the weeping trees, the oak with "fern-like leaves," certain garden-flowers, etc. But none of them have persisted in free nature, or evolved into permanent types.

On the other hand, wherever enduring types have arisen, we find traces of a gradual origin by successive stages, even if, at first sight, their origin may appear to have been sudden. This is the case with SEASONAL DIMORPHISM, the first known cases of which exhibited marked differences between the two generations, the winter and the summer brood. Take for instance the much discussed and studied form Vanessa (Araschnia) levana-prorsa. Here the differences between the two forms are so great and so apparently disconnected, that one might almost believe it to be a sudden mutation, were it not that old transition-stages can be called forth by particular temperatures, and we know other butterflies, as for instance our Garden Whites, in which the differences between the two generations are not nearly so marked; indeed, they are so little apparent that they are scarcely likely to be noticed except by experts. Thus here again there are small initial steps, some of which, indeed, must be regarded as adaptations, such as the green-sprinkled or lightly tinted under-surface which gives them a deceptive resemblance to parsley or to Cardamine leaves.

Even if saltatory variations do occur, we cannot assume that these HAVE EVER LED TO FORMS WHICH ARE CAPABLE OF SURVIVAL UNDER THE CONDITIONS OF WILD LIFE. Experience has shown that in plants which have suddenly varied the power of persistence is diminished. Korschinksky attributes to them weaknesses of organisation in general; "they bloom late, ripen few of their seeds, and show great sensitiveness to cold." These are not the characters which make for success in the struggle for existence.

We must briefly refer here to the views—much discussed in the last decade—of H. de Vries, who believes that the roots of transformation must be sought for in SALTATORY VARIATIONS ARISING FROM INTERNAL CAUSES, and distinguishes such MUTATIONS, as he has called them, from ordinary individual variations, in that they breed true, that is, with strict inbreeding they are handed on pure to the next generation. I have elsewhere endeavoured to point out the weaknesses of this theory ("Vortrage uber Descendenztheorie", Jena, 1904, II. 269. English Translation London, 1904, II. page 317.), and I am the less inclined to return to it here that it now appears (See Poulton, "Essays on Evolution", Oxford, 1908, pages xix-xxii.) that the far-reaching conclusions drawn by de Vries from his observations on the Evening Primrose, Oenothera lamarckiana, rest upon a very insecure foundation. The plant from which de Vries saw numerous "species"—his "mutations"—arise was not, as he assumed, a WILD SPECIES that had been introduced to Europe from America, but was probably a hybrid form which was first discovered in the Jardin des Plantes in Paris, and which does not appear to exist anywhere in America as a wild species.

This gives a severe shock to the "Mutation theory," for the other ACTUALLY WILD species with which de Vries experimented showed no "mutations" but yielded only negative results.

Thus we come to the conclusion that Darwin ("Origin of Species" (6th edition), pages 176 et seq.) was right in regarding transformations as taking place by minute steps, which, if useful, are augmented in the course of innumerable generations, because their possessors more frequently survive in the struggle for existence.

(b) SELECTION-VALUE OF THE INITIAL STEPS.

Is it possible that the significant deviations which we know as "individual variations" can form the beginning of a process of selection? Can they decide which is to perish and which to survive? To use a phrase of Romanes, can they have SELECTION-VALUE?

Darwin himself answered this question, and brought together many excellent examples to show that differences, apparently insignificant because very small, might be of decisive importance for the life of the possessor. But it is by no means enough to bring forward cases of this kind, for the question is not merely whether finished adaptations have selection-value, but whether the first beginnings of these, and whether the small, I might almost say minimal increments, which have led up from these beginnings to the perfect adaptation, have also had selection-value. To this question even one who, like myself, has been for many years a convinced adherent of the theory of selection, can only reply: WE MUST ASSUME SO, BUT WE CANNOT PROVE IT IN ANY CASE. It is not upon demonstrative evidence that we rely when we champion the doctrine of selection as a scientific truth; we base our argument on quite other grounds. Undoubtedly there are many apparently insignificant features, which can nevertheless be shown to be adaptations—for instance, the thickness of the basin-shaped shell of the limpets that live among the breakers on the shore. There can be no doubt that the thickness of these shells, combined with their flat form, protects the animals from the force of the waves breaking upon them,—but how have they become so thick? What proportion of thickness was sufficient to decide that of two variants of a limpet one should survive, the other be eliminated? We can say nothing more than that we infer from the present state of the shell, that it must have varied in regard to differences in shell-thickness, and that these differences must have had selection-value,—no proof therefore, but an assumption which we must show to be convincing.

For a long time the marvellously complex RADIATE and LATTICE-WORK skeletons of Radiolarians were regarded as a mere outflow of "Nature's infinite wealth of form," as an instance of a purely morphological character with no biological significance. But recent investigations have shown that these, too, have an adaptive significance (Hacker). The same thing has been shown by Schutt in regard to the lowly unicellular plants, the Peridineae, which abound alike on the surface of the ocean and in its depths. It has been shown that the long skeletal processes which grow out from these organisms have significance not merely as a supporting skeleton, but also as an extension of the superficial area, which increases the contact with the water-particles, and prevents the floating organisms from sinking. It has been established that the processes are considerably shorter in the colder layers of the ocean, and that they may be twelve times as long (Chun, "Reise der Valdivia", Leipzig, 1904.) in the warmer layers, thus corresponding to the greater or smaller amount of friction which takes place in the denser and less dense layers of the water.

The Peridineae of the warmer ocean layers have thus become long-rayed, those of the colder layers short-rayed, not through the direct effect of friction on the protoplasm, but through processes of selection, which favoured the longer rays in warm water, since they kept the organism afloat, while those with short rays sank and were eliminated. If we put the question as to selection-value in this case, and ask how great the variations in the length of processes must be in order to possess selection-value; what can we answer except that these variations must have been minimal, and yet sufficient to prevent too rapid sinking and consequent elimination? Yet this very case would give the ideal opportunity for a mathematical calculation of the minimal selection-value, although of course it is not feasible from lack of data to carry out the actual calculation.

But even in organisms of more than microscopic size there must frequently be minute, even microscopic differences which set going the process of selection, and regulate its progress to the highest possible perfection.

Many tropical trees possess thick, leathery leaves, as a protection against the force of the tropical rain drops. The DIRECT influence of the rain cannot be the cause of this power of resistance, for the leaves, while they were still thin, would simply have been torn to pieces. Their toughness must therefore be referred to selection, which would favour the trees with slightly thicker leaves, though we cannot calculate with any exactness how great the first stages of increase in thickness must have been. Our hypothesis receives further support from the fact that, in many such trees, the leaves are drawn out into a beak-like prolongation (Stahl and Haberlandt) which facilitates the rapid falling off of the rain water, and also from the fact that the leaves, while they are still young, hang limply down in bunches which offer the least possible resistance to the rain. Thus there are here three adaptations which can only be interpreted as due to selection. The initial stages of these adaptations must undoubtedly have had selection-value.

But even in regard to this case we are reasoning in a circle, not giving "proofs," and no one who does not wish to believe in the selection-value of the initial stages can be forced to do so. Among the many pieces of presumptive evidence a particularly weighty one seems to me to be THE SMALLNESS OF THE STEPS OF PROGRESS which we can observe in certain cases, as for instance in leaf-imitation among butterflies, and in mimicry generally. The resemblance to a leaf, for instance of a particular Kallima, seems to us so close as to be deceptive, and yet we find in another individual, or it may be in many others, a spot added which increases the resemblance, and which could not have become fixed unless the increased deceptiveness so produced had frequently led to the overlooking of its much persecuted possessor. But if we take the selection-value of the initial stages for granted, we are confronted with the further question which I myself formulated many years ago: How does it happen THAT THE NECESSARY BEGINNINGS OF A USEFUL VARIATION ARE ALWAYS PRESENT? How could insects which live upon or among green leaves become all green, while those that live on bark become brown? How have the desert animals become yellow and the Arctic animals white? Why were the necessary variations always present? How could the green locust lay brown eggs, or the privet caterpillar develop white and lilac-coloured lines on its green skin?

It is of no use answering to this that the question is wrongly formulated (Plate, "Selektionsprinzip u. Probleme der Artbildung" (3rd edition), Leipzig, 1908.) and that it is the converse that is true; that the process of selection takes place in accordance with the variations that present themselves. This proposition is undeniably true, but so also is another, which apparently negatives it: the variation required has in the majority of cases actually presented itself. Selection cannot solve this contradiction; it does not call forth the useful variation, but simply works upon it. The ultimate reason why one and the same insect should occur in green and in brown, as often happens in caterpillars and locusts, lies in the fact that variations towards brown presented themselves, and so also did variations towards green: THE KERNEL OF THE RIDDLE LIES IN THE VARYING, and for the present we can only say, that small variations in different directions present themselves in every species. Otherwise so many different kinds of variations could not have arisen. I have endeavoured to explain this remarkable fact by means of the intimate processes that must take place within the germ-plasm, and I shall return to the problem when dealing with "germinal selection."

We have, however, to make still greater demands on variation, for it is not enough that the necessary variation should occur in isolated individuals, because in that case there would be small prospect of its being preserved, notwithstanding its utility. Darwin at first believed, that even single variations might lead to transformation of the species, but later he became convinced that this was impossible, at least without the cooperation of other factors, such as isolation and sexual selection.

In the case of the GREEN CATERPILLARS WITH BRIGHT LONGITUDINAL STRIPES, numerous individuals exhibiting this useful variation must have been produced to start with. In all higher, that is, multicellular organisms, the germ-substance is the source of all transmissible variations, and this germ-plasm is not a simple substance but is made up of many primary constituents. The question can therefore be more precisely stated thus: How does it come about that in so many cases the useful variations present themselves in numbers just where they are required, the white oblique lines in the leaf-caterpillar on the under surface of the body, the accompanying coloured stripes just above them? And, further, how has it come about that in grass caterpillars, not oblique but longitudinal stripes, which are more effective for concealment among grass and plants, have been evolved? And finally, how is it that the same Hawk-moth caterpillars, which to-day show oblique stripes, possessed longitudinal stripes in Tertiary times? We can read this fact from the history of their development, and I have before attempted to show the biological significance of this change of colour. ("Studien zur Descendenz-Theorie" II., "Die Enstehung der Zeichnung bei den Schmetterlings-raupen," Leipzig, 1876.)

For the present I need only draw the conclusion that one and the same caterpillar may exhibit the initial stages of both, and that it depends on the manner in which these marking elements are INTENSIFIED and COMBINED by natural selection whether whitish longitudinal or oblique stripes should result. In this case then the "useful variations" were actually "always there," and we see that in the same group of Lepidoptera, e.g. species of Sphingidae, evolution has occurred in both directions according to whether the form lived among grass or on broad leaves with oblique lateral veins, and we can observe even now that the species with oblique stripes have longitudinal stripes when young, that is to say, while the stripes have no biological significance. The white places in the skin which gave rise, probably first as small spots, to this protective marking could be combined in one way or another according to the requirements of the species. They must therefore either have possessed selection-value from the first, or, if this was not the case at their earliest occurrence, there must have been SOME OTHER FACTORS which raised them to the point of selection-value. I shall return to this in discussing germinal selection. But the case may be followed still farther, and leads us to the same alternative on a still more secure basis.

24

Many years ago I observed in caterpillars of Smerinthus populi (the poplar hawk-moth), which also possess white oblique stripes, that certain individuals showed RED SPOTS above these stripes; these spots occurred only on certain segments, and never flowed together to form continuous stripes. In another species (Smerinthus tiliae) similar blood-red spots unite to form a line-like coloured seam in the last stage of larval life, while in S. ocellata rust-red spots appear in individual caterpillars, but more rarely than in S. Populi, and they show no tendency to flow together.

Thus we have here the origin of a new character, arising from small beginnings, at least in S. tiliae, in which species the coloured stripes are a normal specific character. In the other species, S. populi and S. ocellata, we find the beginnings of the same variation, in one more rarely than in the other, and we can imagine that, in the course of time, in these two species, coloured lines over the oblique stripes will arise. In any case these spots are the elements of variation, out of which coloured lines MAY be evolved, if they are combined in this direction through the agency of natural selection. In S. populi the spots are often small, but sometimes it seems as though several had united to form large spots. Whether a process of selection in this direction will arise in S. populi and S. ocellata, or whether it is now going on cannot be determined, since we cannot tell in advance what biological value the marking might have for these two species. It is conceivable that the spots may have no selection-value as far as these species are concerned, and may therefore disappear again in the course of phylogeny, or, on the other hand, that they may be changed in another direction, for instance towards imitation of the rust-red fungoid patches on poplar and willow leaves. In any case we may regard the smallest spots as the initial stages of variation, the larger as a cumulative summation of these. Therefore either these initial stages must already possess selection-value, or, as I said before: THERE MUST BE SOME OTHER REASON FOR THEIR CUMULATIVE SUMMATION. I should like to give one more example, in which we can infer, though we cannot directly observe, the initial stages.

All the Holothurians or sea-cucumbers have in the skin calcareous bodies of different forms, usually thick and irregular, which make the skin tough and resistant. In a small group of them—the species of Synapta—the calcareous bodies occur in the form of delicate anchors of microscopic size. Up till 1897 these anchors, like many other delicate microscopic structures, were regarded as curiosities, as natural marvels. But a Swedish observer, Oestergren, has recently shown that they have a biological significance: they serve the footless Synapta as auxiliary organs of locomotion, since, when the body swells up in the act of creeping, they press firmly with their tips, which are embedded in the skin, against the substratum on which the animal creeps, and thus prevent slipping backwards. In other Holothurians this slipping is made impossible by the fixing of the tube-feet. The anchors act automatically, sinking their tips towards the ground when the corresponding part of the body thickens, and returning to the original position at an angle of 45 degrees to the upper surface when the part becomes thin again. The arms of the anchor do not lie in the same plane as the shaft, and thus the curve of the arms forms the outermost part of the anchor, and offers no further resistance to the gliding of the animal. Every detail of the anchor, the curved portion, the little teeth at the head, the arms, etc., can be interpreted in the most beautiful way, above all the form of the anchor itself, for the two arms prevent it from swaying round to the side. The position of the anchors, too, is definite and significant; they lie obliquely to the longitudinal axis of the animal, and therefore they act alike whether the animal is creeping backwards or forwards. Moreover, the tips would pierce through the skin if the anchors lay in the longitudinal direction. Synapta burrows in the sand; it first pushes in the thin anterior end, and thickens this again, thus enlarging the hole, then the anterior tentacles displace more sand, the body is worked in a little farther, and the process begins anew. In the first act the anchors are passive, but they begin to take an active share in the forward movement when the body is contracted again. Frequently the animal retains only the posterior end buried in the sand, and then the anchors keep it in position, and make rapid withdrawal possible.

Thus we have in these apparently random forms of the calcareous bodies, complex adaptations in which every little detail as to direction, curve, and pointing is exactly determined. That they have selection-value in their present perfected form is beyond all doubt, since the animals are enabled by means of them to bore rapidly into the ground and so to escape from enemies. We do not know what the initial stages were, but we cannot doubt that the little improvements, which occurred as variations of the originally simple slimy bodies of the Holothurians, were preserved because they already possessed selection-value for the Synaptidae. For such minute microscopic structures whose form is so delicately adapted to the role they have to play in the life of the animal, cannot have arisen suddenly and as a whole, and every new variation of the anchor, that is, in the direction of the development of the two arms, and every curving of the shaft which prevented the tips from projecting at the wrong time, in short, every little adaptation in the modelling of the anchor must have possessed selection-value. And that such minute changes of form fall within the sphere of fluctuating variations, that is to say, THAT THEY OCCUR is beyond all doubt.

In many of the Synaptidae the anchors are replaced by calcareous rods bent in the form of an S, which are said to act in the same way. Others, such as those of the genus Ankyroderma, have anchors which project considerably beyond the skin, and, according to Oestergren, serve "to catch plant-particles and other substances" and so mask the animal. Thus we see that in the Synaptidae the thick and irregular calcareous bodies of the Holothurians have been modified and transformed in various ways in adaptation to the footlessness of these animals, and to the peculiar conditions of their life, and we must conclude that the earlier stages of these changes presented themselves to the processes of selection in the form of microscopic variations. For it is as impossible to think of any origin other than through selection in this case as in the case of the toughness, and the "drip-tips" of tropical leaves. And as these last could not have been produced directly by the beating of the heavy rain-drops upon them, so the calcareous anchors of Synapta cannot have been produced directly by the friction of the sand and mud at the bottom of the sea, and, since they are parts whose function is PASSIVE the Lamarckian factor of use and disuse does not come into question. The conclusion is unavoidable, that the microscopically small variations of the calcareous bodies in the ancestral forms have been intensified and accumulated in a particular direction, till they have led to the formation of the anchor. Whether this has taken place by the action of natural selection alone, or whether the laws of variation and the intimate processes within the germ-plasm have cooperated will become clear in the discussion of germinal selection. This whole process of adaptation has obviously taken place within the time that has elapsed since this group of sea-cucumbers lost their tube-feet, those characteristic organs of locomotion which occur in no group except the Echinoderms, and yet have totally disappeared in the Synaptidae. And after all what would animals that live in sand and mud do with tube-feet?

(c) COADAPTATION.

Darwin pointed out that one of the essential differences between artificial and natural selection lies in the fact that the former can modify only a few characters, usually only one at a time, while Nature preserves in the struggle for existence all the variations of a species, at the same time and in a purely mechanical way, if they possess selection-value.

Herbert Spencer, though himself an adherent of the theory of selection, declared in the beginning of the nineties that in his opinion the range of this principle was greatly over-estimated, if the great changes which have taken place in so many organisms in the course of ages are to be interpreted as due to this process of selection alone, since no transformation of any importance can be evolved by itself; it is always accompanied by a host of secondary changes. He gives the familiar example of the Giant Stag of the Irish peat, the enormous antlers of which required not only a much stronger skull cap, but also greater strength of the sinews, muscles, nerves and bones of the whole anterior half of the animal, if their mass was not to weigh down the animal altogether. It is inconceivable, he says, that so many processes of selection should take place SIMULTANEOUSLY, and we are therefore forced to fall back on the Lamarckian factor of the use and disuse of functional parts. And how, he asks, could natural selection follow two opposite directions of evolution in different parts of the body at the same time, as for instance in the case of the kangaroo, in which the forelegs must have become shorter, while the hind legs and the tail were becoming longer and stronger?

Spencer's main object was to substantiate the validity of the Lamarckian principle, the cooperation of which with selection had been doubted by many. And it does seem as though this principle, if it operates in nature at all, offers a ready and simple explanation of all such secondary variations. Not only muscles, but nerves, bones, sinews, in short all tissues which function actively, increase in strength in proportion as they are used, and conversely they decrease when the claims on them diminish. All the parts, therefore, which depend on the part that varied first, as for instance the enlarged antlers of the Irish Elk, must have been increased or decreased in strength, in exact proportion to the claims made upon them,—just as is actually the case.

But beautiful as this explanation would be, I regard it as untenable, because it assumes the TRANSMISSIBILITY OF FUNCTIONAL MODIFICATIONS (so-called "acquired" characters), and this is not only undemonstrable, but is scarcely theoretically conceivable, for the secondary variations which accompany or follow the first as correlative variations, occur also in cases in which the animals concerned are sterile and THEREFORE CANNOT TRANSMIT ANYTHING TO THEIR DESCENDANTS. This is true of WORKER BEES, and particularly of ANTS, and I shall here give a brief survey of the present state of the problem as it appears to me.

Much has been written on both sides of this question since the published controversy on the subject in the nineties between Herbert Spencer and myself. I should like to return to the matter in detail, if the space at my disposal permitted, because it seems to me that the arguments I advanced at that time are equally cogent to-day, notwithstanding all the objections that have since been urged against them. Moreover, the matter is by no means one of subordinate interest; it is the very kernel of the whole question of the reality and value of the principle of selection. For if selection alone does not suffice to explain "HARMONIOUS ADAPTATION" as I have called Spencer's COADAPTATION, and if we require to call in the aid of the Lamarckian factor it would be questionable whether selection could explain any adaptations whatever. In this particular case—of worker bees—the Lamarckian factor may be excluded altogether, for it can be demonstrated that here at any rate the effects of use and disuse cannot be transmitted.

But if it be asked why we are unwilling to admit the cooperation of the Darwinian factor of selection and the Lamarckian factor, since this would afford us an easy and satisfactory explanation of the phenomena, I answer: BECAUSE THE LAMARCKIAN PRINCIPLE IS FALLACIOUS, AND BECAUSE BY ACCEPTING IT WE CLOSE THE WAY TOWARDS DEEPER INSIGHT. It is not a spirit of combativeness or a desire for self-vindication that induces me to take the field once more against the Lamarckian principle, it is the conviction that the progress of our knowledge is being obstructed by the acceptance of this fallacious principle, since the facile explanation it apparently affords prevents our seeking after a truer explanation and a deeper analysis.

The workers in the various species of ants are sterile, that is to say, they take no regular part in the reproduction of the species, although individuals among them may occasionally lay eggs. In addition to this they have lost the wings, and the receptaculum seminis, and their compound eyes have degenerated to a few facets. How could this last change have come about through disuse, since the eyes of workers are exposed to light in the same way as are those of the sexual insects and thus in this particular case are not liable to "disuse" at all? The same is true of the receptaculum seminis, which can only have been disused as far as its glandular portion and its stalk are concerned, and also of the wings, the nerves tracheae and epidermal cells of which could not cease to function until the whole wing had degenerated, for the chitinous skeleton of the wing does not function at all in the active sense.

But, on the other hand, the workers in all species have undergone modifications in a positive direction, as, for instance, the greater development of brain. In many species large workers have evolved,—the so-called SOLDIERS, with enormous jaws and teeth, which defend the colony,—and in others there are SMALL workers which have taken over other

27

special functions, such as the rearing of the young Aphides. This kind of division of the workers into two castes occurs among several tropical species of ants, but it is also present in the Italian species, Colobopsis truncata. Beautifully as the size of the jaws could be explained as due to the increased use made of them by the "soldiers," or the enlarged brain as due to the mental activities of the workers, the fact of the infertility of these forms is an insurmountable obstacle to accepting such an explanation. Neither jaws nor brain can have been evolved on the Lamarckian principle.

The problem of coadaptation is no easier in the case of the ant than in the case of the Giant Stag. Darwin himself gave a pretty illustration to show how imposing the difference between the two kinds of workers in one species would seem if we translated it into human terms. In regard to the Driver ants (Anomma) we must picture to ourselves a piece of work, "for instance the building of a house, being carried on by two kinds of workers, of which one group was five feet four inches high, the other sixteen feet high." ("Origin of Species" (6th edition), page 232.)

Although the ant is a small animal as compared with man or with the Irish Elk, the "soldier" with its relatively enormous jaws is hardly less heavily burdened than the Elk with its antlers, and in the ant's case, too, a strengthening of the skeleton, of the muscles, the nerves of the head, and of the legs must have taken place parallel with the enlargement of the jaws. HARMONIOUS ADAPTATION (coadaptation) has here been active in a high degree, and yet these "soldiers" are sterile! There thus remains nothing for it but to refer all their adaptations, positive and negative alike, to processes of selection which have taken place in the rudiments of the workers within the egg and sperm-cells of their parents. There is no way out of the difficulty except the one Darwin pointed out. He himself did not find the solution of the riddle at once. At first he believed that the case of the workers among social insects presented "the most serious special difficulty" in the way of his theory of natural selection; and it was only after it had become clear to him, that it was not the sterile insects themselves but their parents that were selected, according as they produced more or less well adapted workers, that he was able to refer to this very case of the conditions among ants "IN ORDER TO SHOW THE POWER OF NATURAL SELECTION" ("Origin of Species", page 233; see also edition 1, page 242.). He explains his view by a simple but interesting illustration. Gardeners have produced, by means of long continued artificial selection, a variety of Stock, which bears entirely double, and therefore infertile flowers (Ibid. page 230.). Nevertheless the variety continues to be reproduced from seed, because in addition to the double and infertile flowers, the seeds always produce a certain number of single, fertile blossoms, and these are used to reproduce the double variety. These single and fertile plants correspond "to the males and females of an ant-colony, the infertile plants, which are regularly produced in large numbers, to the neuter workers of the colony."

This illustration is entirely apt, the only difference between the two cases consisting in the fact that the variation in the flower is not a useful, but a disadvantageous one, which can only be preserved by artificial selection on the part of the gardener, while the transformations that have taken place parallel with the sterility of the ants are useful, since they procure for the colony an advantage in the struggle for existence, and they are therefore preserved by natural selection. Even the sterility itself in this case is not disadvantageous, since the fertility of the true females has at the same time considerably increased. We may therefore regard the sterile forms of ants, which have gradually been adapted in several directions to varying functions, AS A CERTAIN PROOF that selection really takes place in the germ-cells of the fathers and mothers of the workers, and that SPECIAL COMPLEXES OF PRIMORDIA (IDS) are present in the workers and in the males and females, and these complexes contain the primordia of the individual parts (DETERMINANTS). But since all living entities vary, the determinants must also vary, now in a favourable, now in an unfavourable direction. If a female produces eggs, which contain favourably varying determinants in the worker-ids, then these eggs will give rise to workers modified in the favourable direction, and if this happens with many females, the colony concerned will contain a better kind of worker than other colonies.

I digress here in order to give an account of the intimate processes, which, according to my view, take place within the germ-plasm, and which I have called "GERMINAL SELECTION." These processes are of importance since they form the roots of variation, which in its turn is the root of natural selection. I cannot here do more than give a brief outline of the theory in order to show how the Darwin-Wallace theory of selection has gained support from it.

With others, I regard the minimal amount of substance which is contained within the nucleus of the germ-cells, in the form of rods, bands, or granules, as the GERM-SUBSTANCE or GERM-PLASM, and I call the individual granules IDS. There is always a multiplicity of such ids present in the nucleus, either occurring individually, or united in the form of rods or bands (chromosomes). Each id contains the primary constituents of a WHOLE individual, so that several ids are concerned in the development of a new individual.

In every being of complex structure thousands of primary constituents must go to make up a single id; these I call DETERMINANTS, and I mean by this name very small individual particles, far below the limits of microscopic visibility, vital units which feed, grow, and multiply by division. These determinants control the parts of the developing embryo,—in what manner need not here concern us. The determinants differ among themselves, those of a muscle are differently constituted from those of a nerve-cell or a glandular cell, etc., and every determinant is in its turn made up of minute vital units, which I call BIOPHORS, or the bearers of life. According to my view, these determinants not only assimilate, like every other living unit, but they VARY in the course of their growth, as every living unit does; they may vary qualitatively if the elements of which they are composed vary, they may grow and divide more or less rapidly, and their variations give rise to CORRESPONDING variations of the organ, cell, or cell-group which they determine. That they are undergoing ceaseless fluctuations in regard to size and quality seems to me the inevitable consequence of their unequal nutrition; for although the germ-cell as a whole usually receives sufficient nutriment, minute fluctuations in the amount carried to different parts within the germ-plasm cannot fail to occur.

Now, if a determinant, for instance of a sensory cell, receives for a considerable time more abundant nutriment than before, it will grow more rapidly—become bigger, and divide more quickly, and, later, when the id concerned develops into an embryo, this sensory cell will become stronger than in the parents, possibly even twice as strong. This is an instance of a HEREDITARY INDIVIDUAL VARIATION, arising from the germ.

The nutritive stream which, according to our hypothesis, favours the determinant N by chance, that is, for reasons unknown to us, may remain strong for a considerable time, or may decrease again; but even in the latter case it is conceivable that the ascending movement of the determinant may continue, because the strengthened determinant now ACTIVELY nourishes itself more abundantly,—that is to say, it attracts the nutriment to itself, and to a certain extent withdraws it from its fellow-determinants. In this way, it may—as it seems to me—get into PERMANENT UPWARD MOVEMENT, AND ATTAIN A DEGREE OF STRENGTH FROM WHICH THERE IS NO FALLING BACK. Then positive or negative selection sets in, favouring the variations which are advantageous, setting aside those which are disadvantageous.

In a similar manner a DOWNWARD variation of the determinants may take place, if its progress be started by a diminished flow of nutriment. The determinants which are weakened by this diminished flow will have less affinity for attracting nutriment because of their diminished strength, and they will assimilate more feebly and grow more slowly, unless chance streams of nutriment help them to recover themselves. But, as will presently be shown, a change of direction cannot take place at EVERY stage of the degenerative process. If a certain critical stage of downward progress be passed, even favourable conditions of food-supply will no longer suffice permanently to change the direction of the variation. Only two cases are conceivable; if the determinant corresponds to a USEFUL organ, only its removal can bring back the germ-plasm to its former level; therefore personal selection removes the id in question, with its determinants, from the germ-plasm, by causing the elimination of the individual in the struggle for existence. But there is another conceivable case; the determinants concerned may be those of an organ which has become USELESS, and they will then continue unobstructed, but with exceeding slowness, along the downward path, until the organ becomes vestigial, and finally disappears altogether.

The fluctuations of the determinants hither and thither may thus be transformed into a lasting ascending or descending movement; and THIS IS THE CRUCIAL POINT OF THESE GERMINAL PROCESSES.

This is not a fantastic assumption; we can read it in the fact of the degeneration of disused parts. USELESS ORGANS ARE THE ONLY ONES WHICH ARE NOT HELPED TO ASCEND AGAIN BY PERSONAL SELECTION, AND THEREFORE IN THEIR CASE ALONE CAN WE FORM ANY IDEA OF HOW THE PRIMARY CONSTITUENTS BEHAVE, WHEN THEY ARE SUBJECT SOLELY TO INTRA-GERMINAL FORCES.

The whole determinant system of an id, as I conceive it, is in a state of continual fluctuation upwards and downwards. In most cases the fluctuations will counteract one another, because the passive streams of nutriment soon change, but in many cases the limit from which a return is possible will be passed, and then the determinants concerned will continue to vary in the same direction, till they attain positive or negative selection-value. At this stage personal selection intervenes and sets aside the variation if it is disadvantageous, or favours—that is to say, preserves—it if it is advantageous. Only THE DETERMINANT OF A USELESS ORGAN IS UNINFLUENCED BY PERSONAL SELECTION, and, as experience shows, it sinks downwards; that is, the organ that corresponds to it degenerates very slowly but uninterruptedly till, after what must obviously be an immense stretch of time, it disappears from the germ-plasm altogether.

Thus we find in the fact of the degeneration of disused parts the proof that not all the fluctuations of a determinant return to equilibrium again, but that, when the movement has attained to a certain strength, it continues IN THE SAME DIRECTION. We have entire certainty in regard to this as far as the downward progress is concerned, and we must assume it also in regard to ascending variations, as the phenomena of artificial selection certainly justify us in doing. If the Japanese breeders were able to lengthen the tail feathers of the cock to six feet, it can only have been because the determinants of the tail-feathers in the germ-plasm had already struck out a path of ascending variation, and this movement was taken advantage of by the breeder, who continually selected for reproduction the individuals in which the ascending variation was most marked. For all breeding depends upon the unconscious selection of germinal variations.

Of course these germinal processes cannot be proved mathematically, since we cannot actually see the play of forces of the passive fluctuations and their causes. We cannot say how great these fluctuations are, and how quickly or slowly, how regularly or irregularly they change. Nor do we know how far a determinant must be strengthened by the passive flow of the nutritive stream if it is to be beyond the danger of unfavourable variations, or how far it must be weakened passively before it loses the power of recovering itself by its own strength. It is no more possible to bring forward actual proofs in this case than it was in regard to the selection-value of the initial stages of an adaptation. But if we consider that all heritable variations must have their roots in the germ-plasm, and further, that when personal selection does not intervene, that is to say, in the case of parts which have become useless, a degeneration of the part, and therefore also of its determinant must inevitably take place; then we must conclude that processes such as I have assumed are running their course within the germ-plasm, and we can do this with as much certainty as we were able to infer, from the phenomena of adaptation, the selection-value of their initial stages. The fact of the degeneration of disused parts seems to me to afford irrefutable proof that the fluctuations within the germ-plasm ARE THE REAL ROOT OF ALL HEREDITARY VARIATION, and the preliminary condition for the occurrence of the Darwin-Wallace factor of selection. Germinal selection supplies the stones out of which personal selection builds her temples and palaces: ADAPTATIONS. The importance for the theory of the process of degeneration of disused parts cannot be over-estimated, especially when it occurs in sterile animal forms, where we are free from the doubt as to the alleged LAMARCKIAN FACTOR which is apt to confuse our ideas in regard to other cases.

If we regard the variation of the many determinants concerned in the transformation of the female into the sterile worker as having come about through the gradual transformation of the ids into worker-ids, we shall see that the germ-plasm of the sexual ants must contain three kinds of ids, male, female, and worker ids, or if the workers have diverged into soldiers and nest-builders, then four kinds. We understand that the worker-ids arose because their determinants struck out a useful path of variation, whether upward or downward, and that they continued in this path until the highest attainable degree of utility of the parts determined was reached. But in addition to the organs of positive or negative selection-value, there were some which were indifferent as far as the success and especially the functional capacity of the workers was concerned: wings, ovarian tubes, receptaculum seminis, a number of the facets of the eye, perhaps even the whole eye. As to the ovarian tubes it is possible that their degeneration was an advantage for the workers, in saving energy, and if so selection would favour the degeneration; but how could the presence of eyes diminish the usefulness of the workers to the colony? or the minute receptaculum seminis, or even the wings? These parts have therefore degenerated BECAUSE THEY WERE OF NO FURTHER VALUE TO THE INSECT. But if selection did not influence the setting aside of these parts because they were neither of advantage nor of disadvantage to the species, then the Darwinian factor of selection is here confronted with a puzzle which it cannot solve alone, but which at once becomes clear when germinal selection is added. For the determinants of organs that have no further value for the organism, must, as we have already explained, embark on a gradual course of retrograde development.

In ants the degeneration has gone so far that there are no wing-rudiments present in ANY species, as is the case with so many butterflies, flies, and locusts, but in the larvae the imaginal discs of the wings are still laid down. With regard to the ovaries, degeneration has reached different levels in different species of ants, as has been shown by the researches of my former pupil, Elizabeth Bickford. In many species there are twelve ovarian tubes, and they decrease from that number to one; indeed, in one species no ovarian tube at all is present. So much at least is certain from what has been said, that in this case EVERYTHING depends on the fluctuations of the elements of the germ-plasm. Germinal selection, here as elsewhere, presents the variations of the determinants, and personal selection favours or rejects these, or,—if it be a question of organs which have become useless,—it does not come into play at all, and allows the descending variation free course.

It is obvious that even the problem of COADAPTATION IN STERILE ANIMALS can thus be satisfactorily explained. If the determinants are oscillating upwards and downwards in continual fluctuation, and varying more pronouncedly now in one direction now in the other, useful variations of every determinant will continually present themselves anew, and may, in the course of generations, be combined with one another in various ways. But there is one character of the determinants that greatly facilitates this complex process of selection, that, after a certain limit has been reached, they go on varying in the same direction. From this it follows that development along a path once struck out may proceed without the continual intervention of personal selection. This factor only operates, so to speak, at the beginning, when it selects the determinants which are varying in the right direction, and again at the end, when it is necessary to put a check upon further variation. In addition to this, enormously long periods have been available for all these adaptations, as the very gradual transition stages between females and workers in many species plainly show, and thus this process of transformation loses the marvellous and mysterious character that seemed at the first glance to invest it, and takes rank, without any straining, among the other processes of selection. It seems to me that, from the facts that sterile animal forms can adapt themselves to new vital functions, their superfluous parts degenerate, and the parts more used adapt themselves in an ascending direction, those less used in a descending direction, we must draw the conclusion that harmonious adaptation here comes about WITHOUT

31

THE COOPERATION OF THE LAMARCKIAN PRINCIPLE. This conclusion once established, however, we have no reason to refer the thousands of cases of harmonious adaptation, which occur in exactly the same way among other animals or plants, to a principle, the ACTIVE INTERVENTION OF WHICH IN THE TRANSFORMATION OF SPECIES IS NOWHERE PROVED. WE DO NOT REQUIRE IT TO EXPLAIN THE FACTS, AND THEREFORE WE MUST NOT ASSUME IT.

The fact of coadaptation, which was supposed to furnish the strongest argument against the principle of selection, in reality yields the clearest evidence in favour of it. We MUST assume it, BECAUSE NO OTHER POSSIBILITY OF EXPLANATION IS OPEN TO US, AND BECAUSE THESE ADAPTATIONS ACTUALLY EXIST, THAT IS TO SAY, HAVE REALLY TAKEN PLACE. With this conviction I attempted, as far back as 1894, when the idea of germinal selection had not yet occurred to me, to make "harmonious adaptation" (coadaptation) more easily intelligible in some way or other, and so I was led to the idea, which was subsequently expounded in detail by Baldwin, and Lloyd Morgan, and also by Osborn, and Gulick as ORGANIC SELECTION. It seemed to me that it was not necessary that all the germinal variations required for secondary variations should have occurred SIMULTANEOUSLY, since, for instance, in the case of the stag, the bones, muscles, sinews, and nerves would be incited by the increasing heaviness of the antlers to greater activity in THE INDIVIDUAL LIFE, and so would be strengthened. The antlers can only have increased in size by very slow degrees, so that the muscles and bones may have been able to keep pace with their growth in the individual life, until the requisite germinal variations presented themselves. In this way a disharmony between the increasing weight of the antlers and the parts which support and move them would be avoided, since time would be given for the appropriate germinal variations to occur, and so to set agoing the HEREDITARY variation of the muscles, sinews, and bones. ("The Effect of External Influences upon Development", Romanes Lecture, Oxford, 1894.)

I still regard this idea as correct, but I attribute less importance to "organic selection" than I did at that time, in so far that I do not believe that it ALONE could effect complex harmonious adaptations. Germinal selection now seems to me to play the chief part in bringing about such adaptations. Something the same is true of the principle I have called "Panmixia". As I became more and more convinced, in the course of years, that the LAMARCKIAN PRINCIPLE ought not to be called in to explain the dwindling of disused parts, I believed that this process might be simply explained as due to the cessation of the conservative effect of natural selection. I said to myself that, from the moment in which a part ceases to be of use, natural selection withdraws its hand from it, and then it must inevitably fall from the height of its adaptiveness, because inferior variants would have as good a chance of persisting as better ones, since all grades of fitness of the part in question would be mingled with one another indiscriminately. This is undoubtedly true, as Romanes pointed out ten years before I did, and this mingling of the bad with the good probably does bring about a deterioration of the part concerned. But it cannot account for the steady diminution, which always occurs when a part is in process of becoming rudimentary, and which goes on until it ultimately disappears altogether. The process of dwindling cannot therefore be explained as due to panmixia alone; we can only find a sufficient explanation in germinal selection.

IV. DERIVATIVES OF THE THEORY OF SELECTION.

The impetus in all directions given by Darwin through his theory of selection has been an immeasurable one, and its influence is still felt. It falls within the province of the historian of science to enumerate all the ideas which, in the last quarter of the nineteenth century, grew out of Darwin's theories, in the endeavour to penetrate more deeply into the problem of the evolution of the organic world. Within the narrow limits to which this paper is restricted, I cannot attempt to discuss any of these.

V. ARGUMENTS FOR THE REALITY OF THE PROCESSES OF SELECTION.

(a) SEXUAL SELECTION.

Sexual selection goes hand in hand with natural selection. From the very first I have regarded sexual selection as affording an extremely important and interesting corroboration of natural selection, but, singularly enough, it is precisely against this theory that an adverse judgment has been pronounced in so many quarters, and it is only quite recently, and probably in proportion as the wealth of facts in proof of it penetrates into a wider circle, that we seem to be approaching a more general recognition of this side of the problem of adaptation. Thus Darwin's words in his preface to the second edition (1874) of his book, "The Descent of Man and Sexual Selection", are being justified: "My conviction as to the operation of natural selection remains unshaken," and further, "If naturalists were to become more familiar with the idea of sexual selection, it would, I think, be accepted to a much greater extent, and already it is fully and favourably accepted by many competent judges." Darwin was able to speak thus because he was already acquainted with an immense mass of facts, which, taken together, yield overwhelming evidence of the validity of the principle of sexual selection.

NATURAL SELECTION chooses out for reproduction the individuals that are best equipped for the struggle for existence, and it does so at every stage of development; it thus improves the species in all its stages and forms. SEXUAL SELECTION operates only on individuals that are already capable of reproduction, and does so only in relation to the attainment of reproduction. It arises from the rivalry of one sex, usually the male, for the possession of the other, usually the female. Its influence can therefore only DIRECTLY affect one sex, in that it equips it better for attaining possession of the other. But the effect may extend indirectly to the female sex, and thus the whole species may be modified, without, however, becoming any more capable of resistance in the struggle for existence, for sexual selection only gives rise to adaptations which are likely to give their possessor the victory over rivals in the struggle for possession of the female, and which are therefore peculiar to the wooing sex: the manifold "secondary sexual characters." The diversity of these characters is so great that I cannot here attempt to give anything approaching a complete treatment of them, but I should like to give a sufficient number of examples to make the principle itself, in its various modes of expression, quite clear.

One of the chief preliminary postulates of sexual selection is the unequal number of individuals in the two sexes, for if every male immediately finds his mate there can be no competition for the possession of the female. Darwin has shown that, for the most part, the inequality between the sexes is due simply to the fact that there are more males than females, and therefore the males must take some pains to secure a mate. But the inequality does not always depend on the numerical preponderance of the males, it is often due to polygamy; for, if one male claims several females, the number of females in proportion to the rest of the males will be reduced. Since it is almost always the males that are the wooers, we must expect to find the occurrence of secondary sexual characters chiefly among them, and to find it especially frequent in polygamous species. And this is actually the case.

If we were to try to guess—without knowing the facts—what means the male animals make use of to overcome their rivals in the struggle for the possession of the female, we might name many kinds of means, but it would be difficult to suggest any which is not actually employed in some animal group or other. I begin with the mere difference in strength, through which the male of many animals is so sharply distinguished from the female, as, for instance, the lion, walrus, "sea-elephant," and others. Among these the males fight violently for the possession of the female, who falls to the victor in the combat. In this simple case no one can doubt the operation of selection, and there is just as little room for doubt as to the selection-value of the initial stages of the variation. Differences in bodily strength are apparent even among human beings, although in their case the struggle for the possession of the female is no longer decided by bodily strength alone.

Combats between male animals are often violent and obstinate, and the employment of the natural weapons of the species in this way has led to perfecting of these, e.g. the tusks of the boar, the antlers of the stag, and the enormous, antler-like jaws of the stag-beetle. Here again it is impossible to doubt that variations in these organs presented themselves, and that these were considerable enough to be decisive in combat, and so to lead to the improvement of the weapon.

Among many animals, however, the females at first withdraw from the males; they are coy, and have to be sought out, and sometimes held by force. This tracking and grasping of the females by the males has given rise to many different characters in the latter, as, for instance, the larger eyes of the male bee, and especially of the males of the Ephemerids (May-flies), some species of which show, in addition to the usual compound eyes, large, so-called turban-eyes, so that the whole head is covered with seeing surfaces. In these species the females are very greatly in the minority (1-100), and it is easy to understand that a keen competition for them must take place, and that, when the insects of both sexes are floating freely in the air, an unusually wide range of vision will carry with it a decided advantage. Here again the actual adaptations are in accordance with the preliminary postulates of the theory. We do not know the stages through which the eye has passed to its present perfected state, but, since the number of simple eyes (facets) has become very much greater in the male than in the female, we may assume that their increase is due to a gradual duplication of the determinants of the ommatidium in the germ-plasm, as I have already indicated in regard to sense-organs in general. In this case, again, the selection-value of the initial stages hardly admits of doubt; better vision DIRECTLY secures reproduction.

In many cases THE ORGAN OF SMELL shows a similar improvement. Many lower Crustaceans (Daphnidae) have better developed organs of smell in the male sex. The difference is often slight and amounts only to one or two olfactory filaments, but certain species show a difference of nearly a hundred of these filaments (Leptodora). The same thing occurs among insects.

We must briefly consider the clasping or grasping organs which have developed in the males among many lower Crustaceans, but here natural selection plays its part along with sexual selection, for the union of the sexes is an indispensable condition for the maintenance of the species, and as Darwin himself pointed out, in many cases the two forms of selection merge into each other. This fact has always seemed to me to be a proof of natural selection, for, in regard to sexual selection, it is quite obvious that the victory of the best-equipped could have brought about the improvement only of the organs concerned, the factors in the struggle, such as the eye and the olfactory organ.

We come now to the EXCITANTS; that is, to the group of sexual characters whose origin through processes of selection has been most frequently called in question. We may cite the LOVE-CALLS produced by many male insects, such as crickets and cicadas. These could only have arisen in animal groups in which the female did not rapidly flee from the male, but was inclined to accept his wooing from the first. Thus, notes like the chirping of the male cricket serve to entice the females. At first they were merely the signal which showed the presence of a male in the neighbourhood, and the female was gradually enticed nearer and nearer by the continued chirping. The male that could make himself heard to the greatest distance would obtain the largest following, and would transmit the beginnings, and, later, the improvement of his voice to the greatest number of descendants. But sexual excitement in the female became associated with the hearing of the love-call, and then the sound-producing organ of the male began to improve, until it attained to the emission of the long-drawn-out soft notes of the mole-cricket or the maenad-like cry of the cicadas. I cannot here follow the process of development in detail, but will call attention to the fact that the original purpose of the voice, the announcing of the male's presence, became subsidiary, and the exciting of the female became the chief goal to be aimed at. The loudest singers awakened the strongest excitement, and the improvement resulted as a matter of course. I conceive of the origin of bird-song in a somewhat similar manner, first as a means of enticing, then of exciting the female.

One more kind of secondary sexual character must here be mentioned: the odour which emanates from so many animals at the breeding season. It is possible that this odour also served at first merely to give notice of the presence of individuals of the other sex, but it soon became an excitant, and as the individuals which caused the greatest degree of excitement were preferred, it reached as high a pitch of perfection as was possible to it. I shall confine myself here to the comparatively recently discovered fragrance of butterflies. Since Fritz Muller found out that certain Brazilian butterflies gave off fragrance "like a flower," we have become acquainted with many such cases, and we now know that in all lands, not only many diurnal Lepidoptera but nocturnal ones also give off a delicate odour, which is agreeable even to man. The ethereal oil to which this fragrance is due is secreted by the skin-cells, usually of the wing, as I showed soon after the discovery of the SCENT-SCALES. This is the case in the males; the females have no SPECIAL scent-scales recognisable as such by their form, but they must, nevertheless, give off an extremely delicate fragrance, although our imperfect organ of smell cannot perceive it, for the males become aware of the presence of a female, even at night, from a long distance off, and gather round her. We may therefore conclude, that both sexes have long given forth a very delicate perfume, which announced their presence to others of the same species, and that in many species (NOT IN ALL) these small beginnings became, in the males, particularly strong scent-scales of characteristic form (lute, brush, or lyre-shaped). At first these scales were scattered over the surface of the wing, but gradually they concentrated themselves, and formed broad, velvety bands, or strong, prominent brushes, and they attained their highest pitch of evolution when they became enclosed within pits or folds of the skin, which could be opened to let the delicious fragrance stream forth suddenly towards the female. Thus in this case also we see that characters, the original use of which was to bring the sexes together, and so to maintain the species, have been evolved in the males into means for exciting the female. And we can hardly doubt, that the females are most readily enticed to yield to the butterfly that sends out the strongest fragrance,—that is to say, that excites them to the highest degree. It is a pity that our organs of smell are not fine enough to examine the fragrance of male Lepidoptera in general, and to compare it with other perfumes which attract these insects. (See Poulton, "Essays on Evolution", 1908, pages 316, 317.) As far as we can perceive them they resemble the fragrance of flowers, but there are Lepidoptera whose scent suggests musk. A smell of musk is also given off by several plants: it is a sexual excitant in the musk-deer, the musk-sheep, and the crocodile.

As far as we know, then, it is perfumes similar to those of flowers that the male Lepidoptera give off in order to entice their mates, and this is a further indication that animals, like plants, can to a large extent meet the claims made upon them by life, and produce the adaptations which are most purposive,—a further proof, too, of my proposition that the useful variations, so to speak, are ALWAYS THERE. The flowers developed the perfumes which entice their visitors, and the male Lepidoptera developed the perfumes which entice and excite their mates.

There are many pretty little problems to be solved in this connection, for there are insects, such as some flies, that are attracted by smells which are unpleasant to us, like those from decaying flesh and carrion. But there are also certain flowers, some orchids for instance, which give forth no very agreeable odour, but one which is to us repulsive and disgusting; and we should therefore expect that the males of such insects would give off a smell unpleasant to us, but there is no case known to me in which this has been demonstrated.

In cases such as we have discussed, it is obvious that there is no possible explanation except through selection. This brings us to the last kind of secondary sexual characters, and the one in regard to which doubt has been most frequently expressed,—decorative colours and decorative forms, the brilliant plumage of the male pheasant, the humming-birds, and the bird of Paradise, as well as the bright colours of many species of butterfly, from the beautiful blue of our little Lycaenidae to the magnificent azure of the large Morphinae of Brazil. In a great many cases, though not by any means in all, the male butterflies are "more beautiful" than the females, and in the Tropics in particular they shine and glow in the most superb colours. I really see no reason why we should doubt the power of sexual selection, and I myself stand wholly on Darwin's side. Even though we certainly cannot assume that the females exercise a conscious choice of the "handsomest" mate, and deliberate like the judges in a court of justice over the perfections of their wooers, we have no reason to doubt that distinctive forms (decorative feathers) and colours have a particularly exciting effect upon the female, just as certain odours have among animals of so many different groups, including the butterflies. The doubts which existed for a considerable time, as a result of fallacious experiments, as to whether the colours of flowers really had any influence in attracting butterflies have now been set at rest through a series of more careful investigations; we now know that the colours of flowers are there on account of the butterflies, as Sprengel first showed, and that the blossoms of Phanerogams are selected in relation to them, as Darwin pointed out.

Certainly it is not possible to bring forward any convincing proof of the origin of decorative colours through sexual selection, but there are many weighty arguments in favour of it, and these form a body of presumptive evidence so strong that it almost amounts to certainty.

In the first place, there is the analogy with other secondary sexual characters. If the song of birds and the chirping of the cricket have been evolved through sexual selection, if the penetrating odours of male animals,—the crocodile, the musk-deer, the beaver, the carnivores, and, finally, the flower-like fragrances of the butterflies have been evolved to their present pitch in this way, why should decorative colours have arisen in some other way? Why should the eye be less sensitive to SPECIFICALLY MALE colours and other VISIBLE signs ENTICING TO THE FEMALE, than the olfactory sense to specifically male odours, or the sense of hearing to specifically male sounds? Moreover, the decorative feathers of birds are almost always spread out and displayed before the female during courtship. I have elsewhere ("The Evolution Theory", London, 1904, I. page 219.) pointed out that decorative colouring and sweet-scentedness may replace one another in Lepidoptera as well as in flowers, for just as some modestly coloured flowers (mignonette and violet) have often a strong perfume, while strikingly coloured ones are sometimes quite devoid of fragrance, so we find that the most beautiful and gaily-coloured of our native Lepidoptera, the species of Vanessa, have no scent-scales, while these are often markedly developed in grey nocturnal Lepidoptera. Both attractions may, however, be combined in butterflies, just as in flowers. Of course, we cannot explain why both means of attraction should exist in one genus, and only one of them in another, since we do not know the minutest details of the conditions of life of the genera concerned. But from the sporadic distribution of scent-scales in Lepidoptera, and from their occurrence or absence in nearly related species, we may conclude that fragrance is a relatively MODERN acquirement, more recent than brilliant colouring.

One thing in particular that stamps decorative colouring as a product of selection is ITS GRADUAL INTENSIFICATION by the addition of new spots, which we can quite well observe, because in many cases the colours have been first acquired by the males, and later transmitted to the females by inheritance. The scent-scales are never thus transmitted, probably for the same reason that the decorative colours of many birds are often not transmitted to the females: because with these they would be exposed to too great elimination by enemies. Wallace was the first to point out that in species with concealed nests the beautiful feathers of the male occurred in the female also, as in the parrots, for instance, but this is not the case in species which brood on an exposed nest. In the parrots one can often observe that the general brilliant colouring of the male is found in the female, but that certain spots of colour are absent, and these have probably been acquired comparatively recently by the male and have not yet been transmitted to the female.

Isolation of the group of individuals which is in process of varying is undoubtedly of great value in sexual selection, for even a solitary conspicuous variation will become dominant much sooner in a small isolated colony, than among a large number of members of a species.

Anyone who agrees with me in deriving variations from germinal selection will regard that process as an essential aid towards explaining the selection of distinctive courtship-characters, such as coloured spots, decorative feathers, horny outgrowths in birds and reptiles, combs, feather-tufts, and the like, since the beginnings of these would be presented

36

with relative frequency in the struggle between the determinants within the germ-plasm. The process of transmission of decorative feathers to the female results, as Darwin pointed out and illustrated by interesting examples, in the COLOUR-TRANSFORMATION OF A WHOLE SPECIES, and this process, as the phyletically older colouring of young birds shows, must, in the course of thousands of years, have repeated itself several times in a line of descent.

If we survey the wealth of phenomena presented to us by secondary sexual characters, we can hardly fail to be convinced of the truth of the principle of sexual selection. And certainly no one who has accepted natural selection should reject sexual selection, for, not only do the two processes rest upon the same basis, but they merge into one another, so that it is often impossible to say how much of a particular character depends on one and how much on the other form of selection.

(b) NATURAL SELECTION.

An actual proof of the theory of sexual selection is out of the question, if only because we cannot tell when a variation attains to selection-value. It is certain that a delicate sense of smell is of value to the male moth in his search for the female, but whether the possession of one additional olfactory hair, or of ten, or of twenty additional hairs leads to the success of its possessor we are unable to tell. And we are groping even more in the dark when we discuss the excitement caused in the female by agreeable perfumes, or by striking and beautiful colours. That these do make an impression is beyond doubt; but we can only assume that slight intensifications of them give any advantage, and we MUST assume this SINCE OTHERWISE SECONDARY SEXUAL CHARACTERS REMAIN INEXPLICABLE.

The same thing is true in regard to natural selection. It is not possible to bring forward any actual proof of the selection-value of the initial stages, and the stages in the increase of variations, as has been already shown. But the selection-value of a finished adaptation can in many cases be statistically determined. Cesnola and Poulton have made valuable experiments in this direction. The former attached forty-five individuals of the green, and sixty-five of the brown variety of the praying mantis (Mantis religiosa), by a silk thread to plants, and watched them for seventeen days. The insects which were on a surface of a colour similar to their own remained uneaten, while twenty-five green insects on brown parts of plants had all disappeared in eleven days.

The experiments of Poulton and Sanders ("Report of the British Association" (Bristol, 1898), London, 1899, pages 906-909.) were made with 600 pupae of Vanessa urticae, the "tortoise-shell butterfly." The pupae were artificially attached to nettles, tree-trunks, fences, walls, and to the ground, some at Oxford, some at St Helens in the Isle of Wight. In the course of a month 93 per cent of the pupae at Oxford were killed, chiefly by small birds, while at St Helens 68 per cent perished. The experiments showed very clearly that the colour and character of the surface on which the pupa rests—and thus its own conspicuousness—are of the greatest importance. At Oxford only the four pupae which were fastened to nettles emerged; all the rest—on bark, stones and the like—perished. At St Helens the elimination was as follows: on fences where the pupae were conspicuous, 92 per cent; on bark, 66 per cent; on walls, 54 per cent; and among nettles, 57 per cent. These interesting experiments confirm our views as to protective coloration, and show further, THAT THE RATIO OF ELIMINATION IN THE SPECIES IS A VERY HIGH ONE, AND THAT THEREFORE SELECTION MUST BE VERY KEEN.

We may say that the process of selection follows as a logical necessity from the fulfilment of the three preliminary postulates of the theory: variability, heredity, and the struggle for existence, with its enormous ratio of elimination in all species. To this we must add a fourth factor, the INTENSIFICATION of variations which Darwin established as a fact, and which we are now able to account for theoretically on the basis of germinal selection. It may be objected that there is considerable uncertainty about this LOGICAL proof, because of our inability to demonstrate the selection-value of the initial stages and the individual stages of increase. We have therefore to fall back on PRESUMPTIVE

EVIDENCE. This is to be found in THE INTERPRETATIVE VALUE OF THE THEORY. Let us consider this point in greater detail.

In the first place, it is necessary to emphasise what is often overlooked, namely, that the theory not only explains the TRANSFORMATIONS of species, it also explains THEIR REMAINING THE SAME; in addition to the principle of varying, it contains within itself that of PERSISTING. It is part of the essence of selection, that it not only causes a part to VARY till it has reached its highest pitch of adaptation, but that it MAINTAINS IT AT THIS PITCH. THIS CONSERVING INFLUENCE OF NATURAL SELECTION is of great importance, and was early recognised by Darwin; it follows naturally from the principle of the survival of the fittest.

We understand from this how it is that a species which has become fully adapted to certain conditions of life ceases to vary, but remains "constant," as long as the conditions of life FOR IT remain unchanged, whether this be for thousands of years, or for whole geological epochs. But the most convincing proof of the power of the principle of selection lies in the innumerable multitude of phenomena which cannot be explained in any other way. To this category belong all structures which are only PASSIVELY of advantage to the organism, because none of these can have arisen by the alleged LAMARCKIAN PRINCIPLE. These have been so often discussed that we need do no more than indicate them here. Until quite recently the sympathetic coloration of animals—for instance, the whiteness of Arctic animals—was referred, at least in part, to the DIRECT influence of external factors, but the facts can best be explained by referring them to the processes of selection, for then it is unnecessary to make the gratuitous assumption that many species are sensitive to the stimulus of cold and that others are not. The great majority of Arctic land-animals, mammals and birds, are white, and this proves that they were all able to present the variation which was most useful for them. The sable is brown, but it lives in trees, where the brown colouring protects and conceals it more effectively. The musk-sheep (Ovibos moschatus) is also brown, and contrasts sharply with the ice and snow, but it is protected from beasts of prey by its gregarious habit, and therefore it is of advantage to be visible from as great a distance as possible. That so many species have been able to give rise to white varieties does not depend on a special sensitiveness of the skin to the influence of cold, but to the fact that Mammals and Birds have a general tendency to vary towards white. Even with us, many birds—starlings, blackbirds, swallows, etc.—occasionally produce white individuals, but the white variety does not persist, because it readily falls a victim to the carnivores. This is true of white fawns, foxes, deer, etc. The whiteness, therefore, arises from internal causes, and only persists when it is useful. A great many animals living in a GREEN ENVIRONMENT have become clothed in green, especially insects, caterpillars, and Mantidae, both persecuted and persecutors.

That it is not the direct effect of the environment which calls forth the green colour is shown by the many kinds of caterpillar which rest on leaves and feed on them, but are nevertheless brown. These feed by night and betake themselves through the day to the trunk of the tree, and hide in the furrows of the bark. We cannot, however, conclude from this that they were UNABLE to vary towards green, for there are Arctic animals which are white only in winter and brown in summer (Alpine hare, and the ptarmigan of the Alps), and there are also green leaf-insects which remain green only while they are young and difficult to see on the leaf, but which become brown again in the last stage of larval life, when they have outgrown the leaf. They then conceal themselves by day, sometimes only among withered leaves on the ground, sometimes in the earth itself. It is interesting that in one genus, Chaerocampa, one species is brown in the last stage of larval life, another becomes brown earlier, and in many species the last stage is not wholly brown, a part remaining green. Whether this is a case of a double adaptation, or whether the green is being gradually crowded out by the brown, the fact remains that the same species, even the same individual, can exhibit both variations. The case is the same with many of the leaf-like Orthoptera, as, for instance, the praying mantis (Mantis religiosa) which we have already mentioned.

But the best proofs are furnished by those often-cited cases in which the insect bears a deceptive resemblance to another object. We now know many such cases, such as the numerous imitations of green or withered leaves, which are brought about in the most diverse ways, sometimes by mere variations in the form of the insect and in its colour, sometimes by an elaborate marking, like that which occurs in the Indian leaf-butterflies, Kallima inachis. In the single butterfly-genus Anaea, in the woods of South America, there are about a hundred species which are all gaily coloured on the upper surface, and on the reverse side exhibit the most delicate imitation of the colouring and pattern of a leaf, generally without any indication of the leaf-ribs, but extremely deceptive nevertheless. Anyone who has seen only one such butterfly may doubt whether many of the insignificant details of the marking can really be of advantage to the insect. Such details are for instance the apparent holes and splits in the apparently dry or half-rotten leaf, which are usually due to the fact that the scales are absent on a circular or oval patch so that the colourless wing-membrane lies bare, and one can look through the spot as through a window. Whether the bird which is seeking or pursuing the butterflies takes these holes for dewdrops, or for the work of a devouring insect, does not affect the question; the mirror-like spot undoubtedly increases the general deceptiveness, for the same thing occurs in many leaf-butterflies, though not in all, and in some cases it is replaced in quite a peculiar manner. In one species of Anaea (A. divina), the resting butterfly looks exactly like a leaf out of the outer edge of which a large semicircular piece has been eaten, possibly by a caterpillar; but if we look more closely it is obvious that there is no part of the wing absent, and that the semicircular piece is of a clear, pale yellow colour, while the rest of the wing is of a strongly contrasted dark brown.

But the deceptive resemblance may be caused in quite a different manner. I have often speculated as to what advantage the brilliant white C could give to the otherwise dusky-coloured "Comma butterfly" (Grapta C. album). Poulton's recent observations ("Proc. Ent. Soc"., London, May 6, 1903.) have shown that this represents the imitation of a crack such as is often seen in dry leaves, and is very conspicuous because the light shines through it.

The utility obviously lies in presenting to the bird the very familiar picture of a broken leaf with a clear shining slit, and we may conclude, from the imitation of such small details, that the birds are very sharp observers and that the smallest deviation from the usual arrests their attention and incites them to closer investigation. It is obvious that such detailed—we might almost say such subtle—deceptive resemblances could only have come about in the course of long ages through the acquirement from time to time of something new which heightened the already existing resemblance.

In face of facts like these there can be no question of chance, and no one has succeeded so far in finding any other explanation to replace that by selection. For the rest, the apparent leaves are by no means perfect copies of a leaf; many of them only represent the torn or broken piece, or the half or two-thirds of a leaf, but then the leaves themselves frequently do not present themselves to the eye as a whole, but partially concealed among other leaves. Even those butterflies which, like the species of Kallima and Anaea, represent the whole of a leaf with stalk, ribs, apex, and the whole breadth, are not actual copies which would satisfy a botanist; there is often much wanting. In Kallima the lateral ribs of the leaf are never all included in the markings; there are only two or three on the left side and at most four or five on the right, and in many individuals these are rather obscure, while in others they are comparatively distinct. This furnishes us with fresh evidence in favour of their origin through processes of selection, for a botanically perfect picture could not arise in this way; there could only be a fixing of such details as heightened the deceptive resemblance.

Our postulate of origin through selection also enables us to understand why the leaf-imitation is on the lower surface of the wing in the diurnal Lepidoptera, and on the upper surface in the nocturnal forms, corresponding to the attitude of the wings in the resting position of the two groups.

The strongest of all proofs of the theory, however, is afforded by cases of true "mimicry," those adaptations discovered by Bates in 1861, consisting in the imitation of one species by another, which becomes more and more like its model. The model is always a species that enjoys some special protection from enemies, whether because it is unpleasant to taste, or because it is in some way dangerous.

It is chiefly among insects and especially among butterflies that we find the greatest number of such cases. Several of these have been minutely studied, and every detail has been investigated, so that it is difficult to understand how there can still be disbelief in regard to them. If the many and exact observations which have been carefully collected and critically discussed, for instance by Poulton ("Essays on Evolution", 1889-1907, Oxford, 1908, passim, e.g. page 269.) were thoroughly studied, the arguments which are still frequently urged against mimicry would be found untenable; we can hardly hope to find more convincing proof of the actuality of the processes of selection than these cases put into our hands. The preliminary postulates of the theory of mimicry have been disputed, for instance, that diurnal butterflies are persecuted and eaten by birds, but observations specially directed towards this point in India, Africa, America and Europe have placed it beyond all doubt. If it were necessary I could myself furnish an account of my own observations on this point.

In the same way it has been established by experiment and observation in the field that in all the great regions of distribution there are butterflies which are rejected by birds and lizards, their chief enemies, on account of their unpleasant smell or taste. These butterflies are usually gaily and conspicuously coloured and thus—as Wallace first interpreted it—are furnished with an easily recognisable sign: a sign of unpalatableness or WARNING COLOURS. If they were not thus recognisable easily and from a distance, they would frequently be pecked at by birds, and then rejected because of their unpleasant taste; but as it is, the insect-eaters recognise them at once as unpalatable booty and ignore them. Such IMMUNE (The expression does not refer to all the enemies of this butterfly; against ichneumon-flies, for instance, their unpleasant smell usually gives no protection.) species, wherever they occur, are imitated by other palatable species, which thus acquire a certain degree of protection.

It is true that this explanation of the bright, conspicuous colours is only a hypothesis, but its foundations,—unpalatableness, and the liability of other butterflies to be eaten,—are certain, and its consequences—the existence of mimetic palatable forms—confirm it in the most convincing manner. Of the many cases now known I select one, which is especially remarkable, and which has been thoroughly investigated, Papilio dardanus (merope), a large, beautiful, diurnal butterfly which ranges from Abyssinia throughout the whole of Africa to the south coast of Cape Colony.

The males of this form are everywhere ALMOST the same in colour and in form of wings, save for a few variations in the sparse black markings on the pale yellow ground. But the females occur in several quite different forms and colourings, and one of these only, the Abyssinian form, is like the male, while the other three or four are MIMETIC, that is to say, they copy a butterfly of quite a different family the Danaids, which are among the IMMUNE forms. In each region the females have thus copied two or three different immune species. There is much that is interesting to be said in regard to these species, but it would be out of keeping with the general tenor of this paper to give details of this very complicated case of polymorphism in P. dardanus. Anyone who is interested in the matter will find a full and exact statement of the case in as far as we know it, in Poulton's "Essays on Evolution" (pages 373-375). (Professor Poulton has corrected some wrong descriptions which I had unfortunately overlooked in the Plates of my book "Vortrage uber Descendenztheorie", and which refer to Papilio dardanus (merope). These mistakes are of no importance as far as and understanding of the mimicry-theory is concerned, but I hope shortly to be able to correct them in a later edition.) I need only add that three different mimetic female forms have been reared from the eggs of a single female in South Africa. The resemblance of these forms to their immune models goes so far that even the details of the LOCAL forms of the models are copied by the mimetic species.

It remains to be said that in Madagascar a butterfly, Papilio meriones, occurs, of which both sexes are very similar in form and markings to the non-mimetic male of P. dardanus, so that it probably represents the ancestor of this latter species.

In face of such facts as these every attempt at another explanation must fail. Similarly all the other details of the case fulfil the preliminary postulates of selection, and leave no room for any other interpretation. That the males do not take on the protective colouring is easily explained, because they are in general more numerous, and the females are more important for the preservation of the species, and must also live longer in order to deposit their eggs. We find the same state of things in many other species, and in one case (Elymnias undularis) in which the male is also mimetically coloured, it copies quite a differently coloured immune species from the model followed by the female. This is quite intelligible when we consider that if there were TOO MANY false immune types, the birds would soon discover that there were palatable individuals among those with unpalatable warning colours. Hence the imitation of different immune species by Papilio dardanus!

I regret that lack of space prevents my bringing forward more examples of mimicry and discussing them fully. But from the case of Papilio dardanus alone there is much to be learnt which is of the highest importance for our understanding of transformations. It shows us chiefly what I once called, somewhat strongly perhaps, THE OMNIPOTENCE OF NATURAL SELECTION in answer to an opponent who had spoken of its "inadequacy." We here see that one and the same species is capable of producing four or five different patterns of colouring and marking; thus the colouring and marking are not, as has often been supposed, a necessary outcome of the specific nature of the species, but a true adaptation, which cannot arise as a direct effect of climatic conditions, but solely through what I may call the sorting out of the variations produced by the species, according to their utility. That caterpillars may be either green or brown is already something more than could have been expected according to the old conception of species, but that one and the same butterfly should be now pale yellow, with black; now red with black and pure white; now deep black with large, pure white spots; and again black with a large ochreous-yellow spot, and many small white and yellow spots; that in one sub-species it may be tailed like the ancestral form, and in another tailless like its Danaid model,—all this shows a far-reaching capacity for variation and adaptation that wide never have expected if we did not see the facts before us. How it is possible that the primary colour-variations should thus be intensified and combined remains a puzzle even now; we are reminded of the modern three-colour printing,—perhaps similar combinations of the primary colours take place in this case; in any case the direction of these primary variations is determined by the artist whom we know as natural selection, for there is no other conceivable way in which the model could affect the butterfly that is becoming more and more like it. The same climate surrounds all four forms of female; they are subject to the same conditions of nutrition. Moreover, Papilio dardanus is by no means the only species of butterfly which exhibits different kinds of colour-pattern on its wings. Many species of the Asiatic genus Elymnias have on the upper surface a very good imitation of an immune Euploeine (Danainae), often with a steel-blue ground-colour, while the under surface is well concealed when the butterfly is at rest,—thus there are two kinds of protective coloration each with a different meaning! The same thing may be observed in many non-mimetic butterflies, for instance in all our species of Vanessa, in which the under side shows a grey-brown or brownish-black protective coloration, but we do not yet know with certainty what may be the biological significance of the gaily coloured upper surface.

In general it may be said that mimetic butterflies are comparatively rare species, but there are exceptions, for instance Limenitis archippus in North America, of which the immune model (Danaida plexippus) also occurs in enormous numbers.

In another mimicry-category the imitators are often more numerous than the models, namely in the case of the imitation of DANGEROUS INSECTS by harmless species. Bees and wasps are dreaded for their sting, and they are copied by harmless flies of the genera Eristalis and Syrphus, and these mimics often occur in swarms about flowering plants

41

without damage to themselves or to their models; they are feared and are therefore left unmolested.

In regard also to the FAITHFULNESS OF THE COPY the facts are quite in harmony with the theory, according to which the resemblance must have arisen and increased BY DEGREES. We can recognise this in many cases, for even now the mimetic species show very VARYING DEGREES OF RESEMBLANCE to their immune model. If we compare, for instance, the many different imitators of Danaida chrysippus we find that, with their brownish-yellow ground-colour, and the position and size, and more or less sharp limitation of their clear marginal spots, they have reached very different degrees of nearness to their model. Or compare the female of Elymnias undularis with its model Danaida genutia; there is a general resemblance, but the marking of the Danaida is very roughly imitated in Elymnias.

Another fact that bears out the theory of mimicry is, that even when the resemblance in colour-pattern is very great, the WING-VENATION, which is so constant, and so important in determining the systematic position of butterflies, is never affected by the variation. The pursuers of the butterfly have no time to trouble about entomological intricacies.

I must not pass over a discovery of Poulton's which is of great theoretical importance— that mimetic butterflies may reach the same effect by very different means. ("Journ. Linn. Soc. London (Zool.)", Vol. XXVI. 1898, pages 598-602.) Thus the glass-like transparency of the wing of a certain Ithomiine (Methona) and its Pierine mimic (Dismorphia orise) depends on a diminution in the size of the scales; in the Danaine genus Ituna it is due to the fewness of the scales, and in a third imitator, a moth (Castnia linus var. heliconoides) the glass-like appearance of the wing is due neither to diminution nor to absence of scales, but to their absolute colourlessness and transparency, and to the fact that they stand upright. In another moth mimic (Anthomyza) the arrangement of the transparent scales is normal. Thus it is not some unknown external influence that has brought about the transparency of the wing in these five forms, as has sometimes been supposed. Nor is it a hypothetical INTERNAL evolutionary tendency, for all three vary in a different manner. The cause of this agreement can only lie in selection, which preserves and intensifies in each species the favourable variations that present themselves. The great faithfulness of the copy is astonishing in these cases, for it is not THE WHOLE wing which is transparent; certain markings are black in colour, and these contrast sharply with the glass-like ground. It is obvious that the pursuers of these butterflies must be very sharp-sighted, for otherwise the agreement between the species could never have been pushed so far. The less the enemies see and observe, the more defective must the imitation be, and if they had been blind, no visible resemblance between the species which required protection could ever have arisen.

A seemingly irreconcilable contradiction to the mimicry theory is presented in the following cases, which were known to Bates, who, however, never succeeded in bringing them into line with the principle of mimicry.

In South America there are, as we have already said, many mimics of the immune Ithomiinae (or as Bates called them Heliconidae). Among these there occur not merely species which are edible, and thus require the protection of a disguise, but others which are rejected on account of their unpalatableness. How could the Ithomiine dress have developed in their case, and of what use is it, since the species would in any case be immune? In Eastern Brazil, for instance, there are four butterflies, which bear a most confusing resemblance to one another in colour, marking, and form of wing, and all four are unpalatable to birds. They belong to four different genera and three sub-families, and we have to inquire: Whence came this resemblance and what end does it serve? For a long time no satisfactory answer could be found, but Fritz Muller (In "Kosmos", 1879, page 100.), seventeen years after Bates, offered a solution to the riddle, when he pointed out that young birds could not have an instinctive knowledge of the unpalatableness of the Ithomiines, but must learn by experience which species were edible and which inedible. Thus each young bird must have tasted at least one individual of each inedible species and discovered its unpalatability, before it learnt to avoid, and thus to spare the species. But if the four species resemble each other very closely the bird will regard them all as of the same kind, and avoid them all. Thus there developed a process of selection which resulted in the survival of the Ithomiine-like individuals, and in so great an increase of resemblance between the four species, that they are difficult to distinguish one from another even in a collection. The advantage for the four species, living side by side as they do e.g. in Bahia, lies in the fact that only one individual from the MIMICRY-RING ("inedible association") need be tasted by a young bird, instead of at least four individuals, as would otherwise be the case. As the number of young birds is great, this makes a considerable difference in the ratio of elimination.

These interesting mimicry-rings (trusts), which have much significance for the theory, have been the subject of numerous and careful investigations, and at least their essential features are now fully established. Muller took for granted, without making any investigations, that young birds only learn by experience to distinguish between different kinds of victims. But Lloyd Morgan's ("Habit and Instinct", London, 1896.) experiments with young birds proved that this is really the case, and at the same time furnished an additional argument against the LAMARCKIAN PRINCIPLE.

In addition to the mimicry-rings first observed in South America, others have been described from Tropical India by Moore, and by Poulton and Dixey from Africa, and we may expect to learn many more interesting facts in this connection. Here again the preliminary postulates of the theory are satisfied. And how much more that would lead to the same conclusion might be added!

As in the case of mimicry many species have come to resemble one another through processes of selection, so we know whole classes of phenomena in which plants and animals have become adapted to one another, and have thus been modified to a considerable degree. I refer particularly to the relation between flowers and insects; but as there is an article on "The Biology of Flowers" in this volume, I need not discuss the subject, but will confine myself to pointing out the significance of these remarkable cases for the theory of selection. Darwin has shown that the originally inconspicuous blossoms of the phanerogams were transformed into flowers through the visits of insects, and that, conversely, several large orders of insects have been gradually modified by their association with flowers, especially as regards the parts of their body actively concerned. Bees and butterflies in particular have become what they are through their relation to flowers. In this case again all that is apparently contradictory to the theory can, on closer investigation, be beautifully interpreted in corroboration of it. Selection can give rise only to what is of use to the organism actually concerned, never to what is of use to some other organism, and we must therefore expect to find that in flowers only characters of use to THEMSELVES have arisen, never characters which are of use to insects only, and conversely that in the insects characters useful to them and not merely to the plants would have originated. For a long time it seemed as if an exception to this rule existed in the case of the fertilisation of the yucca blossoms by a little moth, Pronuba yuccasella. This little moth has a sickle-shaped appendage to its mouth-parts which occurs in no other Lepidopteron, and which is used for pushing the yellow pollen into the opening of the pistil, thus fertilising the flower. Thus it appears as if a new structure, which is useful only to the plant, has arisen in the insect. But the difficulty is solved as soon as we learn that the moth lays its eggs in the fruit-buds of the Yucca, and that the larvae, when they emerge, feed on the developing seeds. In effecting the fertilisation of the flower the moth is at the same time making provision for its own offspring, since it is only after fertilisation that the seeds begin to develop. There is thus nothing to prevent our referring this structural adaptation in Pronuba yuccasella to processes of selection, which have gradually transformed the maxillary palps of the female into the sickle-shaped instrument for collecting the pollen, and which have at the same time developed in the insect the instinct to press the pollen into the pistil.

In this domain, then, the theory of selection finds nothing but corroboration, and it would be impossible to substitute for it any other explanation, which, now that the facts are so well known, could be regarded as a serious rival to it. That selection is a factor, and a very powerful factor in the evolution of organisms, can no longer be doubted. Even although we cannot bring forward formal proofs of it IN DETAIL, cannot calculate definitely the size of the variations which present themselves, and their selection-value, cannot, in short, reduce the whole process to a mathematical formula, yet we must assume selection, because it is the only possible explanation applicable to whole classes of phenomena, and because, on the other hand, it is made up of factors which we know can be proved actually to exist, and which, IF they exist, must of logical necessity cooperate in the manner required by the theory. WE MUST ACCEPT IT BECAUSE THE PHENOMENA OF EVOLUTION AND ADAPTATION MUST HAVE A NATURAL BASIS, AND BECAUSE IT IS THE ONLY POSSIBLE EXPLANATION OF THEM. (This has been discussed in many of my earlier works. See for instance "The All-Sufficiency of Natural Selection, a reply to Herbert Spencer", London, 1893.)

Many people are willing to admit that selection explains adaptations, but they maintain that only a part of the phenomena are thus explained, because everything does not depend upon adaptation. They regard adaptation as, so to speak, a special effort on the part of Nature, which she keeps in readiness to meet particularly difficult claims of the external world on organisms. But if we look at the matter more carefully we shall find that adaptations are by no means exceptional, but that they are present everywhere in such enormous numbers, that it would be difficult in regard to any structure whatever, to prove that adaptation had NOT played a part in its evolution.

How often has the senseless objection been urged against selection that it can create nothing, it can only reject. It is true that it cannot create either the living substance or the variations of it; both must be given. But in rejecting one thing it preserves another, intensifies it, combines it, and in this way CREATES what is new. EVERYTHING in organisms depends on adaptation; that is to say, everything must be admitted through the narrow door of selection, otherwise it can take no part in the building up of the whole. But, it is asked, what of the direct effect of external conditions, temperature, nutrition, climate and the like? Undoubtedly these can give rise to variations, but they too must pass through the door of selection, and if they cannot do this they are rejected, eliminated from the constitution of the species.

It may, perhaps, be objected that such external influences are often of a compelling power, and that every animal MUST submit to them, and that thus selection has no choice and can neither select nor reject. There may be such cases; let us assume for instance that the effect of the cold of the Arctic regions was to make all the mammals become black; the result would be that they would all be eliminated by selection, and that no mammals would be able to live there at all. But in most cases a certain percentage of animals resists these strong influences, and thus selection secures a foothold on which to work, eliminating the unfavourable variation, and establishing a useful colouring, consistent with what is required for the maintenance of the species.

Everything depends upon adaptation! We have spoken much of adaptation in colouring, in connection with the examples brought into prominence by Darwin, because these are conspicuous, easily verified, and at the same time convincing for the theory of selection. But is it only desert and polar animals whose colouring is determined through adaptation? Or the leaf-butterflies, and the mimetic species, or the terrifying markings, and "warning-colours" and a thousand other kinds of sympathetic colouring? It is, indeed, never the colouring alone which makes up the adaptation; the structure of the animal plays a part, often a very essential part, in the protective disguise, and thus MANY variations may cooperate towards ONE common end. And it is to be noted that it is by no means only external parts that are changed; internal parts are ALWAYS modified at the same time—for instance, the delicate elements of the nervous system on which depend the INSTINCT of the insect to hold its wings, when at rest, in a perfectly definite position, which, in the leaf-butterfly, has the effect of bringing the two pieces on which the marking occurs on the anterior and posterior wing into the same direction, and thus displaying as a whole the fine curve of the midrib on the seeming leaf. But the wing-holding instinct is not regulated in the same way in all leaf-butterflies; even our indigenous species of Vanessa, with their protective ground-colouring, have quite a distinctive way of holding their wings so that the greater part of the anterior wing is covered by the posterior when the butterfly is at rest. But the protective colouring appears on the posterior wing and on the tip of the anterior, TO PRECISELY THE DISTANCE TO WHICH IT IS LEFT UNCOVERED. This occurs, as Standfuss has shown, in different degree in our two most nearly allied species, the uncovered portion being smaller in V. urticae than in V. polychloros. In this case, as in most leaf-butterflies, the holding of the wing was probably the primary character; only after that was thoroughly established did the protective marking develop. In any case, the instinctive manner of holding the wings is associated with the protective colouring, and must remain as it is if the latter is to be effective. How greatly instincts may change, that is to say, may be adapted, is shown by the case of the Noctuid "shark" moth, Xylina vetusta. This form bears a most deceptive resemblance to a piece of rotten wood, and the appearance is greatly increased by the modification of the innate impulse to flight common to so many animals, which has here been transformed into an almost contrary instinct. This moth does not fly away from danger, but "feigns death," that is, it draws antennae, legs and wings close to the body, and remains perfectly motionless. It may be touched, picked up, and thrown down again, and still it does not move. This remarkable instinct must surely have developed simultaneously with the wood-colouring; at all events, both cooperating variations are now present, and prove that both the external and the most minute internal structure have undergone a process of adaptation.

The case is the same with all structural variations of animal parts, which are not absolutely insignificant. When the insects acquired wings they must also have acquired the mechanism with which to move them—the musculature, and the nervous apparatus necessary for its automatic regulation. All instincts depend upon compound reflex mechanisms and are just as indispensable as the parts they have to set in motion, and all may have arisen through processes of selection if the reasons which I have elsewhere given for this view are correct. ("The Evolution Theory", London, 1904, page 144.)

Thus there is no lack of adaptations within the organism, and particularly in its most important and complicated parts, so that we may say that there is no actively functional organ that has not undergone a process of adaptation relative to its function and the requirements of the organism. Not only is every gland structurally adapted, down to the very minutest histological details, to its function, but the function is equally minutely adapted to the needs of the body. Every cell in the mucous lining of the intestine is exactly regulated in its relation to the different nutritive substances, and behaves in quite a different way towards the fats, and towards nitrogenous substances, or peptones.

I have elsewhere called attention to the many adaptations of the whale to the surrounding medium, and have pointed out—what has long been known, but is not universally admitted, even now—that in it a great number of important organs have been transformed in adaptation to the peculiar conditions of aquatic life, although the ancestors of the whale must have lived, like other hair-covered mammals, on land. I cited a number of these transformations—the fish-like form of the body, the hairlessness of the skin, the transformation of the fore-limbs to fins, the disappearance of the hind-limbs and the development of a tail fin, the layer of blubber under the skin, which affords the protection from cold necessary to a warm-blooded animal, the disappearance of the ear-muscles and the auditory passages, the displacement of the external nares to the forehead for the greater security of the breathing-hole during the brief appearance at the surface, and certain remarkable changes in the respiratory and circulatory organs which enable the animal to remain for a long time under water. I might have added many more, for the list of adaptations in the whale to aquatic life is by no means exhausted; they are found in the histological structure and in the minutest combinations in the nervous system. For it is obvious that a tail-fin must be used in quite a different way from a tail, which serves as a fly-brush in hoofed animals, or as an aid to springing in the kangaroo or as a climbing organ; it will require quite different reflex-mechanisms and nerve-combinations in the motor centres.

I used this example in order to show how unnecessary it is to assume a special internal evolutionary power for the phylogenesis of species, for this whole order of whales is, so to speak, MADE UP OF ADAPTATIONS; it deviates in many essential respects from the usual mammalian type, and all the deviations are adaptations to aquatic life. But if precisely the most essential features of the organisation thus depend upon adaptation, what is left for a phyletic force to do, since it is these essential features of the structure it would have to determine? There are few people now who believe in a phyletic evolutionary power, which is not made up of the forces known to us—adaptation and heredity—but the conviction that EVERY part of an organism depends upon adaptation has not yet gained a firm footing. Nevertheless, I must continue to regard this conception as the correct one, as I have long done.

I may be permitted one more example. The feather of a bird is a marvellous structure, and no one will deny that as a whole it depends upon adaptation. But what part of it DOES NOT depend upon adaptation? The hollow quill, the shaft with its hard, thin, light cortex, and the spongy substance within it, its square section compared with the round section of the quill, the flat barbs, their short, hooked barbules which, in the flight-feathers, hook into one another with just sufficient firmness to resist the pressure of the air at each wing-beat, the lightness and firmness of the whole apparatus, the elasticity of the vane, and so on. And yet all this belongs to an organ which is only passively functional, and therefore can have nothing to do with the LAMARCKIAN PRINCIPLE. Nor can the feather have arisen through some magical effect of temperature, moisture, electricity, or specific nutrition, and thus selection is again our only anchor of safety.

But—it will be objected—the substance of which the feather consists, this peculiar kind of horny substance, did not first arise through selection in the course of the evolution of the birds, for it formed the covering of the scales of their reptilian ancestors. It is quite true that a similar substance covered the scales of the Reptiles, but why should it not have arisen among them through selection? Or in what other way could it have arisen, since scales are also passively useful parts? It is true that if we are only to call adaptation what has been acquired by the species we happen to be considering, there would remain a great deal that could not be referred to selection; but we are postulating an evolution which has stretched back through aeons, and in the course of which innumerable adaptations took place, which had not merely ephemeral persistence in a genus, a family or a class, but which was continued into whole Phyla of animals, with continual fresh adaptations to the special conditions of each species, family, or class, yet with persistence of the fundamental elements. Thus the feather, once acquired, persisted in all birds, and the vertebral column, once gained by adaptation in the lowest forms, has persisted in all the Vertebrates, from Amphioxus upwards, although with constant readaptation to the conditions of each particular group. Thus everything we can see in animals is adaptation, whether of to-day, or of yesterday, or of ages long gone by; every kind of cell, whether glandular, muscular, nervous, epidermic, or skeletal, is adapted to absolutely definite and specific functions, and every organ which is composed of these different kinds of cells contains them in the proper proportions, and in the particular arrangement which best serves the function of the organ; it is thus adapted to its function.

All parts of the organism are tuned to one another, that is, THEY ARE ADAPTED TO ONE ANOTHER, and in the same way THE ORGANISM AS A WHOLE IS ADAPTED TO THE CONDITIONS OF ITS LIFE, AND IT IS SO AT EVERY STAGE OF ITS EVOLUTION.

But all adaptations CAN be referred to selection; the only point that remains doubtful is whether they all MUST be referred to it.

However that may be, whether the LAMARCKIAN PRINCIPLE is a factor that has cooperated with selection in evolution, or whether it is altogether fallacious, the fact remains, that selection is the cause of a great part of the phyletic evolution of organisms on our earth. Those who agree with me in rejecting the LAMARCKIAN PRINCIPLE will regard selection as the only GUIDING factor in evolution, which creates what is new out of the transmissible variations, by ordering and arranging these, selecting them in relation to their number and size, as the architect does his building-stones so that a particular style must result. ("Variation under Domestication", 1875 II. pages 426, 427.) But the building-stones themselves, the variations, have their basis in the influences which cause variation in those vital units which are handed on from one generation to another, whether, taken together they form the WHOLE organism, as in Bacteria and other low forms of life, or only a germ-substance, as in unicellular and multicellular organisms. (The Author and Editor are indebted to Professor Poulton for kindly assisting in the revision of the proof of this Essay.)

IV. VARIATION. By HUGO DE VRIES.

Professor of Botany in the University of Amsterdam.

I. DIFFERENT KINDS OF VARIABILITY.

Before Darwin, little was known concerning the phenomena of variability. The fact, that hardly two leaves on a tree were exactly the same, could not escape observation: small deviations of the same kind were met with everywhere, among individuals as well as among the organs of the same plant. Larger aberrations, spoken of as monstrosities, were for a long time regarded as lying outside the range of ordinary phenomena. A special branch of inquiry, that of Teratology, was devoted to them, but it constituted a science by itself, sometimes connected with morphology, but having scarcely any bearing on the processes of evolution and heredity.

Darwin was the first to take a broad survey of the whole range of variations in the animal and vegetable kingdoms. His theory of Natural Selection is based on the fact of variability. In order that this foundation should be as strong as possible he collected all the facts, scattered in the literature of his time, and tried to arrange them in a scientific way. He succeeded in showing that variations may be grouped along a line of almost continuous gradations, beginning with simple differences in size and ending with monstrosities. He was struck by the fact that, as a rule, the smaller the deviations, the more frequently they appear, very abrupt breaks in characters being of rare occurrence.

Among these numerous degrees of variability Darwin was always on the look out for those which might, with the greatest probability, be considered as affording material for natural selection to act upon in the development of new species. Neither of the extremes complied with his conceptions. He often pointed out, that there are a good many small fluctuations, which in this respect must be absolutely useless. On the other hand, he strongly combated the belief, that great changes would be necessary to explain the origin of species. Some authors had propounded the idea that highly adapted organs, e.g. the wings of a bird, could not have been developed in any other way than by a comparatively sudden modification of a well defined and important kind. Such a conception would allow of great breaks or discontinuity in the evolution of highly differentiated animals and plants, shortening the time for the evolution of the whole organic kingdom and getting over numerous difficulties inherent in the theory of slow and gradual progress. It would, moreover, account for the genetic relation of the larger groups of both animals and plants. It would, in a word, undoubtedly afford an easy means of simplifying the problem of descent with modification.

Darwin, however, considered such hypotheses as hardly belonging to the domain of science; they belong, he said, to the realm of miracles. That species have a capacity for change is admitted by all evolutionists; but there is no need to invoke modifications other than those represented by ordinary variability. It is well known that in artificial selection this tendency to vary has given rise to numerous distinct races, and there is no reason for denying that it can do the same in nature, by the aid of natural selection. On both lines an advance may be expected with equal probability.

His main argument, however, is that the most striking and most highly adapted modifications may be acquired by successive variations. Each of these may be slight, and they may affect different organs, gradually adapting them to the same purpose. The direction of the adaptations will be determined by the needs in the struggle for life, and natural selection will simply exclude all such changes as occur on opposite or deviating lines. In this way, it is not variability itself which is called upon to explain beautiful adaptations, but it is quite sufficient to suppose that natural selection has operated during long periods in the same way. Eventually, all the acquired characters, being transmitted together, would appear to us, as if they had all been simultaneously developed.

Correlations must play a large part in such special evolutions: when one part is modified, so will be other parts. The distribution of nourishment will come in as one of the causes, the reactions of different organs to the same external influences as another. But no doubt the more effective cause is that of the internal correlations, which, however, are still but dimly understood. Darwin repeatedly laid great stress on this view, although a definite proof of its correctness could not be given in his time. Such proof requires the direct observation of a mutation, and it should be stated here that even the first observations made in this direction have clearly confirmed Darwin's ideas. The new evening primroses which have sprung in my garden from the old form of Oenothera Lamarckiana, and which have evidently been derived from it, in each case, by a single mutation, do not differ from their parent species in one character only, but in almost all their organs and qualities. Oenothera gigas, for example, has stouter stems and denser foliage; the leaves are larger and broader; its thick flower-buds produce gigantic flowers, but only small fruits with large seeds. Correlative changes of this kind are seen in all my new forms, and they lend support to the view that in the gradual development of highly adapted structures, analogous correlations may have played a large part. They easily explain large deviations from an original type, without requiring the assumption of too many steps.

Monstrosities, as their name implies, are widely different in character from natural species; they cannot, therefore, be adduced as evidence in the investigation of the origin of species. There is no doubt that they may have much in common as regards their manner of origin, and that the origin of species, once understood, may lead to a better understanding of the monstrosities. But the reverse is not true, at least not as regards the main lines of development. Here, it is clear, monstrosities cannot have played a part of any significance.

Reversions, or atavistic changes, would seem to give a better support to the theory of descent through modifications. These have been of paramount importance on many lines of evolution of the animal as well as of the vegetable kingdom. It is often assumed that monocotyledons are descended from some lower group of dicotyledons, probably allied to that which includes the buttercup family. On this view the monocotyledons must be assumed to have lost the cambium and all its influence on secondary growth, the differentiation of the flower into calyx and corolla, the second cotyledon or seed-leaf and several other characters. Losses of characters such as these may have been the result of abrupt changes, but this does not prove that the characters themselves have been produced with equal suddenness. On the contrary, Darwin shows very convincingly that a modification may well be developed by a series of steps, and afterwards suddenly disappear. Many monstrosities, such as those represented by twisted stems, furnish direct proofs in support of this view, since they are produced by the loss of one character and this loss implies secondary changes in a large number of other organs and qualities.

Darwin criticises in detail the hypothesis of great and abrupt changes and comes to the conclusion that it does not give even a shadow of an explanation of the origin of species. It is as improbable as it is unnecessary.

Sports and spontaneous variations must now be considered. It is well known that they have produced a large number of fine horticultural varieties. The cut-leaved maple and many other trees and shrubs with split leaves are known to have been produced at a single step; this is true in the case of the single-leaf strawberry plant and of the laciniate variety of the greater celandine: many white flowers, white or yellow berries and numerous other forms had a similar origin. But changes such as these do not come under the head of adaptations, as they consist for the most part in the loss of some quality or organ belonging to the species from which they were derived. Darwin thinks it impossible to attribute to this cause the innumerable structures, which are so well adapted to the habits of life of each species. At the present time we should say that such adaptations require progressive modifications, which are additions to the stock of qualities already possessed by the ancestors, and cannot, therefore, be explained on the ground of a supposed analogy with sports, which are for the most part of a retrogressive nature.

Excluding all these more or less sudden changes, there remains a long series of gradations of variability, but all of these are not assumed by Darwin to be equally fit for the production of new species. In the first place, he disregards all mere temporary variations, such as size, albinism, etc.; further, he points out that very many species have almost certainly been produced by steps, not greater, and probably not very much smaller, than those separating closely related varieties. For varieties are only small species. Next comes the question of polymorphic species: their occurrence seems to have been a source of much doubt and difficulty in Darwin's mind, although at present it forms one of the main supports of the prevailing explanation of the origin of new species. Darwin simply states that this kind of variability seems to be of a peculiar nature; since polymorphic species are now in a stable condition their occurrence gives no clue as to the mode of origin of new species. Polymorphic species are the expression of the result of previous variability acting on a large scale; but they now simply consist of more or less numerous elementary species, which, as far as we know, do not at present exhibit a larger degree of variability than any other more uniform species. The vernal whitlow-grass (Draba verna) and the wild pansy are the best known examples; both have spread over almost the whole of Europe and are split up into hundreds of elementary forms. These sub-species show no signs of any extraordinary degree of variability, when cultivated under conditions necessary for the exclusion of inter-crossing. Hooker has shown, in the case of some ferns distributed over still wider areas, that the extinction of some of the intermediate forms in such groups would suffice to justify the elevation of the remaining types to the rank of distinct species. Polymorphic species may now be regarded as the link which unites ordinary variability with the historical production of species. But it does not appear that they had this significance for Darwin; and, in fact, they exhibit no phenomena which could explain the processes by which one species has been derived from another. By thus narrowing the limits of the species-producing variability Darwin was led to regard small deviations as the source from which natural selection derives material upon which to act. But even these are not all of the same type, and Darwin was well aware of the fact.

It should here be pointed out that in order to be selected, a change must first have been produced. This proposition, which now seems self-evident, has, however, been a source of much difference of opinion among Darwin's followers. The opinion that natural selection produces changes in useful directions has prevailed for a long time. In other words, it was assumed that natural selection, by the simple means of singling out, could induce small and useful changes to increase and to reach any desired degree of deviation from the original type. In my opinion this view was never actually held by Darwin. It is in contradiction with the acknowledged aim of all his work,—the explanation of the origin of species by means of natural forces and phenomena only. Natural selection acts as a sieve; it does not single out the best variations, but it simply destroys the larger number of those which are, from some cause or another, unfit for their present environment. In this way it keeps the strains up to the required standard, and, in special circumstances, may even improve them.

Returning to the variations which afford the material for the sieving-action of natural selection, we may distinguish two main kinds. It is true that the distinction between these was not clear at the time of Darwin, and that he was unable to draw a sharp line between them. Nevertheless, in many cases, he was able to separate them, and he often discussed the question which of the two would be the real source of the differentiation of species. Certain variations constantly occur, especially such as are connected with size, weight, colour, etc. They are usually too small for natural selection to act upon, having hardly any influence in the struggle for life: others are more rare, occurring only from time to time, perhaps once or twice in a century, perhaps even only once in a thousand years. Moreover, these are of another type, not simply affecting size, number or weight, but bringing about something new, which may be useful or not. Whenever the variation is useful natural selection will take hold of it and preserve it; in other cases the variation may either persist or disappear.

In his criticism of miscellaneous objections brought forward against the theory of natural selection after the publication of the first edition of "The Origin of Species", Darwin stated his view on this point very clearly:—"The doctrine of natural selection or the survival of the fittest, which implies that when variations or individual differences of a beneficial nature happen to arise, these will be preserved." ("Origin of Species" (6th edition), page 169, 1882.) In this sentence the words "HAPPEN TO ARISE" appear to me of prominent significance. They are evidently due to the same general conception which prevailed in Darwin's Pangenesis hypothesis. (Cf. de Vries, "Intracellulare Pangenesis", page 73, Jena, 1889, and "Die Mutationstheorie", I. page 63. Leipzig, 1901.)

A distinction is indicated between ordinary fluctuations which are always present, and such variations as "happen to arise" from time to time. ((I think it right to point out that the interpretation of this passage from the "Origin" by Professor de Vries is not accepted as correct either by Mr Francis Darwin or by myself. We do not believe that Darwin intended to draw any distinction between TWO TYPES of variation; the words "when variations or individual differences of a beneficial nature happen to arise" are not in our opinion meant to imply a distinction between ordinary fluctuations and variations which "happen to arise," but we believe that "or" is here used in the sense of ALIAS. With the permission of Professor de Vries, the following extract is quoted from a letter in which he replied to the objection raised to his reading of the passage in question:

"As to your remarks on the passage on page 6, I agree that it is now impossible to see clearly how far Darwin went in his distinction of the different kinds of variability. Distinctions were only dimly guessed at by him. But in our endeavour to arrive at a true conception of his view I think that the chapter on Pangenesis should be our leading guide, and that we should try to interpret the more difficult passages by that chapter. A careful and often repeated study of the Pangenesis hypothesis has convinced me that Darwin, when he wrote that chapter, was well aware that ordinary variability has nothing to do with evolution, but that other kinds of variation were necessary. In some chapters he comes nearer to a clear distinction than in others. To my mind the expression 'happen to arise' is the sharpest indication of his inclining in this direction. I am quite convinced that numerous expressions in his book become much clearer when looked at in this way."

The statement in this passage that "Darwin was well aware that ordinary variability has nothing to do with evolution, but that other kinds of variation were necessary" is contradicted by many passages in the "Origin". A.C.S.)) The latter afford the material for natural selection to act upon on the broad lines of organic development, but the first do not. Fortuitous variations are the species-producing kind, which the theory requires; continuous fluctuations constitute, in this respect, a useless type.

Of late, the study of variability has returned to the recognition of this distinction. Darwin's variations, which from time to time happen to arise, are MUTATIONS, the opposite type being commonly designed fluctuations. A large mass of facts, collected during the last few decades, has confirmed this view, which in Darwin's time could only be expressed with much reserve, and everyone knows that Darwin was always very careful in statements of this kind.

From the same chapter I may here cite the following paragraph: "Thus as I am inclined to believe, morphological differences,... such as the arrangement of the leaves, the divisions of the flower or of the ovarium, the position of the ovules, etc.—first appeared in many cases as fluctuating variations, which sooner or later became constant through the nature of the organism and of the surrounding conditions... but NOT THROUGH NATURAL SELECTION (The italics are mine (H. de V.).); for as these morphological characters do not affect the welfare of the species, any slight deviation in them could not have been governed or accumulated through this latter agency." ("Origin of Species" (6th edition), page 176.) We thus see that in Darwin's opinion, all small variations had not the same importance. In favourable circumstances some could become constant, but others could not.

Since the appearance of the first edition of "The Origin of Species" fluctuating variability has been thoroughly studied by Quetelet. He discovered the law, which governs all phenomena of organic life falling under this head. It is a very simple law, and states that individual variations follow the laws of probability. He proved it, in the first place, for the size of the human body, using the measurements published for Belgian recruits; he then extended it to various other measurements of parts of the body, and finally concluded that it must be of universal validity for all organic beings. It must hold true for all characters in man, physical as well as intellectual and moral qualities; it must hold true for the plant kingdom as well as for the animal kingdom; in short, it must include the whole living world.

Quetelet's law may be most easily studied in those cases where the variability relates to measure, number and weight, and a vast number of facts have since confirmed its exactness and its validity for all kinds of organisms, organs and qualities. But if we examine it more closely, we find that it includes just those minute variations, which, as Darwin repeatedly pointed out, have often no significance for the origin of species. In the phenomena, described by Quetelet's law nothing "happens to arise"; all is governed by the common law, which states that small deviations from the mean type are frequent, but that larger aberrations are rare, the rarer as they are larger. Any degree of variation will be found to occur, if only the number of individuals studied is large enough: it is even possible to calculate before hand, how many specimens must be compared in order to find a previously fixed degree of deviation.

The variations, which from time to time happen to appear, are evidently not governed by this law. They cannot, as yet, be produced at will: no sowings of thousands or even of millions of plants will induce them, although by such means the chance of their occurring will obviously be increased. But they are known to occur, and to occur suddenly and abruptly. They have been observed especially in horticulture, where they are ranged in the large and ill-defined group called sports. Korschinsky has collected all the evidence which horticultural literature affords on this point. (S. Korschinsky, "Heterogenesis und Evolution", "Flora", Vol. LXXXIX. pages 240-363, 1901.) Several cases of the first appearance of a horticultural novelty have been recorded: this has always happened in the same way; it appeared suddenly and unexpectedly without any definite relation to previously existing variability. Dwarf types are one of the commonest and most favourite varieties of flowering plants; they are not originated by a repeated selection of the smallest specimens, but appear at once, without intermediates and without any previous indication. In many instances they are only about half the height of the original type, thus constituting obvious novelties. So it is in other cases described by Korschinsky: these sports or mutations are now recognised to be the main source of varieties of horticultural plants.

As already stated, I do not pretend that the production of horticultural novelties is the prototype of the origin of new species in nature. I assume that they are, as a rule, derived from the parent species by the loss of some organ or quality, whereas the main lines of the evolution of the animal and vegetable kingdom are of course determined by progressive changes. Darwin himself has often pointed out this difference. But the saltatory origin of horticultural novelties is as yet the simplest parallel for natural mutations, since it relates to forms and phenomena, best known to the general student of evolution.

The point which I wish to insist upon is this. The difference between small and ever present fluctuations and rare and more sudden variations was clear to Darwin, although the facts known at his time were too meagre to enable a sharp line to be drawn between these two great classes of variability. Since Darwin's time evidence, which proves the correctness of his view, has accumulated with increasing rapidity. Fluctuations constitute one type; they are never absent and follow the law of chance, but they do not afford the material from which to build new species. Mutations, on the other hand, only happen to occur from time to time. They do not necessarily produce greater changes than fluctuations, but such as may become, or rather are from their very nature, constant. It is this constancy which is the mark of specific characters, and on this basis every new specific character may be assumed to have arisen by mutation.

Some authors have tried to show that the theory of mutation is opposed to Darwin's views. But this is erroneous. On the contrary, it is in fullest harmony with the great principle laid down by Darwin. In order to be acted upon by that complex of environmental forces, which Darwin has called natural selection, the changes must obviously first be there. The manner in which they are produced is of secondary importance and has hardly any bearing on the theory of descent with modification. ("Life and Letters" II. 125.)

A critical survey of all the facts of variability of plants in nature as well as under cultivation has led me to the conviction, that Darwin was right in stating that those rare beneficial variations, which from time to time happen to arise,—the now so-called mutations—are the real source of progress in the whole realm of the organic world.

II. EXTERNAL AND INTERNAL CAUSES OF VARIABILITY.

All phenomena of animal and plant life are governed by two sets of causes; one of these is external, the other internal. As a rule the internal causes determine the nature of a phenomenon—what an organism can do and what it cannot do. The external causes, on the other hand, decide when a certain variation will occur, and to what extent its features may be developed.

As a very clear and wholly typical instance I cite the cocks-combs (Celosia). This race is distinguished from allied forms by its faculty of producing the well-known broad and much twisted combs. Every single individual possesses this power, but all individuals do not exhibit it in its most complete form. In some cases this faculty may not be exhibited at the top of the main stem, although developed in lateral branches: in others it begins too late for full development. Much depends upon nourishment and cultivation, but almost always the horticulturist has to single out the best individuals and to reject those which do not come up to the standard.

The internal causes are of a historical nature. The external ones may be defined as nourishment and environment. In some cases nutrition is the main factor, as, for instance, in fluctuating variability, but in natural selection environment usually plays the larger part.

The internal or historical causes are constant during the life-time of a species, using the term species in its most limited sense, as designating the so-called elementary species or the units out of which the ordinary species are built up. These historical causes are simply the specific characters, since in the origin of a species one or more of these must have been changed, thus producing the characters of the new type. These changes must, of course, also be due partly to internal and partly to external causes.

In contrast to these changes of the internal causes, the ordinary variability which is exhibited during the life-time of a species is called fluctuating variability. The name mutations or mutating variability is then given to the changes in the specific characters. It is desirable to consider these two main divisions of variability separately.

In the case of fluctuations the internal causes, as well as the external ones, are often apparent. The specific characters may be designated as the mean about which the observed forms vary. Almost every character may be developed to a greater or a less degree, but the variations of the single characters producing a small deviation from the mean are usually the commonest. The limits of these fluctuations may be called wide or narrow, according to the way we look at them, but in numerous cases the extreme on the favoured side hardly surpasses double the value of that on the other side. The degree of this development, for every individual and for every organ, is dependent mainly on nutrition. Better nourishment or an increased supply of food produces a higher development; only it is not always easy to determine which direction is the fuller and which is the poorer one. The differences among individuals grown from different seeds are described as examples of individual variability, but those which may be observed on the same plant, or on cuttings, bulbs or roots derived from one individual are referred to as cases of partial variability. Partial variability, therefore, determines the differences among the flowers, fruits, leaves or branches of one individual: in the main, it follows the same laws as individual variability, but the position of a branch on a plant also determines its strength, and the part it may take in the nourishment of the whole. Composite flowers and umbels therefore have, as a rule, fewer rays on weak branches than on the strong main ones. The number of carpels in the fruits of poppies becomes very small on the weak lateral branches, which are produced towards the autumn, as well as on crowded, and therefore on weakened individuals. Double flowers follow the same rule, and numerous other instances could easily be adduced.

Mutating variability occurs along three main lines. Either a character may disappear, or, as we now say, become latent; or a latent character may reappear, reproducing thereby a character which was once prominent in more or less remote ancestors. The third and most interesting case is that of the production of quite new characters which never existed in the ancestors. Upon this progressive mutability the main development of the animal and vegetable kingdom evidently depends. In contrast to this, the two other cases are called retrogressive and degressive mutability. In nature retrogressive mutability plays a large part; in agriculture and in horticulture it gives rise to numerous varieties, which have in the past been preserved, either on account of their usefulness or beauty, or simply as fancy-types. In fact the possession of numbers of varieties may be considered as the main character of domesticated animals and cultivated plants.

In the case of retrogressive and degressive mutability the internal cause is at once apparent, for it is this which causes the disappearance or reappearance of some character. With progressive mutations the case is not so simple, since the new character must first be produced and then displayed. These two processes are theoretically different, but they may occur together or after long intervals. The production of the new character I call premutation, and the displaying mutation. Both of course must have their external as well as their internal causes, as I have repeatedly pointed out in my work on the Mutation Theory. ("Die Mutationstheorie", 2 vols., Leipzig, 1901.)

It is probable that nutrition plays as important a part among the external causes of mutability as it does among those of fluctuating variability. Observations in support of this view, however, are too scanty to allow of a definite judgment. Darwin assumed an accumulative influence of external causes in the case of the production of new varieties or species. The accumulation might be limited to the life-time of a single individual, or embrace that of two or more generations. In the end a degree of instability in the equilibrium of one or more characters might be attained, great enough for a character to give way under a small shock produced by changed conditions of life. The character would then be thrown over from the old state of equilibrium into a new one.

Characters which happen to be in this state of unstable equilibrium are called mutable. They may be either latent or active, being in the former case derived from old active ones or produced as new ones (by the process, designated premutation). They may be inherited in this mutable condition during a long series of generations. I have shown that in the case of the evening primrose of Lamarck this state of mutability must have existed for at least half a century, for this species was introduced from Texas into England about the year 1860, and since then all the strains derived from its first distribution over the several countries of Europe show the same phenomena in producing new forms. The production of the dwarf evening primrose, or Oenothera nanella, is assumed to be due to one of the factors, which determines the tall stature of the parent form, becoming latent; this would, therefore, afford an example of retrogressive mutation. Most of the other types of my new mutants, on the other hand, seem to be due to progressive mutability.

The external causes of this curious period of mutability are as yet wholly unknown and can hardly be guessed at, since the origin of the Oenothera Lamarckiana is veiled in mystery. The seeds, introduced into England about 1860, were said to have come from Texas, but whether from wild or from cultivated plants we do not know. Nor has the species been recorded as having been observed in the wild condition. This, however, is nothing peculiar. The European types of Oenothera biennis and O. muricata are in the same condition. The first is said to have been introduced from Virginia, and the second from Canada, but both probably from plants cultivated in the gardens of these countries. Whether the same elementary species are still growing on those spots is unknown, mainly because the different sub-species of the species mentioned have not been systematically studied and distinguished.

The origin of new species, which is in part the effect of mutability, is, however, due mainly to natural selection. Mutability provides the new characters and new elementary species. Natural selection, on the other hand, decides what is to live and what to die. Mutability seems to be free, and not restricted to previously determined lines. Selection, however, may take place along the same main lines in the course of long geological epochs, thus directing the development of large branches of the animal and vegetable kingdom. In natural selection it is evident that nutrition and environment are the main factors. But it is probable that, while nutrition may be one of the main causes of mutability, environment may play the chief part in the decisions ascribed to natural selection. Relations to neighbouring plants and to injurious or useful animals, have been considered the most important determining factors ever since the time when Darwin pointed out their prevailing influence.

From this discussion of the main causes of variability we may derive the proposition that the study of every phenomenon in the field of heredity, of variability, and of the origin of new species will have to be considered from two standpoints; on one hand we have the internal causes, on the other the external ones. Sometimes the first are more easily detected, in other cases the latter are more accessible to investigation. But the complete elucidation of any phenomenon of life must always combine the study of the influence of internal with that of external causes.

III. POLYMORPHIC VARIABILITY IN CEREALS.

One of the propositions of Darwin's theory of the struggle for life maintains that the largest amount of life can be supported on any area, by great diversification or divergence in the structure and constitution of its inhabitants. Every meadow and every forest affords a proof of this thesis. The numerical proportion of the different species of the flora is always changing according to external influences. Thus, in a given meadow, some species will flower abundantly in one year and then almost disappear, until, after a series of years, circumstances allow them again to multiply rapidly. Other species, which have taken their places, will then become rare. It follows from this principle, that notwithstanding the constantly changing conditions, a suitable selection from the constituents of a meadow will ensure a continued high production. But, although the principle is quite clear, artificial selection has, as yet, done very little towards reaching a really high standard.

The same holds good for cereals. In ordinary circumstances a field will give a greater yield, if the crop grown consists of a number of sufficiently differing types. Hence it happens that almost all older varieties of wheat are mixtures of more or less diverging forms. In the same variety the numerical composition will vary from year to year, and in oats this may, in bad years, go so far as to destroy more than half of the harvest, the wind-oats (Avena fatua), which scatter their grain to the winds as soon as it ripens, increasing so rapidly that they assume the dominant place. A severe winter, a cold spring and other extreme conditions of life will destroy one form more completely than another, and it is evident that great changes in the numerical composition of the mixture may thus be brought about.

This mixed condition of the common varieties of cereals was well known to Darwin. For him it constituted one of the many types of variability. It is of that peculiar nature to which, in describing other groups, he applies the term polymorphy. It does not imply that the single constituents of the varieties are at present really changing their characters. On the other hand, it does not exclude the possibility of such changes. It simply states that observation shows the existence of different forms; how these have originated is a question which it does not deal with. In his well-known discussion of the variability of cereals, Darwin is mainly concerned with the question, whether under cultivation they have undergone great changes or only small ones. The decision ultimately depends on the question, how many forms have originally been taken into cultivation. Assuming five or six initial species, the variability must be assumed to have been very large, but on the assumption that there were between ten and fifteen types, the necessary range of variability is obviously much smaller. But in regard to this point, we are of course entirely without historical data.

Few of the varieties of wheat show conspicuous differences, although their number is great. If we compare the differentiating characters of the smaller types of cereals with those of ordinary wild species, even within the same genus or family, they are obviously much less marked. All these small characters, however, are strictly inherited, and this fact makes it very probable that the less obvious constituents of the mixtures in ordinary fields must be constant and pure as long as they do not intercross. Natural crossing is in most cereals a phenomenon of rare occurrence, common enough to admit of the production of all possible hybrid combinations, but requiring the lapse of a long series of years to reach its full effect.

Darwin laid great stress on this high amount of variability in the plants of the same variety, and illustrated it by the experience of Colonel Le Couteur ("On the Varieties, Properties, and Classification of Wheat", Jersey, 1837.) on his farm on the isle of Jersey, who cultivated upwards of 150 varieties of wheat, which he claimed were as pure as those of any other agriculturalist. But Professor La Gasca of Madrid, who visited him, drew attention to aberrant ears, and pointed out, that some of them might be better yielders than the majority of plants in the crop, whilst others might be poor types. Thence he concluded that the isolation of the better ones might be a means of increasing his crops. Le Couteur seems to have considered the constancy of such smaller types after isolation as absolutely probable, since he did not even discuss the possibility of their being variable or of their yielding a changeable or mixed progeny. This curious fact proves that he considered the types, discovered in his fields by La Gasca to be of the same kind as his other varieties, which until that time he had relied upon as being pure and uniform. Thus we see, that for him, the variability of cereals was what we now call polymorphy. He looked through his fields for useful aberrations, and collected twenty-three new types of wheat. He was, moreover, clear about one point, which, on being rediscovered after half a century, has become the starting-point for the new Swedish principle of selecting agricultural plants. It was the principle of single-ear sowing, instead of mixing the grains of all the selected ears together. By sowing each ear on a separate plot he intended not only to multiply them, but also to compare their value. This comparison ultimately led him to the choice of some few valuable sorts, one of which, the "Bellevue de Talavera," still holds its place among the prominent sorts of wheat cultivated in France. This variety seems to be really a uniform type, a quality very useful under favourable conditions of cultivation, but which seems to have destroyed its capacity for further improvement by selection.

The principle of single-ear sowing, with a view to obtain pure and uniform strains without further selection, has, until a few years ago, been almost entirely lost sight of. Only a very few agriculturists have applied it: among these are Patrick Shirreff ("Die Verbesserung der Getreide-Arten", translated by R. Hesse, Halle, 1880.) in Scotland and Willet M. Hays ("Wheat, varieties, breeding, cultivation", Univ. Minnesota, Agricultural Experimental Station, Bull. no. 62, 1899.) in Minnesota. Patrick Shirreff observed the fact, that in large fields of cereals, single plants may from time to time be found with larger ears, which justify the expectation of a far greater yield. In the course of about twenty-five years he isolated in this way two varieties of wheat and two of oats. He simply multiplied them as fast as possible, without any selection, and put them on the market.

Hays was struck by the fact that the yield of wheat in Minnesota was far beneath that in the neighbouring States. The local varieties were Fife and Blue Stem. They gave him, on inspection, some better specimens, "phenomenal yielders" as he called them. These were simply isolated and propagated, and, after comparison with the parent-variety and with some other selected strains of less value, were judged to be of sufficient importance to be tested by cultivation all over the State of Minnesota. They have since almost supplanted the original types, at least in most parts of the State, with the result that the total yield of wheat in Minnesota is said to have been increased by about a million dollars yearly.

Definite progress in the method of single-ear sowing has, however, been made only recently. It had been foreshadowed by Patrick Shirreff, who after the production of the four varieties already mentioned, tried to carry out his work on a larger scale, by including numerous minor deviations from the main type. He found by doing so that the chances of obtaining a better form were sufficiently increased to justify the trial. But it was Nilsson who discovered the almost inexhaustible polymorphy of cereals and other agricultural crops and made it the starting-point for a new and entirely trustworthy method of the highest utility. By this means he has produced during the last fifteen years a number of new and valuable races, which have already supplanted the old types on numerous farms in Sweden and which are now being introduced on a large scale into Germany and other European countries.

It is now twenty years since the station at Svalof was founded. During the first period of its work, embracing about five years, selection was practised on the principle which was then generally used in Germany. In order to improve a race a sample of the best ears was carefully selected from the best fields of the variety. These ears were considered as representatives of the type under cultivation, and it was assumed that by sowing their grains on a small plot a family could be obtained, which could afterwards be improved by a continuous selection. Differences between the collected ears were either not observed or disregarded. At Svalof this method of selection was practised on a far larger scale than on any German farm, and the result was, broadly speaking, the same. This may be stated in the following words: improvement in a few cases, failure in all the others. Some few varieties could be improved and yielded excellent new types, some of which have since been introduced into Swedish agriculture and are now prominent races in the southern and middle parts of the country. But the station had definite aims, and among them was the improvement of the Chevalier barley. This, in Middle Sweden, is a fine brewer's barley, but liable to failure during unfavourable summers on account of its slender stems. It was selected with a view of giving it stiffer stems, but in spite of all the care and work bestowed upon it no satisfactory result was obtained.

This experience, combined with a number of analogous failures, could not fail to throw doubt upon the whole method. It was evident that good results were only exceptions, and that in most cases the principle was not one that could be relied upon. The exceptions might be due to unknown causes, and not to the validity of the method; it became therefore of much more interest to search for the causes than to continue the work along these lines.

In the year 1892 a number of different varieties of cereals were cultivated on a large scale and a selection was again made from them. About two hundred samples of ears were chosen, each apparently constituting a different type. Their seeds were sown on separate plots and manured and treated as much as possible in the same manner. The plots were small and arranged in rows so as to facilitate the comparison of allied types. During the whole period of growth and during the ripening of the ears the plots were carefully studied and compared: they were harvested separately; ears and kernels were counted and weighed, and notes were made concerning layering, rust and other cereal pests.

The result of this experiment was, in the main, no distinct improvement. Nilsson was especially struck by the fact that the plots, which should represent distinct types, were far from uniform. Many of them were as multiform as the fields from which the parent-ears were taken. Others showed variability in a less degree, but in almost all of them it was clear that a pure race had not been obtained. The experiment was a fair one, inasmuch as it demonstrated the polymorphic variability of cereals beyond all doubt and in a degree hitherto unsuspected; but from the standpoint of the selectionist it was a failure. Fortunately there were, however, one or two exceptions. A few lots showed a perfect uniformity in regard to all the stalks and ears: these were small families. This fact suggested the idea that each might have been derived from a single ear. During the selection in the previous summer, Nilsson had tried to find as many ears as possible of each new type which he recognised in his fields. But the variability of his crops was so great, that he was rarely able to include more than two or three ears in the same group, and, in a few cases, he found only one representative of the supposed type. It might, therefore, be possible that those small uniform plots were the direct progeny of ears, the grains of which had not been mixed with those from other ears before sowing. Exact records had, of course, been kept of the chosen samples, and the number of ears had been noted in each case. It was, therefore, possible to answer the question and it was found that those plots alone were uniform on which the kernels of one single ear only had been sown. Nilsson concluded that the mixture of two or more ears in a single sowing might be the cause of the lack of uniformity in the progeny. Apparently similar ears might be different in their progeny.

Once discovered, this fact was elevated to the rank of a leading principle and tested on as large a scale as possible. The fields were again carefully investigated and every single ear, which showed a distinct divergence from the main type in one character or another, was selected. A thousand samples were chosen, but this time each sample consisted of one ear only. Next year, the result corresponded to the expectation. Uniformity prevailed almost everywhere; only a few lots showed a discrepancy, which might be ascribed to the accidental selection of hybrid ears. It was now clear that the progeny of single ears was, as a rule, pure, whereas that of mixed ears was impure. The single-ear selection or single-ear sowing, which had fallen into discredit in Germany and elsewhere in Europe, was rediscovered. It proved to be the only trustworthy principle of selection. Once isolated, such single-parent races are constant from seed and remain true to their type. No further selection is needed; they have simply to be multiplied and their real value tested.

Patrick Shirreff, in his early experiments, Le Couteur, Hays and others had observed the rare occurrence of exceptionally good yielders and the value of their isolation to the agriculturist. The possibility of error in the choice of such striking specimens and the necessity of judging their value by their progeny were also known to these investigators, but they had not the slightest idea of all the possibilities suggested by their principle. Nilsson, who is a botanist as well as an agriculturist, discovered that, besides these exceptionally good yielders, every variety of a cereal consists of hundreds of different types, which find the best conditions for success when grown together, but which, after isolation, prove to be constant. Their preference for mixed growth is so definite, that once isolated, their claims on manure and treatment are found to be much higher than those of the original mixed variety. Moreover, the greatest care is necessary to enable them to retain their purity, and as soon as they are left to themselves they begin to deteriorate through accidental crosses and admixtures and rapidly return to the mixed condition.

Reverting now to Darwin's discussion of the variability of cereals, we may conclude that subsequent investigation has proved it to be exactly of the kind which he describes. The only difference is that in reality it reaches a degree, quite unexpected by Darwin and his contemporaries. But it is polymorphic variability in the strictest sense of the word. How the single constituents of a variety originate we do not see. We may assume, and there can hardly be a doubt about the truth of the assumption, that a new character, once produced, will slowly but surely be combined through accidental crosses with a large number of previously existing types, and so will tend to double the number of the constituents of the variety. But whether it first appears suddenly or whether it is only slowly evolved we cannot determine. It would, of course, be impossible to observe either process in such a mixture. Only cultures of pure races, of single-parent races as we have called them, can afford an opportunity for this kind of observation. In the fields of Svalof new and unexpected qualities have recently been seen, from time to time, to appear suddenly. These characters are as distinct as the older ones and appear to be constant from the moment of their origin.

Darwin has repeatedly insisted that man does not cause variability. He simply selects the variations given to him by the hand of nature. He may repeat this process in order to accumulate different new characters in the same family, thus producing varieties of a higher order. This process of accumulation would, if continued for a longer time, lead to the augmentation of the slight differences characteristic of varieties into the greater differences characteristic of species and genera. It is in this way that horticultural and agricultural experience contribute to the problem of the conversion of varieties into species, and to the explanation of the admirable adaptations of each organism to its complex conditions of life. In the long run new forms, distinguished from their allies by quite a number of new characters, would, by the extermination of the older intermediates, become distinct species.

Thus we see that the theory of the origin of species by means of natural selection is quite independent of the question, how the variations to be selected arise. They may arise slowly, from simple fluctuations, or suddenly, by mutations; in both cases natural selection will take hold of them, will multiply them if they are beneficial, and in the course of time accumulate them, so as to produce that great diversity of organic life, which we so highly admire.

Darwin has left the decision of this difficult and obviously subordinate point to his followers. But in his Pangenesis hypothesis he has given us the clue for a close study and ultimate elucidation of the subject under discussion.

V. HEREDITY AND VARIATION IN MODERN LIGHTS. By W. Bateson, M.A., F.R.S.

Professor of Biology in the University of Cambridge.

Darwin's work has the property of greatness in that it may be admired from more aspects than one. For some the perception of the principle of Natural Selection stands out as his most wonderful achievement to which all the rest is subordinate. Others, among whom I would range myself, look up to him rather as the first who plainly distinguished, collected, and comprehensively studied that new class of evidence from which hereafter a true understanding of the process of Evolution may be developed. We each prefer our own standpoint of admiration; but I think that it will be in their wider aspect that his labours will most command the veneration of posterity.

A treatise written to advance knowledge may be read in two moods. The reader may keep his mind passive, willing merely to receive the impress of the writer's thought; or he may read with his attention strained and alert, asking at every instant how the new knowledge can be used in a further advance, watching continually for fresh footholds by which to climb higher still. Of Shelley it has been said that he was a poet for poets: so Darwin was a naturalist for naturalists. It is when his writings are used in the critical and more exacting spirit with which we test the outfit for our own enterprise that we learn their full value and strength. Whether we glance back and compare his performance with the efforts of his predecessors, or look forward along the course which modern research is disclosing, we shall honour most in him not the rounded merit of finite accomplishment, but the creative power by which he inaugurated a line of discovery endless in variety and extension. Let us attempt thus to see his work in true perspective between the past from which it grew, and the present which is its consequence. Darwin attacked the problem of Evolution by reference to facts of three classes: Variation; Heredity; Natural Selection. His work was not as the laity suppose, a sudden and unheralded revelation, but the first fruit of a long and hitherto barren controversy. The occurrence of variation from type, and the hereditary transmission of such variation had of course been long familiar to practical men, and inferences as to the possible bearing of those phenomena on the nature of specific difference had been from time to time drawn by naturalists. Maupertuis, for example, wrote

"Ce qui nous reste a examiner, c'est comment d'un seul individu, il a pu naitre tant d'especes si differentes." And again "La Nature contient le fonds de toutes ces varietes: mais le hasard ou l'art les mettent en oeuvre. C'est ainsi que ceux dont l'industrie s'applique a satisfaire le gout des curieux, sont, pour ainsi dire, creatures d'especes nouvelles." ("Venus Physique, contenant deux Dissertations, l'une sur l'origine des Hommes et des Animaux: Et l'autre sur l'origine des Noirs" La Haye, 1746, pages 124 and 129. For an introduction to the writings of Maupertuis I am indebted to an article by Professor Lovejoy in "Popular Sci. Monthly", 1902.)

Such passages, of which many (though few so emphatic) can be found in eighteenth century writers, indicate a true perception of the mode of Evolution. The speculations hinted at by Buffon (For the fullest account of the views of these pioneers of Evolution, see the works of Samuel Butler, especially "Evolution, Old and New" (2nd edition) 1882. Butler's claims on behalf of Buffon have met with some acceptance; but after reading what Butler has said, and a considerable part of Buffon's own works, the word "hinted" seems to me a sufficiently correct description of the part he played. It is interesting to note that in the chapter on the Ass, which contains some of his evolutionary passages, there is a reference to "plusieurs idees tres-elevees sur la generation" contained in the Letters of Maupertuis.), developed by Erasmus Darwin, and independently proclaimed above all by Lamarck, gave to the doctrine of descent a wide renown. The uniformitarian teaching which Lyell deduced from geological observation had gained acceptance. The facts of geographical distribution (See especially W. Lawrence, "Lectures on Physiology", London, 1823, pages 213 f.) had been shown to be obviously inconsistent with the Mosaic legend. Prichard, and Lawrence, following the example of Blumenbach, had successfully demonstrated that the races of Man could be regarded as different forms of one species, contrary to the opinion up till then received. These treatises all begin, it is true, with a profound obeisance to the sons of Noah, but that performed, they continue on strictly modern lines. The question of the mutability of species was thus prominently raised.

Those who rate Lamarck no higher than did Huxley in his contemptuous phrase "buccinator tantum," will scarcely deny that the sound of the trumpet had carried far, or that its note was clear. If then there were few who had already turned to evolution with positive conviction, all scientific men must at least have known that such views had been promulgated; and many must, as Huxley says, have taken up his own position of "critical expectancy." (See the chapter contributed to the "Life and Letters of Charles Darwin" II. page 195. I do not clearly understand the sense in which Darwin wrote (Autobiography, ibid. I. page 87): "It has sometimes been said that the success of the "Origin" proved 'that the subject was in the air,' or 'that men's minds were prepared for it.' I do not think that this is strictly true, for I occasionally sounded not a few naturalists, and never happened to come across a single one who seemed to doubt about the permanence of species." This experience may perhaps have been an accident due to Darwin's isolation. The literature of the period abounds with indications of "critical expectancy." A most interesting expression of that feeling is given in the charming account of the "Early Days of Darwinism" by Alfred Newton, "Macmillan's Magazine", LVII. 1888, page 241. He tells how in 1858 when spending a dreary summer in Iceland, he and his friend, the ornithologist John Wolley, in default of active occupation, spent their days in discussion. "Both of us taking a keen interest in Natural History, it was but reasonable that a question, which in those days was always coming up wherever two or more naturalists were gathered together, should be continually recurring. That question was, 'What is a species?' and connected therewith was the other question, 'How did a species begin?'... Now we were of course fairly well acquainted with what had been published on these subjects." He then enumerates some of these publications, mentioning among others T. Vernon Wollaston's "Variation of Species"—a work which has in my opinion never been adequately appreciated. He proceeds: "Of course we never arrived at anything like a solution of these problems, general or special, but we felt very strongly that a solution ought to be found, and that quickly, if the study of Botany and Zoology was to make any great advance." He then describes how on his return home he received the famous number of the "Linnean Journal" on a certain evening. "I sat up late that night to read it; and never shall I forget the impression it made upon me. Herein was contained a perfectly simple solution of all the difficulties which had been troubling me for months past... I went to bed satisfied that a solution had been found.")

Why, then, was it, that Darwin succeeded where the rest had failed? The cause of that success was two-fold. First, and obviously, in the principle of Natural Selection he had a suggestion which would work. It might not go the whole way, but it was true as far as it went. Evolution could thus in great measure be fairly represented as a consequence of demonstrable processes. Darwin seldom endangers the mechanism he devised by putting on it strains much greater than it can bear. He at least was under no illusion as to the omnipotence of Selection; and he introduces none of the forced pleading which in recent years has threatened to discredit that principle.

For example, in the latest text of the "Origin" ("Origin", (6th edition (1882), page 421.)) we find him saying:

"But as my conclusions have lately been much misrepresented, and it has been stated that I attribute the modification of species exclusively to natural selection, I may be permitted to remark that in the first edition of this work, and subsequently, I placed in a most conspicuous position—namely, at the close of the Introduction—the following words: 'I am convinced that natural selection has been the main but not the exclusive means of modification.'"

But apart from the invention of this reasonable hypothesis, which may well, as Huxley estimated, "be the guide of biological and psychological speculation for the next three or four generations," Darwin made a more significant and imperishable contribution. Not for a few generations, but through all ages he should be remembered as the first who showed clearly that the problems of Heredity and Variation are soluble by observation, and laid down the course by which we must proceed to their solution. (Whatever be our estimate of the importance of Natural Selection, in this we all agree. Samuel Butler, the most brilliant, and by far the most interesting of Darwin's opponents—whose works are at length emerging from oblivion—in his Preface (1882) to the 2nd edition of "Evolution, Old and New", repeats his earlier expression of homage to one whom he had come to regard as an enemy: "To the end of time, if the question be asked, 'Who taught people to believe in Evolution?' the answer must be that it was Mr. Darwin. This is true, and it is hard to see what palm of higher praise can be awarded to any philosopher.") The moment of inspiration did not come with the reading of Malthus, but with the opening of the "first note-book on Transmutation of Species." ("Life and Letters", I. pages 276 and 83.) Evolution is a process of Variation and Heredity. The older writers, though they had some vague idea that it must be so, did not study Variation and Heredity. Darwin did, and so begat not a theory, but a science.

The extent to which this is true, the scientific world is only beginning to realise. So little was the fact appreciated in Darwin's own time that the success of his writings was followed by an almost total cessation of work in that special field. Of the causes which led to this remarkable consequence I have spoken elsewhere. They proceeded from circumstances peculiar to the time; but whatever the causes there is no doubt that this statement of the result is historically exact, and those who make it their business to collect facts elucidating the physiology of Heredity and Variation are well aware that they will find little to reward their quest in the leading scientific Journals of the Darwinian epoch.

In those thirty years the original stock of evidence current and in circulation even underwent a process of attrition. As in the story of the Eastern sage who first wrote the collected learning of the universe for his sons in a thousand volumes, and by successive compression and burning reduced them to one, and from this by further burning distilled the single ejaculation of the Faith, "There is no god but God and Mohamed is the Prophet of God," which was all his maturer wisdom deemed essential:—so in the books of that period do we find the corpus of genetic knowledge dwindle to a few prerogative instances, and these at last to the brief formula of an unquestioned creed.

And yet in all else that concerns biological science this period was, in very truth, our Golden Age, when the natural history of the earth was explored as never before; morphology and embryology were exhaustively ransacked; the physiology of plants and animals began to rival chemistry and physics in precision of method and in the rapidity of its advances; and the foundations of pathology were laid.

In contrast with this immense activity elsewhere the neglect which befel the special physiology of Descent, or Genetics as we now call it, is astonishing. This may of course be interpreted as meaning that the favoured studies seemed to promise a quicker return for effort, but it would be more true to say that those who chose these other pursuits did so without making any such comparison; for the idea that the physiology of Heredity and Variation was a coherent science, offering possibilities of extraordinary discovery, was not present to their minds at all. In a word, the existence of such a science was well nigh forgotten. It is true that in ancillary periodicals, as for example those that treat of entomology or horticulture, or in the writings of the already isolated systematists (This isolation of the systematists is the one most melancholy sequela of Darwinism. It seems an irony that we should read in the peroration to the "Origin" that when the Darwinian view is accepted "Systematists will be able to pursue their labours as at present; but they will not be incessantly haunted by the shadowy doubt whether this or that form be a true species. This, I feel sure, and I speak after experience, will be no slight relief. The endless disputes whether or not some fifty species of British brambles are good species will cease." "Origin", 6th edition (1882), page 425. True they have ceased to attract the attention of those who lead opinion, but anyone who will turn to the literature of systematics will find that they have not ceased in any other sense. Should there not be something disquieting in the fact that among the workers who come most into contact with specific differences, are to be found the only men who have failed to be persuaded of the unreality of those differences?), observations with this special bearing were from time to time related, but the class of fact on which Darwin built his conceptions of Heredity and Variation was not seen in the highways of biology. It formed no part of the official curriculum of biological students, and found no place among the subjects which their teachers were investigating.

During this period nevertheless one distinct advance was made, that with which Weismann's name is prominently connected. In Darwin's genetic scheme the hereditary transmission of parental experience and its consequences played a considerable role. Exactly how great that role was supposed to be, he with his habitual caution refrained from specifying, for the sufficient reason that he did not know. Nevertheless much of the process of Evolution, especially that by which organs have become degenerate and rudimentary, was certainly attributed by Darwin to such inheritance, though since belief in the inheritance of acquired characters fell into disrepute, the fact has been a good deal overlooked. The "Origin" without "use and disuse" would be a materially different book. A certain vacillation is discernible in Darwin's utterances on this question, and the fact gave to the astute Butler an opportunity for his most telling attack. The discussion which best illustrates the genetic views of the period arose in regard to the production of the rudimentary condition of the wings of many beetles in the Madeira group of islands, and by comparing passages from the "Origin" (6th edition pages 109 and 401. See Butler, "Essays on Life, Art, and Science", page 265, reprinted 1908, and "Evolution, Old and New", chapter XXII. (2nd edition), 1882.) Butler convicts Darwin of saying first that this condition was in the main the result of Selection, with disuse aiding, and in another place that the main cause of degeneration was disuse, but that Selection had aided. To Darwin however I think the point would have seemed one of dialectics merely. To him the one paramount purpose was to show that somehow an Evolution by means of Variation and Heredity might have brought about the facts observed, and whether they had come to pass in the one way or the other was a matter of subordinate concern.

To us moderns the question at issue has a diminished significance. For over all such debates a change has been brought by Weismann's challenge for evidence that use and disuse have any transmitted effects at all. Hitherto the transmission of many acquired characteristics had seemed to most naturalists so obvious as not to call for demonstration. (W. Lawrence was one of the few who consistently maintained the contrary opinion. Prichard, who previously had expressed himself in the same sense, does not, I believe repeat these views in his later writings, and there are signs that he came to believe in the transmission of acquired habits. See Lawrence, "Lect. Physiol." 1823, pages 436-437, 447 Prichard, Edin. Inaug. Disp. 1808 (not seen by me), quoted ibid. and "Nat. Hist. Man", 1843, pages 34 f.) Weismann's demand for facts in support of the main proposition revealed at once that none having real cogency could be produced. The time-honoured examples were easily shown to be capable of different explanations. A few certainly remain which cannot be so summarily dismissed, but—though it is manifestly impossible here to do justice to such a subject—I think no one will dispute that these residual and doubtful phenomena, whatever be their true nature, are not of a kind to help us much in the interpretation of any of those complex cases of adaptation which on the hypothesis of unguided Natural Selection are especially difficult to understand. Use and disuse were invoked expressly to help us over these hard places; but whatever changes can be induced in offspring by direct treatment of the parents, they are not of a kind to encourage hope of real assistance from that quarter. It is not to be denied that through the collapse of this second line of argument the Selection hypothesis has had to take an increased and perilous burden. Various ways of meeting the difficulty have been proposed, but these mostly resolve themselves into improbable attempts to expand or magnify the powers of Natural Selection.

Weismann's interpellation, though negative in purpose, has had a lasting and beneficial effect, for through his thorough demolition of the old loose and distracting notions of inherited experience, the ground has been cleared for the construction of a true knowledge of heredity based on experimental fact.

In another way he made a contribution of a more positive character, for his elaborate speculations as to the genetic meaning of cytological appearances have led to a minute investigation of the visible phenomena occurring in those divisions by which germ-cells arise. Though the particular views he advocated have very largely proved incompatible with the observed facts of heredity, yet we must acknowledge that it was chiefly through the stimulus of Weismann's ideas that those advances in cytology were made; and though the doctrine of the continuity of germ-plasm cannot be maintained in the form originally propounded, it is in the main true and illuminating. (It is interesting to see how nearly Butler was led by natural penetration, and from absolutely opposite conclusions, back to this underlying truth: "So that each ovum when impregnate should be considered not as descended from its ancestors, but as being a continuation of the personality of every ovum in the chain of its ancestry, which every ovum IT ACTUALLY IS quite as truly as the octogenarian IS the same identity with the ovum from which he has been developed. This process cannot stop short of the primordial cell, which again will probably turn out to be but a brief resting-place. We therefore prove each one of us to BE ACTUALLY the primordial cell which never died nor dies, but has differentiated itself into the life of the world, all living beings whatever, being one with it and members one of another," "Life and Habit", 1878, page 86.) Nevertheless in the present state of knowledge we are still as a rule quite unable to connect cytological appearances with any genetic consequence and save in one respect (obviously of extreme importance—to be spoken of later) the two sets of phenomena might, for all we can see, be entirely distinct.

I cannot avoid attaching importance to this want of connection between the nuclear phenomena and the features of bodily organisation. All attempts to investigate Heredity by cytological means lie under the disadvantage that it is the nuclear changes which can alone be effectively observed. Important as they must surely be, I have never been persuaded that the rest of the cell counts for nothing. What we know of the behaviour and variability of chromosomes seems in my opinion quite incompatible with the belief that they alone govern form, and are the sole agents responsible in heredity. (This view is no doubt contrary to the received opinion. I am however interested to see it lately maintained by Driesch ("Science and Philosophy of the Organism", London, 1907, page 233), and from the recent observations of Godlewski it has received distinct experimental support.)

If, then, progress was to be made in Genetics, work of a different kind was required. To learn the laws of Heredity and Variation there is no other way than that which Darwin himself followed, the direct examination of the phenomena. A beginning could be made by collecting fortuitous observations of this class, which have often thrown a suggestive light, but such evidence can be at best but superficial and some more penetrating instrument of research is required. This can only be provided by actual experiments in breeding.

The truth of these general considerations was becoming gradually clear to many of us when in 1900 Mendel's work was rediscovered. Segregation, a phenomenon of the utmost novelty, was thus revealed. From that moment not only in the problem of the origin of species, but in all the great problems of biology a new era began. So unexpected was the discovery that many naturalists were convinced it was untrue, and at once proclaimed Mendel's conclusions as either altogether mistaken, or if true, of very limited application. Many fantastic notions about the workings of Heredity had been asserted as general principles before: this was probably only another fancy of the same class.

Nevertheless those who had a preliminary acquaintance with the facts of Variation were not wholly unprepared for some such revelation. The essential deduction from the discovery of segregation was that the characters of living things are dependent on the presence of definite elements or factors, which are treated as units in the processes of Heredity. These factors can thus be recombined in various ways. They act sometimes separately, and sometimes they interact in conjunction with each other, producing their various effects. All this indicates a definiteness and specific order in heredity, and therefore in variation. This order cannot by the nature of the case be dependent on Natural Selection for its existence, but must be a consequence of the fundamental chemical and physical nature of living things. The study of Variation had from the first shown that an orderliness of this kind was present. The bodies and the properties of living things are cosmic, not chaotic. No matter how low in the scale we go, never do we find the slightest hint of a diminution in that all-pervading orderliness, nor can we conceive an organism existing for a moment in any other state. Moreover not only does this order prevail in normal forms, but again and again it is to be seen in newly-sprung varieties, which by general consent cannot have been subjected to a prolonged Selection. The discovery of Mendelian elements admirably coincided with and at once gave a rationale of these facts. Genetic Variation is then primarily the consequence of additions to, or omissions from, the stock of elements which the species contains. The further investigation of the species-problem must thus proceed by the analytical method which breeding experiments provide.

In the nine years which have elapsed since Mendel's clue became generally known, progress has been rapid. We now understand the process by which a polymorphic race maintains its polymorphism. When a family consists of dissimilar members, given the numerical proportions in which these members are occurring, we can represent their composition symbolically and state what types can be transmitted by the various members. The difficulty of the "swamping effects of intercrossing" is practically at an end. Even the famous puzzle of sex-limited inheritance is solved, at all events in its more regular manifestations, and we know now how it is brought about that the normal sisters of a colour-blind man can transmit the colour-blindness while his normal brothers cannot transmit it.

We are still only on the fringe of the inquiry. It can be seen extending and ramifying in many directions. To enumerate these here would be impossible. A whole new range of possibilities is being brought into view by study of the interrelations between the simple factors. By following up the evidence as to segregation, indications have been obtained which can only be interpreted as meaning that when many factors are being simultaneously redistributed among the germ-cells, certain of them exert what must be described as a repulsion upon other factors. We cannot surmise whither this discovery may lead.

In the new light all the old problems wear a fresh aspect. Upon the question of the nature of Sex, for example, the bearing of Mendelian evidence is close. Elsewhere I have shown that from several sets of parallel experiments the conclusion is almost forced upon us that, in the types investigated, of the two sexes the female is to be regarded as heterozygous in sex, containing one unpaired dominant element, while the male is similarly homozygous in the absence of that element. (In other words, the ova are each EITHER female, OR male (i.e. non-female), but the sperms are all non-female.) It is not a little remarkable that on this point—which is the only one where observations of the nuclear processes of gameto-genesis have yet been brought into relation with the visible characteristics of the organisms themselves—there should be diametrical opposition between the results of breeding experiments and those derived from cytology.

Those who have followed the researches of the American school will be aware that, after it had been found in certain insects that the spermatozoa were of two kinds according as they contained or did not contain the accessory chromosome, E.B. Wilson succeeded in proving that the sperms possessing this accessory body were destined to form FEMALES on fertilisation, while sperms without it form males, the eggs being apparently indifferent. Perhaps the most striking of all this series of observations is that lately made by T.H. Morgan (Morgan, "Proc. Soc. Exp. Biol. Med." V. 1908, and von Baehr, "Zool. Anz." XXXII. page 507, 1908.), since confirmed by von Baehr, that in a Phylloxeran two kinds of spermatids are formed, respectively with and without an accessory (in this case, DOUBLE) chromosome. Of these, only those possessing the accessory body become functional spermatozoa, the others degenerating. We have thus an elucidation of the puzzling fact that in these forms fertilisation results in the formation of FEMALES only. How the males are formed—for of course males are eventually produced by the parthenogenetic females—we do not know.

If the accessory body is really to be regarded as bearing the factor for femaleness, then in Mendelian terms female is DD and male is DR. The eggs are indifferent and the spermatozoa are each male, OR female. But according to the evidence derived from a study of the sex-limited descent of certain features in other animals the conclusion seems equally clear that in them female must be regarded as DR and male as RR. The eggs are thus each either male or female and the spermatozoa are indifferent. How this contradictory evidence is to be reconciled we do not yet know. The breeding work concerns fowls, canaries, and the Currant moth (Abraxas grossulariata). The accessory chromosome has been now observed in most of the great divisions of insects (As Wilson has proved, the unpaired body is not a universal feature even in those orders in which it has been observed. Nearly allied types may differ. In some it is altogether unpaired. In others it is paired with a body of much smaller size, and by selection of various types all gradations can be demonstrated ranging to the condition in which the members of the pair are indistinguishable from each other.), except, as it happens, Lepidoptera. At first sight it seems difficult to suppose that a feature apparently so fundamental as sex should be differently constituted in different animals, but that seems at present the least improbable inference. I mention these two groups of facts as illustrating the nature and methods of modern genetic work. We must proceed by minute and specific analytical investigation. Wherever we look we find traces of the operation of precise and specific rules.

In the light of present knowledge it is evident that before we can attack the Species-problem with any hope of success there are vast arrears to be made up. He would be a bold man who would now assert that there was no sense in which the term Species might not have a strict and concrete meaning in contradistinction to the term Variety. We have been taught to regard the difference between species and variety as one of degree. I think it unlikely that this conclusion will bear the test of further research. To Darwin the question, What is a variation? presented no difficulties. Any difference between parent and offspring was a variation. Now we have to be more precise. First we must, as de Vries has shown, distinguish real, genetic, variation from FLUCTUATIONAL variations, due to environmental and other accidents, which cannot be transmitted. Having excluded these sources of error the variations observed must be expressed in terms of the factors to which they are due before their significance can be understood. For example, numbers of the variations seen under domestication, and not a few witnessed in nature, are simply the consequence of some ingredient being in an unknown way omitted from the composition of the varying individual. The variation may on the contrary be due to the addition of some new element, but to prove that it is so is by no means an easy matter. Casual observation is useless, for though these latter variations will always be dominants, yet many dominant characteristics may arise from another cause, namely the meeting of complementary factors, and special study of each case in two generations at least is needed before these two phenomena can be distinguished.

When such considerations are fully appreciated it will be realised that medleys of most dissimilar occurrences are all confused together under the term Variation. One of the first objects of genetic analysis is to disentangle this mass of confusion.

To those who have made no study of heredity it sometimes appears that the question of the effect of conditions in causing variation is one which we should immediately investigate, but a little thought will show that before any critical inquiry into such possibilities can be attempted, a knowledge of the working of heredity under conditions as far as possible uniform must be obtained. At the time when Darwin was writing, if a plant brought into cultivation gave off an albino variety, such an event was without hesitation ascribed to the change of life. Now we see that albino GAMETES, germs, that is to say, which are destitute of the pigment-forming factor, may have been originally produced by individuals standing an indefinite number of generations back in the ancestry of the actual albino, and it is indeed almost certain that the variation to which the appearance of the albino is due cannot have taken place in a generation later than that of the grandparents. It is true that when a new DOMINANT appears we should feel greater confidence that we were witnessing the original variation, but such events are of extreme rarity, and no such case has come under the notice of an experimenter in modern times, as far as I am aware. That they must have appeared is clear enough. Nothing corresponding to the Brown-breasted Game fowl is known wild, yet that colour is a most definite dominant, and at some moment since Gallus bankiva was domesticated, the element on which that special colour depends must have at least once been formed in the germ-cell of a fowl; but we need harder evidence than any which has yet been produced before we can declare that this novelty came through over-feeding, or change of climate, or any other disturbance consequent on domestication. When we reflect on the intricacies of genetic problems as we must now conceive them there come moments when we feel almost thankful that the Mendelian principles were unknown to Darwin. The time called for a bold pronouncement, and he made it, to our lasting profit and delight. With fuller knowledge we pass once more into a period of cautious expectation and reserve.

In every arduous enterprise it is pleasanter to look back at difficulties overcome than forward to those which still seem insurmountable, but in the next stage there is nothing to be gained by disguising the fact that the attributes of living things are not what we used to suppose. If they are more complex in the sense that the properties they display are throughout so regular (I have in view, for example, the marvellous and specific phenomena of regeneration, and those discovered by the students of "Entwicklungsmechanik". The circumstances of its occurrence here preclude any suggestion that this regularity has been brought about by the workings of Selection. The attempts thus to represent the phenomena have resulted in mere parodies of scientific reasoning.) that the Selection of minute random variations is an unacceptable account of the origin of their diversity, yet by virtue of that very regularity the problem is limited in scope and thus simplified.

To begin with, we must relegate Selection to its proper place. Selection permits the viable to continue and decides that the non-viable shall perish; just as the temperature of our atmosphere decides that no liquid carbon shall be found on the face of the earth: but we do not suppose that the form of the diamond has been gradually achieved by a process of Selection. So again, as the course of descent branches in the successive generations, Selection determines along which branch Evolution shall proceed, but it does not decide what novelties that branch shall bring forth. "La Nature contient le fonds de toutes ces varietes, mais le hazard ou l'art les mettent en oeuvre," as Maupertuis most truly said.

Not till knowledge of the genetic properties of organisms has attained to far greater completeness can evolutionary speculations have more than a suggestive value. By genetic experiment, cytology and physiological chemistry aiding, we may hope to acquire such knowledge. In 1872 Nathusius wrote ("Vortrage uber Viehzucht und Rassenerkenntniss", page 120, Berlin, 1872.): "Das Gesetz der Vererbung ist noch nicht erkannt; der Apfel ist noch nicht vom Baum der Erkenntniss gefallen, welcher, der Sage nach, Newton auf den rechten Weg zur Ergrundung der Gravitationsgesetze fuhrte." We cannot pretend that the words are not still true, but in Mendelian analysis the seeds of that apple-tree at last are sown.

If we were asked what discovery would do most to forward our inquiry, what one bit of knowledge would more than any other illuminate the problem, I think we may give the answer without hesitation. The greatest advance that we can foresee will be made when it is found possible to connect the geometrical phenomena of development with the chemical. The geometrical symmetry of living things is the key to a knowledge of their regularity, and the forces which cause it. In the symmetry of the dividing cell the basis of that resemblance we call Heredity is contained. To imitate the morphological phenomena of life we have to devise a system which can divide. It must be able to divide, and to segment as—grossly—a vibrating plate or rod does, or as an icicle can do as it becomes ribbed in a continuous stream of water; but with this distinction, that the distribution of chemical differences and properties must simultaneously be decided and disposed in orderly relation to the pattern of the segmentation. Even if a model which would do this could be constructed it might prove to be a useful beginning.

This may be looking too far ahead. If we had to choose some one piece of more proximate knowledge which we would more especially like to acquire, I suppose we should ask for the secret of interracial sterility. Nothing has yet been discovered to remove the grave difficulty, by which Huxley in particular was so much oppressed, that among the many varieties produced under domestication—which we all regard as analogous to the species seen in nature—no clear case of interracial sterility has been demonstrated. The phenomenon is probably the only one to which the domesticated products seem to afford no parallel. No solution of the difficulty can be offered which has positive value, but it is perhaps worth considering the facts in the light of modern ideas. It should be observed that we are not discussing incompatibility of two species to produce offspring (a totally distinct phenomenon), but the sterility of the offspring which many of them do produce.

When two species, both perfectly fertile severally, produce on crossing a sterile progeny, there is a presumption that the sterility is due to the development in the hybrid of some substance which can only be formed by the meeting of two complementary factors. That some such account is correct in essence may be inferred from the well-known observation that if the hybrid is not totally sterile but only partially so, and thus is able to form some good germ-cells which develop into new individuals, the sterility of these daughter-individuals is sensibly reduced or may be entirely absent. The fertility once re-established, the sterility does not return in the later progeny, a fact strongly suggestive of segregation. Now if the sterility of the cross-bred be really the consequence of the meeting of two complementary factors, we see that the phenomenon could only be produced among the divergent offspring of one species by the acquisition of at least TWO new factors; for if the acquisition of a single factor caused sterility the line would then end. Moreover each factor must be separately acquired by distinct individuals, for if both were present together, the possessors would by hypothesis be sterile. And in order to imitate the case of species each of these factors must be acquired by distinct breeds. The factors need not, and probably would not, produce any other perceptible effects; they might, like the colour-factors present in white flowers, make no difference in the form or other characters. Not till the cross was actually made between the two complementary individuals would either factor come into play, and the effects even then might be unobserved until an attempt was made to breed from the cross-bred.

Next, if the factors responsible for sterility were acquired, they would in all probability be peculiar to certain individuals and would not readily be distributed to the whole breed. Any member of the breed also into which BOTH the factors were introduced would drop out of the pedigree by virtue of its sterility. Hence the evidence that the various domesticated breeds say of dogs or fowls when mated together produce fertile offspring, is beside the mark. The real question is, Do they ever produce sterile offspring? I think the evidence is clearly that sometimes they do, oftener perhaps than is commonly supposed. These suggestions are quite amenable to experimental tests. The most obvious way to begin is to get a pair of parents which are known to have had any sterile offspring, and to find the proportions in which these steriles were produced. If, as I anticipate, these proportions are found to be definite, the rest is simple.

In passing, certain other considerations may be referred to. First, that there are observations favouring the view that the production of totally sterile cross-breds is seldom a universal property of two species, and that it may be a matter of individuals, which is just what on the view here proposed would be expected. Moreover, as we all know now, though incompatibility may be dependent to some extent on the degree to which the species are dissimilar, no such principle can be demonstrated to determine sterility or fertility in general. For example, though all our Finches can breed together, the hybrids are all sterile. Of Ducks some species can breed together without producing the slightest sterility; others have totally sterile offspring, and so on. The hybrids between several genera of Orchids are perfectly fertile on the female side, and some on the male side also, but the hybrids produced between the Turnip (Brassica napus) and the Swede (Brassica campestris), which, according to our estimates of affinity should be nearly allied forms, are totally sterile. (See Sutton, A.W., "Journ. Linn. Soc." XXXVIII. page 341, 1908.) Lastly, it may be recalled that in sterility we are almost certainly considering a meristic phenomenon. FAILURE TO DIVIDE is, we may feel fairly sure, the immediate "cause" of the sterility. Now, though we know very little about the heredity of meristic differences, all that we do know points to the conclusion that the less-divided is dominant to the more-divided, and we are thus justified in supposing that there are factors which can arrest or prevent cell-division. My conjecture therefore is that in the case of sterility of cross-breds we see the effect produced by a complementary pair of such factors. This and many similar problems are now open to our analysis.

The question is sometimes asked, Do the new lights on Variation and Heredity make the process of Evolution easier to understand? On the whole the answer may be given that they do. There is some appearance of loss of simplicity, but the gain is real. As was said above, the time is not ripe for the discussion of the origin of species. With faith in Evolution unshaken—if indeed the word faith can be used in application to that which is certain—we look on the manner and causation of adapted differentiation as still wholly mysterious. As Samuel Butler so truly said: "To me it seems that the 'Origin of Variation,' whatever it is, is the only true 'Origin of Species'" ("Life and Habit", London, page 263, 1878.), and of that Origin not one of us knows anything. But given Variation—and it is given: assuming further that the variations are not guided into paths of adaptation—and both to the Darwinian and to the modern school this hypothesis appears to be sound if unproven—an evolution of species proceeding by definite steps is more, rather than less, easy to imagine than an evolution proceeding by the accumulation of indefinite and insensible steps. Those who have lost themselves in contemplating the miracles of Adaptation (whether real or spurious) have not unnaturally fixed their hopes rather on the indefinite than on the definite changes. The reasons are obvious. By suggesting that the steps through which an adaptive mechanism arose were indefinite and insensible, all further trouble is spared. While it could be said that species arise by an insensible and imperceptible process of variation, there was clearly no use in tiring ourselves by trying to perceive that process. This labour-saving counsel found great favour. All that had to be done to develop evolution-theory was to discover the good in everything, a task which, in the complete absence of any control or test whereby to check the truth of the discovery, is not very onerous. The doctrine "que tout est au mieux" was therefore preached with fresh vigour, and examples of that illuminating principle were discovered with a facility that Pangloss himself might have envied, till at last even the spectators wearied of such dazzling performances.

But in all seriousness, why should indefinite and unlimited variation have been regarded as a more probable account of the origin of Adaptation? Only, I think, because the obstacle was shifted one plane back, and so looked rather less prominent. The abundance of Adaptation, we all grant, is an immense, almost an unsurpassable difficulty in all non-Lamarckian views of Evolution; but if the steps by which that adaptation arose were fortuitous, to imagine them insensible is assuredly no help. In one most important respect indeed, as has often been observed, it is a multiplication of troubles. For the smaller the steps, the less could Natural Selection act upon them. Definite variations—and of the occurrence of definite variations in abundance we have now the most convincing proof—have at least the obvious merit that they can make and often do make a real difference in the chances of life.

There is another aspect of the Adaptation problem to which I can only allude very briefly. May not our present ideas of the universality and precision of Adaptation be greatly exaggerated? The fit of organism to its environment is not after all so very close—a proposition unwelcome perhaps, but one which could be illustrated by very copious evidence. Natural Selection is stern, but she has her tolerant moods.

We have now most certain and irrefragable proof that much definiteness exists in living things apart from Selection, and also much that may very well have been preserved and so in a sense constituted by Selection. Here the matter is likely to rest. There is a passage in the sixth edition of the "Origin" which has I think been overlooked. On page 70 Darwin says "The tuft of hair on the breast of the wild turkey-cock cannot be of any use, and it is doubtful whether it can be ornamental in the eyes of the female bird." This tuft of hair is a most definite and unusual structure, and I am afraid that the remark that it "cannot be of any use" may have been made inadvertently; but it may have been intended, for in the first edition the usual qualification was given and must therefore have been deliberately excised. Anyhow I should like to think that Darwin did throw over that tuft of hair, and that he felt relief when he had done so. Whether however we have his great authority for such a course or not, I feel quite sure that we shall be rightly interpreting the facts of nature if we cease to expect to find purposefulness wherever we meet with definite structures or patterns. Such things are, as often as not, I suspect rather of the nature of tool-marks, mere incidents of manufacture, benefiting their possessor not more than the wire-marks in a sheet of paper, or the ribbing on the bottom of an oriental plate renders those objects more attractive in our eyes.

If Variation may be in any way definite, the question once more arises, may it not be definite in direction? The belief that it is has had many supporters, from Lamarck onwards, who held that it was guided by need, and others who, like Nageli, while laying no emphasis on need, yet were convinced that there was guidance of some kind. The latter view under the name of "Orthogenesis," devised I believe by Eimer, at the present day commends itself to some naturalists. The objection to such a suggestion is of course that no fragment of real evidence can be produced in its support. On the other hand, with the experimental proof that variation consists largely in the unpacking and repacking of an original complexity, it is not so certain as we might like to think that the order of these events is not pre-determined. For instance the original "pack" may have been made in such a way that at the nth division of the germ-cells of a Sweet Pea a colour-factor might be dropped, and that at the n plus n prime division the hooded variety be given off, and so on. I see no ground whatever for holding such a view, but in fairness the possibility should not be forgotten, and in the light of modern research it scarcely looks so absurdly improbable as before.

No one can survey the work of recent years without perceiving that evolutionary orthodoxy developed too fast, and that a great deal has got to come down; but this satisfaction at least remains, that in the experimental methods which Mendel inaugurated, we have means of reaching certainty in regard to the physiology of Heredity and Variation upon which a more lasting structure may be built.

68

VI. THE MINUTE STRUCTURE OF
CELLS IN RELATION TO
HEREDITY. By Eduard Strasburger.

Professor of Botany in the University of
Bonn.

Since 1875 an unexpected insight has been gained into the internal structure of cells. Those who are familiar with the results of investigations in this branch of Science are convinced that any modern theory of heredity must rest on a basis of cytology and cannot be at variance with cytological facts. Many histological discoveries, both such as have been proved correct and others which may be accepted as probably well founded, have acquired a fundamental importance from the point of view of the problems of heredity.

My aim is to describe the present position of our knowledge of Cytology. The account must be confined to essentials and cannot deal with far-reaching and controversial questions. In cases where difference of opinion exists, I adopt my own view for which I hold myself responsible. I hope to succeed in making myself intelligible even without the aid of illustrations: in order to convey to the uninitiated an adequate idea of the phenomena connected with the life of a cell, a greater number of figures would be required than could be included within the scope of this article.

So long as the most eminent investigators (As for example the illustrious Wilhelm Hofmeister in his "Lehre von der Pflanzenzelle" (1867).) believed that the nucleus of a cell was destroyed in the course of each division and that the nuclei of the daughter-cells were produced de novo, theories of heredity were able to dispense with the nucleus. If they sought, as did Charles Darwin, who showed a correct grasp of the problem in the enunciation of his Pangenesis hypothesis, for histological connecting links, their hypotheses, or at least the best of them, had reference to the cell as a whole. It was known to Darwin that the cell multiplied by division and was derived from a similar pre-existing cell. Towards 1870 it was first demonstrated that cell-nuclei do not arise de novo, but are invariably the result of division of pre-existing nuclei. Better methods of investigation rendered possible a deeper insight into the phenomena accompanying cell and nuclear divisions and at the same time disclosed the existence of remarkable structures. The work of O. Butschli, O. Hertwig, W. Flemming H. Fol and of the author of this article (For further reference to literature, see my article on "Die Ontogenie der Zelle seit 1875", in the "Progressus Rei Botanicae", Vol. I. page 1, Jena, 1907.), have furnished conclusive evidence in favour of these facts. It was found that when the reticular framework of a nucleus prepares to divide, it separates into single segments. These then become thicker and denser, taking up with avidity certain stains, which are used as aids to investigation, and finally form longer or shorter, variously bent, rodlets of uniform thickness. In these organs which, on account of their special property of absorbing certain stains, were styled Chromosomes (By W. Waldeyer in 1888.), there may usually be recognised a separation into thicker and thinner discs; the former are often termed Chromomeres. (Discovered by W. Pfitzner in 1880.) In the course of division of the nucleus, the single rows of chromomeres in the chromosomes are doubled and this produces a band-like flattening and leads to the longitudinal splitting by which each chromosome is divided into two exactly equal halves. The nuclear membrane then disappears and fibrillar cell-plasma or cytoplasm invades the nuclear area. In animal cells these fibrillae in the cytoplasm centre on definite bodies (Their existence and their multiplication by fission were demonstrated by E. van Beneden and Th. Boveri in 1887.), which it is customary to speak of as Centrosomes. Radiating lines in the adjacent cell-plasma suggest that these bodies

69

constitute centres of force. The cells of the higher plants do not possess such individualised centres; they have probably disappeared in the course of phylogenetic development: in spite of this, however, in the nuclear division-figures the fibrillae of the cell-plasma are seen to radiate from two opposite poles. In both animal and plant cells a fibrillar bipolar spindle is formed, the fibrillae of which grasp the longitudinally divided chromosomes from two opposite sides and arrange them on the equatorial plane of the spindle as the so-called nuclear or equatorial plate. Each half-chromosome is connected with one of the spindle poles only and is then drawn towards that pole. (These important facts, suspected by W. Flemming in 1882, were demonstrated by E. Heuser, L. Guignard, E. van Beneden, M. Nussbaum, and C. Rabl.)

The formation of the daughter-nuclei is then effected. The changes which the daughter-chromosomes undergo in the process of producing the daughter-nuclei repeat in the reverse order the changes which they went through in the course of their progressive differentiation from the mother-nucleus. The division of the cell-body is completed midway between the two daughter-nuclei. In animal cells, which possess no chemically differentiated membrane, separation is effected by simple constriction, while in the case of plant cells provided with a definite wall, the process begins with the formation of a cytoplasmic separating layer.

The phenomena observed in the course of the division of the nucleus show beyond doubt that an exact halving of its substance is of the greatest importance. (First shown by W. Roux in 1883.) Compared with the method of division of the nucleus, that of the cytoplasm appears to be very simple. This led to the conception that the cell-nucleus must be the chief if not the sole carrier of hereditary characters in the organism. It is for this reason that the detailed investigation of fertilisation phenomena immediately followed researches into the nucleus. The fundamental discovery of the union of two nuclei in the sexual act was then made (By O. Hertwig in 1875.) and this afforded a new support for the correct conception of the nuclear functions. The minute study of the behaviour of the other constituents of sexual cells during fertilisation led to the result, that the nucleus alone is concerned with handing on hereditary characters (This was done by O. Hertwig and the author of this essay simultaneously in 1884.) from one generation to another. Especially important, from the point of view of this conclusion, is the study of fertilisation in Angiosperms (Flowering plants); in these plants the male sexual cells lose their cell-body in the pollen-tube and the nucleus only—the sperm-nucleus—reaches the egg. The cytoplasm of the male sexual cell is therefore not necessary to ensure a transference of hereditary characters from parents to offspring. I lay stress on the case of the Angiosperms because researches recently repeated with the help of the latest methods failed to obtain different results. As regards the descendants of angiospermous plants, the same laws of heredity hold good as for other sexually differentiated organisms; we may, therefore, extend to the latter what the Angiosperms so clearly teach us.

The next advance in the hitherto rapid progress in our knowledge of nuclear division was delayed, because it was not at once recognised that there are two absolutely different methods of nuclear division. All such nuclear divisions were united under the head of indirect or mitotic divisions; these were also spoken of as karyo-kineses, and were distinguished from the direct or amitotic divisions which are characterised by a simple constriction of the nuclear body. So long as the two kinds of indirect nuclear division were not clearly distinguished, their correct interpretation was impossible. This was accomplished after long and laborious research, which has recently been carried out and with results which should, perhaps, be regarded as provisional.

Soon after the new study of the nucleus began, investigators were struck by the fact that the course of nuclear division in the mother-cells, or more correctly in the grandmother-cells, of spores, pollen-grains, and embryo-sacs of the more highly organised plants and in the spermatozoids and eggs of the higher animals, exhibits similar phenomena, distinct from those which occur in the somatic cells.

In the nuclei of all those cells which we may group together as gonotokonts (At the suggestion of J.P. Lotsy in 1904.) (i.e. cells concerned in reproduction) there are fewer chromosomes than in the adjacent body-cells (somatic cells). It was noticed also that there is a peculiarity characteristic of the gonotokonts, namely the occurrence of two nuclear divisions rapidly succeeding one another. It was afterwards recognised that in the first stage of nuclear division in the gonotokonts the chromosomes unite in pairs: it is these chromosome-pairs, and not the two longitudinal halves of single chromosomes, which form the nuclear plate in the equatorial plane of the nuclear spindle. It has been proposed to call these pairs gemini. (J.E.S. Moore and A.L. Embleton, "Proc. Roy. Soc." London, Vol. LXXVII. page 555, 1906; V. Gregoire, 1907.) In the course of this division the spindle-fibrillae attach themselves to the gemini, i.e. to entire chromosomes and direct them to the points where the new daughter-nuclei are formed, that is to those positions towards which the longitudinal halves of the chromosomes travel in ordinary nuclear divisions. It is clear that in this way the number of chromosomes which the daughter-nuclei contain, as the result of the first stage in division in the gonotokonts, will be reduced by one half, while in ordinary divisions the number of chromosomes always remains the same. The first stage in the division of the nucleus in the gonotokonts has therefore been termed the reduction division. (In 1887 W. Flemming termed this the heterotypic form of nuclear division.) This stage in division determines the conditions for the second division which rapidly ensues. Each of the paired chromosomes of the mother-nucleus has already, as in an ordinary nuclear division, completed the longitudinal fission, but in this case it is not succeeded by the immediate separation of the longitudinal halves and their allotment to different nuclei. Each chromosome, therefore, takes its two longitudinal halves into the same daughter-nucleus. Thus, in each daughter-nucleus the longitudinal halves of the chromosomes are present ready for the next stage in the division; they only require to be arranged in the nuclear plate and then distributed among the granddaughter-nuclei. This method of division, which takes place with chromosomes already split, and which have only to provide for the distribution of their longitudinal halves to the next nuclear generation, has been called homotypic nuclear division. (The name was proposed by W. Flemming in 1887; the nature of this type of division was, however, not explained until later.)

Reduction division and homotypic nuclear division are included together under the term allotypic nuclear division and are distinguished from the ordinary or typical nuclear division. The name Meiosis (By J. Bretland Farmer and J.E.S. Moore in 1905.) has also been proposed for these two allotypic nuclear divisions. The typical divisions are often spoken of as somatic.

Observers who were actively engaged in this branch of recent histological research soon noticed that the chromosomes of a given organism are differentiated in definite numbers from the nuclear network in the course of division. This is especially striking in the gonotokonts, but it applies also to the somatic tissues. In the latter, one usually finds twice as many chromosomes as in the gonotokonts. Thus the conclusion was gradually reached that the doubling of chromosomes, which necessarily accompanies fertilisation, is maintained in the product of fertilisation, to be again reduced to one half in the gonotokonts at the stage of reduction-division. This enabled us to form a conception as to the essence of true alternation of generations, in which generations containing single and double chromosomes alternate with one another.

The single-chromosome generation, which I will call the HAPLOID, must have been the primitive generation in all organisms; it might also persist as the only generation. Every sexual differentiation in organisms, which occurred in the course of phylogenetic development, was followed by fertilisation and therefore by the creation of a diploid or double-chromosome product. So long as the germination of the product of fertilisation, the zygote, began with a reducing process, a special DIPLOID generation was not represented. This, however, appeared later as a product of the further evolution of the zygote, and the reduction division was correspondingly postponed. In animals, as in plants, the diploid generation attained the higher development and gradually assumed the dominant position. The haploid generation suffered a proportional reduction, until it finally ceased to have an independent existence and became restricted to the role of producing the sexual products within the body of the diploid generation. Those who do not possess the necessary special

71

knowledge are unable to realise what remains of the first haploid generation in a phanerogamic plant or in a vertebrate animal. In Angiosperms this is actually represented only by the short developmental stages which extend from the pollen mother-cells to the sperm-nucleus of the pollen-tube, and from the embryo-sac mother-cell to the egg and the endosperm tissue. The embryo-sac remains enclosed in the diploid ovule, and within this from the fertilised egg is formed the embryo which introduces the new diploid generation. On the full development of the diploid embryo of the next generation, the diploid ovule of the preceding diploid generation is separated from the latter as a ripe seed. The uninitiated sees in the more highly organised plants only a succession of diploid generations. Similarly all the higher animals appear to us as independent organisms with diploid nuclei only. The haploid generation is confined in them to the cells produced as the result of the reduction division of the gonotokonts; the development of these is completed with the homotypic stage of division which succeeds the reduction division and produces the sexual products.

The constancy of the numbers in which the chromosomes separate themselves from the nuclear network during division gave rise to the conception that, in a certain degree, chromosomes possess individuality. Indeed the most careful investigations (Particularly those of V. Gregoire and his pupils.) have shown that the segments of the nuclear network, which separate from one another and condense so as to produce chromosomes for a new division, correspond to the segments produced from the chromosomes of the preceding division. The behaviour of such nuclei as possess chromosomes of unequal size affords confirmatory evidence of the permanence of individual chromosomes in corresponding sections of an apparently uniform nuclear network. Moreover at each stage in division chromosomes with the same differences in size reappear. Other cases are known in which thicker portions occur in the substance of the resting nucleus, and these agree in number with the chromosomes. In this network, therefore, the individual chromosomes must have retained their original position. But the chromosomes cannot be regarded as the ultimate hereditary units in the nuclei, as their number is too small. Moreover, related species not infrequently show a difference in the number of their chromosomes, whereas the number of hereditary units must approximately agree. We thus picture to ourselves the carriers of hereditary characters as enclosed in the chromosomes; the transmitted fixed number of chromosomes is for us only the visible expression of the conception that the number of hereditary units which the chromosomes carry must be also constant. The ultimate hereditary units may, like the chromosomes themselves, retain a definite position in the resting nucleus. Further, it may be assumed that during the separation of the chromosomes from one another and during their assumption of the rod-like form, the hereditary units become aggregated in the chromomeres and that these are characterised by a constant order of succession. The hereditary units then grow, divide into two and are uniformly distributed by the fission of the chromosomes between their longitudinal halves.

As the contraction and rod-like separation of the chromosomes serve to isnure the transmission of all hereditary units in the products of division of a nucleus, so, on the other hand, the reticular distension of each chromosome in the so-called resting nucleus may effect a separation of the carriers of hereditary units from each other and facilitate the specific activity of each of them.

In the stages preliminary to their division, the chromosomes become denser and take up a substance which increases their staining capacity; this is called chromatin. This substance collects in the chromomeres and may form the nutritive material for the carriers of hereditary units which we now believe to be enclosed in them. The chromatin cannot itself be the hereditary substance, as it afterwards leaves the chromosomes, and the amount of it is subject to considerable variation in the nucleus, according to its stage of development. Conjointly with the materials which take part in the formation of the nuclear spindle and other processes in the cell, the chromatin accumulates in the resting nucleus to form the nucleoli.

Naturally connected with the conclusion that the nuclei are the carriers of hereditary characters in the organism, is the question whether enucleate organisms can also exist.

72

Phylogenetic considerations give an affirmative answer to this question. The differentiation into nucleus and cytoplasm represents a division of labour in the protoplast. A study of organisms which belong to the lowest class of the organic world teaches us how this was accomplished. Instead of well-defined nuclei, scattered granules have been described in the protoplasm of several of these organisms (Bacteria, Cyanophyceae, Protozoa.), characterised by the same reactions as nuclear material, provided also with a nuclear network, but without a limiting membrane. (This is the result of the work of R. Hertwig and of the most recently published investigations.) Thus the carriers of hereditary characters may originally have been distributed in the common protoplasm, afterwards coming together and eventually assuming a definite form as special organs of the cell. It may be also assumed that in the protoplasm and in the primitive types of nucleus, the carriers of the same hereditary unit were represented in considerable quantity; they became gradually differentiated to an extent commensurate with newly acquired characters. It was also necessary that, in proportion as this happened, the mechanism of nuclear division must be refined. At first processes resembling a simple constriction would suffice to provide for the distribution of all hereditary units to each of the products of division, but eventually in both organic kingdoms nuclear division, which alone insured the qualitative identity of the products of division, became a more marked feature in the course of cell-multiplication.

Where direct nuclear division occurs by constriction in the higher organisms, it does not result in the halving of hereditary units. So far as my observations go, direct nuclear division occurs in the more highly organised plants only in cells which have lost their specific functions. Such cells are no longer capable of specific reproduction. An interesting case in this connection is afforded by the internodal cells of the Characeae, which possess only vegetative functions. These cells grow vigorously and their cytoplasm increases, their growth being accompanied by a correspondingly direct multiplication of the nuclei. They serve chiefly to nourish the plant, but, unlike the other cells, they are incapable of producing any offspring. This is a very instructive case, because it clearly shows that the nuclei are not only carriers of hereditary characters, but that they also play a definite part in the metabolism of the protoplasts.

Attention was drawn to the fact that during the reducing division of nuclei which contain chromosomes of unequal size, gemini are constantly produced by the pairing of chromosomes of the same size. This led to the conclusion that the pairing chromosomes are homologous, and that one comes from the father, the other from the mother. (First stated by T.H. Montgomery in 1901 and by W.S. Sutton in 1902.) This evidently applies also to the pairing of chromosomes in those reduction-divisions in which differences in size do not enable us to distinguish the individual chromosomes. In this case also each pair would be formed by two homologous chromosomes, the one of paternal, the other of maternal origin. When the separation of these chromosomes and their distribution to both daughter-nuclei occur a chromosome of each kind is provided for each of these nuclei. It would seem that the components of each pair might pass to either pole of the nuclear spindle, so that the paternal and maternal chromosomes would be distributed in varying proportion between the daughter-nuclei; and it is not impossible that one daughter-nucleus might occasionally contain paternal chromosomes only and its sister-nucleus exclusively maternal chromosomes.

The fact that in nuclei containing chromosomes of various sizes, the chromosomes which pair together in reduction-division are always of equal size, constitutes a further and more important proof of their qualitative difference. This is supported also by ingenious experiments which led to an unequal distribution of chromosomes in the products of division of a sea-urchin's egg, with the result that a difference was induced in their further development. (Demonstrated by Th. Boveri in 1902.)

The recently discovered fact that in diploid nuclei the chromosomes are arranged in pairs affords additional evidence in favour of the unequal value of the chromosomes. This is still more striking in the case of chromosomes of different sizes. It has been shown that in the first division-figure in the nucleus of the fertilised egg the chromosomes of corresponding size form pairs. They appear with this arrangement in all subsequent nuclear divisions in the diploid generation. The longitudinal fissions of the chromosomes provide for the unaltered preservation of this condition. In the reduction nucleus of the gonotokonts the homologous chromosomes being near together need not seek out one another; they are ready to form gemini. The next stage is their separation to the haploid daughter-nuclei, which have resulted from the reduction process.

Peculiar phenomena in the reduction nucleus accompany the formation of gemini in both organic kingdoms. (This has been shown more particularly by the work of L. Guignard, M. Mottier, J.B. Farmer, C.B. Wilson, V. Hacker and more recently by V. Gregoire and his pupil C.A. Allen, by the researches conducted in the Bonn Botanical Institute, and by A. and K.E. Schreiner.) Probably for the purpose of entering into most intimate relation, the pairs are stretched to long threads in which the chromomeres come to lie opposite one another. (C.A. Allen, A. and K.E. Schreiner, and Strasburger.) It seems probable that these are homologous chromomeres, and that the pairs afterwards unite for a short time, so that an exchange of hereditary units is rendered possible. (H. de Vries and Strasburger.) This cannot be actually seen, but certain facts of heredity point to the conclusion that this occurs. It follows from these phenomena that any exchange which may be effected must be one of homologous carriers of hereditary units only. These units continue to form exchangeable segments after they have undergone unequal changes; they then constitute allelotropic pairs. We may thus calculate what sum of possible combinations the exchange of homologous hereditary units between the pairing chromosomes provides for before the reduction division and the subsequent distribution of paternal and maternal chromosomes in the haploid daughter-nuclei. These nuclei then transmit their characters to the sexual cells, the conjugation of which in fertilization again produces the most varied combinations. (A. Weismann gave the impulse to these ideas in his theory on "Amphimixis".) In this way all the cooperations which the carriers of hereditary characters are capable of in a species are produced; this must give it an appreciable advantage in the struggle for life.

The admirers of Charles Darwin must deeply regret that he did not live to see the results achieved by the new Cytology. What service would they have been to him in the presentation of his hypothesis of Pangenesis; what an outlook into the future would they have given to his active mind!

The Darwinian hypothesis of Pangenesis rests on the conception that all inheritable properties are represented in the cells by small invisible particles or gemmules and that these gemmules increase by division. Cytology began to develop on new lines some years after the publication in 1868 of Charles Darwin's "Provisional hypothesis of Pangenesis" ("Animals and Plants under Domestication", London, 1868, Chapter XXVII.), and when he died in 1882 it was still in its infancy. Darwin would have soon suggested the substitution of the nuclei for his gemmules. At least the great majority of present-day investigators in the domain of cytology have been led to the conclusion that the nucleus is the carrier of hereditary characters, and they also believe that hereditary characters are represented in the nucleus as distinct units. Such would be Darwin's gemmules, which in conformity with the name of his hypothesis may be called pangens (So called by H. de Vries in 1889.): these pangens multiply by division. All recently adopted views may be thus linked on to this part of Darwin's hypothesis. It is otherwise with Darwin's conception to which Pangenesis owes its name, namely the view that all cells continually give off gemmules, which migrate to other places in the organism, where they unite to form reproductive cells. When Darwin foresaw this possibility, the continuity of the germinal substance was still unknown (Demonstrated by Nussbaum in 1880, by Sachs in 1882, and by Weismann in 1885.), a fact which excludes a transference of gemmules.

But even Charles Darwin's genius was confined within finite boundaries by the state of science in his day.

It is not my province to deal with other theories of development which followed from Darwin's Pangenesis, or to discuss their histological probabilities. We can, however, affirm that Charles Darwin's idea that invisible gemmules are the carriers of hereditary characters and that they multiply by division has been removed from the position of a provisional hypothesis to that of a well-founded theory. It is supported by histology, and the results of experimental work in heredity, which are now assuming extraordinary prominence, are in close agreement with it.

VII. "THE DESCENT OF MAN". By G. Schwalbe.

Professor of Anatomy in the University of Strassburg.

The problem of the origin of the human race, of the descent of man, is ranked by Huxley in his epoch-making book "Man's Place in Nature", as the deepest with which biology has to concern itself, "the question of questions,"—the problem which underlies all others. In the same brilliant and lucid exposition, which appeared in 1863, soon after the publication of Darwin's "Origin of Species", Huxley stated his own views in regard to this great problem. He tells us how the idea of a natural descent of man gradually grew up in his mind, it was especially the assertions of Owen in regard to the total difference between the human and the simian brain that called forth strong dissent from the great anatomist Huxley, and he easily succeeded in showing that Owen's supposed differences had no real existence; he even established, on the basis of his own anatomical investigations, the proposition that the anatomical differences between the Marmoset and the Chimpanzee are much greater than those between the Chimpanzee and Man.

But why do we thus introduce the study of Darwin's "Descent of Man", which is to occupy us here, by insisting on the fact that Huxley had taken the field in defence of the descent of man in 1863, while Darwin's book on the subject did not appear till 1871? It is in order that we may clearly understand how it happened that from this time onwards Darwin and Huxley followed the same great aim in the most intimate association.

Huxley and Darwin working at the same Problema maximum! Huxley fiery, impetuous, eager for battle, contemptuous of the resistance of a dull world, or energetically triumphing over it. Darwin calm, weighing every problem slowly, letting it mature thoroughly,—not a fighter, yet having the greater and more lasting influence by virtue of his immense mass of critically sifted proofs. Darwin's friend, Huxley, was the first to do him justice, to understand his nature, and to find in it the reason why the detailed and carefully considered book on the descent of man made its appearance so late. Huxley, always generous, never thought of claiming priority for himself. In enthusiastic language he tells how Darwin's immortal work, "The Origin of Species", first shed light for him on the problem of the descent of man; the recognition of a vera causa in the transformation of species illuminated his thoughts as with a flash. He was now content to leave what perplexed him, what he could not yet solve, as he says himself, "in the mighty hands of Darwin." Happy in the bustle of strife against old and deep-rooted prejudices, against intolerance and superstition, he wielded his sharp weapons on Darwin's behalf; wearing Darwin's armour he joyously overthrew adversary after adversary. Darwin spoke of Huxley as his "general agent." ("Life and Letters of Thomas Henry Huxley", Vol. I. page 171, London, 1900.) Huxley says of himself "I am Darwin's bulldog." (Ibid. page 363.)

Thus Huxley openly acknowledged that it was Darwin's "Origin of Species" that first set the problem of the descent of man in its true light, that made the question of the origin of the human race a pressing one. That this was the logical consequence of his book Darwin himself had long felt. He had been reproached with intentionally shirking the application of his theory to Man. Let us hear what he says on this point in his autobiography: "As soon as I had become, in the year 1837 or 1838, convinced that species were mutable productions, I could not avoid the belief that man must come under the same law. Accordingly I collected notes on the subject for my own satisfaction, and not for a long time with any intention of publishing. Although in the 'Origin of Species' the derivation of any particular species is never discussed, yet I thought it best, in order THAT NO HONOURABLE MAN SHOULD ACCUSE ME OF CONCEALING MY VIEWS (No italics in original.), to add that by the work 'light would be thrown on the origin of man and his history.' It would have been useless and injurious to the success of the book to have paraded, without giving any evidence, my conviction with respect to his origin." ("Life and Letters of Charles Darwin", Vol. 1. page 93.)

In a letter written in January, 1860, to the Rev. L. Blomefield, Darwin expresses himself in similar terms. "With respect to man, I am very far from wishing to obtrude my belief; but I thought it dishonest to quite conceal my opinion." (Ibid. Vol. II. page 263.)

The brief allusion in the "Origin of Species" is so far from prominent and so incidental that it was excusable to assume that Darwin had not touched upon the descent of man in this work. It was solely the desire to have his mass of evidence sufficiently complete, solely Darwin's great characteristic of never publishing till he had carefully weighed all aspects of his subject for years, solely, in short, his most fastidious scientific conscience that restrained him from challenging the world in 1859 with a book in which the theory of the descent of man was fully set forth. Three years, frequently interrupted by ill-health, were needed for the actual writing of the book ("Life and Letters", Vol. I. page 94.): the first edition, which appeared in 1871, was followed in 1874 by a much improved second edition, the preparation of which he very reluctantly undertook. (Ibid. Vol. III. page 175.)

This, briefly, is the history of the work, which, with the "Origin of Species", marks an epoch in the history of biological sciences—the work with which the cautious, peace-loving investigator ventured forth from his contemplative life into the arena of strife and unrest, and laid himself open to all the annoyances that deep-rooted belief and prejudice, and the prevailing tendency of scientific thought at the time could devise.

Darwin did not take this step lightly. Of great interest in this connection is a letter written to Wallace on Dec. 22, 1857 (Ibid. Vol. II. page 109.), in which he says "You ask whether I shall discuss 'man.' I think I shall avoid the whole subject, as so surrounded with prejudices; though I fully admit that it is the highest and most interesting problem for the naturalist." But his conscientiousness compelled him to state briefly his opinion on the subject in the "Origin of Species" in 1859. Nevertheless he did not escape reproaches for having been so reticent. This is unmistakably apparent from a letter to Fritz Muller dated February 22 (1869?), in which he says: "I am thinking of writing a little essay on the Origin of Mankind, as I have been taunted with concealing my opinions." (Ibid. Vol. III. page 112.)

It might be thought that Darwin behaved thus hesitatingly, and was so slow in deciding on the full publication of his collected material in regard to the descent of man, because he had religious difficulties to overcome.

But this was not the case, as we can see from his admirable confession of faith, the publication of which we owe to his son Francis. (Ibid. Vol. I. pages 304-317.) Whoever wishes really to understand the lofty character of this great man should read these immortal lines in which he unfolds to us in simple and straightforward words the development of his conception of the universe. He describes how, though he was still quite orthodox during his voyage round the world on board the "Beagle", he came gradually to see, shortly afterwards (1836-1839) that the Old Testament was no more to be trusted than the Sacred Books of the Hindoos; the miracles by which Christianity is supported, the discrepancies between the accounts in the different Gospels, gradually led him to disbelieve in Christianity as a divine revelation. "Thus," he writes ("Life and Letters", Vol. 1. page 309.), "disbelief crept over me at a very slow rate, but was at last complete. The rate was so slow that I felt no distress." But Darwin was too modest to presume to go beyond the limits laid down by science. He wanted nothing more than to be able to go, freely and unhampered by belief in authority or in the Bible, as far as human knowledge could lead him. We learn this from the concluding words of his chapter on religion: "The mystery of the beginning of all things is insoluble by us; and I for one must be content to remain an Agnostic." (Loc. cit. page 313.)

Darwin was always very unwilling to give publicity to his views in regard to religion. In a letter to Asa Gray on May 22, 1860 (Ibid. Vol. II. page 310.), he declares that it is always painful to him to have to enter into discussion of religious problems. He had, he said, no intention of writing atheistically.

Finally, let us cite one characteristic sentence from a letter from Darwin to C. Ridley (Ibid. Vol. III. page. 236. ("C. Ridley," Mr Francis Darwin points out to me, should be H.N. Ridley. A.C.S.)) (Nov. 28, 1878.) A clergyman, Dr Pusey, had asserted that Darwin had written the "Origin of Species" with some relation to theology. Darwin writes emphatically, "Many years ago, when I was collecting facts for the 'Origin', my belief in what is called a personal God was as firm as that of Dr Pusey himself, and as to the eternity of matter I never troubled myself about such insoluble questions." The expression "many years ago" refers to the time of his voyage round the world, as has already been pointed out. Darwin means by this utterance that the views which had gradually developed in his mind in regard to the origin of species were quite compatible with the faith of the Church.

If we consider all these utterances of Darwin in regard to religion and to his outlook on life (Weltanschauung), we shall see at least so much, that religious reflection could in no way have influenced him in regard to the writing and publishing of his book on "The Descent of Man". Darwin had early won for himself freedom of thought, and to this freedom he remained true to the end of his life, uninfluenced by the customs and opinions of the world around him.

Darwin was thus inwardly fortified and armed against the host of calumnies, accusations, and attacks called forth by the publication of the "Origin of Species", and to an even greater extent by the appearance of the "Descent of Man". But in his defence he could rely on the aid of a band of distinguished auxiliaries of the rarest ability. His faithful confederate, Huxley, was joined by the botanist Hooker, and, after longer resistance, by the famous geologist Lyell, whose "conversion" afforded Darwin peculiar satisfaction. All three took the field with enthusiasm in defence of the natural descent of man. From Wallace, on the other hand, though he shared with him the idea of natural selection, Darwin got no support in this matter. Wallace expressed himself in a strange manner. He admitted everything in regard to the morphological descent of man, but maintained, in a mystic way, that something else, something of a spiritual nature must have been added to what man inherited from his animal ancestors. Darwin, whose esteem for Wallace was extraordinarily high, could not understand how he could give utterance to such a mystical view in regard to man; the idea seemed to him so "incredibly strange" that he thought some one else must have added these sentences to Wallace's paper.

Even now there are thinkers who, like Wallace, shrink from applying to man the ultimate consequences of the theory of descent. The idea that man is derived from ape-like forms is to them unpleasant and humiliating.

So far I have been depicting the development of Darwin's work on the descent of man. In what follows I shall endeavour to give a condensed survey of the contents of the book.

It must at once be said that the contents of Darwin's work fall into two parts, dealing with entirely different subjects. "The Descent of Man" includes a very detailed investigation in regard to secondary sexual characters in the animal series, and on this investigation Darwin founded a new theory, that of sexual selection. With astonishing patience he gathered together an immense mass of material, and showed, in regard to Arthropods and Vertebrates, the wide distribution of secondary characters, which develop almost exclusively in the male, and which enable him, on the one hand, to get the better of his rivals in the struggle for the female by the greater perfection of his weapons, and on the other hand, to offer greater allurements to the female through the higher development of decorative characters, of song, or of scent-producing glands. The best equipped males will thus crowd out the less well-equipped in the matter of reproduction, and thus the relevant characters will be increased and perfected through sexual selection. It is, of course, a necessary assumption that these secondary sexual characters may be transmitted to the female, although perhaps in rudimentary form.

As we have said, this theory of sexual selection takes up a great deal of space in Darwin's book, and it need only be considered here in so far as Darwin applied it to the descent of man. To this latter problem the whole of Part I is devoted, while Part III contains a discussion of sexual selection in relation to man, and a general summary. Part II treats of sexual selection in general, and may be disregarded in our present study. Moreover, many interesting details must necessarily be passed over in what follows, for want of space.

The first part of the "Descent of Man" begins with an enumeration of the proofs of the animal descent of man taken from the structure of the human body. Darwin chiefly emphasises the fact that the human body consists of the same organs and of the same tissues as those of the other mammals; he shows also that man is subject to the same diseases and tormented by the same parasites as the apes. He further dwells on the general agreement exhibited by young, embryonic forms, and he illustrates this by two figures placed one above the other, one representing a human embryo, after Eaker, the other a dog embryo, after Bischoff. ("Descent of Man" (Popular Edition, 1901), fig. 1, page 14.)

Darwin finds further proofs of the animal origin of man in the reduced structures, in themselves extremely variable, which are either absolutely useless to their possessors, or of so little use that they could never have developed under existing conditions. Of such vestiges he enumerates: the defective development of the panniculus carnosus (muscle of the skin) so widely distributed among mammals, the ear-muscles, the occasional persistence of the animal ear-point in man, the rudimentary nictitating membrane (plica semilunaris) in the human eye, the slight development of the organ of smell, the general hairiness of the human body, the frequently defective development or entire absence of the third molar (the wisdom tooth), the vermiform appendix, the occasional reappearance of a bony canal (foramen supracondyloideum) at the lower end of the humerus, the rudimentary tail of man (the so-called taillessness), and so on. Of these rudimentary structures the occasional occurrence of the animal ear-point in man is most fully discussed. Darwin's attention was called to this interesting structure by the sculptor Woolner. He figures such a case observed in man, and also the head of an alleged orang-foetus, the photograph of which he received from Nitsche.

Darwin's interpretation of Woolner's case as having arisen through a folding over of the free edge of a pointed ear has been fully borne out by my investigations on the external ear. (G. Schwalbe, "Das Darwin'sche Spitzohr beim menschlichen Embryo", "Anatom. Anzeiger", 1889, pages 176-189, and other papers.) In particular, it was established by these investigations that the human foetus, about the middle of its embryonic life, possesses a pointed ear somewhat similar to that of the monkey genus Macacus. One of Darwin's statements in regard to the head of the orang-foetus must be corrected. A LARGE ear with a point is shown in the photograph ("Descent of Man", fig.3, page 24.), but it can easily be demonstrated—and Deniker has already pointed this out—that the figure is not that of an orang-foetus at all, for that form has much smaller ears with no point; nor can it be a gibbon-foetus, as Deniker supposes, for the gibbon ear is also without a point. I myself regard it as that of a Macacus-embryo. But this mistake, which is due to Nitsche, in no way affects the fact recognised by Darwin, that ear-forms showing the point characteristic of the animal ear occur in man with extraordinary frequency.

Finally, there is a discussion of those rudimentary structures which occur only in ONE sex, such as the rudimentary mammary glands in the male, the vesicula prostatica, which corresponds to the uterus of the female, and others. All these facts tell in favour of the common descent of man and all other vertebrates. The conclusion of this section is characteristic: "IT IS ONLY OUR NATURAL PREJUDICE, AND THAT ARROGANCE WHICH MADE OUR FOREFATHERS DECLARE THAT THEY WERE DESCENDED FROM DEMI-GODS, WHICH LEADS US TO DEMUR TO THIS CONCLUSION. BUT THE TIME WILL BEFORE LONG COME, WHEN IT WILL BE

THOUGHT WONDERFUL THAT NATURALISTS, WHO WERE WELL ACQUAINTED WITH THE COMPARATIVE STRUCTURE AND DEVELOPMENT OF MAN, AND OTHER MAMMALS, SHOULD HAVE BELIEVED THAT EACH WAS THE WORK OF A SEPARATE ACT OF CREATION." (Ibid. page 36.)

In the second chapter there is a more detailed discussion, again based upon an extraordinary wealth of facts, of the problem as to the manner in which, and the causes through which, man evolved from a lower form. Precisely the same causes are here suggested for the origin of man, as for the origin of species in general. Variability, which is a necessary assumption in regard to all transformations, occurs in man to a high degree. Moreover, the rapid multiplication of the human race creates conditions which necessitate an energetic struggle for existence, and thus afford scope for the intervention of natural selection. Of the exercise of ARTIFICIAL selection in the human race, there is nothing to be said, unless we cite such cases as the grenadiers of Frederick William I, or the population of ancient Sparta. In the passages already referred to and in those which follow, the transmission of acquired characters, upon which Darwin does not dwell, is taken for granted. In man, direct effects of changed conditions can be demonstrated (for instance in regard to bodily size), and there are also proofs of the influence exerted on his physical constitution by increased use or disuse. Reference is here made to the fact, established by Forbes, that the Quechua-Indians of the high plateaus of Peru show a striking development of lungs and thorax, as a result of living constantly at high altitudes.

Such special forms of variation as arrests of development (microcephalism) and reversion to lower forms are next discussed. Darwin himself felt ("Descent of Man", page 54.) that these subjects are so nearly related to the cases mentioned in the first chapter, that many of them might as well have been dealt with there. It seems to me that it would have been better so, for the citation of additional instances of reversion at this place rather disturbs the logical sequence of his ideas as to the conditions which have brought about the evolution of man from lower forms. The instances of reversion here discussed are microcephalism, which Darwin wrongly interpreted as atavistic, supernumerary mammae, supernumerary digits, bicornuate uterus, the development of abnormal muscles, and so on. Brief mention is also made of correlative variations observed in man.

Darwin next discusses the question as to the manner in which man attained to the erect position from the state of a climbing quadruped. Here again he puts the influence of Natural Selection in the first rank. The immediate progenitors of man had to maintain a struggle for existence in which success was to the more intelligent, and to those with social instincts. The hand of these climbing ancestors, which had little skill and served mainly for locomotion, could only undergo further development when some early member of the Primate series came to live more on the ground and less among trees.

A bipedal existence thus became possible, and with it the liberation of the hand from locomotion, and the one-sided development of the human foot. The upright position brought about correlated variations in the bodily structure; with the free use of the hand it became possible to manufacture weapons and to use them; and this again resulted in a degeneration of the powerful canine teeth and the jaws, which were then no longer necessary for defence. Above all, however, the intelligence immediately increased, and with it skull and brain. The nakedness of man, and the absence of a tail (rudimentariness of the tail vertebrae) are next discussed. Darwin is inclined to attribute the nakedness of man, not to the action of natural selection on ancestors who originally inhabited a tropical land, but to sexual selection, which, for aesthetic reasons, brought about the loss of the hairy covering in man, or primarily in woman. An interesting discussion of the loss of the tail, which, however, man shares with the anthropoid apes, some other monkeys and lemurs, forms the conclusion of the almost superabundant material which Darwin worked up in the second chapter. His object was to show that some of the most distinctive human characters are in all probability directly or indirectly due to natural selection. With characteristic modesty he adds ("Descent of Man", page 92.): "Hence, if I have erred in giving to natural selection great power, which I am very far from admitting, or in having exaggerated its power, which is in itself probable, I have at least, as I hope, done good service in aiding to overthrow the dogma of separate creations." At the end of the chapter he touches upon the objection as to man's helpless and defenceless condition. Against this he urges his intelligence and social instincts.

The two following chapters contain a detailed discussion of the objections drawn from the supposed great differences between the mental powers of men and animals. Darwin at once admits that the differences are enormous, but not that any fundamental difference between the two can be found. Very characteristic of him is the following passage: "In what manner the mental powers were first developed in the lowest organisms, is as hopeless an enquiry as how life itself first originated. These are problems for the distant future, if they are ever to be solved by man." (Ibid. page 100.)

After some brief observations on instinct and intelligence, Darwin brings forward evidence to show that the greater number of the emotional states, such as pleasure and pain, happiness and misery, love and hate are common to man and the higher animals. He goes on to give various examples showing that wonder and curiosity, imitation, attention, memory and imagination (dreams of animals), can also be observed in the higher mammals, especially in apes. In regard even to reason there are no sharply defined limits. A certain faculty of deliberation is characteristic of some animals, and the more thoroughly we know an animal the more intelligence we are inclined to credit it with. Examples are brought forward of the intelligent and deliberate actions of apes, dogs and elephants. But although no sharply defined differences exist between man and animals, there is, nevertheless, a series of other mental powers which are characteristics usually regarded as absolutely peculiar to man. Some of these characteristics are examined in detail, and it is shown that the arguments drawn from them are not conclusive. Man alone is said to be capable of progressive improvement; but against this must be placed as something analogous in animals, the fact that they learn cunning and caution through long continued persecution. Even the use of tools is not in itself peculiar to man (monkeys use sticks, stones and twigs), but man alone fashions and uses implements DESIGNED FOR A SPECIAL PURPOSE. In this connection the remarks taken from Lubbock in regard to the origin and gradual development of the earliest flint implements will be read with interest; these are similar to the observations on modern eoliths, and their bearing on the development of the stone-industry. It is interesting to learn from a letter to Hooker ("Life and Letters", Vol. II. page 161, June 22, 1859.), that Darwin himself at first doubted whether the stone implements discovered by Boucher de Perthes were really of the nature of tools. With the relentless candour as to himself which characterised him, he writes four years later in a letter to Lyell in regard to this view of Boucher de Perthes' discoveries: "I know something about his errors, and looked at his book many years ago, and am ashamed to think that I concluded the whole was rubbish! Yet he has done for man something like what Agassiz did for glaciers." (Ibid. Vol. III. page 15, March 17, 1863.)

To return to Darwin's further comparisons between the higher mental powers of man and animals. He takes much of the force from the argument that man alone is capable of abstraction and self-consciousness by his own observations on dogs. One of the main differences between man and animals, speech, receives detailed treatment. He points out that various animals (birds, monkeys, dogs) have a large number of different sounds for different emotions, that, further, man produces in common with animals a whole series of inarticulate cries combined with gestures, and that dogs learn to understand whole sentences of human speech. In regard to human language, Darwin expresses a view contrary to that held by Max Muller ("Descent of Man", page 132.): "I cannot doubt that language owes its origin to the imitation and modification of various natural sounds, the voices of other animals, and man's own instinctive cries, aided by signs and gestures." The development of actual language presupposes a higher degree of intelligence than is found in any kind of ape. Darwin remarks on this point (Ibid. pages 136, 137.): "The fact of the higher apes not using their vocal organs for speech no doubt depends on their intelligence not having been sufficiently advanced."

The sense of beauty, too, has been alleged to be peculiar to man. In refutation of this assertion Darwin points to the decorative colours of birds, which are used for display. And to the last objection, that man alone has religion, that he alone has a belief in God, it is answered "that numerous races have existed, and still exist, who have no idea of one or more gods, and who have no words in their languages to express such an idea." (Ibid. page 143.)

The result of the investigations recorded in this chapter is to show that, great as the difference in mental powers between man and the higher animals may be, it is undoubtedly only a difference "of degree and not of kind." ("Descent of Man", page 193.)

In the fourth chapter Darwin deals with the MORAL SENSE or CONSCIENCE, which is the most important of all differences between man and animals. It is a result of social instincts, which lead to sympathy for other members of the same society, to non-egoistic actions for the good of others. Darwin shows that social tendencies are found among many animals, and that among these love and kin-sympathy exist, and he gives examples of animals (especially dogs) which may exhibit characters that we should call moral in man (e.g. disinterested self-sacrifice for the sake of others). The early ape-like progenitors of the human race were undoubtedly social. With the increase of intelligence the moral sense develops farther; with the acquisition of speech public opinion arises, and finally, moral sense becomes habit. The rest of Darwin's detailed discussions on moral philosophy may be passed over.

The fifth chapter may be very briefly summarised. In it Darwin shows that the intellectual and moral faculties are perfected through natural selection. He inquires how it can come about that a tribe at a low level of evolution attains to a higher, although the best and bravest among them often pay for their fidelity and courage with their lives without leaving any descendants. In this case it is the sentiment of glory, praise and blame, the admiration of others, which bring about the increase of the better members of the tribe. Property, fixed dwellings, and the association of families into a community are also indispensable requirements for civilisation. In the longer second section of the fifth chapter Darwin acts mainly as recorder. On the basis of numerous investigations, especially those of Greg, Wallace, and Galton, he inquires how far the influence of natural selection can be demonstrated in regard to civilised nations. In the final section, which deals with the proofs that all civilised nations were once barbarians, Darwin again uses the results gained by other investigators, such as Lubbock and Tylor. There are two sets of facts which prove the proposition in question. In the first place, we find traces of a former lower state in the customs and beliefs of all civilised nations, and in the second place, there are proofs to show that savage races are independently able to raise themselves a few steps in the scale of civilisation, and that they have thus raised themselves.

In the sixth chapter of the work, Morphology comes into the foreground once more. Darwin first goes back, however, to the argument based on the great difference between the mental powers of the highest animals and those of man. That this is only quantitative, not qualitative, he has already shown. Very instructive in this connection is the reference to the enormous difference in mental powers in another class. No one would draw from the fact that the cochineal insect (Coccus) and the ant exhibit enormous differences in their mental powers, the conclusion that the ant should therefore be regarded as something quite distinct, and withdrawn from the class of insects altogether.

Darwin next attempts to establish the SPECIFIC genealogical tree of man, and carefully weighs the differences and resemblances between the different families of the Primates. The erect position of man is an adaptive character, just as are the various characters referable to aquatic life in the seals, which, notwithstanding these, are ranked as a mere family of the Carnivores. The following utterance is very characteristic of Darwin ("Descent of Man", page 231.): "If man had not been his own classifier, he would never have thought of founding a separate order for his own reception." In numerous characters not mentioned in systematic works, in the features of the face, in the form of the nose, in the structure of the external ear, man resembles the apes. The arrangement of the hair in man has also much in common with the apes; as also the occurrence of hair on the forehead of the human embryo, the beard, the convergence of the hair of the upper and under arm towards the elbow, which occurs not only in the anthropoid apes, but also in some American monkeys. Darwin here adopts Wallace's explanation of the origin of the ascending direction of the hair in the forearm of the orang,—that it has arisen through the habit of holding the hands over the head in rain. But this explanation cannot be maintained when we consider that this disposition of the hair is widely distributed among the most different mammals, being found in the dog, in the sloth, and in many of the lower monkeys.

After further careful analysis of the anatomical characters Darwin reaches the conclusion that the New World monkeys (Platyrrhine) may be excluded from the genealogical tree altogether, but that man is an offshoot from the Old World monkeys (Catarrhine) whose progenitors existed as far back as the Miocene period. Among these Old World monkeys the forms to which man shows the greatest resemblance are the anthropoid apes, which, like him, possess neither tail nor ischial callosities. The platyrrhine and catarrhine monkeys have their primitive ancestor among extinct forms of the Lemuridae. Darwin also touches on the question of the original home of the human race and supposes that it may have been in Africa, because it is there that man's nearest relatives, the gorilla and the chimpanzee, are found. But he regards speculation on this point as useless. It is remarkable that, in this connection, Darwin regards the loss of the hair-covering in man as having some relation to a warm climate, while elsewhere he is inclined to make sexual selection responsible for it. Darwin recognises the great gap between man and his nearest relatives, but similar gaps exist at other parts of the mammalian genealogical tree: the allied forms have become extinct. After the extermination of the lower races of mankind, on the one hand, and of the anthropoid apes on the other, which will undoubtedly take place, the gulf will be greater than ever, since the baboons will then bound it on the one side, and the white races on the other. Little weight need be attached to the lack of fossil remains to fill up this gap, since the discovery of these depends upon chance. The last part of the chapter is devoted to a discussion of the earlier stages in the genealogy of man. Here Darwin accepts in the main the genealogical tree, which had meantime been published by Haeckel, who traces the pedigree back through Monotremes, Reptiles, Amphibians, and Fishes, to Amphioxus.

Then follows an attempt to reconstruct, from the atavistic characters, a picture of our primitive ancestor who was undoubtedly an arboreal animal. The occurrence of rudiments of parts in one sex which only come to full development in the other is next discussed. This state of things Darwin regards as derived from an original hermaphroditism. In regard to the mammary glands of the male he does not accept the theory that they are vestigial, but considers them rather as not fully developed.

The last chapter of Part I deals with the question whether the different races of man are to be regarded as different species, or as sub-species of a race of monophyletic origin. The striking differences between the races are first emphasised, and the question of the fertility or infertility of hybrids is discussed. That fertility is the more usual is shown by the excessive fertility of the hybrid population of Brazil. This, and the great variability of the distinguishing characters of the different races, as well as the fact that all grades of transition stages are found between these, while considerable general agreement exists, tell in favour of the unity of the races and lead to the conclusion that they all had a common primitive ancestor.

Darwin therefore classifies all the different races as sub-species of ONE AND THE SAME SPECIES. Then follows an interesting inquiry into the reasons for the extinction of human races. He recognises as the ultimate reason the injurious effects of a change of the conditions of life, which may bring about an increase in infantile mortality, and a diminished fertility. It is precisely the reproductive system, among animals also, which is most susceptible to changes in the environment.

The final section of this chapter deals with the formation of the races of mankind. Darwin discusses the question how far the direct effect of different conditions of life, or the inherited effects of increased use or disuse may have brought about the characteristic differences between the different races. Even in regard to the origin of the colour of the skin he rejects the transmitted effects of an original difference of climate as an explanation. In so doing he is following his tendency to exclude Lamarckian explanations as far as possible. But here he makes gratuitous difficulties from which, since natural selection fails, there is no escape except by bringing in the principle of sexual selection, to which, he regarded it as possible, skin-colouring, arrangement of hair, and form of features might be traced. But with his characteristic conscientiousness he guards himself thus: "I do not intend to assert that sexual selection will account for all the differences between the races." ("Descent of Man", page 308.)

I may be permitted a remark as to Darwin's attitude towards Lamarck. While, at an earlier stage, when he was engaged in the preliminary labours for his immortal work, "The Origin of Species", Darwin expresses himself very forcibly against the views of Lamarck, speaking of Lamarckian "nonsense," ("Life and Letters", Vol. II. page 23.), and of Lamarck's "absurd, though clever work" (Loc. cit. page 39.) and expressly declaring, "I attribute very little to the direct action of climate, etc." (Loc. cit. (1856), page 82.) yet in later life he became more and more convinced of the influence of external conditions. In 1876, that is, two years after the appearance of the second edition of "The Descent of Man", he writes with his usual candid honesty: "In my opinion the greatest error which I have committed, has been not allowing sufficient weight to the direct action of the environment, i.e. food, climate, etc. independently of natural selection." (Ibid. Vol. III. page 159.) It is certain from this change of opinion that, if he had been able to make up his mind to issue a third edition of "The Descent of Man", he would have ascribed a much greater influence to the effect of external conditions in explaining the different characters of the races of man than he did in the second edition. He would also undoubtedly have attributed less influence to sexual selection as a factor in the origin of the different bodily characteristics, if indeed he would not have excluded it altogether.

In Part III of the "Descent" two additional chapters are devoted to the discussion of sexual selection in relation to man. These may be very briefly referred to. Darwin here seeks to show that sexual selection has been operative on man and his primitive progenitor. Space fails me to follow out his interesting arguments. I can only mention that he is inclined to trace back hairlessness, the development of the beard in man, and the characteristic colour of the different human races to sexual selection. Since bareness of the skin could be no advantage, but rather a disadvantage, this character cannot have been brought about by natural selection. Darwin also rejected a direct influence of climate as a cause of the origin of the skin-colour. I have already expressed the opinion, based on the development of his views as shown in his letters, that in a third edition Darwin would probably have laid more stress on the influence of external environment. He himself feels that there are gaps in his proofs here, and says in self-criticism: "The views here advanced, on the part which sexual selection has played in the history of man, want scientific precision." ("Descent of Man", page 924.) I need here only point out that it is impossible to explain the graduated stages of skin-colour by sexual selection, since it would have produced races sharply defined by their colour and not united to other races by transition stages, and this, it is well known, is not the case. Moreover, the fact established by me ("Die Hautfarbe des Menschen", "Mitteilungen der Anthropologischen Gesellschaft in Wien", Vol. XXXIV. pages 331-352.), that in all races the ventral side of the trunk is paler than the dorsal side, and the inner surface of the extremities paler than the outer side, cannot be explained by sexual selection in the Darwinian sense.

With this I conclude my brief survey of the rich contents of Darwin's book. I may be permitted to conclude by quoting the magnificent final words of "The Descent of Man": "We must, however, acknowledge, as it seems to me, that man, with all his noble qualities, with sympathy which feels for the most debased, with benevolence which extends not only to other men but to the humblest living creature, with his god-like intellect which has penetrated into the movements and constitution of the solar system—with all these exalted powers—Man still bears in his bodily frame the indelible stamp of his lowly origin." (Ibid. page 947.)

What has been the fate of Darwin's doctrines since his great achievement? How have they been received and followed up by the scientific and lay world? And what do the successors of the mighty hero and genius think now in regard to the origin of the human race?

At the present time we are incomparably more favourably placed than Darwin was for answering this question of all questions. We have at our command an incomparably greater wealth of material than he had at his disposal. And we are more fortunate than he in this respect, that we now know transition-forms which help to fill up the gap, still great, between the lowest human races and the highest apes. Let us consider for a little the more essential additions to our knowledge since the publication of "The Descent of Man".

Since that time our knowledge of animal embryos has increased enormously. While Darwin was obliged to content himself with comparing a human embryo with that of a dog, there are now available the youngest embryos of monkeys of all possible groups (Orang, Gibbon, Semnopithecus, Macacus), thanks to Selenka's most successful tour in the East Indies in search of such material. We can now compare corresponding stages of the lower monkeys and of the Anthropoid apes with human embryos, and convince ourselves of their great resemblance to one another, thus strengthening enormously the armour prepared by Darwin in defence of his view on man's nearest relatives. It may be said that Selenka's material fils up the blanks in Darwin's array of proofs in the most satisfactory manner.

The deepening of our knowledge of comparative anatomy also gives us much surer foundations than those on which Darwin was obliged to build. Just of late there have been many workers in the domain of the anatomy of apes and lemurs, and their investigations extend to the most different organs. Our knowledge of fossil apes and lemurs has also become much wider and more exact since Darwin's time: the fossil lemurs have been especially worked up by Cope, Forsyth Major, Ameghino, and others. Darwin knew very little about fossil monkeys. He mentions two or three anthropoid apes as occurring in the Miocene of Europe ("Descent of Man", page 240.), but only names Dryopithecus, the largest form from the Miocene of France. It was erroneously supposed that this form was related to Hylobates. We now know not only a form that actually stands near to the gibbon (Pliopithecus), and remains of other anthropoids (Pliohylobates and the fossil chimpanzee, Palaeopithecus), but also several lower catarrhine monkeys, of which Mesopithecus, a form nearly related to the modern Sacred Monkeys (a species of Semnopithecus) and found in strata of the Miocene period in Greece, is the most important. Quite recently, too, Ameghino's investigations have made us acquainted with fossil monkeys from South America (Anthropops, Homunculus), which, according to their discoverer, are to be regarded as in the line of human descent.

What Darwin missed most of all—intermediate forms between apes and man—has been recently furnished. (E. Dubois, as is well known, discovered in 1893, near Trinil in Java, in the alluvial deposits of the river Bengawan, an important form represented by a skull-cap, some molars, and a femur. His opinion—much disputed as it has been—that in this form, which he named Pithecanthropus, he has found a long-desired transition-form is shared by the present writer. And although the geological age of these fossils, which, according to Dubois, belong to the uppermost Tertiary series, the Pliocene, has recently been fixed at a later date (the older Diluvium)), the MORPHOLOGICAL VALUE of these interesting remains, that is, the intermediate position of Pithecanthropus, still holds good. Volz says with justice ("Das geologische Alter der Pithecanthropus-Schichten bei Trinil, Ost-Java". "Neues Jahrb. f.Mineralogie". Festband, 1907.), that even if Pithecanthropus is not THE missing link, it is undoubtedly A missing link.

As on the one hand there has been found in Pithecanthropus a form which, though intermediate between apes and man, is nevertheless more closely allied to the apes, so on the other hand, much progress has been made since Darwin's day in the discovery and description of the older human remains. Since the famous roof of a skull and the bones of the extremities belonging to it were found in 1856 in the Neandertal near Dusseldorf, the most varied judgments have been expressed in regard to the significance of the remains and of the skull in particular. In Darwin's "Descent of Man" there is only a passing allusion to them ("Descent of Man", page 82.) in connection with the discussion of the skull-capacity, although the investigations of Schaaffhausen, King, and Huxley were then known. I believe I have shown, in a series of papers, that the skull in question belongs to a form different from any of the races of man now living, and, with King and Cope, I regard it as at least a different species from living man, and have therefore designated it Homo primigenius. The form unquestionably belongs to the older Diluvium, and in the later Diluvium human forms already appear, which agree in all essential points with existing human races.

As far back as 1886 the value of the Neandertal skull was greatly enhanced by Fraipont's discovery of two skulls and skeletons from Spy in Belgium. These are excellently described by their discoverer ("La race humaine de Neanderthal ou de Canstatt en Belgique". "Arch. de Biologie", VII. 1887.), and are regarded as belonging to the same group of forms as the Neandertal remains. In 1899 and the following years came the discovery by Gorjanovic-Kramberger of different skeletal parts of at least ten individuals in a cave near Krapina in Croatia. (Gorjanovic-Kramberger "Der diluviale Mensch von Krapina in Kroatien", 1906.) It is in particular the form of the lower jaw which is different from that of all recent races of man, and which clearly indicates the lowly position of Homo primigenius, while, on the other hand, the long-known skull from Gibraltar, which I ("Studien zur Vorgeschichte des Menschen", 1906, pages 154 ff.) have referred to Homo primigenius, and which has lately been examined in detail by Sollas ("On the cranial and facial characters of the Neandertal Race". "Trans. R. Soc." London, vol. 199, 1908, page 281.), has made us acquainted with the surprising shape of the eye-orbit, of the nose, and of the whole upper part of the face. Isolated lower jaws found at La Naulette in Belgium, and at Malarnaud in France, increase our material which is now as abundant as could be desired. The most recent discovery of all is that of a skull dug up in August of this year (1908) by Klaatsch and Hauser in the lower grotto of the Le Moustier in Southern France, but this skull has not yet been fully described. Thus Homo primigenius must also be regarded as occupying a position in the gap existing between the highest apes and the lowest human races, Pithecanthropus, standing in the lower part of it, and Homo primigenius in the higher, near man. In order to prevent misunderstanding, I should like here to emphasise that in arranging this structural series—anthropoid apes, Pithecanthropus, Homo primigenius, Homo sapiens—I have no intention of establishing it as a direct genealogical series. I shall have something to say in regard to the genetic relations of these forms, one to another, when discussing the different theories of descent current at the present day. ((Since this essay was written Schoetensack has discovered near Heidelberg and briefly described an exceedingly interesting lower jaw from rocks between the Pliocene and Diluvial beds. This exhibits interesting differences from the forms of lower jaw of Homo primigenius. (Schoetensack "Der Unterkiefer des Homo heidelbergensis". Leipzig, 1908.) G.S.))

In quite a different domain from that of morphological relationship, namely in the physiological study of the blood, results have recently been gained which are of the highest importance to the doctrine of descent. Uhlenhuth, Nuttall, and others have established the fact that the blood-serum of a rabbit which has previously had human blood injected into it, forms a precipitate with human blood. This biological reaction was tried with a great variety of mammalian species, and it was found that those far removed from man gave no precipitate under these conditions. But as in other cases among mammals all nearly related forms yield an almost equally marked precipitate, so the serum of a rabbit treated with human blood and then added to the blood of an anthropoid ape gives ALMOST as marked a precipitate as in human blood; the reaction to the blood of the lower Eastern monkeys is weaker, that to the Western monkeys weaker still; indeed in this last case there is only a slight clouding after a considerable time and no actual precipitate. The blood of the Lemuridae (Nuttall) gives no reaction or an extremely weak one, that of the other mammals none whatever. We have in this not only a proof of the literal blood-relationship between man and apes, but the degree of relationship with the different main groups of apes can be determined beyond possibility of mistake.

Finally, it must be briefly mentioned that in regard to remains of human handicraft also, the material at our disposal has greatly increased of late years, that, as a result of this, the opinions of archaeologists have undergone many changes, and that, in particular, their views in regard to the age of the human race have been greatly influenced. There is a tendency at the present time to refer the origin of man back to Tertiary times. It is true that no remains of Tertiary man have been found, but flints have been discovered which, according to the opinion of most investigators, bear traces either of use, or of very primitive workmanship. Since Rutot's time, following Mortillet's example, investigators have called these "eoliths," and they have been traced back by Verworn to the Miocene of the Auvergne, and by Rutot even to the upper Oligocene. Although these eoliths are even nowadays the subject of many different views, the preoccupation with them has kept the problem of the age of the human race continually before us.

Geology, too, has made great progress since the days of Darwin and Lyell, and has endeavoured with satisfactory results to arrange the human remains of the Diluvial period in chronological order (Penck). I do not intend to enter upon the question of the primitive home of the human race; since the space at my disposal will not allow of my touching even very briefly upon all the departments of science which are concerned in the problem of the descent of man. How Darwin would have rejoiced over each of the discoveries here briefly outlined! What use he would have made of the new and precious material, which would have prevented the discouragement from which he suffered when preparing the second edition of "The Descent of Man"! But it was not granted to him to see this progress towards filling up the gaps in his edifice of which he was so painfully conscious.

He did, however, have the satisfaction of seeing his ideas steadily gaining ground, notwithstanding much hostility and deep-rooted prejudice. Even in the years between the appearance of "The Origin of Species" and of the first edition of the "Descent", the idea of a natural descent of man, which was only briefly indicated in the work of 1859, had been eagerly welcomed in some quarters. It has been already pointed out how brilliantly Huxley contributed to the defence and diffusion of Darwin's doctrines, and how in "Man's Place in Nature" he has given us a classic work as a foundation for the doctrine of the descent of man. As Huxley was Darwin's champion in England, so in Germany Carl Vogt in particular, made himself master of the Darwinian ideas. But above all it was Haeckel who, in energy, eagerness for battle, and knowledge may be placed side by side with Huxley, who took over the leadership in the controversy over the new conception of the universe. As far back as 1866, in his "Generelle Morphologie", he had inquired minutely into the question of the descent of man, and not content with urging merely the general theory of descent from lower animal forms, he drew up for the first time genealogical trees showing the close relationships of the different animal groups; the last of these illustrated the relationships of Mammals, and among them of all groups of the Primates, including man. It was Haeckel's genealogical trees that formed the basis of the special discussion of the relationships of man, in the sixth chapter of Darwin's "Descent of Man".

In the last section of this essay I shall return to Haeckel's conception of the special descent of man, the main features of which he still upholds, and rightly so. Haeckel has contributed more than any one else to the spread of the Darwinian doctrine.

I can only allow myself a few words as to the spread of the theory of the natural descent of man in other countries. The Parisian anthropological school, founded and guided by the genius of Broca, took up the idea of the descent of man, and made many notable contributions to it (Broca, Manouvrier, Mahoudeau, Deniker and others). In England itself Darwin's work did not die. Huxley took care of that, for he, with his lofty and unprejudiced mind, dominated and inspired English biology until his death on June 29, 1895. He had the satisfaction shortly before his death of learning of Dubois' discovery, which he illustrated by a humorous sketch. ("Life and Letters of Thomas Henry Huxley", Vol. II. page 394.) But there are still many followers in Darwin's footsteps in England. Keane has worked at the special genealogical tree of the Primates; Keith has inquired which of the anthropoid apes has the greatest number of characters in common with man; Morris concerns himself with the evolution of man in general, especially with his acquisition of the erect position. The recent discoveries of Pithecanthropus and Homo primigenius are being vigorously discussed;

but the present writer is not in a position to form an opinion of the extent to which the idea of descent has penetrated throughout England generally.

In Italy independent work in the domain of the descent of man is being produced, especially by Morselli; with him are associated, in the investigation of related problems, Sergi and Giuffrida-Ruggeri. From the ranks of American investigators we may single out in particular the eminent geologist Cope, who championed with much decision the idea of the specific difference of Homo neandertalensis (primigenius) and maintained a more direct descent of man from the fossil Lemuridae. In South America too, in Argentina, new life is stirring in this department of science. Ameghino in Buenos Ayres has awakened the fossil primates of the Pampas formation to new life; he even believes that in Tetraprothomo, represented by a femur, he has discovered a direct ancestor of man. Lehmann-Nitsche is working at the other side of the gulf between apes and men, and he describes a remarkable first cervical vertebra (atlas) from Monte Hermoso as belonging to a form which may bear the same relation to Homo sapiens in South America as Homo primigenius does in the Old World. After a minute investigation he establishes a human species Homo neogaeus, while Ameghino ascribes this atlas vertebra to his Tetraprothomo.

Thus throughout the whole scientific world there is arising a new life, an eager endeavour to get nearer to Huxley's problema maximum, to penetrate more deeply into the origin of the human race. There are to-day very few experts in anatomy and zoology who deny the animal descent of man in general. Religious considerations, old prejudices, the reluctance to accept man, who so far surpasses mentally all other creatures, as descended from "soulless" animals, prevent a few investigators from giving full adherence to the doctrine. But there are very few of these who still postulate a special act of creation for man. Although the majority of experts in anatomy and zoology accept unconditionally the descent of man from lower forms, there is much diversity of opinion among them in regard to the special line of descent.

In trying to establish any special hypothesis of descent, whether by the graphic method of drawing up genealogical trees or otherwise, let us always bear in mind Darwin's words ("Descent of Man", page 229.) and use them as a critical guiding line: "As we have no record of the lines of descent, the pedigree can be discovered only by observing the degrees of resemblance between the beings which are to be classed." Darwin carries this further by stating "that resemblances in several unimportant structures, in useless and rudimentary organs, or not now functionally active, or in an embryological condition, are by far the most serviceable for classification." (Loc. cit.) It has also to be remembered that NUMEROUS separate points of agreement are of much greater importance than the amount of similarity or dissimilarity in a few points.

The hypotheses as to descent current at the present day may be divided into two main groups. The first group seeks for the roots of the human race not among any of the families of the apes—the anatomically nearest forms—nor among their very similar but less specialised ancestral forms, the fossil representatives of which we can know only in part, but, setting the monkeys on one side, it seeks for them lower down among the fossil Eocene Pseudo-lemuridae or Lemuridae (Cope), or even among the primitive pentadactylous Eocene forms, which may either have led directly to the evolution of man (Adloff), or have given rise to an ancestral form common to apes and men (Klaatsch (Klaatsch in his last publications speaks in the main only of an ancestral form common to men and anthropoid apes.), Giuffrida-Ruggeri). The common ancestral form, from which man and apes are thus supposed to have arisen independently, may explain the numerous resemblances which actually exist between them. That is to say, all the characters upon which the great structural resemblance between apes and man depends must have been present in their common ancestor. Let us take an example of such a common character. The bony external ear-passage is in general as highly developed in the lower Eastern monkeys and the anthropoid apes as in man. This character must, therefore, have already been present in the common primitive form. In that case it is not easy to understand why the Western monkeys have not also inherited the character, instead of possessing only a tympanic ring. But it becomes more intelligible if we assume that forms with a primitive tympanic ring were the original type, and that from these were evolved, on the one hand, the existing New World monkeys with persistent tympanic ring, and on the other an ancestral form common to the lower Old World monkeys, the anthropoid apes and man. For man shares with these the character in question, and it is also one of the "unimportant" characters required by Darwin. Thus we have two divergent lines arising from the ancestral form, the Western monkeys (Platyrrhine) on the one hand, and an ancestral form common to the lower Eastern monkeys, the anthropoid apes, and man, on the other. But considerations similar to those which showed it to be impossible that man should have developed from an ancestor common to him and the monkeys, yet outside of and parallel with these, may be urged also against the likelihood of a parallel evolution of the lower Eastern monkeys, the anthropoid apes, and man. The anthropoid apes have in common with man many characters which are not present in the lower Old World monkeys. These characters must therefore have been present in the ancestral form common to the three groups. But here, again, it is difficult to understand why the lower Eastern monkeys should not also have inherited these characters. As this is not the case, there remains no alternative but to assume divergent evolution from an indifferent form. The lower Eastern monkeys are carrying on the evolution in one direction—I might almost say towards a blind alley—while anthropoids and men have struck out a progressive path, at first in common, which explains the many points of resemblance between them, without regarding man as derived directly from the anthropoids. Their many striking points of agreement indicate a common descent, and cannot be explained as phenomena of convergence.

I believe I have shown in the above sketch that a theory which derives man directly from lower forms without regarding apes as transition-types leads ad absurdum. The close structural relationship between man and monkeys can only be understood if both are brought into the same line of evolution. To trace man's line of descent directly back to the old Eocene mammals, alongside of, but with no relation to these very similar forms, is to abandon the method of exact comparison, which, as Darwin rightly recognised, alone justifies us in drawing up genealogical trees on the basis of resemblances and differences. The farther down we go the more does the ground slip from beneath our feet. Even the Lemuridae show very numerous divergent conditions, much more so the Eocene mammals (Creodonta, Condylarthra), the chief resemblance of which to man consists in the possession of pentadactylous hands and feet! Thus the farther course of the line of descent disappears in the darkness of the ancestry of the mammals. With just as much reason we might pass by the Vertebrates altogether, and go back to the lower Invertebrates, but in that case it would be much easier to say that man has arisen independently, and has evolved, without relation to any animals, from the lowest primitive form to his present isolated and dominant position. But this would be to deny all value to classification, which must after all be the ultimate basis of a genealogical tree. We can, as Darwin rightly observed, only infer the line of descent from the degree of resemblance between single forms. If we regard man as directly derived from primitive forms very far back, we have no way of explaining the many points of agreement between him and the monkeys in general, and the anthropoid apes in particular. These must remain an inexplicable marvel.

I have thus, I trust, shown that the first class of special theories of descent, which assumes that man has developed, parallel with the monkeys, but without relation to them, from very low primitive forms cannot be upheld, because it fails to take into account the close structural affinity of man and monkeys. I cannot but regard this hypothesis as lamentably retrograde, for it makes impossible any application of the facts that have been discovered in the course of the anatomical and embryological study of man and monkeys, and indeed prejudges investigations of that class as pointless. The whole method is perverted; an unjustifiable theory of descent is first formulated with the aid of the imagination, and then we are asked to declare that all structural relations between man and monkeys, and between the different groups of the latter, are valueless,—the fact being that they are the only true basis on which a genealogical tree can be constructed.

So much for this most modern method of classification, which has probably found adherents because it would deliver us from the relationship to apes which many people so much dislike. In contrast to it we have the second class of special hypotheses of descent, which keeps strictly to the nearest structural relationships. This is the only basis that justifies the drawing up of a special hypothesis of descent. If this fundamental proposition be recognised, it will be admitted that the doctrine of special descent upheld by Haeckel, and set forth in Darwin's "Descent of Man", is still valid to-day. In the genealogical tree, man's place is quite close to the anthropoid apes; these again have as their nearest relatives the lower Old World monkeys, and their progenitors must be sought among the less differentiated Platyrrhine monkeys, whose most important characters have been handed on to the present day New World monkeys. How the different genera are to be arranged within the general scheme indicated depends in the main on the classificatory value attributed to individual characters. This is particularly true in regard to Pithecanthropus, which I consider as the root of a branch which has sprung from the anthropoid ape root and has led up to man; the latter I have designated the family of the Hominidae.

For the rest, there are, as we have said, various possible ways of constructing the narrower genealogy within the limits of this branch including men and apes, and these methods will probably continue to change with the accumulation of new facts. Haeckel himself has modified his genealogical tree of the Primates in certain details since the publication of his "Generelle Morphologie" in 1866, but its general basis remains the same. (Haeckel's latest genealogical tree is to be found in his most recent work, "Unsere Ahnenreihe". Jena, 1908.) All the special genealogical trees drawn up on the lines laid down by Haeckel and Darwin—and that of Dubois may be specially mentioned—are based, in general, on the close relationship of monkeys and men, although they may vary in detail. Various hypotheses have been formulated on these lines, with special reference to the evolution of man. "Pithecanthropus" is regarded by some authorities as the direct ancestor of man, by others as a side-track failure in the attempt at the evolution of man. The problem of the monophyletic or polyphyletic origin of the human race has also been much discussed. Sergi (Sergi G. "Europa", 1908.) inclines towards the assumption of a polyphyletic origin of the three main races of man, the African primitive form of which has given rise also to the gorilla and chimpanzee, the Asiatic to the Orang, the Gibbon, and Pithecanthropus. Kollmann regards existing human races as derived from small primitive races (pigmies), and considers that Homo primigenius must have arisen in a secondary and degenerative manner.

But this is not the place, nor have I the space to criticise the various special theories of descent. One, however, must receive particular notice. According to Ameghino, the South American monkeys (Pitheculites) from the oldest Tertiary of the Pampas are the forms from which have arisen the existing American monkeys on the one hand, and on the other, the extinct South American Homunculidae, which are also small forms. From these last, anthropoid apes and man have, he believes, been evolved. Among the progenitors of man, Ameghino reckons the form discovered by him (Tetraprothomo), from which a South American primitive man, Homo pampaeus, might be directly evolved, while on the other hand all the lower Old World monkeys may have arisen from older fossil South American forms (Clenialitidae), the distribution of which may be explained by the bridge formerly existing between South America and Africa, as may be the derivation of all existing human races from Homo pampaeus. (See Ameghino's latest paper, "Notas preliminares sobre el Tetraprothomo argentinus", etc. "Anales del Museo nacional de Buenos Aires", XVI. pages 107-242, 1907.) The fossil forms discovered by Ameghino deserve the most minute investigation, as does also the fossil man from South America of which Lehmann-Nitsche ("Nouvelles recherches sur la formation pampeenne et l'homme fossile de la Republique Argentine". "Rivista del Museo de la Plata", T. XIV. pages 193-488.) has made a thorough study.

It is obvious that, notwithstanding the necessity for fitting man's line of descent into the genealogical tree of the Primates, especially the apes, opinions in regard to it differ greatly in detail. This could not be otherwise, since the different Primate forms, especially the fossil forms, are still far from being exhaustively known. But one thing remains certain,—the idea of the close relationship between man and monkeys set forth in Darwin's "Descent of Man". Only those who deny the many points of agreement, the sole basis of classification, and thus of a natural genealogical tree, can look upon the position of Darwin and Haeckel as antiquated, or as standing on an insufficient foundation. For such a genealogical tree is nothing more than a summarised representation of what is known in regard to the degree of resemblance between the different forms.

Darwin's work in regard to the descent of man has not been surpassed; the more we immerse ourselves in the study of the structural relationships between apes and man, the more is our path illumined by the clear light radiating from him, and through his calm and deliberate investigation, based on a mass of material in the accumulation of which he has never had an equal. Darwin's fame will be bound up for all time with the unprejudiced investigation of the question of all questions, the descent of the human race.

VIII. CHARLES DARWIN AS AN ANTHROPOLOGIST. By Ernst Haeckel.

Professor of Zoology in the University of Jena.

The great advance that anthropology has made in the second half of the nineteenth century is due in the first place, to Darwin's discovery of the origin of man. No other problem in the whole field of research is so momentous as that of "Man's place in nature," which was justly described by Huxley (1863) as the most fundamental of all questions. Yet the scientific solution of this problem was impossible until the theory of descent had been established.

It is now a hundred years since the great French biologist Jean Lamarck published his "Philosophie Zoologique". By a remarkable coincidence the year in which that work was issued, 1809, was the year of the birth of his most distinguished successor, Charles Darwin. Lamarck had already recognised that the descent of man from a series of other Vertebrates—that is, from a series of Ape-like Primates—was essentially involved in the general theory of transformation which he had erected on a broad inductive basis; and he had sufficient penetration to detect the agencies that had been at work in the evolution of the erect bimanous man from the arboreal and quadrumanous ape. He had, however, few empirical arguments to advance in support of his hypothesis, and it could not be established until the further development of the biological sciences—the founding of comparative embryology by Baer (1828) and of the cell-theory by Schleiden and Schwann (1838), the advance of physiology under Johannes Muller (1833), and the enormous progress of palaeontology and comparative anatomy between 1820 and 1860—provided this necessary foundation. Darwin was the first to coordinate the ample results of these lines of research. With no less comprehensiveness than discrimination he consolidated them as a basis of a modified theory of descent, and associated with them his own theory of natural selection, which we take to be distinctive of "Darwinism" in the stricter sense. The illuminating truth of these cumulative arguments was so great in every branch of biology that, in spite of the most vehement opposition, the battle was won within a single decade, and Darwin secured the general admiration and recognition that had been denied to his forerunner, Lamarck, up to the hour of his death (1829).

Before, however, we consider the momentous influence that Darwinism has had in anthropology, we shall find it useful to glance at its history in the course of the last half century, and notice the various theories that have contributed to its advance. The first attempt to give extensive expression to the reform of biology by Darwin's work will be found in my "Generelle Morphologie" (1866) ("Generelle Morphologie der Organismen", 2 vols., Berlin, 1866.) which was followed by a more popular treatment of the subject in my "Naturliche Schopfungsgeschichte" (1868) (English translation; "The History of Creation", London, 1876.), a compilation from the earlier work. In the first volume of the "Generelle Morphologie" I endeavoured to show the great importance of evolution in settling the fundamental questions of biological philosophy, especially in regard to comparative anatomy. In the second volume I dealt broadly with the principle of evolution, distinguishing ontogeny and phylogeny as its two coordinate main branches, and associating the two in the Biogenetic Law. The Law may be formulated thus: "Ontogeny (embryology or the development of the individual) is a concise and compressed recapitulation of phylogeny (the palaeontological or genealogical series) conditioned by laws of heredity and adaptation." The "Systematic introduction to general evolution," with which the second volume of the "Generelle Morphologie" opens, was the first attempt to draw up a natural system of organisms (in harmony with the principles of Lamarck and Darwin) in the form of a hypothetical pedigree, and was provisionally set forth in eight genealogical tables.

In the nineteenth chapter of the "Generelle Morphologie"—a part of which has been republished, without any alteration, after a lapse of forty years—I made a critical study of Lamarck's theory of descent and of Darwin's theory of selection, and endeavoured to bring the complex phenomena of heredity and adaptation under definite laws for the first time. Heredity I divided into conservative and progressive: adaptation into indirect (or potential) and direct (or actual). I then found it possible to give some explanation of the correlation of the two physiological functions in the struggle for life (selection), and to indicate the important laws of divergence (or differentiation) and complexity (or division of labour), which are the direct and inevitable outcome of selection. Finally, I marked off dysteleology as the science of the aimless (vestigial, abortive, atrophied, and useless) organs and parts of the body. In all this I worked from a strictly monistic standpoint, and sought to explain all biological phenomena on the mechanical and naturalistic lines that had long been recognised in the study of inorganic nature. Then (1866), as now, being convinced of the unity of nature, the fundamental identity of the agencies at work in the inorganic and the organic worlds, I discarded vitalism, teleology, and all hypotheses of a mystic character.

It was clear from the first that it was essential, in the monistic conception of evolution, to distinguish between the laws of conservative and progressive heredity. Conservative heredity maintains from generation to generation the enduring characters of the species. Each organism transmits to its descendants a part of the morphological and physiological qualities that it has received from its parents and ancestors. On the other hand, progressive heredity brings new characters to the species—characters that were not found in preceding generations. Each organism may transmit to its offspring a part of the morphological and physiological features that it has itself acquired, by adaptation, in the course of its individual career, through the use or disuse of particular organs, the influence of environment, climate, nutrition, etc. At that time I gave the name of "progressive heredity" to this inheritance of acquired characters, as a short and convenient expression, but have since changed the term to "transformative heredity" (as distinguished from conservative). This term is preferable, as inherited regressive modifications (degeneration, retrograde metamorphisis, etc.) come under the same head.

Transformative heredity—or the transmission of acquired characters—is one of the most important principles in evolutionary science. Unless we admit it most of the facts of comparative anatomy and physiology are inexplicable. That was the conviction of Darwin no less than of Lamarck, of Spencer as well as Virchow, of Huxley as well as Gegenbaur, indeed of the great majority of speculative biologists. This fundamental principle was for the first time called in question and assailed in 1885 by August Weismann of Freiburg, the eminent zoologist to whom the theory of evolution owes a great deal of valuable support, and who has attained distinction by his extension of the theory of selection. In explanation of the phenomena of heredity he introduced a new theory, the "theory of the continuity of the germ-plasm." According to him the living substance in all organisms consists of two quite distinct kinds of plasm, somatic and germinal. The permanent germ-plasm, or the active substance of the two germ-cells (egg-cell and sperm-cell), passes unchanged through a series of generations, and is not affected by environmental influences. The environment modifies only the soma-plasm, the organs and tissues of the body. The modifications that these parts undergo through the influence of the environment or their own activity (use and habit), do not affect the germ-plasm, and cannot therefore be transmitted.

This theory of the continuity of the germ-plasm has been expounded by Weismann during the last twenty-four years in a number of able volumes, and is regarded by many biologists, such as Mr Francis Galton, Sir E. Ray Lankester, and Professor J. Arthur Thomson (who has recently made a thoroughgoing defence of it in his important work "Heredity" (London, 1908.)), as the most striking advance in evolutionary science. On the other hand, the theory has been rejected by Herbert Spencer, Sir W. Turner, Gegenbaur, Kolliker, Hertwig, and many others. For my part I have, with all respect for the distinguished Darwinian, contested the theory from the first, because its whole foundation seems to me erroneous, and its deductions do not seem to be in accord with the main facts of comparative morphology and physiology. Weismann's theory in its entirety is a finely conceived molecular hypothesis, but it is devoid of empirical basis. The notion of the absolute and permanent independence of the germ-plasm, as distinguished from the soma-plasm, is purely speculative; as is also the theory of germinal selection. The determinants, ids, and idants, are purely hypothetical elements. The experiments that have been devised to demonstrate their existence really prove nothing.

It seems to me quite improper to describe this hypothetical structure as "Neodarwinism." Darwin was just as convinced as Lamarck of the transmission of acquired characters and its great importance in the scheme of evolution. I had the good fortune to visit Darwin at Down three times and discuss with him the main principles of his system, and on each occasion we were fully agreed as to the incalculable importance of what I call transformative inheritance. It is only proper to point out that Weismann's theory of the germ-plasm is in express contradiction to the fundamental principles of Darwin and Lamarck. Nor is it more acceptable in what one may call its "ultradarwinism"—the idea that the theory of selection explains everything in the evolution of the organic world. This belief in the "omnipotence of natural selection" was not shared by Darwin himself. Assuredly, I regard it as of the utmost value, as the process of natural selection through the struggle for life affords an explanation of the mechanical origin of the adapted organisation. It solves the great problem: how could the finely adapted structure of the animal or plant body be formed unless it was built on a preconceived plan? It thus enables us to dispense with the teleology of the metaphysician and the dualist, and to set aside the old mythological and poetic legends of creation. The idea had occurred in vague form to the great Empedocles 2000 years before the time of Darwin, but it was reserved for modern research to give it ample expression. Nevertheless, natural selection does not of itself give the solution of all our evolutionary problems. It has to be taken in conjunction with the transformism of Lamarck, with which it is in complete harmony.

The monumental greatness of Charles Darwin, who surpasses every other student of science in the nineteenth century by the loftiness of his monistic conception of nature and the progressive influence of his ideas, is perhaps best seen in the fact that not one of his many successors has succeeded in modifying his theory of descent in any essential point or in discovering an entirely new standpoint in the interpretation of the organic world. Neither Nageli nor Weismann, neither De Vries nor Roux, has done this. Nageli, in his "Mechanisch-Physiologische Theorie der Abstammungslehre" (Munich, 1884.), which is to a great extent in agreement with Weismann, constructed a theory of the idioplasm, that represents it (like the germ-plasm) as developing continuously in a definite direction from internal causes. But his internal "principle of progress" is at the bottom just as teleological as the vital force of the Vitalists, and the micellar structure of the idioplasm is just as hypothetical as the "dominant" structure of the germ-plasm. In 1889 Moritz Wagner sought to explain the origin of species by migration and isolation, and on that basis constructed a special "migration-theory." This, however, is not out of harmony with the theory of selection. It merely elevates one single factor in the theory to a predominant position. Isolation is only a special case of selection, as I had pointed out in the fifteenth chapter of my "Natural history of creation". The "mutation-theory" of De Vries ("Die Mutationstheorie", Leipzig, 1903.), that would explain the origin of species by sudden and saltatory variations rather than by gradual modification, is regarded by many botanists as a great step in advance, but it is generally rejected by zoologists. It affords no explanation of the facts of adaptation, and has no causal value.

Much more important than these theories is that of Wilhelm Roux ("Der Kampf der Theile im Organismus", Leipzig, 1881.) of "the struggle of parts within the organism, a supplementation of the theory of mechanical adaptation." He explains the functional autoformation of the purposive structure by a combination of Darwin's principle of selection with Lamarck's idea of transformative heredity, and applies the two in conjunction to the facts of histology. He lays stress on the significance of functional adaptation, which I had described in 1866, under the head of cumulative adaptation, as the most important factor in evolution. Pointing out its influence in the cell-life of the tissues, he puts "cellular selection" above "personal selection," and shows how the finest conceivable adaptations in the structure of the tissue may be brought about quite mechanically, without preconceived plan. This "mechanical teleology" is a valuable extension of Darwin's monistic principle of selection to the whole field of cellular physiology and histology, and is wholly destructive of dualistic vitalism.

The most important advance that evolution has made since Darwin and the most valuable amplification of his theory of selection is, in my opinion, the work of Richard Semon: "Die Mneme als erhaltendes Prinzip im Wechsel des organischen Geschehens" (Leipzig, 1904.). He offers a psychological explanation of the facts of heredity by reducing them to a process of (unconscious) memory. The physiologist Ewald Hering had shown in 1870 that memory must be regarded as a general function of organic matter, and that we are quite unable to explain the chief vital phenomena, especially those of reproduction and inheritance, unless we admit this unconscious memory. In my essay "Die Perigenesis der Plastidule" (Berlin, 1876.) I elaborated this far-reaching idea, and applied the physical principle of transmitted motion to the plastidules, or active molecules of plasm. I concluded that "heredity is the memory of the plastidules, and variability their power of comprehension." This "provisional attempt to give a mechanical explanation of the elementary processes of evolution" I afterwards extended by showing that sensitiveness is (as Carl Nageli, Ernst Mach, and Albrecht Rau express it) a general quality of matter. This form of panpsychism finds its simplest expression in the "trinity of substance."

To the two fundamental attributes that Spinoza ascribed to substance—Extension (matter as occupying space) and Cogitation (energy, force)—we now add the third fundamental quality of Psychoma (sensitiveness, soul). I further elaborated this trinitarian conception of substance in the nineteenth chapter of my "Die Lebenswunder" (1904) ("Wonders of Life", London, 1904.), and it seems to me well calculated to afford a monistic solution of many of the antitheses of philosophy.

This important Mneme-theory of Semon and the luminous physiological experiments and observations associated with it not only throw considerable light on transformative inheritance, but provide a sound physiological foundation for the biogenetic law. I had endeavoured to show in 1874, in the first chapter of my "Anthropogenie" (English translation; "The Evolution of Man", 2 volumes, London, 1879 and 1905.), that this fundamental law of organic evolution holds good generally, and that there is everywhere a direct causal connection between ontogeny and phylogeny. "Phylogenesis is the mechanical cause of ontogenesis"; in other words, "The evolution of the stem or race is—in accordance with the laws of heredity and adaptation—the real cause of all the changes that appear, in a condensed form, in the development of the individual organism from the ovum, in either the embryo or the larva."

93

It is now fifty years since Charles Darwin pointed out, in the thirteenth chapter of his epoch-making "Origin of Species", the fundamental importance of embryology in connection with his theory of descent:

"The leading facts in embryology, which are second to none in importance, are explained on the principle of variations in the many descendants from some one ancient progenitor, having appeared at a not very early period of life, and having been inherited at a corresponding period." ("Origin of Species" (6th edition), page 396.)

He then shows that the striking resemblance of the embryos and larvae of closely related animals, which in the mature stage belong to widely different species and genera, can only be explained by their descent from a common progenitor. Fritz Muller made a closer study of these important phenomena in the instructive instance of the Crustacean larva, as given in his able work "Fur Darwin" (1864). (English translation; "Facts and Arguments for Darwin", London, 1869.) I then, in 1872, extended the range so as to include all animals (with the exception of the unicellular Protozoa) and showed, by means of the theory of the Gastraea, that all multicellular, tissue-forming animals—all the Metazoa—develop in essentially the same way from the primary germ-layers. I conceived the embryonic form, in which the whole structure consists of only two layers of cells, and is known as the gastrula, to be the ontogenetic recapitulation, maintained by tenacious heredity, of a primitive common progenitor of all the Metazoa, the Gastraea. At a later date (1895) Monticelli discovered that this conjectural ancestral form is still preserved in certain primitive Coelenterata—Pemmatodiscus, Kunstleria, and the nearly-related Orthonectida.

The general application of the biogenetic law to all classes of animals and plants has been proved in my "Systematische Phylogenie". (3 volumes, Berlin, 1894-96.) It has, however, been frequently challenged, both by botanists and zoologists, chiefly owing to the fact that many have failed to distinguish its two essential elements, palingenesis and cenogenesis. As early as 1874 I had emphasised, in the first chapter of my "Evolution of Man", the importance of discriminating carefully between these two sets of phenomena:

"In the evolutionary appreciation of the facts of embryology we must take particular care to distinguish sharply and clearly between the primary, palingenetic evolutionary processes and the secondary, cenogenetic processes. The palingenetic phenomena, or embryonic RECAPITULATIONS, are due to heredity, to the transmission of characters from one generation to another. They enable us to draw direct inferences in regard to corresponding structures in the development of the species (e.g. the chorda or the branchial arches in all vertebrate embryos). The cenogenetic phenomena, on the other hand, or the embryonic VARIATIONS, cannot be traced to inheritance from a mature ancestor, but are due to the adaptation of the embryo or the larva to certain conditions of its individual development (e.g. the amnion, the allantois, and the vitelline arteries in the embryos of the higher vertebrates). These cenogenetic phenomena are later additions; we must not infer from them that there were corresponding processes in the ancestral history, and hence they are apt to mislead."

The fundamental importance of these facts of comparative anatomy, atavism, and the rudimentary organs, was pointed out by Darwin in the first part of his classic work, "The Descent of Man and Selection in Relation to Sex" (1871). ("Descent of Man" (Popular Edition), page 927.) In the "General summary and conclusion" (chapter XXI.) he was able to say, with perfect justice: "He who is not content to look, like a savage, at the phenomena of nature as disconnected, cannot any longer believe that man is the work of a separate act of creation. He will be forced to admit that the close resemblance of the embryo of man to that, for instance, of a dog—the construction of his skull, limbs, and whole frame on the same plan with that of other mammals, independently of the uses to which the parts may be put—the occasional reappearance of various structures, for instance of several muscles, which man does not normally possess, but which are common to the Quadrumana—and a crowd of analogous facts—all point in the plainest manner to the conclusion that man is the co-descendant with other mammals of a common progenitor."

These few lines of Darwin's have a greater scientific value than hundreds of those so-called "anthropological treatises," which give detailed descriptions of single organs, or mathematical tables with series of numbers and what are claimed to be "exact analyses," but are devoid of synoptic conclusions and a philosophical spirit.

Charles Darwin is not generally recognised as a great anthropologist, nor does the school of modern anthropologists regard him as a leading authority. In Germany, especially, the great majority of the members of the anthropological societies took up an attitude of hostility to him from the very beginning of the controversy in 1860. "The Descent of Man" was not merely rejected, but even the discussion of it was forbidden on the ground that it was "unscientific."

The centre of this inveterate hostility for thirty years—especially after 1877—was Rudolph Virchow of Berlin, the leading investigator in pathological anatomy, who did so much for the reform of medicine by his establishment of cellular pathology in 1858. As a prominent representative of "exact" or "descriptive" anthropology, and lacking a broad equipment in comparative anatomy and ontogeny, he was unable to accept the theory of descent. In earlier years, and especially during his splendid period of activity at Wurzburg (1848-1856), he had been a consistent free-thinker, and had in a number of able articles (collected in his "Gesammelte Abhandlungen") ("Gesammelte Abhandlungen zur wissenschaftlichen Medizin", Berlin, 1856.) upheld the unity of human nature, the inseparability of body and spirit. In later years at Berlin, where he was more occupied with political work and sociology (especially after 1866), he abandoned the positive monistic position for one of agnosticism and scepticism, and made concessions to the dualistic dogma of a spiritual world apart from the material frame.

In the course of a Scientific Congress at Munich in 1877 the conflict of these antithetic views of nature came into sharp relief. At this memorable Congress I had undertaken to deliver the first address (September 18th) on the subject of "Modern evolution in relation to the whole of science." I maintained that Darwin's theory not only solved the great problem of the origin of species, but that its implications, especially in regard to the nature of man, threw considerable light on the whole of science, and on anthropology in particular. The discovery of the real origin of man by evolution from a long series of mammal ancestors threw light on his place in nature in every aspect, as Huxley had already shown in his excellent lectures of 1863. Just as all the organs and tissues of the human body had originated from those of the nearest related mammals, certain ape-like forms, so we were bound to conclude that his mental qualities also had been derived from those of his extinct primate ancestor.

This monistic view of the origin and nature of man, which is now admitted by nearly all who have the requisite acquaintance with biology, and approach the subject without prejudice, encountered a sharp opposition at that time. The opposition found its strongest expression in an address that Virchow delivered at Munich four days afterwards (September 22nd), on "The freedom of science in the modern State." He spoke of the theory of evolution as an unproved hypothesis, and declared that it ought not to be taught in the schools, because it was dangerous to the State. "We must not," he said, "teach that man has descended from the ape or any other animal." When Darwin, usually so lenient in his judgment, read the English translation of Virchow's speech, he expressed his disapproval in strong terms. But the great authority that Virchow had—an authority well founded in pathology and sociology—and his prestige as President of the German Anthropological Society, had the effect of preventing any member of the Society from raising serious opposition to him for thirty years. Numbers of journals and treatises repeated his dogmatic statement: "It is quite certain that man has descended neither from the ape nor from any other animal." In this he persisted till his death in 1902. Since that time the whole position of German anthropology has changed. The question is no longer whether man was created by a distinct supernatural act or evolved from other mammals, but to which line of the animal hierarchy we must look for the actual series of ancestors. The interested reader will find an account of this "battle of Munich" (1877) in my three Berlin lectures (April, 1905) ("Der

Kampf um die Entwickelungs-Gedanken". (English translation; "Last Words on Evolution", London, 1906.))

The main points in our genealogical tree were clearly recognised by Darwin in the sixth chapter of the "Descent of Man". Lowly organised fishes, like the lancelet (Amphioxus), are descended from lower invertebrates resembling the larvae of an existing Tunicate (Appendicularia). From these primitive fishes were evolved higher fishes of the ganoid type and others of the type of Lepidosiren (Dipneusta). It is a very small step from these to the Amphibia:

"In the class of mammals the steps are not difficult to conceive which led from the ancient Monotremata to the ancient Marsupials; and from these to the early progenitors of the placental mammals. We may thus ascend to the Lemuridae; and the interval is not very wide from these to the Simiadae. The Simiadae then branched off into two great stems, the New World and Old World monkeys; and from the latter, at a remote period, Man, the wonder and glory of the Universe, proceeded." ("Descent of Man" (Popular Edition), page 255.)

In these few lines Darwin clearly indicated the way in which we were to conceive our ancestral series within the vertebrates. It is fully confirmed by all the arguments of comparative anatomy and embryology, of palaeontology and physiology; and all the research of the subsequent forty years has gone to establish it. The deep interest in geology which Darwin maintained throughout his life and his complete knowledge of palaeontology enabled him to grasp the fundamental importance of the palaeontological record more clearly than anthropologists and zoologists usually do.

There has been much debate in subsequent decades whether Darwin himself maintained that man was descended from the ape, and many writers have sought to deny it. But the lines I have quoted verbatim from the conclusion of the sixth chapter of the "Descent of Man" (1871) leave no doubt that he was as firmly convinced of it as was his great precursor Jean Lamarck in 1809. Moreover, Darwin adds, with particular explicitness, in the "general summary and conclusion" (chapter XXI.) of that standard work ("Descent of Man", page 930.):

"By considering the embryological structure of man—the homologies which he presents with the lower animals,—the rudiments which he retains,—and the reversions to which he is liable, we can partly recall in imagination the former condition of our early progenitors; and can approximately place them in their proper place in the zoological series. We thus learn that man is descended from a hairy, tailed quadruped, probably arboreal in its habits, and an inhabitant of the Old World. This creature, if its whole structure had been examined by a naturalist, would have been classed amongst the Quadrumana, as surely as the still more ancient progenitor of the Old and New World monkeys."

These clear and definite lines leave no doubt that Darwin—so critical and cautious in regard to important conclusions—was quite as firmly convinced of the descent of man from the apes (the Catarrhinae, in particular) as Lamarck was in 1809 and Huxley in 1863.

It is to be noted particularly that, in these and other observations on the subject, Darwin decidedly assumes the monophyletic origin of the mammals, including man. It is my own conviction that this is of the greatest importance. A number of difficult questions in regard to the development of man, in respect of anatomy, physiology, psychology, and embryology, are easily settled if we do not merely extend our progonotaxis to our nearest relatives, the anthropoid apes and the tailed monkeys from which these have descended, but go further back and find an ancestor in the group of the Lemuridae, and still further back to the Marsupials and Monotremata. The essential identity of all the Mammals in point of anatomical structure and embryonic development—in spite of their astonishing differences in external appearance and habits of life—is so palpably significant that modern zoologists are agreed in the hypothesis that they have all sprung from a common root, and that this root may be sought in the earlier Palaeozoic Amphibia.

The fundamental importance of this comparative morphology of the Mammals, as a sound basis of scientific anthropology, was recognised just before the beginning of the nineteenth century, when Lamarck first emphasised (1794) the division of the animal kingdom into Vertebrates and Invertebrates. Even thirteen years earlier (1781), when Goethe made a close study of the mammal skeleton in the Anatomical Institute at Jena, he was intensely interested to find that the composition of the skull was the same in man as in the other mammals. His discovery of the os intermaxillare in man (1784), which was contradicted by most of the anatomists of the time, and his ingenious "vertebral theory of the skull," were the splendid fruit of his morphological studies. They remind us how Germany's greatest philosopher and poet was for many years ardently absorbed in the comparative anatomy of man and the mammals, and how he divined that their wonderful identity in structure was no mere superficial resemblance, but pointed to a deep internal connection. In my "Generelle Morphologie" (1866), in which I published the first attempts to construct phylogenetic trees, I have given a number of remarkable theses of Goethe, which may be called "phyletic prophecies." They justify us in regarding him as a precursor of Darwin.

In the ensuing forty years I have made many conscientious efforts to penetrate further along that line of anthropological research that was opened up by Goethe, Lamarck, and Darwin. I have brought together the many valuable results that have constantly been reached in comparative anatomy, physiology, ontogeny, and palaeontology, and maintained the effort to reform the classification of animals and plants in an evolutionary sense. The first rough drafts of pedigrees that were published in the "Generelle Morphologie" have been improved time after time in the ten editions of my "Naturaliche Schopfungsgeschichte" (1868-1902). (English translation; "The History of Creation", London, 1876.) A sounder basis for my phyletic hypotheses, derived from a discriminating combination of the three great records— morphology, ontogeny, and palaeontology—was provided in the three volumes of my "Systematische Phylogenie" (Berlin, 1894-96.) (1894 Protists and Plants, 1895 Vertebrates, 1896 Invertebrates). In my "Anthropogenie" (Leipzig, 1874, 5th edition 1905. English translation; "The Evolution of Man", London, 1905.) I endeavoured to employ all the known facts of comparative ontogeny (embryology) for the purpose of completing my scheme of human phylogeny (evolution). I attempted to sketch the historical development of each organ of the body, beginning with the most elementary structures in the germ-layers of the Gastraea. At the same time I drew up a corrected statement of the most important steps in the line of our ancestral series.

At the fourth International Congress of Zoology at Cambridge (August 26th, 1898) I delivered an address on "Our present knowledge of the Descent of Man." It was translated into English, enriched with many valuable notes and additions, by my friend and pupil in earlier days Dr Hans Gadow (Cambridge), and published under the title: "The Last Link; our present knowledge of the Descent of Man". (London, 1898.) The determination of the chief animal forms that occur in the line of our ancestry is there restricted to thirty types, and these are distributed in six main groups.

The first half of this "Progonotaxis hominis," which has no support from fossil evidence, comprises three groups: (i) Protista (unicellular organisms, 1-5: (ii) Invertebrate Metazoa (Coelenteria 6-8, Vermalia 9-11): (iii) Monorrhine Vertebrates (Acrania 12-13, Cyclostoma 14-15). The second half, which is based on fossil records, also comprises three groups: (iv) Palaeozoic cold-blooded Craniota (Fishes 16-18, Amphibia 19, Reptiles 20: (v) Mesozoic Mammals (Monotrema 21, Marsupialia 22, Mallotheria 23): (vi) Cenozoic Primates (Lemuridae 24-25, Tailed Apes 26-27, Anthropomorpha 28-30). An improved and enlarged edition of this hypothetic "Progonotaxis hominis" was published in 1908, in my essay "Unsere Ahnenreihe". ("Festschrift zur 350-jahrigen Jubelfeier der Thuringer Universitat Jena". Jena, 1908.)

If I have succeeded in furthering, in some degree, by these anthropological works, the solution of the great problem of Man's place in nature, and particularly in helping to trace the definite stages in our ancestral series, I owe the success, not merely to the vast progress that biology has made in the last half century, but largely to the luminous example of the great investigators who have applied themselves to the problem, with so much assiduity and genius, for a century and a quarter—I mean Goethe and Lamarck, Gegenbaur and Huxley, but, above all, Charles Darwin. It was the great genius of Darwin that first brought together the scattered material of biology and shaped it into that symmetrical temple of scientific knowledge, the theory of descent. It was Darwin who put the crown on the edifice by his theory of natural selection. Not until this broad inductive law was firmly established was it possible to vindicate the special conclusion, the descent of man from a series of other Vertebrates. By his illuminating discovery Darwin did more for anthropology than thousands of those writers, who are more specifically titled anthropologists, have done by their technical treatises. We may, indeed, say that it is not merely as an exact observer and ingenious experimenter, but as a distinguished anthropologist and far-seeing thinker, that Darwin takes his place among the greatest men of science of the nineteenth century.

To appreciate fully the immortal merit of Darwin in connection with anthropology, we must remember that not only did his chief work, "The Origin of Species", which opened up a new era in natural history in 1859, sustain the most virulent and widespread opposition for a lengthy period, but that even thirty years later, when its principles were generally recognised and adopted, the application of them to man was energetically contested by many high scientific authorities. Even Alfred Russel Wallace, who discovered the principle of natural selection independently in 1858, did not concede that it was applicable to the higher mental and moral qualities of man. Dr Wallace still holds a spiritualist and dualist view of the nature of man, contending that he is composed of a material frame (descended from the apes) and an immortal immaterial soul (infused by a higher power). This dual conception, moreover, is still predominant in the wide circles of modern theology and metaphysics, and has the general and influential adherence of the more conservative classes of society.

In strict contradiction to this mystical dualism, which is generally connected with teleology and vitalism, Darwin always maintained the complete unity of human nature, and showed convincingly that the psychological side of man was developed, in the same way as the body, from the less advanced soul of the anthropoid ape, and, at a still more remote period, from the cerebral functions of the older vertebrates. The eighth chapter of the "Origin of Species", which is devoted to instinct, contains weighty evidence that the instincts of animals are subject, like all other vital processes, to the general laws of historic development. The special instincts of particular species were formed by adaptation, and the modifications thus acquired were handed on to posterity by heredity; in their formation and preservation natural selection plays the same part as in the transformation of every other physiological function. The higher moral qualities of civilised man have been derived from the lower mental functions of the uncultivated barbarians and savages, and these in turn from the social instincts of the mammals. This natural and monistic psychology of Darwin's was afterwards more fully developed by his friend George Romanes in his excellent works "Mental Evolution in Animals" and "Mental Evolution in Man". (London, 1885; 1888.)

Many valuable and most interesting contributions to this monistic psychology of man were made by Darwin in his fine work on "The Descent of Man and Selection in Relation to Sex", and again in his supplementary work, "The Expression of the Emotions in Man and Animals". To understand the historical development of Darwin's anthropology one must read his life and the introduction to "The Descent of Man". From the moment that he was convinced of the truth of the principle of descent—that is to say, from his thirtieth year, in 1838—he recognised clearly that man could not be excluded from its range. He recognised as a logical necessity the important conclusion that "man is the co-descendant with other species of some ancient, lower, and extinct form." For many years he gathered notes and arguments in support of this thesis, and for the purpose of showing the probable line of man's ancestry. But in the first edition of "The Origin of Species" (1859) he restricted himself to the single line, that by this work "light would be thrown on the origin of man and his history." In the fifty years that have elapsed since that time the science of the origin and nature of man has made astonishing progress, and we are now fairly agreed in a monistic conception of nature that regards the whole universe, including man, as a wonderful unity, governed by unalterable and eternal laws. In my philosophical book "Die Weltratsel" (1899) ("The Riddle of the Universe", London, 1900.) and in the supplementary volume "Die Lebenswunder" (1904) "The Wonders of Life", London, (1904.), I have endeavoured to show that this pure monism is securely established, and that the admission of the all-powerful rule of the same principle of evolution throughout the universe compels us to formulate a single supreme law—the all-embracing "Law of Substance," or the united laws of the constancy of matter and the conservation of energy. We should never have reached this supreme general conception if Charles Darwin—a "monistic philosopher" in the true sense of the word—had not prepared the way by his theory of descent by natural selection, and crowned the great work of his life by the association of this theory with a naturalistic anthropology.

IX. SOME PRIMITIVE THEORIES OF THE ORIGIN OF MAN.

By J.G. FRAZER. Fellow of Trinity College, Cambridge.

On a bright day in late autumn a good many years ago I had ascended the hill of Panopeus in Phocis to examine the ancient Greek fortifications which crest its brow. It was the first of November, but the weather was very hot; and when my work among the ruins was done, I was glad to rest under the shade of a clump of fine holly-oaks, to inhale the sweet refreshing perfume of the wild thyme which scented all the air, and to enjoy the distant prospects, rich in natural beauty, rich too in memories of the legendary and historic past. To the south the finely-cut peak of Helicon peered over the low intervening hills. In the west loomed the mighty mass of Parnassus, its middle slopes darkened by pine-woods like shadows of clouds brooding on the mountain-side; while at its skirts nestled the ivy-mantled walls of Daulis overhanging the deep glen, whose romantic beauty accords so well with the loves and sorrows of Procne and Philomela, which Greek tradition associated with the spot. Northwards, across the broad plain to which the hill of Panopeus descends, steep and bare, the eye rested on the gap in the hills through which the Cephissus winds his tortuous way to flow under grey willows, at the foot of barren stony hills, till his turbid waters lose themselves, no longer in the vast reedy swamps of the now vanished Copaic Lake, but in the darkness of a cavern in the limestone rock. Eastward, clinging to the slopes of the bleak range of which the hill of Panopeus forms part, were the ruins of Chaeronea, the birthplace of Plutarch; and out there in the plain was fought the disastrous battle which laid Greece at the feet of Macedonia. There, too, in a later age East and West met in deadly conflict, when the Roman armies under Sulla defeated the Asiatic hosts of Mithridates. Such was the landscape spread out before me on one of those farewell autumn days of almost pathetic splendour, when the departing summer seems to linger fondly, as if loth to resign to winter the enchanted mountains of Greece. Next day the scene had changed: summer was gone. A grey November mist hung low on the hills which only yesterday had shone resplendent in the sun, and under its melancholy curtain the dead flat of the Chaeronean plain, a wide treeless expanse shut in by desolate slopes, wore an aspect of chilly sadness befitting the battlefield where a nation's freedom was lost.

But crowded as the prospect from Panopeus is with memories of the past, the place itself, now so still and deserted, was once the scene of an event even more ancient and memorable, if Greek story-tellers can be trusted. For here, they say, the sage Prometheus created our first parents by fashioning them, like a potter, out of clay. (Pausanias X. 4.4. Compare Apollodorus, "Bibliotheca", I. 7. 1; Ovid, "Metamorph." I. 82 sq.; Juvenal, "Sat". XIV. 35. According to another version of the tale, this creation of mankind took place not at Panopeus, but at Iconium in Lycaonia. After the original race of mankind had been destroyed in the great flood of Deucalion, the Greek Noah, Zeus commanded Prometheus and Athena to create men afresh by moulding images out of clay, breathing the winds into them, and making them live. See "Etymologicum Magnum", s.v. "Ikonion", pages 470 sq. It is said that Prometheus fashioned the animals as well as men, giving to each kind of beast its proper nature. See Philemon, quoted by Stobaeus, "Florilegium" II. 27. The creation of man by Prometheus is figured on ancient works of art. See J. Toutain, "Etudes de Mythologie et d'Histoire des Religions Antiques" (Paris, 1909), page 190. According to Hesiod ("Works and Days", 60 sqq.) it was Hephaestus who at the bidding of Zeus moulded the first woman out of moist earth.) The very spot where he did so can still be seen. It is a forlorn little glen or rather hollow behind the hill of Panopeus, below the ruined but still stately walls and towers which crown the grey rocks of the summit. The glen, when I visited it that hot day after the long drought of summer, was quite dry; no water trickled down its bushy sides, but in the bottom I found a reddish crumbling earth, a relic perhaps of the clay out of which the potter Prometheus moulded the Greek Adam and Eve. In a volume dedicated to the honour of one who has done more than any other in modern times to shape the ideas of mankind as to their origin it may not be out of place to recall this crude Greek notion of the creation of the human race, and to compare or contrast it with other rudimentary speculations of primitive peoples on the same subject, if only for the sake of marking the interval which divides the childhood from the maturity of science.

The simple notion that the first man and woman were modelled out of clay by a god or other superhuman being is found in the traditions of many peoples. This is the Hebrew belief recorded in Genesis: "The Lord God formed man of the dust of the ground, and breathed into his nostrils the breath of life; and man became a living soul." (Genesis ii.7.) To the Hebrews this derivation of our species suggested itself all the more naturally because in their language the word for "ground" (adamah) is in form the feminine of the word for man (adam). (S.R. Driver and W.H.Bennett, in their commentaries on Genesis ii. 7.) From various allusions in Babylonian literature it would seem that the Babylonians also conceived man to have been moulded out of clay. (H. Zimmern, in E. Schrader's "Die Keilinschriften und das Alte Testament" 3 (Berlin, 1902), page 506.) According to Berosus, the Babylonian priest whose account of creation has been preserved in a Greek version, the god Bel cut off his own head, and the other gods caught the flowing blood, mixed it with earth, and fashioned men out of the bloody paste; and that, they said, is why men are so wise, because their mortal clay is tempered with divine blood. (Eusebius, "Chronicon", ed. A. Schoene, Vol. I. (Berlin, 1875), col. 16.) In Egyptian mythology Khnoumou, the Father of the gods, is said to have moulded men out of clay. (G. Maspero, "Histoire Ancienne des Peuples de l'Orient Classique", I. (Paris, 1895), page 128.) We cannot doubt that such crude conceptions of the origin of our race were handed down to the civilised peoples of antiquity by their savage or barbarous forefathers. Certainly stories of the same sort are known to be current among savages and barbarians.

Thus the Australian blacks in the neighbourhood of Melbourne said that Pund-jel, the creator, cut three large sheets of bark with his big knife. On one of these he placed some clay and worked it up with his knife into a proper consistence. He then laid a portion of the clay on one of the other pieces of bark and shaped it into a human form; first he made the feet, then the legs, then the trunk, the arms, and the head. Thus he made a clay man on each of the two pieces of bark; and being well pleased with them he danced round them for joy. Next he took stringy bark from the Eucalyptus tree, made hair of it, and stuck it on the heads of his clay men. Then he looked at them again, was pleased with his work, and again danced round them for joy. He then lay down on them, blew his breath hard into their mouths, their noses, and their navels; and presently they stirred, spoke, and rose up as full-grown men. (R. Brough Smyth, "The Aborigines of Victoria" (Melbourne, 1878), I. 424. This and many of the following legends of creation have been already cited by me in a note on Pausanias X. 4. 4 ("Pausanias's Description of Greece, translated with a Commentary" (London, 1898), Vol V. pages 220 sq.).) The Maoris of New Zealand say that Tiki made man after his own image. He took red clay, kneaded it, like the Babylonian Bel, with his own blood, fashioned it in human form, and gave the image breath. As he had made man in his own likeness he called him Tiki-ahua or Tiki's likeness. (R. Taylor "Te Ika A Maui, or New Zealand and its Inhabitants", Second Edition (London, 1870), page 117. Compare E. Shortland, "Maori Religion and Mythology" (London, 1882), pages 21 sq.) A very generally received tradition in Tahiti was that the first human pair was made by Taaroa, the chief god. They say that after he had formed the world he created man out of red earth, which was also the food of mankind until bread-fruit was produced. Further, some say that one day Taaroa called for the man by name, and when he came he made him fall asleep. As he slept, the

100

creator took out one of his bones (ivi) and made a woman of it, whom he gave to the man to be his wife, and the pair became the progenitors of mankind. This narrative was taken down from the lips of the natives in the early years of the mission to Tahiti. The missionary who records it observes: "This always appeared to me a mere recital of the Mosaic account of creation, which they had heard from some European, and I never placed any reliance on it, although they have repeatedly told me it was a tradition among them before any foreigner arrived. Some have also stated that the woman's name was Ivi, which would be by them pronounced as if written "Eve". "Ivi" is an aboriginal word, and not only signifies a bone, but also a widow, and a victim slain in war. Notwithstanding the assertion of the natives, I am disposed to think that "Ivi", or Eve, is the only aboriginal part of the story, as far as it respects the mother of the human race. (W. Ellis, "Polynesian Researches", Second Edition (London, 1832), I. 110 sq. "Ivi" or "iwi" is the regular word for "bone" in the various Polynesian languages. See E. Tregear, "The Maori-Polynesian Comparative Dictionary" (Wellington, New Zealand, 1891), page 109.) However, the same tradition has been recorded in other parts of Polynesia besides Tahiti. Thus the natives of Fakaofo or Bowditch Island say that the first man was produced out of a stone. After a time he bethought him of making a woman. So he gathered earth and moulded the figure of a woman out of it, and having done so he took a rib out of his left side and thrust it into the earthen figure, which thereupon started up a live woman. He called her Ivi (Eevee) or "rib" and took her to wife, and the whole human race sprang from this pair. (G. Turner, "Samoa" (London, 1884), pages 267 sq.) The Maoris also are reported to believe that the first woman was made out of the first man's ribs. (J.L. Nicholas, "Narrative of a Voyage to New Zealand" (London, 1817), I. 59, who writes "and to add still more to this strange coincidence, the general term for bone is 'Hevee'.") This wide diffusion of the story in Polynesia raises a doubt whether it is merely, as Ellis thought, a repetition of the Biblical narrative learned from Europeans. In Nui, or Netherland Island, it was the god Aulialia who made earthen models of a man and woman, raised them up, and made them live. He called the man Tepapa and the woman Tetata. (G. Turner, "Samoa", pages 300 sq.)

In the Pelew Islands they say that a brother and sister made men out of clay kneaded with the blood of various animals, and that the characters of these first men and of their descendants were determined by the characters of the animals whose blood had been kneaded with the primordial clay; for instance, men who have rat's blood in them are thieves, men who have serpent's blood in them are sneaks, and men who have cock's blood in them are brave. (J. Kubary, "Die Religion der Pelauer", in A. Bastian's "Allerlei aus Volks- und Menschenkunde" (Berlin, 1888), I. 3, 56.) According to a Melanesian legend, told in Mota, one of the Banks Islands, the hero Qat moulded men of clay, the red clay from the marshy river-side at Vanua Lava. At first he made men and pigs just alike, but his brothers remonstrated with him, so he beat down the pigs to go on all fours and made men walk upright. Qat fashioned the first woman out of supple twigs, and when she smiled he knew she was a living woman. (R.H. Codrington, "The Melanesians" (Oxford, 1891), page 158.) A somewhat different version of the Melanesian story is told at Lakona, in Santa Maria. There they say that Qat and another spirit ("vui") called Marawa both made men. Qat made them out of the wood of dracaena-trees. Six days he worked at them, carving their limbs and fitting them together. Then he allowed them six days to come to life. Three days he hid them away, and three days more he worked to make them live. He set them up and danced to them and beat his drum, and little by little they stirred, till at last they could stand all by themselves. Then Qat divided them into pairs and called each pair husband and wife. Marawa also made men out of a tree, but it was a different tree, the tavisoviso. He likewise worked at them six days, beat his drum, and made them live, just as Qat did. But when he saw them move, he dug a pit and buried them in it for six days, and then, when he scraped away the earth to see what they were doing, he found them all rotten and stinking. That was the origin of death. (R.H. Codrington op. cit., pages 157 sq.)

The inhabitants of Noo-Hoo-roa, in the Kei Islands say that their ancestors were fashioned out of clay by the supreme god, Dooadlera, who breathed life into the clay figures. (C.M. Pleyte, "Ethnographische Beschrijving der Kei-Eilanden", "Tijdschrift van het

Nederlandsch Aardrijkskundig Genootschap", Tweede Serie X. (1893), page 564.) The aborigines of Minahassa, in the north of Celebes, say that two beings called Wailan Wangko and Wangi were alone on an island, on which grew a cocoa-nut tree. Said Wailan Wangko to Wangi, "Remain on earth while I climb up the tree." Said Wangi to Wailan Wangko, "Good." But then a thought occurred to Wangi and he climbed up the tree to ask Wailan Wangko why he, Wangi, should remain down there all alone. Said Wailan Wangko to Wangi, "Return and take earth and make two images, a man and a woman." Wangi did so, and both images were men who could move but could not speak. So Wangi climbed up the tree to ask Wailan Wangko, "How now? The two images are made, but they cannot speak." Said Wailan Wangko to Wangi, "Take this ginger and go and blow it on the skulls and the ears of these two images, that they may be able to speak; call the man Adam and the woman Ewa." (N. Graafland "De Minahassa" (Rotterdam, 1869), I. pages 96 sq.) In this narrative the names of the man and woman betray European influence, but the rest of the story may be aboriginal. The Dyaks of Sakarran in British Borneo say that the first man was made by two large birds. At first they tried to make men out of trees, but in vain. Then they hewed them out of rocks, but the figures could not speak. Then they moulded a man out of damp earth and infused into his veins the red gum of the kumpang-tree. After that they called to him and he answered; they cut him and blood flowed from his wounds. (Horsburgh, quoted by H. Ling Roth, "The Natives of Sarawak and of British North Borneo" (London, 1896), I. pages 299 sq. Compare The Lord Bishop of Labuan, "On the Wild Tribes of the North-West Coast of Borneo," "Transactions of the Ethnological Society of London", New Series, II. (1863), page 27.)

The Kumis of South-Eastern India related to Captain Lewin, the Deputy Commissioner of Hill Tracts, the following tradition of the creation of man. "God made the world and the trees and the creeping things first, and after that he set to work to make one man and one woman, forming their bodies of clay; but each night, on the completion of his work, there came a great snake, which, while God was sleeping, devoured the two images. This happened twice or thrice, and God was at his wit's end, for he had to work all day, and could not finish the pair in less than twelve hours; besides, if he did not sleep, he would be no good," said Captain Lewin's informant. "If he were not obliged to sleep, there would be no death, nor would mankind be afflicted with illness. It is when he rests that the snake carries us off to this day. Well, he was at his wit's end, so at last he got up early one morning and first made a dog and put life into it, and that night, when he had finished the images, he set the dog to watch them, and when the snake came, the dog barked and frightened it away. This is the reason at this day that when a man is dying the dogs begin to howl; but I suppose God sleeps heavily now-a-days, or the snake is bolder, for men die all the same." (Capt. T.H. Lewin, "Wild Races of South-Eastern India" (London, 1870), pages 224-26.) The Khasis of Assam tell a similar tale. (A. Bastian, "Volkerstamme am Brahmaputra und verwandtschaftliche Nachbarn" (Berlin, 1883), page 8; Major P.R.T. Gurdon, "The Khasis" (London, 1907), page 106.)

The Ewe-speaking tribes of Togo-land, in West Africa, think that God still makes men out of clay. When a little of the water with which he moistens the clay remains over, he pours it on the ground and out of that he makes the bad and disobedient people. When he wishes to make a good man he makes him out of good clay; but when he wishes to make a bad man, he employs only bad clay for the purpose. In the beginning God fashioned a man and set him on the earth; after that he fashioned a woman. The two looked at each other and began to laugh, whereupon God sent them into the world. (J. Spieth, "Die Ewe-Stamme, Material zur Kunde des Ewe-Volkes in Deutsch-Togo" (Berlin, 1906), pages 828, 840.) The Innuit or Esquimaux of Point Barrow, in Alaska, tell of a time when there was no man in the land, till a spirit named "a se lu", who resided at Point Barrow, made a clay man, set him up on the shore to dry, breathed into him and gave him life. ("Report of the International Expedition to Point Barrow" (Washington, 1885), page 47.) Other Esquimaux of Alaska relate how the Raven made the first woman out of clay to be a companion to the first man; he fastened water-grass to the back of the head to be hair, flapped his wings over the clay figure, and it arose, a beautiful young woman. (E.W. Nelson, "The Eskimo about Bering

102

Strait", "Eighteenth Annual Report of the Bureau of American Ethnology", Part I. (Washington, 1899), page 454.) The Acagchemem Indians of California said that a powerful being called Chinigchinich created man out of clay which he found on the banks of a lake; male and female created he them, and the Indians of the present day are their descendants. (Friar Geronimo Boscana, "Chinigchinich", appended to (A. Robinson's) "Life in California" (New York, 1846), page 247.) A priest of the Natchez Indians in Louisiana told Du Pratz "that God had kneaded some clay, such as that which potters use and had made it into a little man; and that after examining it, and finding it well formed, he blew up his work, and forthwith that little man had life, grew, acted, walked, and found himself a man perfectly well shaped." As to the mode in which the first woman was created, the priest had no information, but thought she was probably made in the same way as the first man; so Du Pratz corrected his imperfect notions by reference to Scripture. (M. Le Page Du Pratz, "The History of Louisiana" (London, 1774), page 330.) The Michoacans of Mexico said that the great god Tucapacha first made man and woman out of clay, but that when the couple went to bathe in a river they absorbed so much water that the clay of which they were composed all fell to pieces. Then the creator went to work again and moulded them afresh out of ashes, and after that he essayed a third time and made them of metal. This last attempt succeeded. The metal man and woman bathed in the river without falling to pieces, and by their union they became the progenitors of mankind. (A. de Herrera, "General History of the vast Continent and Islands of America", translated into English by Capt. J. Stevens (London, 1725, 1726), III. 254; Brasseur de Bourbourg, "Histoire des Nations Civilisees du Mexique et de l'Amerique-Centrale" (Paris, 1857—1859), III. 80 sq; compare id. I. 54 sq.)

According to a legend of the Peruvian Indians, which was told to a Spanish priest in Cuzco about half a century after the conquest, it was in Tiahuanaco that man was first created, or at least was created afresh after the deluge. "There (in Tiahuanaco)," so runs the legend, "the Creator began to raise up the people and nations that are in that region, making one of each nation of clay, and painting the dresses that each one was to wear; those that were to wear their hair, with hair, and those that were to be shorn, with hair cut. And to each nation was given the language, that was to be spoken, and the songs to be sung, and the seeds and food that they were to sow. When the Creator had finished painting and making the said nations and figures of clay, he gave life and soul to each one, as well men as women, and ordered that they should pass under the earth. Thence each nation came up in the places to which he ordered them to go." (E.J. Payne, "History of the New World called America", I. (Oxford, 1892), page 462.)

These examples suffice to prove that the theory of the creation of man out of dust or clay has been current among savages in many parts of the world. But it is by no means the only explanation which the savage philosopher has given of the beginnings of human life on earth. Struck by the resemblances which may be traced between himself and the beasts, he has often supposed, like Darwin himself, that mankind has been developed out of lower forms of animal life. For the simple savage has none of that high notion of the transcendant dignity of man which makes so many superior persons shrink with horror from the suggestion that they are distant cousins of the brutes. He on the contrary is not too proud to own his humble relations; indeed his difficulty often is to perceive the distinction between him and them. Questioned by a missionary, a Bushman of more than average intelligence "could not state any difference between a man and a brute—he did not know but a buffalo might shoot with bows and arrows as well as man, if it had them." (Reverend John Campbell, "Travels in South Africa" (London, 1822, II. page 34.) When the Russians first landed on one of the Alaskan islands, the natives took them for cuttle-fish "on account of the buttons on their clothes." (I. Petroff, "Report on the Population, Industries, and Resources of Alaska", page 145.) The Giliaks of the Amoor think that the outward form and size of an animal are only apparent; in substance every beast is a real man, just like a Giliak himself, only endowed with an intelligence and strength, which often surpass those of mere ordinary human beings. (L. Sternberg, "Die Religion der Giljaken", "Archiv fur Religionswissenschaft", VIII. (1905), page 248.) The Borororos, an Indian tribe of Brazil, will have it that they are parrots of a gorgeous red plumage which live in their native forests.

Accordingly they treat the birds as their fellow-tribesmen, keeping them in captivity, refusing to eat their flesh, and mourning for them when they die. (K. von den Steinen, "Unter den Naturvolkern Zentral-Brasiliens" (Berlin, 1894), pages 352 sq., 512.))

This sense of the close relationship of man to the lower creation is the essence of totemism, that curious system of superstition which unites by a mystic bond a group of human kinsfolk to a species of animals or plants. Where that system exists in full force, the members of a totem clan identify themselves with their totem animals in a way and to an extent which we find it hard even to imagine. For example, men of the Cassowary clan in Mabuiag think that cassowaries are men or nearly so. "Cassowary, he all same as relation, he belong same family," is the account they give of their relationship with the long-legged bird. Conversely they hold that they themselves are cassowaries for all practical purposes. They pride themselves on having long thin legs like a cassowary. This reflection affords them peculiar satisfaction when they go out to fight, or to run away, as the case may be; for at such times a Cassowary man will say to himself, "My leg is long and thin, I can run and not feel tired; my legs will go quickly and the grass will not entangle them." Members of the Cassowary clan are reputed to be pugnacious, because the cassowary is a bird of very uncertain temper and can kick with extreme violence. (A.C. Haddon, "The Ethnography of the Western Tribe of Torres Straits", "Journal of the Anthropological Institute", XIX. (1890), page 393; "Reports of the Cambridge Anthropological Expedition to Torres Straits", V. (Cambridge, 1904), pages 166, 184.) So among the Ojibways men of the Bear clan are reputed to be surly and pugnacious like bears, and men of the Crane clan to have clear ringing voices like cranes. (W.W. Warren, "History of the Ojibways", "Collections of the Minnesota Historical Society", V. (Saint Paul, Minn. 1885), pages 47, 49.) Hence the savage will often speak of his totem animal as his father or his brother, and will neither kill it himself nor allow others to do so, if he can help it. For example, if somebody were to kill a bird in the presence of a native Australian who had the bird for his totem, the black might say, "What for you kill that fellow? that my father!" or "That brother belonging to me you have killed; why did you do it?" (E. Palmer, "Notes on some Australian Tribes", "Journal of the Anthropological Institute", XIII. (1884), page 300.) Bechuanas of the Porcupine clan are greatly afflicted if anybody hurts or kills a porcupine in their presence. They say, "They have killed our brother, our master, one of ourselves, him whom we sing of"; and so saying they piously gather the quills of their murdered brother, spit on them, and rub their eyebrows with them. They think they would die if they touched its flesh. In like manner Bechuanas of the Crocodile clan call the crocodile one of themselves, their master, their brother; and they mark the ears of their cattle with a long slit like a crocodile's mouth by way of a family crest. Similarly Bechuanas of the Lion clan would not, like the members of other clans, partake of lion's flesh; for how, say they, could they eat their grandfather? If they are forced in self-defence to kill a lion, they do so with great regret and rub their eyes carefully with its skin, fearing to lose their sight if they neglected this precaution. (T. Arbousset et F. Daumas, "Relation d'un Voyage d'Exploration au Nord-Est de la Colonie du Cap de Bonne-Esperance" (Paris, 1842), pages 349 sq., 422-24.) A Mandingo porter has been known to offer the whole of his month's pay to save the life of a python, because the python was his totem and he therefore regarded the reptile as his relation; he thought that if he allowed the creature to be killed, the whole of his own family would perish, probably through the vengeance to be taken by the reptile kinsfolk of the murdered serpent. (M. le Docteur Tautain, "Notes sur les Croyances et Pratiques Religieuses des Banmanas", "Revue d'Ethnographie", III. (1885), pages 396 sq.; A. Rancon, "Dans la Haute-Gambie, Voyage d'Exploration Scientifique" (Paris, 1894), page 445.)

Sometimes, indeed, the savage goes further and identifies the revered animal not merely with a kinsman but with himself; he imagines that one of his own more or less numerous souls, or at all events that a vital part of himself, is in the beast, so that if it is killed he must die. Thus, the Balong tribe of the Cameroons, in West Africa, think that every man has several souls, of which one is lodged in an elephant, a wild boar, a leopard, or what not. When any one comes home, feels ill, and says, "I shall soon die," and is as good as his word, his friends are of opinion that one of his souls has been shot by a hunter in a wild boar or a leopard, for example, and that that is the real cause of his death. (J. Keller, "Ueber das Land und Volk der Balong", "Deutsches Kolonialblatt", 1 October, 1895, page 484.) A Catholic missionary, sleeping in the hut of a chief of the Fan negroes, awoke in the middle of the night to see a huge black serpent of the most dangerous sort in the act of darting at him. He was about to shoot it when the chief stopped him, saying, "In killing that serpent, it is me that you would have killed. Fear nothing, the serpent is my elangela." (Father Trilles, "Chez les Fang, leurs Moeurs, leur Langue, leur Religion", "Les Missions Catholiques", XXX. (1898), page 322.) At Calabar there used to be some years ago a huge old crocodile which was well known to contain the spirit of a chief who resided in the flesh at Duke Town. Sporting Vice-Consuls, with a reckless disregard of human life, from time to time made determined attempts to injure the animal, and once a peculiarly active officer succeeded in hitting it. The chief was immediately laid up with a wound in his leg. He SAID that a dog had bitten him, but few people perhaps were deceived by so flimsy a pretext. (Miss Mary H. Kingsley, "Travels in West Africa" (London, 1897), pages 538 sq. As to the external or bush souls of human beings, which in this part of Africa are supposed to be lodged in the bodies of animals, see Miss Mary H. Kingsley op. cit. pages 459-461; R. Henshaw, "Notes on the Efik belief in 'bush soul'", "Man", VI.(1906), pages 121 sq.; J. Parkinson, "Notes on the Asaba people (Ibos) of the Niger", "Journal of the Anthropological Institute", XXXVI. (1906), pages 314 sq.) Once when Mr Partridge's canoe-men were about to catch fish near an Assiga town in Southern Nigeria, the natives of the town objected, saying, "Our souls live in those fish, and if you kill them we shall die." (Charles Partridge, "Cross River Natives" (London, 1905), pages 225 sq.) On another occasion, in the same region, an Englishman shot a hippopotamus near a native village. The same night a woman died in the village, and her friends demanded and obtained from the marksman five pounds as compensation for the murder of the woman, whose soul or second self had been in that hippopotamus. (C.H. Robinson, "Hausaland" (London, 1896), pages 36 sq.) Similarly at Ndolo, in the Congo region, we hear of a chief whose life was bound up with a hippopotamus, but he prudently suffered no one to fire at the animal. ("Notes Analytiques sur les Collections Ethnographiques du Musee du Congo", I. (Brussels, 1902-06), page 150.)

Amongst people who thus fail to perceive any sharp line of distinction between beasts and men it is not surprising to meet with the belief that human beings are directly descended from animals. Such a belief is often found among totemic tribes who imagine that their ancestors sprang from their totemic animals or plants; but it is by no means confined to them. Thus, to take instances, some of the Californian Indians, in whose mythology the coyote or prairie-wolf is a leading personage, think that they are descended from coyotes. At first they walked on all fours; then they began to have some members of the human body, one finger, one toe, one eye, one ear, and so on; then they got two fingers, two toes, two eyes, two ears, and so forth; till at last, progressing from period to period, they became perfect human beings. The loss of their tails, which they still deplore, was produced by the habit of sitting upright. (H.R. Schoolcraft, "Indian Tribes of the United States", IV. (Philadelphia, 1856), pages 224 sq.; compare id. V. page 217. The descent of some, not all, Indians from coyotes is mentioned also by Friar Boscana, in (A. Robinson's) "Life in California" (New York, 1846), page 299.) Similarly Darwin thought that "the tail has disappeared in man and the anthropomorphous apes, owing to the terminal portion having been injured by friction during a long lapse of time; the basal and embedded portion having been reduced and modified, so as to become suitable to the erect or semi-erect position." (Charles Darwin, "The Descent of Man", Second Edition (London, 1879), page 60.) The Turtle clam of the Iroquois think that they are descended from real mud turtles which used to live in a pool. One hot summer the pool dried up, and the mud turtles set out to find another. A very fat turtle, waddling after the rest in the heat, was much incommoded by the weight of his shell, till by a great effort he heaved it off altogether. After that he gradually developed into a man and became the progenitor of the Turtle clan. (E.A. Smith, "Myths of the Iroquois", "Second Annual Report of the Bureau of Ethnology" (Washington, 1883), page 77.) The Crawfish band of the Choctaws are in like manner descended from real crawfish, which used to live under ground, only coming up occasionally through the mud to the surface. Once a party of Choctaws smoked them out, taught them the Choctaw language, taught them to walk on two legs, made them cut off their toe nails and pluck the hair from their bodies, after which they adopted them into the tribe. But the rest of their kindred, the crawfish, are crawfish under ground to this day. (Geo. Catlin, "North American Indians" 4 (London, 1844), II. page 128.) The Osage Indians universally believed that they were descended from a male snail and a female beaver. A flood swept the snail down to the Missouri and left him high and dry on the bank, where the sun ripened him into a man. He met and married a beaver maid, and from the pair the tribe of the Osages is descended. For a long time these Indians retained a pious reverence for their animal ancestors and refrained from hunting beavers, because in killing a beaver they killed a brother of the Osages. But when white men came among them and offered high prices for beaver skins, the Osages yielded to the temptation and took the lives of their furry brethren. (Lewis and Clarke, "Travels to the Source of the Missouri River" (London, 1815), I. 12 (Vol. I. pages 44 sq. of the London reprint, 1905).) The Carp clan of the Ootawak Indians are descended from the eggs of a carp which had been deposited by the fish on the banks of a stream and warmed by the sun. ("Lettres Edifiantes et Curieuses", Nouvelle Edition, VI. (Paris, 1781), page 171.) The Crane clan of the Ojibways are sprung originally from a pair of cranes, which after long wanderings settled on the rapids at the outlet of Lake Superior, where they were changed by the Great Spirit into a man and woman. (L.H. Morgan, "Ancient Society" (London, 1877), page 180.) The members of two Omaha clans were originally buffaloes and lived, oddly enough, under water, which they splashed about, making it muddy. And at death all the members of these clans went back to their ancestors the buffaloes. So when one of them lay adying, his friends used to wrap him up in a buffalo skin with the hair outside and say to him, "You came hither from the animals and you are going back thither. Do not face this way again. When you go, continue walking. (J. Owen Dorsey, "Omaha Sociology", "Third Annual Report of the Bureau of Ethnology" (Washington, 1884), pages 229, 233.) The Haida Indians of Queen Charlotte Islands believe that long ago the raven, who is the chief figure in the mythology of North-West America, took a cockle from the beach and married it; the cockle gave birth to a female child, whom the raven took to wife, and from their union the Indians were produced. (G.M. Dawson, "Report on the Queen Charlotte Islands" (Montreal, 1880), pages 149B sq. ("Geological Survey of Canada"); F. Poole, "Queen Charlotte Islands", page 136.) The Delaware Indians called the rattle-snake their grandfather and would on no account destroy one of these reptiles, believing that were they to do so the whole race of rattle-snakes would rise up and bite them. Under the influence of the white man, however, their respect for their grandfather the rattle-snake gradually died away, till at last they killed him without compunction or ceremony whenever they met him. The writer who records the old custom observes that he had often reflected on the curious connection which appears to subsist in the mind of an Indian between man and the brute creation; "all animated nature," says he, "in whatever degree, is in their eyes a great whole, from which they have not yet ventured to separate themselves." (Rev. John Heckewelder, "An Account of the History, Manners, and Customs, of the Indian Nations, who once inhabited Pennsylvania and the Neighbouring States", "Transactions of the Historical and Literary Committee of the American Philosophical Society", I. (Philadelphia, 1819), pages 245, 247, 248.)

Some of the Indians of Peru boasted of being descended from the puma or American lion; hence they adored the lion as a god and appeared at festivals like Hercules dressed in the skins of lions with the heads of the beasts fixed over their own. Others claimed to be sprung from condors and attired themselves in great black and white wings, like that enormous bird. (Garcilasso de la Vega, "First Part of the Royal Commentaries of the Yncas", Vol. I. page 323, Vol. II. page 156 (Markham's translation).) The Wanika of East Africa look upon the hyaena as one of their ancestors or as associated in some way with their origin and destiny. The death of a hyaena is mourned by the whole people, and the greatest funeral ceremonies which they perform are performed for this brute. The wake held over a chief is as nothing compared to the wake held over a hyaena; one tribe only mourns the death of its chief, but all the tribes unite to celebrate the obsequies of a hyaena. (Charles New, "Life, Wanderings, and Labours in Eastern Africa" (London, 1873) page 122.) Some Malagasy families claim to be descended from the babacoote (Lichanotus brevicaudatus), a large lemur of grave appearance and staid demeanour, which lives in the depth of the forest. When they find one of these creatures dead, his human descendants bury it solemnly, digging a grave for it, wrapping it in a shroud, and weeping and lamenting over its carcase. A doctor who had shot a babacoote was accused by the inhabitants of a Betsimisaraka village of having killed "one of their grandfathers in the forest," and to appease their indignation he had to promise not to skin the animal in the village but in a solitary place where nobody could see him. (Father Abinal, "Croyances fabuleuses des Malgaches", "Les Missions Catholiques", XII. (1880), page 526; G.H. Smith, "Some Betsimisaraka superstitions", "The Antananarivo Annual and Madagascar Magazine", No. 10 (Antananarivo, 1886), page 239; H.W. Little, "Madagascar, its History and People" (London, 1884), pages 321 sq; A. van Gennep, "Tabou et Totemisme a Madagascar" (Paris, 1904), pages 214 sqq.) Many of the Betsimisaraka believe that the curious nocturnal animal called the aye-aye (Cheiromys madagascariensis) "is the embodiment of their forefathers, and hence will not touch it, much less do it an injury. It is said that when one is discovered dead in the forest, these people make a tomb for it and bury it with all the forms of a funeral. They think that if they attempt to entrap it, they will surely die in consequence." (G.A. Shaw, "The Aye-aye", "Antananarivo Annual and Madagascar Magazine", Vol. II. (Antananarivo, 1896), pages 201, 203 (Reprint of the Second four Numbers). Compare A. van Gennep, "Tabou et Totemisme a Madagascar", pages 223 sq.) Some Malagasy tribes believe themselves descended from crocodiles and accordingly they deem the formidable reptiles their brothers. If one of these scaly brothers so far forgets the ties of kinship as to devour a man, the chief of the tribe, or in his absence an old man familiar with the tribal customs, repairs at the head of the people to the edge of the water, and summons the family of the culprit to deliver him up to the arm of justice. A hook is then baited and cast into the river or lake. Next day the guilty brother or one of his family is dragged ashore, formally tried, sentenced to death, and executed. The claims of justice being thus satisfied, the dead animal is lamented and buried like a kinsman; a mound is raised over his grave and a stone marks the place of his head. (Father Abinal, "Croyances fabuleuses des Malgaches", "Les Missions Catholiques", XII. (1880), page 527; A. van Gennep, "Tabou et Totemisme a Madagascar", pages 281 sq.)

Amongst the Tshi-speaking tribes of the Gold Coast in West Africa the Horse-mackerel family traces its descent from a real horse-mackerel whom an ancestor of theirs once took to wife. She lived with him happily in human shape on shore till one day a second wife, whom the man had married, cruelly taunted her with being nothing but a fish. That hurt her so much that bidding her husband farewell she returned to her old home in the sea, with her youngest child in her arms, and never came back again. But ever since the Horse-mackerel people have refrained from eating horse-mackerels, because the lost wife and mother was a fish of that sort. (A.B. Ellis, "The Tshi-speaking Peoples of the Gold Coast of West Africa" (London, 1887), pages 208-11. A similar tale is told by another fish family who abstain from eating the fish (appei) from which they take their name (A.B. Ellis op. cit. pages 211 sq.).) Some of the Land Dyaks of Borneo tell a similar tale to explain a similar custom. "There is a fish which is taken in their rivers called a puttin, which they would on no account touch, under the idea that if they did they would be eating their relations. The tradition respecting it is, that a solitary old man went out fishing and caught a puttin, which he dragged out of the water and laid down in his boat. On turning round, he found it had changed into a very pretty little girl. Conceiving the idea she would make, what he had long wished for, a charming wife for his son, he took her home and educated her until she was fit to be married. She consented to be the son's wife cautioning her husband to use her well. Some time after their marriage, however, being out of temper, he struck her, when she screamed, and rushed away into the water; but not without leaving behind her a beautiful daughter, who became afterwards the mother of the race." (The Lord Bishop of Labuan, "On the Wild Tribes of the North-West Coast of Borneo", "Transactions of the Ethnological Society of London", New Series II. (London, 1863), pages 26 sq. Such stories conform to a well-known type which may be called the Swan-Maiden type of story, or Beauty and the Beast, or Cupid and Psyche. The occurrence of stories of this type among totemic peoples, such as the Tshi-speaking negroes of the Gold Coast, who tell them to explain their totemic taboos, suggests that all such tales may have originated in totemism. I shall deal with this question elsewhere.)

Members of a clan in Mandailing, on the west coast of Sumatra, assert that they are descended from a tiger, and at the present day, when a tiger is shot, the women of the clan are bound to offer betel to the dead beast. When members of this clan come upon the tracks of a tiger, they must, as a mark of homage, enclose them with three little sticks. Further, it is believed that the tiger will not attack or lacerate his kinsmen, the members of the clan. (H. Ris, "De Onderafdeeling Klein Mandailing Oeloe en Pahantan en hare Bevolking met uitzondering van de Oeloes", "Bijdragen tot de Tall- Land- en Volkenkunde van Nederlansch-Indie, XLVI." (1896), page 473.) The Battas of Central Sumatra are divided into a number of clans which have for their totems white buffaloes, goats, wild turtle-doves, dogs, cats, apes, tigers, and so forth; and one of the explanations which they give of their totems is that these creatures were their ancestors, and that their own souls after death can transmigrate into the animals. (J.B. Neumann, "Het Pane en Bila-stroomgebied op het eiland Sumatra", "Tijdschrift van het Nederlandsch Aardrijkskundig Genootschap", Tweede Serie, III. Afdeeling, Meer uitgebreide Artikelen, No. 2 (Amsterdam, 1886), pages 311 sq.; id. ib. Tweede Serie, IV. Afdeeling, Meer uitgebreide Artikelen, No. 1 (Amsterdam, 1887), pages 8 sq.) In Amboyna and the neighbouring islands the inhabitants of some villages aver that they are descended from trees, such as the Capellenia moluccana, which had been fertilised by the Pandion Haliaetus. Others claim to be sprung from pigs, octopuses, crocodiles, sharks, and eels. People will not burn the wood of the trees from which they trace their descent, nor eat the flesh of the animals which they regard as their ancestors. Sicknesses of all sorts are believed to result from disregarding these taboos. (J.G.F. Riedel, "De sluik- en kroesharige rassen tusschen Selebes en Papua" (The Hague, 1886), pages 32, 61; G.W.W.C. Baron van Hoevell, "Ambon en meer bepaaldelijk de Oeliasers" (Dordrecht, 1875), page 152.) Similarly in Ceram persons who think they are descended from crocodiles, serpents, iguanas, and sharks will not eat the flesh of these animals. (J.G.F. Riedel op. cit. page 122.) Many other peoples of the Molucca Islands entertain similar beliefs and observe similar taboos. (J.G.F. Riedel "De sluik- en kroesharige rassen tusschen Selebes en Papua" (The Hague, 1886), pages 253, 334, 341, 348, 412, 414, 432.) Again, in Ponape, one of the Caroline Islands, "The different families suppose themselves to stand in a certain relation to animals, and especially to fishes, and believe in their descent from them. They actually name these animals 'mothers'; the creatures are sacred to the family and may not be injured. Great dances, accompanied with the offering of prayers, are performed in their honour. Any person who killed such an animal would expose himself to contempt and punishment, certainly also to the vengeance of the insulted deity." Blindness is commonly supposed to be the consequence of such a sacrilege. (Dr Hahl, "Mittheilungen uber Sitten und rechtliche Verhaltnisse auf Ponape", "Ethnologisches Notizblatt", Vol. II. Heft 2 (Berlin, 1901), page 10.)

Some of the aborigines of Western Australia believe that their ancestors were swans, ducks, or various other species of water-fowl before they were transformed into men. (Captain G. Grey, "A Vocabulary of the Dialects of South Western Australia", Second Edition (London, 1840), pages 29, 37, 61, 63, 66, 71.) The Dieri tribe of Central Australia, who are divided into totemic clans, explain their origin by the following legend. They say that in the beginning the earth opened in the midst of Perigundi Lake, and the totems (murdus or madas) came trooping out one after the other. Out came the crow, and the shell parakeet, and the emu, and all the rest. Being as yet imperfectly formed and without members or organs of sense, they laid themselves down on the sandhills which surrounded the lake then just as they do now. It was a bright day and the totems lay basking in the sunshine, till at last, refreshed and invigorated by it, they stood up as human beings and dispersed in all directions. That is why people of the same totem are now scattered all over the country. You may still see the island in the lake out of which the totems came trooping long ago. (A.W. Howitt, "Native Tribes of South-East Australia" (London, 1904), pages 476, 779 sq.) Another Dieri legend relates how Paralina, one of the Mura-Muras or mythical predecessors of the Dieri, perfected mankind. He was out hunting kangaroos, when he saw

108

four incomplete beings cowering together. So he went up to them, smoothed their bodies, stretched out their limbs, slit up their fingers and toes, formed their mouths, noses, and eyes, stuck ears on them, and blew into their ears in order that they might hear. Having perfected their organs and so produced mankind out of these rudimentary beings, he went about making men everywhere. (A.W. Howitt op. cit., pages 476, 780 sq.) Yet another Dieri tradition sets forth how the Mura-Mura produced the race of man out of a species of small black lizards, which may still be met with under dry bark. To do this he divided the feet of the lizards into fingers and toes, and, applying his forefinger to the middle of their faces, created a nose; likewise he gave them human eyes, mouths and ears. He next set one of them upright, but it fell down again because of its tail; so he cut off its tail and the lizard then walked on its hind legs. That is the origin of mankind. (S. Gason, "The Manners and Customs of the Dieyerie tribe of Australian Aborigines", "Native Tribes of South Australia" (Adelaide, 1879), page 260. This writer fell into the mistake of regarding the Mura-Mura (Mooramoora) as a Good-Spirit instead of as one of the mythical but more or less human predecessors of the Dieri in the country. See A.W. Howitt, "Native Tribes of South-East Australia", pages 475 sqq.)

The Arunta tribe of Central Australia similarly tell how in the beginning mankind was developed out of various rudimentary forms of animal life. They say that in those days two beings called Ungambikula, that is, "out of nothing," or "self-existing," dwelt in the western sky. From their lofty abode they could see, far away to the east, a number of inapertwa creatures, that is, rudimentary human beings or incomplete men, whom it was their mission to make into real men and women. For at that time there were no real men and women; the rudimentary creatures (inapertwa) were of various shapes and dwelt in groups along the shore of the salt water which covered the country. These embryos, as we may call them, had no distinct limbs or organs of sight, hearing, and smell; they did not eat food, and they presented the appearance of human beings all doubled up into a rounded mass, in which only the outline of the different parts of the body could be vaguely perceived. Coming down from their home in the western sky, armed with great stone knives, the Ungambikula took hold of the embryos, one after the other. First of all they released the arms from the bodies, then making four clefts at the end of each arm they fashioned hands and fingers; afterwards legs, feet, and toes were added in the same way. The figure could now stand; a nose was then moulded and the nostrils bored with the fingers. A cut with the knife made the mouth, which was pulled open several times to render it flexible. A slit on each side of the face separated the upper and lower eye-lids, disclosing the eyes, which already existed behind them; and a few strokes more completed the body. Thus out of the rudimentary creatures were formed men and women. These rudimentary creatures or embryos, we are told, "were in reality stages in the transformation of various animals and plants into human beings, and thus they were naturally, when made into human beings, intimately associated with the particular animal or plant, as the case may be, of which they were the transformations—in other words, each individual of necessity belonged to a totem, the name of which was of course that of the animal or plant of which he or she was a transformation." However, it is not said that all the totemic clans of the Arunta were thus developed; no such tradition, for example, is told to explain the origin of the important Witchetty Grub clan. The clans which are positively known, or at least said, to have originated out of embryos in the way described are the Plum Tree, the Grass Seed, the Large Lizard, the Small Lizard, the Alexandra Parakeet, and the Small Rat clans. When the Ungambikula had thus fashioned people of these totems, they circumcised them all, except the Plum Tree men, by means of a fire-stick. After that, having done the work of creation or evolution, the Ungambikula turned themselves into little lizards which bear a name meaning "snappers-up of flies." (Baldwin Spencer and F.J. Gillen, "Native Tribes of Central Australia" (London, 1899), pages 388 sq.; compare id., "Northern Tribes of Central Australia" (London, 1904), page 150.)

This Arunta tradition of the origin of man, as Messrs Spencer and Gillen, who have recorded it, justly observe, "is of considerable interest; it is in the first place evidently a crude attempt to describe the origin of human beings out of non-human creatures who were of various forms; some of them were representatives of animals, others of plants, but in all

109

cases they are to be regarded as intermediate stages in the transition of an animal or plant ancestor into a human individual who bore its name as that of his or her totem." (Baldwin Spencer and F.J. Gillen, "Native Tribes of Central Australia", pages 391 sq.) In a sense these speculations of the Arunta on their own origin may be said to combine the theory of creation with the theory of evolution; for while they represent men as developed out of much simpler forms of life, they at the same time assume that this development was effected by the agency of two powerful beings, whom so far we may call creators. It is well known that at a far higher stage of culture a crude form of the evolutionary hypothesis was propounded by the Greek philosopher Empedocles. He imagined that shapeless lumps of earth and water, thrown up by the subterranean fires, developed into monstrous animals, bulls with the heads of men, men with the heads of bulls, and so forth; till at last, these hybrid forms being gradually eliminated, the various existing species of animals and men were evolved. (E. Zeller, "Die Philosophie der Griechen", I.4 (Leipsic, 1876), pages 718 sq.; H. Ritter et L. Preller, "Historia Philosophiae Graecae et Romanae ex fontium locis contexta" 5, pages 102 sq. H. Diels, "Die Fragmente der Vorsokratiker" 2, I. (Berlin, 1906), pages 190 sqq. Compare Lucretius "De rerum natura", V. 837 sqq.) The theory of the civilised Greek of Sicily may be set beside the similar theory of the savage Arunta of Central Australia. Both represent gropings of the human mind in the dark abyss of the past; both were in a measure grotesque anticipations of the modern theory of evolution.

In this essay I have made no attempt to illustrate all the many various and divergent views which primitive man has taken of his own origin. I have confined myself to collecting examples of two radically different views, which may be distinguished as the theory of creation and the theory of evolution. According to the one, man was fashioned in his existing shape by a god or other powerful being; according to the other he was evolved by a natural process out of lower forms of animal life. Roughly speaking, these two theories still divide the civilised world between them. The partisans of each can appeal in support of their view to a large consensus of opinion; and if truth were to be decided by weighing the one consensus against the other, with "Genesis" in the one scale and "The Origin of Species" in the other, it might perhaps be found, when the scales were finally trimmed, that the balance hung very even between creation and evolution.

X. THE INFLUENCE OF DARWIN ON THE STUDY OF ANIMAL EMBRYOLOGY. By A. Sedgwick, M.A., F.R.S.

Professor of Zoology and Comparative Anatomy in the University of Cambridge.

The publication of "The Origin of Species" ushered in a new era in the study of Embryology. Whereas, before the year 1859 the facts of anatomy and development were loosely held together by the theory of types, which owed its origin to the great anatomists of the preceding generation, to Cuvier, L. Agassiz, J. Muller, and R. Owen, they were now combined together into one organic whole by the theory of descent and by the hypothesis of recapitulation which was deduced from that theory. The view (First clearly enunciated by Fritz Muller in his well-known work, "Fur Darwin", Leipzig, 1864; (English Edition, "Facts for Darwin", 1869).) that a knowledge of embryonic and larval histories would lay bare the secrets of race-history and enable the course of evolution to be traced, and so lead to the discovery of the natural system of classification, gave a powerful stimulus to morphological study in general and to embryological investigation in particular. In Darwin's words:

"Embryology rises greatly in interest, when we look at the embryo as a picture, more or less obscured, of the progenitor, either in its adult or larval state, of all the members of the same great class." ("Origin" (6th edition), page 396.) In the period under consideration the output of embryological work has been enormous. No group of the animal kingdom has escaped exhaustive examination and no effort has been spared to obtain the embryos of isolated and out of the way forms, the development of which might have an important bearing upon questions of phylogeny and classification. Marine zoological stations have been established, expeditions have been sent to distant countries, and the methods of investigation have been greatly improved. The result of this activity has been that the main features of the developmental history of all the most important animals are now known and the curiosity as to developmental processes, so greatly excited by the promulgation of the Darwinian theory, has to a considerable extent been satisfied.

To what extent have the results of this vast activity fulfilled the expectations of the workers who have achieved them? The Darwin centenary is a fitting moment at which to take stock of our position. In this inquiry we shall leave out of consideration the immense and intensely interesting additions to our knowledge of Natural History. These may be said to constitute a capital fund upon which philosophers, poets and men of science will draw for many generations. The interest of Natural History existed long before Darwinian evolution was thought of and will endure without any reference to philosophic speculations. She is a mistress in whose face are beauties and in whose arms are delights elsewhere unattainable. She is and always has been pursued for her own sake without any reference to philosophy, science, or utility.

Darwin's own views of the bearing of the facts of embryology upon questions of wide scientific interest are perfectly clear. He writes ("Origin" (6th edition), page 395.):

"On the other hand it is highly probable that with many animals the embryonic or larval stages show us, more or less completely, the condition of the progenitor of the whole group in its adult state. In the great class of the Crustacea, forms wonderfully distinct from each other, namely, suctorial parasites, cirripedes, entomostraca, and even the malacostraca, appear at first as larvae under the nauplius-form; and as these larvae live and feed in the open sea, and are not adapted for any peculiar habits of life, and from other reasons assigned by Fritz Muller, it is probable that at some very remote period an independent adult animal, resembling the Nauplius, existed, and subsequently produced, along several divergent lines of descent, the above-named great Crustacean groups. So again it is probable, from what we know of the embryos of mammals, birds, fishes, and reptiles, that these animals are the modified descendants of some ancient progenitor, which was furnished in its adult state with branchiae, a swim-bladder, four fin-like limbs, and a long tail, all fitted for an aquatic life.

"As all the organic beings, extinct and recent, which have ever lived, can be arranged within a few great classes; and as all within each class have, according to our theory, been connected together by fine gradations, the best, and, if our collections were nearly perfect, the only possible arrangement, would be genealogical; descent being the hidden bond of connexion which naturalists have been seeking under the term of the Natural System. On this view we can understand how it is that, in the eyes of most naturalists, the structure of the embryo is even more important for classification than that of the adult. In two or more groups of animals, however much they may differ from each other in structure and habits in their adult condition, if they pass through closely similar embryonic stages, we may feel assured that they all are descended from one parent-form, and are therefore closely related. Thus, community in embryonic structure reveals community of descent; but dissimilarity in embryonic development does not prove discommunity of descent, for in one of two groups the developmental stages may have been suppressed, or may have been so greatly modified through adaptation to new habits of life, as to be no longer recognisable. Even in groups, in which the adults have been modified to an extreme degree, community of origin is often revealed by the structure of the larvae; we have seen, for instance, that cirripedes, though externally so like shell-fish, are at once known by their larvae to belong to the great class of crustaceans. As the embryo often shows us more or less plainly the structure of the less modified and ancient progenitor of the group, we can see why ancient and extinct forms so often resemble in their adult state the embryos of existing species of the same class. Agassiz believes this to be a universal law of nature; and we may hope hereafter to see the law proved true. It can, however, be proved true only in those cases in which the ancient state of the progenitor of the group has not been wholly obliterated, either by successive variations having supervened at a very early period of growth, or by such variations having been inherited at an earlier stage than that at which they first appeared. It should also be borne in mind, that the law may be true, but yet, owing to the geological record not extending far enough back in time, may remain for a long period, or for ever, incapable of demonstration. The law will not strictly hold good in those cases in which an ancient form became adapted in its larval state to some special line of life, and transmitted the same larval state to a whole group of descendants; for such larvae will not resemble any still more ancient form in its adult state."

As this passage shows, Darwin held that embryology was of interest because of the light it seems to throw upon ancestral history (phylogeny) and because of the help it would give in enabling us to arrive at a natural system of classification. With regard to the latter point, he quotes with approval the opinion that "the structure of the embryo is even more important for classification than that of the adult." What justification is there for this view? The phase of life chosen for the ordinary anatomical and physiological studies, namely, the adult phase, is merely one of the large number of stages of structure through which the organism passes. By far the greater number of these are included in what is specially called the developmental or (if we include larvae with embryos) embryonic period, for the developmental changes are more numerous and take place with greater rapidity at the beginning of life than in its later periods. As each of these stages is equal in value, for our present purpose, to the adult phase, it clearly follows that if there is anything in the view that the anatomical study of organisms is of importance in determining their mutual relations, the study of the organism in its various embryonic (and larval) stages must have a greater importance than the study of the single and arbitrarily selected stage of life called the adult.

But a deeper reason than this has been assigned for the importance of embryology in classification. It has been asserted, and is implied by Darwin in the passage quoted, that the ancestral history is repeated in a condensed form in the embryonic, and that a study of the latter enables us to form a picture of the stages of structure through which the organism has passed in its evolution. It enables us on this view to reconstruct the pedigrees of animals and so to form a genealogical tree which shall be the true expression of their natural relations.

The real question which we have to consider is to what extent the embryological studies of the last 50 years have confirmed or rendered probable this "theory of recapitulation." In the first place it must be noted that the recapitulation theory is itself a deduction from the theory of evolution. The facts of embryology, particularly of vertebrate embryology, and of larval history receive, it is argued, an explanation on the view that the successive stages of development are, on the whole, records of adult stages of structure which the species has passed through in its evolution. Whether this statement will bear a critical verbal examination I will not now pause to inquire, for it is more important to determine whether any independent facts can be alleged in favour of the theory. If it could be shown, as was stated to be the case by L. Agassiz, that ancient and extinct forms of life present features of structure now only found in embryos, we should have a body of facts of the greatest importance in the present discussion. But as Huxley (See Huxley's "Scientific Memoirs", London, 1898, Vol. I. page 303: "There is no real parallel between the successive forms assumed in the development of the life of the individual at present, and those which have appeared at different epochs in the past." See also his Address to the Geological Society of London (1862) 'On the Palaeontological Evidence of Evolution', ibid. Vol. II. page 512.) has shown and as the whole course of palaeontological and embryological investigation has demonstrated, no such statement can be made. The extinct forms of life are very similar to those now existing and there is nothing specially embryonic about them. So that the facts, as we know them, lend no support to theory.

But there is another class of facts which have been alleged in favour of the theory, viz. the facts which have been included in the generalisation known as the Law of v. Baer. The law asserts that embryos of different species of animals of the same group are more alike than the adults and that, the younger the embryo, the greater are the resemblances. If this law could be established it would undoubtedly be a strong argument in favour of the "recapitulation" explanation of the facts of embryology. But its truth has been seriously disputed. If it were true we should expect to find that the embryos of closely similar species would be indistinguishable from one another, but this is notoriously not the case. It is more difficult to meet the assertion when it is made in the form given above, for here we are dealing with matters of opinion. For instance, no one would deny that the embryo of a dogfish is different from the embryo of a rabbit, but there is room for difference of opinion when it is asserted that the difference is less than the difference between an adult dogfish and an adult rabbit. It would be perfectly true to say that the differences between the embryos concern other organs more than do the differences between the adults, but who is prepared to affirm that the presence of a cephalic coelom and of cranial segments, of external gills, of six gill slits, of the kidney tubes opening into the muscle-plate coelom, of an enormous yolk-sac, of a neurenteric canal, and the absence of any trace of an amnion, of an allantois and of a primitive streak are not morphological facts of as high an import as those implied by the differences between the adults? The generalisation undoubtedly had its origin in the fact that there is what may be called a family resemblance between embryos and larvae, but this resemblance, which is by no means exact, is largely superficial and does not extend to anatomical detail.

It is useless to say, as Weismann has stated ("The Evolution Theory", by A. Weismann, English Translation, Vol. II. page 176, London, 1904.), that "it cannot be disputed that the rudiments [vestiges his translator means] of gill-arches and gill-clefts, which are peculiar to one stage of human ontogeny, give us every ground for concluding that we possessed fish-like ancestors." The question at issue is: did the pharyngeal arches and clefts of mammalian embryos ever discharge a branchial function in an adult ancestor of the mammalia? We cannot therefore, without begging the question at issue in the grossest manner, apply to them the terms "gill-arches" and "gill-clefts". That they are homologous with the "gill-arches" and "gill-clefts" of fishes is true; but there is no evidence to show that they ever discharged a branchial function. Until such evidence is forthcoming, it is beside the point to say that it "cannot be disputed" that they are evidence of a piscine ancestry.

It must, therefore, be admitted that one outcome of the progress of embryological and palaeontological research for the last 50 years is negative. The recapitulation theory originated as a deduction from the evolution theory and as a deduction it still remains.

Let us before leaving the subject apply another test. If the evolution theory and the recapitulation theory are both true, how is it that living birds are not only without teeth but have no rudiments of teeth at any stage of their existence? How is it that the missing digits in birds and mammals, the missing or reduced limb of snakes and whales, the reduced mandibulo-hyoid cleft of elasmobranch fishes are not present or relatively more highly developed in the embryo than in the adult? How is it that when a marked variation, such as an extra digit, or a reduced limb, or an extra segment, makes its appearance, it is not confined to the adult but can be seen all through the development? All the clear evidence we can get tends to show that marked variations, whether of reduction or increase, of organs are manifest during the whole of the development of the organ and do not merely affect the adult. And on reflection we see that it could hardly be otherwise. All such evidence is distinctly at variance with the theory of recapitulation, at least as applied to embryos. In the case of larvae of course the case will be different, for in them the organs are functional, and reduction in the adult will not be accompanied by reduction in the larva unless a change in the conditions of life of the larva enables it to occur.

If after 50 years of research and close examination of the facts of embryology the recapitulation theory is still without satisfactory proof, it seems desirable to take a wider sweep and to inquire whether the facts of embryology cannot be included in a larger category.

As has been pointed out by Huxley, development and life are co-extensive, and it is impossible to point to any period in the life of an organism when the developmental changes cease. It is true that these changes take place more rapidly at the commencement of life, but they are never wholly absent, and those which occur in the later or so-called adult stages of life do not differ in their essence, however much they may differ in their degree, from those which occur during the embryonic and larval periods. This consideration at once brings the changes of the embryonic period into the same category as those of the adult and suggests that an explanation which will account for the one will account for the other. What then is the problem we are dealing with? Surely it is this: Why does an organism as soon as it is established at the fertilisation of the ovum enter upon a cycle of transformations which never cease until death puts an end to them? In other words what is the meaning of that cycle of changes which all organisms present in a greater or less degree and which constitute the very essence of life? It is impossible to give an answer to this question so long as we remain within the precincts of Biology—and it is not my present purpose to penetrate beyond those precincts into the realms of philosophy. We have to do with an ultimate biological fact, with a fundamental property of living matter, which governs and includes all its other properties. How may this property be stated? Thus: it is a property of living matter to react in a remarkable way to external forces without undergoing destruction. The life-cycle, of which the embryonic and larval periods are a part, consists of the orderly interaction between the organism and its environment. The action of the environment produces certain morphological changes in the organism. These changes enable the organism to come into relation with new external forces, to move into what is practically a new environment, which in its turn produces further structural changes in the organism. These in their turn enable, indeed necessitate, the organism to move again into a new environment, and so the process continues until the structural changes are of such a nature that the organism is unable to adapt itself to the environment in which it finds itself. The essential condition of success in this process is that the organism should always shift into the environment to which its new structure is suited—any failure in this leading to the impairment of the organism. In most cases the shifting of the environment is a very gradual process (whether consisting in the very slight and gradual alteration in the relation of the embryo as a whole to the egg-shell or uterine wall, or in the relations of its parts to each other, or in the successive phases of adult life), and the morphological changes in connection with each step of it are but slight. But in some cases jumps are made such as we find in the phenomena known as hatching, birth, and metamorphosis.

This property of reacting to the environment without undergoing destruction is, as has been stated, a fundamental property of organisms. It is impossible to conceive of any matter, to which the term living could be applied, being without it. And with this property of reacting to the environment goes the further property of undergoing a change which alters the relation of the organism to the old environment and places it in a new environment. If this reasoning is correct, it necessarily follows that this property must have been possessed by living matter at its first appearance on the earth. In other words living matter must always have presented a life-cycle, and the question arises what kind of modification has that cycle undergone? Has it increased or diminished in duration and complexity since organisms first appeared on the earth? The current view is that the cycle was at first very short and that it has increased in length by the evolutionary creation of new adult phases, that these new phases are in addition to those already existing and that each of them as it appears takes over from the preceding adult phase the functional condition of the reproductive organs. According to the same view the old adult phases are not obliterated but persist in a more or less modified form as larval stages. It is further supposed that as the life-history lengthens at one end by the addition of new adult phases, it is shortened at the other by the abbreviation of embryonic development and by the absorption of some of the early larval stages into the embryonic period; but on the whole the lengthening process has exceeded that of shortening, so that the whole life-history has, with the progress of evolution, become longer and more complicated.

Now there can be no doubt that the life-history of organisms has been shortened in the way above suggested, for cases are known in which this can practically be seen to occur at the present day. But the process of lengthening by the creation of new stages at the other end of the life-cycle is more difficult to conceive and moreover there is no evidence for its having occurred. This, indeed, may have occurred, as is suggested below, but the evidence we have seems to indicate that evolutionary modification has proceeded by ALTERING and not by SUPERSEDING: that is to say that each stage in the life-history, as we see it to-day, has proceeded from a corresponding stage in a former era by the modification of that stage and not by the creation of a new one. Let me, at the risk of repetition, explain my meaning more fully by taking a concrete illustration. The mandibulo-hyoid cleft (spiracle) of the elasmobranch fishes, the lateral digits of the pig's foot, the hind-limbs of whales, the enlarged digit of the ostrich's foot are supposed to be organs which have been recently modified. This modification is not confined to the final adult stage of the life-history but characterises them throughout the whole of their development. A stage with a reduced spiracle does not proceed in development from a preceding stage in which the spiracle shows no reduction: it is reduced at its first appearance. The same statement may be made of organs which have entirely disappeared in the adult, such as bird's teeth and snake's fore-limbs: the adult stage in which they have disappeared is not preceded by embryonic stages in which the teeth and limbs or rudiments of them are present. In fact the evidence indicates that adult variations of any part are accompanied by precedent variations in the same direction in the embryo. The evidence seems to show, not that a stage is added on at the end of the life-history, but only that some of the stages in the life-history are modified. Indeed, on the wider view of development taken in this essay, a view which makes it coincident with life, one would not expect often to find, even if new stages are added in the course of evolution, that they are added at the end of the series when the organism has passed through its reproductive period. It is possible of course that new stages have been intercalated in the course of the life-history, though it is difficult to see how this has occurred. It is much more likely, if we may judge from available evidence, that every stage has had its counterpart in the ancestral form from which it has been derived by descent with modification. Just as the adult phase of the living form differs, owing to evolutionary modification, from the adult phase of the ancestor from which it has proceeded, so each larval phase will differ for the same reason from the corresponding larval phase in the life-history of the ancestor. Inasmuch as the organism is variable at every stage of its independent existence and is exposed to the action of natural selection there is no reason why it should escape modification at any stage.

If there is any truth in these considerations it would seem to follow that at the dawn of life the life-cycle must have been, either in posse or in esse, at least as long as it is at the present time, and that the peculiarity of passing through a series of stages in which new characters are successively evolved is a primordial quality of living matter.

Before leaving this part of the subject, it is necessary to touch upon another aspect of it. What are these variations in structure which succeed one another in the life-history of an organism? I am conscious that I am here on the threshold of a chamber which contains the clue to some of our difficulties, and that I cannot enter it. Looked at from one point of view they belong to the class of genetic variations, which depend upon the structure or constitution of the protoplasm; but instead of appearing in different zygotes (A zygote is a fertilised ovum, i.e. a new organism resulting from the fusion of an ovum and a spermatozoon.), they are present in the same zygote though at different times in its life-history. They are of the same order as the mutational variations of the modern biologist upon which the appearance of a new character depends. What is a genetic or mutational variation? It is a genetic character which was not present in either of the parents. But these "growth variations" were present in the parents, and in this they differ from mutational variations. But what are genetic characters? They are characters which must appear if any development occurs. They are usually contrasted with "acquired characters," using the expression "acquired character" in the Lamarckian sense. But strictly speaking they ARE acquired characters, for the zygote at first has none of the characters which it subsequently acquires, but only the power of acquiring them in response to the action of the environment. But the characters so acquired are not what we technically understand and what Lamarck meant by "acquired characters." They are genetic characters, as defined above. What then are Lamarck's "acquired characters"? They are variations in genetic characters caused in a particular way. There are, in fact, two kinds of variation in genetic characters depending on the mode of causation. Firstly, there are those variations consequent upon a variation in the constitution of the protoplasm of a particular zygote, and independent of the environment in which the organism develops, save in so far as this simply calls them forth: these are the so-called genetic or mutational variations. Secondly, there are those variations which occur in zygotes of similar germinal constitution and which are caused solely by differences in the environment to which the individuals are respectively exposed: these are the "acquired characters" of Lamarck and of authors generally. In consequence of this double sense in which the term "acquired characters" may be used, great confusion may and does occur. If the protoplasm be compared to a machine, and the external conditions to the hand that works the machine, then it may be said that, as the machine can only work in one way, it can only produce one kind of result (genetic character), but the particular form or quality (Lamarckian "acquired character") of the result will depend upon the hand that works the machine (environment), just as the quality of the sound produced by a fiddle depends entirely upon the hand which plays upon it. It would be improper to apply the term "mutation" to those genetic characters which are not new characters or new variants of old characters, but such genetic characters are of the same nature as those characters to which the term mutation has been applied. It may be noticed in passing that it is very questionable if the modern biologist has acted in the real interests of science in applying the term mutation in the sense in which he has applied it. The genetic characters of organisms come from one of two sources: either they are old characters and are due to the action of what we call inheritance or they are new and are due to what we call variation. If the term mutation is applied to the actual alteration of the machinery of the protoplasm, no objection can be felt to its use; but if it be applied, as it is, to the product of the action of the altered machine, viz. to the new genetic character, it leads to confusion. Inheritance is the persistence of the structure of the machine; characters are the products of the working of the machine; variation in genetic characters is due to the alteration (mutation) in the arrangement of the machinery, while variation in acquired characters (Lamarckian) is due to differences in the mode of working the machinery. The machinery when it starts (in the new zygote) has the power of grinding out certain results, which we call the characters of the organism. These appear at successive intervals of time, and the orderly manifestation of them is what we call the life-history of the organism. This brings us back to the question with which we started this discussion, viz. what is the relation of these variations in structure, which successively appear in an organism and constitute its life-history, to the mutational variations which appear in different organisms of the same brood or species. The question is brought home to us when we ask what is a bud-sport, such as a nectarine appearing on a peach-tree? From one point of view, it is simply a mutation appearing in asexual reproduction; from another it is one of these successional characters ("growth variations") which constitute the life-history of the zygote, for it appears in the same zygote which first produces a peach. Here our analogy of a machine which only works in one way seems to fail us, for these bud-sports do not appear in all parts of the organism, only in certain buds or parts of it, so that one part of the zygotic machine would appear to work differently to another. To discuss this question further would take us too far from our subject. Suffice it to say that we cannot answer it, any more than we can this further question of burning interest at the present day, viz. to what extent and in what manner is the machine itself altered by the particular way in which it is worked. In connection with this question we can only submit one consideration: the zygotic machine can, by its nature, only work once, so that any alteration in it can only be ascertained by studying the replicas of it which are produced in the reproductive organs.

It is a peculiarity that the result which we call the ripening of the generative organs nearly always appears among the final products of the action of the zygotic machine. It is remarkable that this should be the case. What is the reason of it? The late appearance of functional reproductive organs is almost a universal law, and the explanation of it is suggested by expressing the law in another way, viz. that the machine is almost always so constituted that it ceases to work efficiently soon after the reproductive organs have sufficiently discharged their function. Why this should occur we cannot explain: it is an ultimate fact of nature, and cannot be included in any wider category. The period during which the reproductive organs can act may be short as in ephemerids or long as in man and trees, and there is no reason to suppose that their action damages the vital machinery, though sometimes, as in the case of annual plants (Metschnikoff), it may incidentally do so; but, long or short, the cessation of their actions is always a prelude to the end. When they and their action are impaired, the organism ceases to react with precision to the environment, and the organism as a whole undergoes retrogressive changes.

It has been pointed out above that there is reason to believe that at the dawn of life the life-cycle was, EITHER IN ESSE OR IN POSSE, at least as long as it is at the present time. The qualification implied by the words in italics is necessary, for it is clearly possible that the external conditions then existing were not suitable for the production of all the stages of the potential life-history, and that what we call organic evolution has consisted in a gradual evolution of new environments to which the organism's innate capacity of change has enabled it to adapt itself. We have warrant for this possibility in the case of the Axolotl and in other similar cases of neoteny. And these cases further bring home to us the fact, to which I have already referred, that the full development of the functional reproductive organs is nearly always associated with the final stages of the life-history.

On this view of the succession of characters in the life-history of organisms, how shall we explain the undoubted fact that the development of buds hardly ever presents any phenomena corresponding to the embryonic and larval changes? The reason is clearly this, that budding usually occurs after the embryonic stage is past; when the characters of embryonic life have been worked out by the machine. When it takes place at an early stage in embryonic life, as it does in cases of so-called embryonic fission, the product shows, either partly or entirely, phenomena similar to those of embryonic development. The only case known to me in which budding by the adult is accompanied by morphological features similar to those displayed by embryos is furnished by the budding of the medusiform spore-sacs of hydrozoon polyps. But this case is exceptional, for here we have to do with an attempt, which fails, to form a free-swimming organism, the medusa; and the vestiges which appear in the buds are the umbrella-cavity, marginal tentacles, circular canal, etc., of the medusa arrested in development.

But the question still remains, are there no cases in which, as implied by the recapitulation theory, variations in any organ are confined to the period in which the organ is functional and do not affect it in the embryonic stages? The teeth of the whalebone whales may be cited as a case in which this is said to occur; but here the teeth are only imperfectly developed in the embryo and are soon absorbed. They have been affected by the change which has produced their disappearance in the adult, but not to complete extinction. Nor are they now likely to be extinguished, for having become exclusively embryonic they are largely protected from the action of natural selection. This consideration brings up a most important aspect of the question, so far as disappearing organs are concerned. Every organ is laid down at a certain period in the embryo and undergoes a certain course of growth until it obtains full functional development. When for any cause reduction begins, it is affected at all stages of its growth, unless it has functional importance in the larva, and in some cases its life is shortened at one or both ends. In cases, as in that of the whale's teeth, in which it entirely disappears in the adult, the latter part of its life is cut off; in others, the beginning of its life may be deferred. This happens, for instance, with the spiracle of many Elasmobranchs, which makes its appearance after the hyobranchial cleft, not before it as it should do, being anterior to it in position, and as it does in the Amniota in which it shows no reduction in size as compared with the other pharyngeal clefts. In those Elasmobranchs in which it is absent in the adult but present in the embryo (e.g. Carcharias) its life is shortened at both ends. Many more instances of organs, of which the beginning and end have been cut off, might be mentioned; e.g. the muscle-plate coelom of Aves, the primitive streak and the neurenteric canal of amniote blastoderms. In yet other cases in which the reduced organ is almost on the verge of disappearance, it may appear for a moment and disappear more than once in the course of development. As an instance of this striking phenomenon I may mention the neurenteric canal of avine embryos, and the anterior neuropore of Ascidians. Lastly the reduced organ may disappear in the developing stages before it does so in the adult. As an instance of this may be mentioned the mandibular palp of those Crustacea with zoaea larvae. This structure disappears in the larva only to reappear in a reduced form in later stages. In all these cases we are dealing with an organ which, we imagine, attained a fuller functional development at some previous stage in race-history, but in most of them we have no proof that it did so. It may be, and the possibility must not be lost sight of, that these organs never were anything else than functionless and that though they have been got rid of in the adult by elimination in the course of time, they have been able to persist in embryonic stages which are protected from the full action of natural selection. There is no reason to suppose that living matter at its first appearance differed from non-living matter in possessing only properties conducive to its well-being and prolonged existence. No one thinks that the properties of the various forms of inorganic matter are all strictly related to external conditions. Of what use to the diamond is its high specific gravity and high refrangibility, and to gold of its yellow colour and great weight? These substances continue to exist in virtue of other properties than these. It is impossible to suppose that the properties of living matter at its first appearance were all useful to it, for even now after aeons of elimination we find that it possesses many useless organs and that many of its relations to the external world are capable of considerable improvement.

In writing this essay I have purposely refrained from taking a definite position with regard to the problems touched. My desire has been to write a chapter showing the influence of Darwin's work so far as Embryology is concerned, and the various points which come up for consideration in discussing his views. Darwin was the last man who would have claimed finality for any of his doctrines, but he might fairly have claimed to have set going a process of intellectual fermentation which is still very far from completion.

XI. THE PALAEONTOLOGICAL RECORD. By W.B. Scott.

Professor of Geology in the University of Princeton, U.S.A.

I. ANIMALS.

To no branch of science did the publication of "The Origin of Species" prove to be a more vivifying and transforming influence than to Palaeontology. This science had suffered, and to some extent, still suffers from its rather anomalous position between geology and biology, each of which makes claim to its territory, and it was held in strict bondage to the Linnean and Cuvierian dogma that species were immutable entities. There is, however, reason to maintain that this strict bondage to a dogma now abandoned, was not without its good side, and served the purpose of keeping the infant science in leading-strings until it was able to walk alone, and preventing a flood of premature generalisations and speculations.

As Zittel has said: "Two directions were from the first apparent in palaeontological research—a stratigraphical and a biological. Stratigraphers wished from palaeontology mainly confirmation regarding the true order or relative age of zones of rock-deposits in the field. Biologists had, theoretically at least, the more genuine interest in fossil organisms as individual forms of life." (Zittel, "History of Geology and Palaeontology", page 363, London, 1901.) The geological or stratigraphical direction of the science was given by the work of William Smith, "the father of historical geology," in the closing decade of the eighteenth century. Smith was the first to make a systematic use of fossils in determining the order of succession of the rocks which make up the accessible crust of the earth, and this use has continued, without essential change, to the present day. It is true that the theory of evolution has greatly modified our conceptions concerning the introduction of new species and the manner in which palaeontological data are to be interpreted in terms of stratigraphy, but, broadly speaking, the method remains fundamentally the same as that introduced by Smith.

The biological direction of palaeontology was due to Cuvier and his associates, who first showed that fossils were not merely varieties of existing organisms, but belonged to extinct species and genera, an altogether revolutionary conception, which startled the scientific world. Cuvier made careful studies, especially of fossil vertebrates, from the standpoint of zoology and was thus the founder of palaeontology as a biological science. His great work on "Ossements Fossiles" (Paris, 1821) has never been surpassed as a masterpiece of the comparative method of anatomical investigation, and has furnished to the palaeontologist the indispensable implements of research.

On the other hand, Cuvier's theoretical views regarding the history of the earth and its successive faunas and floras are such as no one believes to-day. He held that the earth had been repeatedly devastated by great cataclysms, which destroyed every living thing, necessitating an entirely new creation, thus regarding the geological periods as sharply demarcated and strictly contemporaneous for the whole earth, and each species of animal and plant as confined to a single period. Cuvier's immense authority and his commanding personality dominated scientific thought for more than a generation and marked out the line which the development of palaeontology was to follow. The work was enthusiastically taken up by many very able men in the various European countries and in the United States, but, controlled as it was by the belief in the fixity of species, it remained almost entirely descriptive and consisted in the description and classification of the different groups of fossil organisms. As already intimated, this narrowness of view had its compensations, for it deferred generalisations until some adequate foundations for these had been laid.

Dominant as it was, Cuvier's authority was slowly undermined by the progress of knowledge and the way was prepared for the introduction of more rational conceptions. The theory of "Catastrophism" was attacked by several geologists, most effectively by Sir Charles Lyell, who greatly amplified the principles enunciated by Hutton and Playfair in the preceding century, and inaugurated a new era in geology. Lyell's uniformitarian views of the earth's history and of the agencies which had wrought its changes, had undoubted effect in educating men's minds for the acceptance of essentially similar views regarding the organic world. In palaeontology too the doctrine of the immutability of species, though vehemently maintained and reasserted, was gradually weakening. In reviewing long series of fossils, relations were observed which pointed to genetic connections and yet were interpreted as purely ideal. Agassiz, for example, who never accepted the evolutionary theory, drew attention to facts which could be satisfactorily interpreted only in terms of that theory. Among the fossils he indicated "progressive," "synthetic," "prophetic," and "embryonic" types, and pointed out the parallelism which obtains between the geological succession of ancient animals and the ontogenetic development of recent forms. In Darwin's words: "This view accords admirably well with our theory." ("Origin of Species" (6th edition), page 310.) Of similar import were Owen's views on "generalised types" and "archetypes."

The appearance of "The Origin of Species" in 1859 revolutionised all the biological sciences. From the very nature of the case, Darwin was compelled to give careful consideration to the palaeontological evidence; indeed, it was the palaeontology and modern distribution of animals in South America which first led him to reflect upon the great problem. In his own words: "I had been deeply impressed by discovering in the Pampean formation great fossil animals covered with armour like that on the existing armadillos;

secondly, by the manner in which closely allied animals replace one another in proceeding southward over the Continent; and thirdly, by the South American character of most of the productions of the Galapagos archipelago, and more especially by the manner in which they differ slightly on each island of the group." ("Life and Letters of Charles Darwin", I. page 82.) In the famous tenth and eleventh chapters of the "Origin", the palaeontological evidence is examined at length and the imperfection of the geological record is strongly emphasised. The conclusion is reached, that, in view of this extreme imperfection, palaeontology could not reasonably be expected to yield complete and convincing proof of the evolutionary theory. "I look at the geological record as a history of the world imperfectly kept, and written in a changing dialect; of this history we possess the last volume alone, relating only to two or three countries. Of this volume, only here and there a short chapter has been preserved; and of each page, only here and there a few lines." ("Origin of Species", page 289.) Yet, aside from these inevitable difficulties, he concludes, that "the other great leading facts in palaeontology agree admirably with the theory of descent with modification through variation and natural selection." (Ibid. page 313.)

Darwin's theory gave an entirely new significance and importance to palaeontology. Cuvier's conception of the science had been a limited, though a lofty one. "How glorious it would be if we could arrange the organised products of the universe in their chronological order!... The chronological succession of organised forms, the exact determination of those types which appeared first, the simultaneous origin of certain species and their gradual decay, would perhaps teach us as much about the mysteries of organisation as we can possibly learn through experiments with living organisms." (Zittel op. cit. page 140.) This, however, was rather the expression of a hope for the distant future than an account of what was attainable, and in practice the science remained almost purely descriptive, until Darwin gave it a new standpoint, new problems and an altogether fresh interest and charm. The revolution thus accomplished is comparable only to that produced by the Copernican astronomy.

From the first it was obvious that one of the most searching tests of the evolutionary theory would be given by the advance of palaeontological discovery. However imperfect the geological record might be, its ascertained facts would necessarily be consistent, under any reasonable interpretation, with the demands of a true theory; otherwise the theory would eventually be overwhelmed by the mass of irreconcilable data. A very great stimulus was thus given to geological investigation and to the exploration of new lands. In the last forty years, the examination of North and South America, of Africa and Asia has brought to light many chapters in the history of life, which are astonishingly full and complete. The flood of new material continues to accumulate at such a rate that it is impossible to keep abreast of it, and the very wealth of the collections is a source of difficulty and embarrassment. In modern palaeontology phylogenetic questions and problems occupy a foremost place and, as a result of the labours of many eminent investigators in many lands, it may be said that this science has proved to be one of the most solid supports of Darwin's theory. True, there are very many unsolved problems, and the discouraged worker is often tempted to believe that the fossils raise more questions than they answer. Yet, on the other hand, the whole trend of the evidence is so strongly in favour of the evolutionary doctrine, that no other interpretation seems at all rational.

To present any adequate account of the palaeontological record from the evolutionary standpoint, would require a large volume and a singularly unequal, broken and disjointed history it would be. Here the record is scanty, interrupted, even unintelligible, while there it is crowded with embarrassing wealth of material, but too often these full chapters are separated by such stretches of unrecorded time, that it is difficult to connect them. It will be more profitable to present a few illustrative examples than to attempt an outline of the whole history.

At the outset, the reader should be cautioned not to expect too much, for the task of determining phylogenies fairly bristles with difficulties and encounters many unanswered questions. Even when the evidence seems to be as copious and as complete as could be wished, different observers will put different interpretations upon it, as in the notorious case of the Steinheim shells. (In the Miocene beds of Steinheim, Wurtemberg, occur countless fresh-water shells, which show numerous lines of modification, but these have been very differently interpreted by different writers.) The ludicrous discrepancies which often appear between the phylogenetic "trees" of various writers have cast an undue discredit upon the science and have led many zoologists to ignore palaeontology altogether as unworthy of serious attention. One principal cause of these discrepant and often contradictory results is our ignorance concerning the exact modes of developmental change. What one writer postulates as almost axiomatic, another will reject as impossible and absurd. Few will be found to agree as to how far a given resemblance is offset by a given unlikeness, and so long as the question is one of weighing evidence and balancing probabilities, complete harmony is not to be looked for. These formidable difficulties confront us even in attempting to work out from abundant material a brief chapter in the phylogenetic history of some small and clearly limited group, and they become disproportionately greater, when we extend our view over vast periods of time and undertake to determine the mutual relationships of classes and types. If the evidence were complete and available, we should hardly be able to unravel its infinite complexity, or to find a clue through the mazes of the labyrinth. "Our ideas of the course of descent must of necessity be diagrammatic." (D.H. Scott, "Studies in Fossil Botany", page 524. London, 1900.)

Some of the most complete and convincing examples of descent with modification are to be found among the mammals, and nowhere more abundantly than in North America, where the series of continental formations, running through the whole Tertiary period, is remarkably full. Most of these formations contain a marvellous wealth of mammalian remains and in an unusual state of preservation. The oldest Eocene (Paleocene) has yielded a mammalian fauna which is still of prevailingly Mesozoic character, and contains but few forms which can be regarded as ancestral to those of later times. The succeeding fauna of the lower Eocene proper (Wasatch stage) is radically different and, while a few forms continue over from the Paleocene, the majority are evidently recent immigrants from some region not yet identified. From the Wasatch onward, the development of many phyla may be traced in almost unbroken continuity, though from time to time the record is somewhat obscured by migrations from the Old World and South America. As a rule, however, it is easy to distinguish between the immigrant and the indigenous elements of the fauna.

From their gregarious habits and individual abundance, the history of many hoofed animals is preserved with especial clearness. So well known as to have become a commonplace, is the phylogeny of the horses, which, contrary to all that would have been expected, ran the greater part of its course in North America. So far as it has yet been traced, the line begins in the lower Eocene with the genus Eohippus, a little creature not much larger than a cat, which has a short neck, relatively short limbs, and in particular, short feet, with four functional digits and a splint-like rudiment in the fore-foot, three functional digits and a rudiment in the hind-foot. The forearm bones (ulna and radius) are complete and separate, as are also the bones of the lower leg (fibula and tibia). The skull has a short face, with the orbit, or eye-socket, incompletely enclosed with bone, and the brain-case is slender and of small capacity. The teeth are short-crowned, the incisors without "mark," or enamel pit, on the cutting edge; the premolars are all smaller and simpler than the molars. The pattern of the upper molars is so entirely different from that seen in the modern horses that, without the intermediate connecting steps, no one would have ventured to derive the later from the earlier plan. This pattern is quadritubercular, with four principal, conical cusps arranged in two transverse pairs, forming a square, and two minute cuspules between each transverse pair, a tooth which is much more pig-like than horse-like. In the lower molars the cusps have already united to form two crescents, one behind the other, forming a pattern which is extremely common in the early representatives of many different families, both of the Perissodactyla and the Artiodactyla. In spite of the manifold differences in all parts of the skeleton between Eohippus and the recent horses, the former has stamped upon it an equine character which is unmistakable, though it can hardly be expressed in words.

Each one of the different Eocene and Oligocene horizons has its characteristic genus of horses, showing a slow, steady progress in a definite direction, all parts of the structure participating in the advance. It is not necessary to follow each of these successive steps of change, but it should be emphasised that the changes are gradual and uninterrupted. The genus Mesohippus, of the middle Oligocene, may be selected as a kind of half-way stage in the long progression. Comparing Mesohippus with Eohippus, we observe that the former is much larger, some species attaining the size of a sheep, and has a relatively longer neck, longer limbs and much more elongate feet, which are tridactyl, and the middle toe is so enlarged that it bears most of the weight, while the lateral digits are very much more slender. The fore-arm bones have begun to co-ossify and the ulna is greatly reduced, while the fibula, though still complete, is hardly more than a thread of bone. The skull has a longer face and a nearly enclosed orbit, and the brain-case is fuller and more capacious, the internal cast of which shows that the brain was richly convoluted. The teeth are still very short-crowned, but the upper incisors plainly show the beginning of the "mark"; the premolars have assumed the molar form, and the upper molars, though plainly derived from those of Eohippus, have made a long stride toward the horse pattern, in that the separate cusps have united to form a continuous outer wall and two transverse crests.

In the lower Miocene the interesting genus Desmatippus shows a further advance in the development of the teeth, which are beginning to assume the long-crowned shape, delaying the formation of roots; a thin layer of cement covers the crowns, and the transverse crests of the upper grinding teeth display an incipient degree of their modern complexity. This tooth-pattern is strictly intermediate between the recent type and the ancient type seen in Mesohippus and its predecessors. The upper Miocene genera, Protohippus and Hipparion are, to all intents and purposes, modern in character, but their smaller size, tridactyl feet and somewhat shorter-crowned teeth are reminiscences of their ancestry.

From time to time, when a land-connection between North America and Eurasia was established, some of the successive equine genera migrated to the Old World, but they do not seem to have gained a permanent footing there until the end of the Miocene or beginning of the Pliocene, eventually diversifying into the horses, asses, and zebras of Africa, Asia and Europe. At about the same period, the family extended its range to South America and there gave rise to a number of species and genera, some of them extremely peculiar. For some unknown reason, all the horse tribe had become extinct in the western hemisphere before the European discovery, but not until after the native race of man had peopled the continents.

In addition to the main stem of equine descent, briefly considered in the foregoing paragraphs, several side-branches were given off at successive levels of the stem. Most of these branches were short-lived, but some of them flourished for a considerable period and ramified into many species.

Apparently related to the horses and derived from the same root-stock is the family of the Palaeotheres, confined to the Eocene and Oligocene of Europe, dying out without descendants. In the earlier attempts to work out the history of the horses, as in the famous essay of Kowalevsky ("Sur l'Anchitherium aurelianense Cuv. et sur l'histoire paleontologique des Chevaux", "Mem. de l'Acad. Imp. des Sc. de St Petersbourg", XX. no. 5, 1873.), the Palaeotheres were placed in the direct line, because the number of adequately known Eocene mammals was then so small, that Cuvier's types were forced into various incongruous positions, to serve as ancestors for unrelated series.

The American family of the Titanotheres may also be distantly related to the horses, but passed through an entirely different course of development. From the lower Eocene to the lower sub-stage of the middle Oligocene the series is complete, beginning with small and rather lightly built animals. Gradually the stature and massiveness increase, a transverse pair of nasal horns make their appearance and, as these increase in size, the canine tusks and incisors diminish correspondingly. Already in the oldest known genus the number of digits had been reduced to four in the fore-foot and three in the hind, but there the reduction stops, for the increasing body-weight made necessary the development of broad and heavy feet. The final members of the series comprise only large, almost elephantine animals, with immensely developed and very various nasal horns, huge and massive heads, and altogether a grotesque appearance. The growth of the brain did not at all keep pace with the increase of the head and body, and the ludicrously small brain may will have been one of the factors which determined the startlingly sudden disappearance and extinction of the group.

Less completely known, but of unusual interest, is the genealogy of the rhinoceros family, which probably, though not certainly, was likewise of American origin. The group in North America at least, comprised three divisions, or sub-families, of very different proportions, appearance and habits, representing three divergent lines from the same stem. Though the relationship between the three lines seems hardly open to question, yet the form ancestral to all of them has not yet been identified. This is because of our still very incomplete knowledge of several perissodactyl genera of the Eocene, any one of which may eventually prove to be the ancestor sought for.

The first sub-family is the entirely extinct group of Hyracodonts, which may be traced in successive modifications through the upper Eocene, lower and middle Oligocene, then disappearing altogether. As yet, the hyracodonts have been found only in North America, and the last genus of the series, Hyracodon, was a cursorial animal. Very briefly stated, the modifications consist in a gradual increase in size, with greater slenderness of proportions, accompanied by elongation of the neck, limbs, and feet, which become tridactyl and very narrow. The grinding teeth have assumed the rhinoceros-like pattern and the premolars resemble the molars in form; on the other hand, the front teeth, incisors and canines, have become very small and are useless as weapons. As the animal had no horns, it was quite defenceless and must have found its safety in its swift running, for Hyracodon displays many superficial resemblances to the contemporary Oligocene horses, and was evidently adapted for speed. It may well have been the competition of the horses which led to the extinction of these cursorial rhinoceroses.

The second sub-family, that of the Amynodonts, followed a totally different course of development, becoming short-legged and short-footed, massive animals, the proportions of which suggest aquatic habits; they retained four digits in the front foot. The animal was well provided with weapons in the large canine tusks, but was without horns. Some members of this group extended their range to the Old World, but they all died out in the middle Oligocene, leaving no successors.

The sub-family of the true rhinoceroses cannot yet be certainly traced farther back than to the base of the middle Oligocene, though some fragmentary remains found in the lower Oligocene are probably also referable to it. The most ancient and most primitive member of this series yet discovered, the genus Trigonias, is unmistakably a rhinoceros, yet much less massive, having more the proportions of a tapir; it had four toes in the front foot, three in the hind, and had a full complement of teeth, except for the lower canines, though the upper canines are about to disappear, and the peculiar modification of the incisors, characteristic of the true rhinoceroses, is already apparent; the skull is hornless. Representatives of this sub-family continue through the Oligocene and Miocene of North America, becoming rare and localised in the Pliocene and then disappearing altogether. In the Old World, on the other hand, where the line appeared almost as early as it did in America, this group underwent a great expansion and ramification, giving rise not only to the Asiatic and African forms, but also to several extinct series.

Turning now to the Artiodactyla, we find still another group of mammals, that of the camels and llamas, which has long vanished from North America, yet took its rise and ran the greater part of its course in that continent. From the lower Eocene onward the history of this series is substantially complete, though much remains to be learned concerning the earlier members of the family. The story is very like that of the horses, to which in many respects it runs curiously parallel. Beginning with very small, five-toed animals, we observe in the successive genera a gradual transformation in all parts of the skeleton, an elongation of the neck, limbs and feet, a reduction of the digits from five to two, and eventually the coalescence of the remaining two digits into a "cannon-bone." The grinding teeth, by equally gradual steps, take on the ruminant pattern. In the upper Miocene the line divides into the two branches of the camels and llamas, the former migrating to Eurasia and the latter to

South America, though representatives of both lines persisted in North America until a very late period. Interesting side-branches of this line have also been found, one of which ended in the upper Miocene in animals which had almost the proportions of the giraffes and must have resembled them in appearance.

The American Tertiary has yielded several other groups of ruminant-like animals, some of which form beautifully complete evolutionary series, but space forbids more than this passing mention of them.

It was in Europe that the Artiodactyla had their principal development, and the upper Eocene, Oligocene and Miocene are crowded with such an overwhelming number and variety of forms that it is hardly possible to marshal them in orderly array and determine their mutual relationships. Yet in this chaotic exuberance of life, certain important facts stand out clearly, among these none is of greater interest and importance than the genealogy of the true Ruminants, or Pecora, which may be traced from the upper Eocene onward. The steps of modification and change are very similar to those through which the camel phylum passed in North America, but it is instructive to note that, despite their many resemblances, the two series can be connected only in their far distant beginnings. The pecoran stock became vastly more expanded and diversified than did the camel line and was evidently more plastic and adaptable, spreading eventually over all the continents except Australia, and forming to-day one of the dominant types of mammals, while the camels are on the decline and not far from extinction. The Pecora successively ramified into the deer, antelopes, sheep, goats and oxen, and did not reach North America till the Miocene, when they were already far advanced in specialisation. To this invasion of the Pecora, or true ruminants, it seems probable that the decline and eventual disappearance of the camels is to be ascribed.

Recent discoveries in Egypt have thrown much light upon a problem which long baffled the palaeontologist, namely, the origin of the elephants. (C.W. Andrews, "On the Evolution of the Proboscidea", "Phil. Trans. Roy. Soc." London, Vol. 196, 1904, page 99.) Early representatives of this order, Mastodons, had appeared almost simultaneously (in the geological sense of that word) in the upper Miocene of Europe and North America, but in neither continent was any more ancient type known which could plausibly be regarded as ancestral to them. Evidently, these problematical animals had reached the northern continents by migrating from some other region, but no one could say where that region lay. The Eocene and Oligocene beds of the Fayoum show us that the region sought for is Africa, and that the elephants form just such a series of gradual modifications as we have found among other hoofed animals. The later steps of the transformation, by which the mastodons lost their lower tusks, and their relatively small and simple grinding teeth acquired the great size and highly complex structure of the true elephants, may be followed in the uppermost Miocene and Pliocene fossils of India and southern Europe.

Egypt has also of late furnished some very welcome material which contributes to the solution of another unsolved problem which had quite eluded research, the origin of the whales. The toothed-whales may be traced back in several more or less parallel lines as far as the lower Miocene, but their predecessors in the Oligocene are still so incompletely known that safe conclusions can hardly be drawn from them. In the middle Eocene of Egypt, however, has been found a small, whale-like animal (Protocetus), which shows what the ancestral toothed-whale was like, and at the same time seems to connect these thoroughly marine mammals with land-animals. Though already entirely adapted to an aquatic mode of life, the teeth, skull and backbone of Protocetus display so many differences from those of the later whales and so many approximations to those of primitive, carnivorous land-mammals, as, in a large degree, to bridge over the gap between the two groups. Thus one of the most puzzling of palaeontological questions is in a fair way to receive a satisfactory answer. The origin of the whalebone-whales and their relations to the toothed-whales cannot yet be determined, since the necessary fossils have not been discovered.

Among the carnivorous mammals, phylogenetic series are not so clear and distinct as among the hoofed animals, chiefly because the carnivores are individually much less abundant, and well-preserved skeletons are among the prizes of the collector. Nevertheless,

124

much has already been learned concerning the mutual relations of the carnivorous families, and several phylogenetic series, notably that of the dogs, are quite complete. It has been made extremely probable that the primitive dogs of the Eocene represent the central stock, from which nearly or quite all the other families branched off, though the origin and descent of the cats have not yet been determined.

It should be clearly understood that the foregoing account of mammalian descent is merely a selection of a few representative cases and might be almost indefinitely extended. Nothing has been said, for example, of the wonderful museum of ancient mammalian life which is entombed in the rocks of South America, especially of Patagonia, and which opens a world so entirely different from that of the northern continents, yet exemplifying the same laws of "descent with modification." Very beautiful phylogenetic series have already been established among these most interesting and marvellously preserved fossils, but lack of space forbids a consideration of them.

The origin of the mammalia, as a class, offers a problem of which palaeontology can as yet present no definitive solution. Many morphologists regard the early amphibia as the ancestral group from which the mammals were derived, while most palaeontologists believe that the mammals are descended from the reptiles. The most ancient known mammals, those from the upper Triassic of Europe and North America, are so extremely rare and so very imperfectly known, that they give little help in determining the descent of the class, but, on the other hand, certain reptilian orders of the Permian period, especially well represented in South Africa, display so many and such close approximations to mammalian structure, as strongly to suggest a genetic relationship. It is difficult to believe that all those likenesses should have been independently acquired and are without phylogenetic significance.

Birds are comparatively rare as fossils and we should therefore look in vain among them for any such long and closely knit series as the mammals display in abundance. Nevertheless, a few extremely fortunate discoveries have made it practically certain that birds are descended from reptiles, of which they represent a highly specialised branch. The most ancient representative of this class is the extraordinary genus Archaeopteryx from the upper Jurassic of Bavaria, which, though an unmistakable bird, retains so many reptilian structures and characteristics as to make its derivation plain. Not to linger over anatomical minutiae, it may suffice to mention the absence of a horny beak, which is replaced by numerous true teeth, and the long lizard-like tail, which is made up of numerous distinct vertebrae, each with a pair of quill-like feathers attached to it. Birds with teeth are also found in the Cretaceous, though in most other respects the birds of that period had attained a substantially modern structure. Concerning the interrelations of the various orders and families of birds, palaeontology has as yet little to tell us.

The life of the Mesozoic era was characterised by an astonishing number and variety of reptiles, which were adapted to every mode of life, and dominated the air, the sea and the land, and many of which were of colossal proportions. Owing to the conditions of preservation which obtained during the Mesozoic period, the history of the reptiles is a broken and interrupted one, so that we can make out many short series, rather than any one of considerable length. While the relations of several reptilian orders can be satisfactorily determined, others still baffle us entirely, making their first known appearance in a fully differentiated state. We can trace the descent of the sea-dragons, the Ichthyosaurs and Plesiosaurs, from terrestrial ancestors, but the most ancient turtles yet discovered show us no closer approximation to any other order than do the recent turtles; and the oldest known Pterosaurs, the flying dragons of the Jurassic, are already fully differentiated. There is, however, no ground for discouragement in this, for the progress of discovery has been so rapid of late years, and our knowledge of Mesozoic life has increased with such leaps and bounds, that there is every reason to expect a solution of many of the outstanding problems in the near future.

Passing over the lower vertebrates, for lack of space to give them any adequate consideration, we may briefly take up the record of invertebrate life. From the overwhelming mass of material it is difficult to make a representative selection and even more difficult to state the facts intelligibly without the use of unduly technical language and without the aid of illustrations.

Several groups of the Mollusca, or shell-fish, yield very full and convincing evidence of their descent from earlier and simpler forms, and of these none is of greater interest than the Ammonites, an extinct order of the cephalopoda. The nearest living ally of the ammonites is the pearly nautilus, the other existing cephalopods, such as the squids, cuttle-fish, octopus, etc., are much more distantly related. Like the nautilus, the ammonites all possess a coiled and chambered shell, but their especial characteristic is the complexity of the "sutures." By sutures is meant the edges of the transverse partitions, or septa, where these join the shell-wall, and their complexity in the fully developed genera is extraordinary, forming patterns like the most elaborate oak-leaf embroidery, while in the nautiloids the sutures form simple curves. In the rocks of the Mesozoic era, wherever conditions of preservation are favourable, these beautiful shells are stored in countless multitudes, of an incredible variety of form, size and ornamentation, as is shown by the fact that nearly 5000 species have already been described. The ammonites are particularly well adapted for phylogenetic studies, because, by removing the successive whorls of the coiled shell, the individual development may be followed back in inverse order, to the microscopic "protoconch," or embryonic shell, which lies concealed in the middle of the coil. Thus the valuable aid of embryology is obtained in determining relationships.

The descent of the ammonites, taken as a group, is simple and clear; they arose as a branch of the nautiloids in the lower Devonian, the shells known as goniatites having zigzag, angulated sutures. Late in the succeeding Carboniferous period appear shells with a truly ammonoid complexity of sutures, and in the Permian their number and variety cause them to form a striking element of the marine faunas. It is in the Mesozoic era, however, that these shells attain their full development; increasing enormously in the Triassic, they culminate in the Jurassic in the number of families, genera and species, in the complexity of the sutures, and in the variety of shell-ornamentation. A slow decline begins in the Cretaceous, ending in the complete extinction of the whole group at the end of that period. As a final phase in the history of the ammonites, there appear many so-called "abnormal" genera, in which the shell is irregularly coiled, or more or less uncoiled, in some forms becoming actually straight. It is interesting to observe that some of these genera are not natural groups, but are "polyphyletic," i.e. are each derived from several distinct ancestral genera, which have undergone a similar kind of degeneration.

In the huge assembly of ammonites it is not yet possible to arrange all the forms in a truly natural classification, which shall express the various interrelations of the genera, yet several beautiful series have already been determined. In these series the individual development of the later general shows transitory stages which are permanent in antecedent genera. To give a mere catalogue of names without figures would not make these series more intelligible.

The Brachiopoda, or "lamp-shells," are a phylum of which comparatively few survive to the present day; their shells have a superficial likeness to those of the bivalved Mollusca, but are not homologous with the latter, and the phylum is really very distinct from the molluscs. While greatly reduced now, these animals were incredibly abundant throughout the Palaeozoic era, great masses of limestone being often composed almost exclusively of their shells, and their variety is in keeping with their individual abundance. As in the case of the ammonites, the problem is to arrange this great multitude of forms in an orderly array that shall express the ramifications of the group according to a genetic system. For many brachiopods, both recent and fossil, the individual development, or ontogeny, has been worked out and has proved to be of great assistance in the problems of classification and phylogeny. Already very encouraging progress has been made in the solution of these problems. All brachiopods form first a tiny, embryonic shell, called the protegulum, which is believed to represent the ancestral form of the whole group, and in the more advanced genera the developmental stages clearly indicate the ancestral genera of the series, the succession of adult forms in time corresponding to the order of the ontogenetic stages. The transformation of the delicate calcareous supports of the arms, often exquisitely preserved, are extremely interesting. Many of the Palaeozoic genera had these supports coiled like a pair of spiral springs, and it has been shown that these genera were derived from types in which the supports were simply shelly loops.

The long extinct class of crustacea known as the Trilobites are likewise very favourable subjects for phylogenetic studies. So far as the known record can inform us, the trilobites are exclusively Palaeozoic in distribution, but their course must have begun long before that era, as is shown by the number of distinct types among the genera of the lower Cambrian. The group reached the acme of abundance and relative importance in the Cambrian and Ordovician; then followed a long, slow decline, ending in complete and final disappearance before the end of the Permian. The newly-hatched and tiny trilobite larva, known as the protaspis, is very near to the primitive larval form of all the crustacea. By the aid of the correlated ontogenetic stages and the succession of the adult forms in the rocks, many phylogenetic series have been established and a basis for the natural arrangement of the whole class has been laid.

Very instructive series may also be observed among the Echinoderms and, what is very rare, we are able in this sub-kingdom to demonstrate the derivation of one class from another. Indeed, there is much reason to believe that the extinct class Cystidea of the Cambrian is the ancestral group, from which all the other Echinoderms, star-fishes, brittle-stars, sea-urchins, feather-stars, etc., are descended.

The foregoing sketch of the palaeontological record is, of necessity, extremely meagre, and does not represent even an outline of the evidence, but merely a few illustrative examples, selected almost at random from an immense body of material. However, it will perhaps suffice to show that the geological record is not so hopelessly incomplete as Darwin believed it to be. Since "The Origin of Species" was written, our knowledge of that record has been enormously extended and we now possess, no complete volumes, it is true, but some remarkably full and illuminating chapters. The main significance of the whole lies in the fact, that JUST IN PROPORTION TO THE COMPLETENESS OF THE RECORD IS THE UNEQUIVOCAL CHARACTER OF ITS TESTIMONY TO THE TRUTH OF THE EVOLUTIONARY THEORY.

The test of a true, as distinguished from a false, theory is the manner in which newly discovered and unanticipated facts arrange themselves under it. No more striking illustration of this can be found than in the contrasted fates of Cuvier's theory and of that of Darwin. Even before Cuvier's death his views had been undermined and the progress of discovery soon laid them in irreparable ruin, while the activity of half-a-century in many different lines of inquiry has established the theory of evolution upon a foundation of ever growing solidity. It is Darwin's imperishable glory that he prescribed the lines along which all the biological sciences were to advance to conquests not dreamed of when he wrote.

XII. THE PALAEONTOLOGICAL
RECORD. By D.H. Scott, F.R.S.

President of the Linnean Society.

II. PLANTS.

There are several points of view from which the subject of the present essay may be regarded. We may consider the fossil record of plants in its bearing: I. on the truth of the doctrine of Evolution; II. on Phylogeny, or the course of Evolution; III. on the theory of

Natural Selection. The remarks which follow, illustrating certain aspects only of an extensive subject, may conveniently be grouped under these three headings.

I. THE TRUTH OF EVOLUTION.

When "The Origin of Species" was written, it was necessary to show that the Geological Record was favourable to, or at least consistent with, the Theory of Descent. The point is argued, closely and fully, in Chapter X. "On the Imperfection of the Geological Record," and Chapter XI. "On the Geological Succession of Organic Beings"; there is, however, little about plants in these chapters. At the present time the truth of Evolution is no longer seriously disputed, though there are writers, like Reinke, who insist, and rightly so, that the doctrine is still only a belief, rather than an established fact of science. (J. Reinke, "Kritische Abstammungslehre", "Wiesner-Festschrift", page 11, Vienna, 1908.) Evidently, then, however little the Theory of Descent may be questioned in our own day, it is desirable to assure ourselves how the case stands, and in particular how far the evidence from fossil plants has grown stronger with time.

As regards direct evidence for the derivation of one species from another, there has probably been little advance since Darwin wrote, at least so we must infer from the emphasis laid on the discontinuity of successive fossil species by great systematic authorities like Grand'Eury and Zeiller in their most recent writings. We must either adopt the mutationist views of those authors (referred to in the last section of this essay) or must still rely on Darwin's explanation of the absence of numerous intermediate varieties. The attempts which have been made to trace, in the Tertiary rocks, the evolution of recent species, cannot, owing to the imperfect character of the evidence, be regarded as wholly satisfactory.

When we come to groups of a somewhat higher order we have an interesting history of the evolution of a recent family in the work, not yet completed, of Kidston and Gwynne-Vaughan on the fossil Osmundaceae. ("Trans. Royal Soc. Edinburgh", Vol. 45, Part III. 1907, Vol. 46, Part II. 1908, Vol. 46, Part III. 1909.) The authors are able, mainly on anatomical evidence, to trace back this now limited group of Ferns, through the Tertiary and Mesozoic to the Permian, and to show, with great probability, how their structure has been derived from that of early Palaeozoic types.

The history of the Ginkgoaceae, now represented only by the isolated maidenhair tree, scarcely known in a wild state, offers another striking example of a family which can be traced with certainty to the older Mesozoic and perhaps further back still. (See Seward and Gowan, "The Maidenhair Tree (Gingko biloba)", "Annals of Botany", Vol. XIV. 1900, page 109; also A. Sprecher "Le Ginkgo biloba", L., Geneva, 1907.)

On the wider question of the derivation of the great groups of plants, a very considerable advance has been made, and, so far as the higher plants are concerned, we are now able to form a far better conception than before of the probable course of evolution. This is a matter of phylogeny, and the facts will be considered under that head; our immediate point is that the new knowledge of the relations between the classes of plants in question materially strengthens the case for the theory of descent. The discoveries of the last few years throw light especially on the relation of the Angiosperms to the Gymnosperms, on that of the Seed-plants generally to the Ferns, and on the interrelations between the various classes of the higher Cryptogams.

That the fossil record has not done still more for Evolution is due to the fact that it begins too late—a point on which Darwin laid stress ("Origin of Species" (6th edition), page 286.) and which has more recently been elaborated by Poulton. ("Essays on Evolution", pages 46 et seq., Oxford, 1908.) An immense proportion of the whole evolutionary history lies behind the lowest fossiliferous rocks, and the case is worse for plants than for animals, as the record for the former begins, for all practical purposes, much higher up in the rocks.

It may be well here to call attention to a question, often overlooked, which has lately been revived by Reinke. (Reinke, loc. cit. page 13.) As all admit, we know nothing of the origin of life; consequently, for all we can tell, it is as probable that life began, on this planet, with many living things, as with one. If the first organic beings were many, they may have been heterogeneous, or at least exposed to different conditions, from their origin; in either case there would have been a number of distinct series from the beginning, and if so we should not be justified in assuming that all organisms are related to one another. There may conceivably be several of the original lines of descent still surviving, or represented among extinct forms—to reverse the remark of a distinguished botanist, there may be several Vegetable Kingdoms! However improbable this may sound, the possibility is one to be borne in mind.

That all VASCULAR plants really belong to one stock seems certain, and here the palaeontological record has materially strengthened the case for a monophyletic history. The Bryophyta are not likely to be absolutely distinct, for their sexual organs, and the stomata of the Mosses strongly suggest community of descent with the higher plants; if this be so it no doubt establishes a certain presumption in favour of a common origin for plants generally, for the gap between "Mosses and Ferns" has been regarded as the widest in the Vegetable Kingdom. The direct evidence of consanguinity is however much weaker when we come to the Algae, and it is conceivable (even if improbable) that the higher plants may have had a distinct ancestry (now wholly lost) from the beginning. The question had been raised in Darwin's time, and he referred to it in these words: "No doubt it is possible, as Mr G.H. Lewes has urged, that at the first commencement of life many different forms were evolved; but if so, we may conclude that only a very few have left modified descendants." ("Origin of Species", page 425.) This question, though it deserves attention, does not immediately affect the subject of the palaeontological record of plants, for there can be no reasonable doubt as to the interrelationship of those groups on which the record at present throws light.

The past history of plants by no means shows a regular progression from the simple to the complex, but often the contrary. This apparent anomaly is due to two causes.

1. The palaeobotanical record is essentially the story of the successive ascendancy of a series of dominant families, each of which attained its maximum, in organisation as well as in extent, and then sank into comparative obscurity, giving place to other families, which under new conditions were better able to take a leading place. As each family ran its downward course, either its members underwent an actual reduction in structure as they became relegated to herbaceous or perhaps aquatic life (this may have happened with the Horsetails and with Isoetes if derived from Lepidodendreae), or the higher branches of the family were crowded out altogether and only the "poor relations" were able to maintain their position by evading the competition of the ascendant races; this is also illustrated by the history of the Lycopod phylum. In either case there would result a lowering of the type of organisation within the group.

2. The course of real progress is often from the complex to the simple. If, as we shall find some grounds for believing, the Angiosperms came from a type with a flower resembling in its complexity that of Mesozoic "Cycads," almost the whole evolution of the flower in the highest plants has been a process of reduction. The stamen, in particular, has undoubtedly become extremely simplified during evolution; in the most primitive known seed-plants it was a highly compound leaf or pinna; its reduction has gone on in the Conifers and modern Cycads, as well as in the Angiosperms, though in different ways and to a varying extent.

The seed offers another striking example; the Palaeozoic seeds (if we leave the seed-like organs of certain Lycopods out of consideration) were always, so far as we know, highly complex structures, with an elaborate vascular system, a pollen-chamber, and often a much-differentiated testa. In the present day such seeds exist only in a few Gymnosperms which retain their ancient characters—in all the higher Spermophytes the structure is very much simplified, and this holds good even in the Coniferae, where there is no countervailing complication of ovary and stigma.

Reduction, in fact, is not always, or even generally, the same thing as degeneration. Simplification of parts is one of the most usual means of advance for the organism as a whole. A large proportion of the higher plants are microphyllous in comparison with the highly megaphyllous fern-like forms from which they appear to have been derived.

Darwin treated the general question of advance in organisation with much caution, saying: "The geological record... does not extend far enough back, to show with unmistakeable clearness that within the known history of the world organisation has largely advanced." ("Origin of Species", page 308.) Further on (Ibid. page 309.) he gives two standards by which advance may be measured: "We ought not solely to compare the highest members of a class at any two periods... but we ought to compare all the members, high and low, at the two periods." Judged by either standard the Horsetails and Club Mosses of the Carboniferous were higher than those of our own day, and the same is true of the Mesozoic Cycads. There is a general advance in the succession of classes, but not within each class.

Darwin's argument that "the inhabitants of the world at each successive period in its history have beaten their predecessors in the race for life, and are, in so far, higher in the scale" ("Origin of Species", page 315.) is unanswerable, but we must remember that "higher in the scale" only means "better adapted to the existing conditions." Darwin points out (Ibid. page 279.) that species have remained unchanged for long periods, probably longer than the periods of modification, and only underwent change when the conditions of their life were altered. Higher organisation, judged by the test of success, is thus purely relative to the changing conditions, a fact of which we have a striking illustration in the sudden incoming of the Angiosperms with all their wonderful floral adaptations to fertilisation by the higher families of Insects.

II. PHYLOGENY.

The question of phylogeny is really inseparable from that of the truth of the doctrine of evolution, for we cannot have historical evidence that evolution has actually taken place without at the same time having evidence of the course it has followed.

As already pointed out, the progress hitherto made has been rather in the way of joining up the great classes of plants than in tracing the descent of particular species or genera of the recent flora. There appears to be a difference in this respect from the Animal record, which tells us so much about the descent of living species, such as the elephant or the horse. The reason for this difference is no doubt to be found in the fact that the later part of the palaeontological record is the most satisfactory in the case of animals and the least so in the case of plants. The Tertiary plant-remains, in the great majority of instances, are impressions of leaves, the conclusions to be drawn from which are highly precarious; until the whole subject of Angiospermous palaeobotany has been reinvestigated, it would be rash to venture on any statements as to the descent of the families of Dicotyledons or Monocotyledons.

Our attention will be concentrated on the following questions, all relating to the phylogeny of main groups of plants: i. The Origin of the Angiosperms. ii. The Origin of the Seed-plants. iii. The Origin of the different classes of the Higher Cryptogamia.

i. THE ORIGIN OF THE ANGIOSPERMS.

The first of these questions has long been the great crux of botanical phylogeny, and until quite recently no light had been thrown upon the difficulty. The Angiosperms are the Flowering Plants, par excellence, and form, beyond comparison, the dominant sub-kingdom in the flora of our own age, including, apart from a few Conifers and Ferns, all the most familiar plants of our fields and gardens, and practically all plants of service to man. All recent work has tended to separate the Angiosperms more widely from the other seed-plants now living, the Gymnosperms. Vast as is the range of organisation presented by the great modern sub-kingdom, embracing forms adapted to every environment, there is yet a marked uniformity in certain points of structure, as in the development of the embryo-sac and its contents, the pollination through the intervention of a stigma, the strange phenomenon of double fertilisation (One sperm fertilising the egg, while the other unites with the embryo-sac nucleus, itself the product of a nuclear fusion, to give rise to a nutritive tissue, the endosperm.), the structure of the stamens, and the arrangement of the parts of the flower. All these points are common to Monocotyledons and Dicotyledons, and separate the Angiosperms collectively from all other plants.

In geological history the Angiosperms first appear in the Lower Cretaceous, and by Upper Cretaceous times had already swamped all other vegetation and seized the dominant position which they still hold. Thus they are isolated structurally from the rest of the Vegetable Kingdom, while historically they suddenly appear, almost in full force, and apparently without intermediaries with other groups. To quote Darwin's vigorous words: "The rapid development, as far as we can judge, of all the higher plants within recent geological times is an abominable mystery." ("More Letters of Charles Darwin", Vol. II. page 20, letter to J.D. Hooker, 1879.) A couple of years later he made a bold suggestion (which he only called an "idle thought") to meet this difficulty. He says: "I have been so astonished at the apparently sudden coming in of the higher phanerogams, that I have sometimes fancied that development might have slowly gone on for an immense period in some isolated continent or large island, perhaps near the South Pole." (Ibid, page 26, letter to Hooker, 1881.) This idea of an Angiospermous invasion from some lost southern land has sometimes been revived since, but has not, so far as the writer is aware, been supported by evidence. Light on the problem has come from a different direction.

The immense development of plants with the habit of Cycads, during the Mesozoic Period up to the Lower Cretaceous, has long been known. The existing Order Cycadaceae is a small family, with 9 genera and perhaps 100 species, occurring in the tropical and sub-tropical zones of both the Old and New World, but nowhere forming a dominant feature in the vegetation. Some few attain the stature of small trees, while in the majority the stem is short, though often living to a great age. The large pinnate or rarely bipinnate leaves give the Cycads a superficial resemblance in habit to Palms. Recent Cycads are dioecious; throughout the family the male fructification is in the form of a cone, each scale of the cone representing a stamen, and bearing on its lower surface numerous pollen-sacs, grouped in sori like the sporangia of Ferns. In all the genera, except Cycas itself, the female fructifications are likewise cones, each carpel bearing two ovules on its margin. In Cycas, however, no female cone is produced, but the leaf-like carpels, bearing from two to six ovules each, are borne directly on the main stem of the plant in rosettes alternating with those of the ordinary leaves—the most primitive arrangement known in any living seed-plant. The whole Order is relatively primitive, as shown most strikingly in its cryptogamic mode of fertilisation, by means of spermatozoids, which it shares with the maidenhair tree alone, among recent seed-plants.

In all the older Mesozoic rocks, from the Trias to the Lower Cretaceous, plants of the Cycad class (Cycadophyta, to use Nathorst's comprehensive name) are extraordinarily abundant in all parts of the world; in fact they were almost as prominent in the flora of those ages as the Dicotyledons are in that of our own day. In habit and to a great extent in anatomy, the Mesozoic Cycadophyta for the most part much resemble the recent Cycadaceae. But, strange to say, it is only in the rarest cases that the fructification has proved to be of the simple type characteristic of the recent family; the vast majority of the abundant fertile specimens yielded by the Mesozoic rocks possess a type of reproductive apparatus far more elaborate than anything known in Cycadaceae or other Gymnosperms. The predominant Mesozoic family, characterised by this advanced reproductive organisation, is known as the Bennettiteae; in habit these plants resembled the more stunted Cycads of the recent flora, but differed from them in the presence of numerous lateral fructifications, like large buds, borne on the stem among the crowded bases of the leaves. The organisation of these fructifications was first worked out on European specimens by Carruthers, Solms-Laubach, Lignier and others, but these observers had only more or less ripe fruits to deal with; the complete structure of the flower has only been elucidated within the last few years by the researches of Wieland on the magnificent American material, derived from the Upper Jurassic and Lower Cretaceous beds of Maryland, Dakota and Wyoming. (G.R. Wieland, "American Fossil Cycads", Carnegie Institution, Washington, 1906.) The word "flower" is used deliberately, for reasons which will be apparent from the following brief description, based on Wieland's observations.

The fructification is attached to the stem by a thick stalk, which, in its upper part, bears a large number of spirally arranged bracts, forming collectively a kind of perianth and

131

completely enclosing the essential organs of reproduction. The latter consist of a whorl of stamens, of extremely elaborate structure, surrounding a central cone or receptacle bearing numerous ovules. The stamens resemble the fertile fronds of a fern; they are of a compound, pinnate form, and bear very large numbers of pollen-sacs, each of which is itself a compound structure consisting of a number of compartments in which the pollen was formed. In their lower part the stamens are fused together by their stalks, like the "monadelphous" stamens of a mallow. The numerous ovules borne on the central receptacle are stalked, and are intermixed with sterile scales; the latter are expanded at their outer ends, which are united to form a kind of pericarp or ovary-wall, only interrupted by the protruding micropyles of the ovules. There is thus an approach to the closed pistil of an Angiosperm, but it is evident that the ovules received the pollen directly. The whole fructification is of large size; in the case of Cycadeoidea dacotensis, one of the species investigated by Wieland, the total length, in the bud condition, is about 12 cm., half of which belongs to the peduncle.

The general arrangement of the organs is manifestly the same as in a typical Angiospermous flower, with a central pistil, a surrounding whorl of stamens and an enveloping perianth; there is, as we have seen, some approach to the closed ovary of an Angiosperm; another point, first discovered nearly 20 years ago by Solms-Laubach in his investigation of a British species, is that the seed was practically "exalbuminous," its cavity being filled by the large, dicotyledonous embryo, whereas in all known Gymnosperms a large part of the sac is occupied by a nutritive tissue, the prothallus or endosperm; here also we have a condition only met with elsewhere among the higher Flowering Plants.

Taking all the characters into account, the indications of affinity between the Mesozoic Cycadophyta and the Angiosperms appear extremely significant, as was recognised by Wieland when he first discovered the hermaphrodite nature of the Bennettitean flower. The Angiosperm with which he specially compared the fossil type was the Tulip tree (Liriodendron) and certainly there is a remarkable analogy with the Magnoliaceous flowers, and with those of related orders such as Ranunculaceae and the Water-lilies. It cannot, of course, be maintained that the Bennettiteae, or any other Mesozoic Cycadophyta at present known, were on the direct line of descent of the Angiosperms, for there are some important points of difference, as, for example, in the great complexity of the stamens, and in the fact that the ovary-wall or pericarp was not formed by the carpels themselves, but by the accompanying sterile scale-leaves. Botanists, since the discovery of the bisexual flowers of the Bennettiteae, have expressed different views as to the nearness of their relation to the higher Flowering Plants, but the points of agreement are so many that it is difficult to resist the conviction that a real relation exists, and that the ancestry of the Angiosperms, so long shrouded in complete obscurity, is to be sought among the great plexus of Cycad-like plants which dominated the flora of the world in Mesozoic times. (On this subject see, in addition to Wieland's great work above cited, F.W. Oliver, "Pteridosperms and Angiosperms", "New Phytologist", Vol. V. 1906; D.H. Scott, "The Flowering Plants of the Mesozoic Age in the Light of Recent Discoveries", "Journal R. Microscop. Soc." 1907, and especially E.A.N. Arber and J. Parkin, "On the Origin of Angiosperms", "Journal Linn. Soc." (Bot.) Vol. XXXVIII. page 29, 1907.)

The great complexity of the Bennettitean flower, the earliest known fructification to which the word "flower" can be applied without forcing the sense, renders it probable, as Wieland and others have pointed out, that the evolution of the flower in Angiosperms has consisted essentially in a process of reduction, and that the simplest forms of flower are not to be regarded as the most primitive. The older morphologists generally took the view that such simple flowers were to be explained as reductions from a more perfect type, and this opinion, though abandoned by many later writers, appears likely to be true when we consider the elaboration of floral structure attained among the Mesozoic Cycadophyta, which preceded the Angiosperms in evolution.

If, as now seems probable, the Angiosperms were derived from ancestors allied to the Cycads, it would naturally follow that the Dicotyledons were first evolved, for their structure has most in common with that of the Cycadophyta. We should then have to regard the

Monocotyledons as a side-line, diverging probably at a very early stage from the main dicotyledonous stock, a view which many botanists have maintained, of late, on other grounds. (See especially Ethel Sargant, "The Reconstruction of a Race of Primitive Angiosperms", "Annals of Botany", Vol. XXII. page 121, 1908.) So far, however, as the palaeontological record shows, the Monocotyledons were little if at all later in their appearance than the Dicotyledons, though always subordinate in numbers. The typical and beautifully preserved Palm-wood from Cretaceous rocks is striking evidence of the early evolution of a characteristic monocotyledonous family. It must be admitted that the whole question of the evolution of Monocotyledons remains to be solved.

Accepting, provisionally, the theory of the cycadophytic origin of Angiosperms, it is interesting to see to what further conclusions we are led. The Bennettiteae, at any rate, were still at the gymnospermous level as regards their pollination, for the exposed micropyles of the ovules were in a position to receive the pollen directly, without the intervention of a stigma. It is thus indicated that the Angiosperms sprang from a gymnospermous source, and that the two great phyla of Seed-plants have not been distinct from the first, though no doubt the great majority of known Gymnosperms, especially the Coniferae, represent branch-lines of their own.

The stamens of the Bennettiteae are arranged precisely as in an angiospermous flower, but in form and structure they are like the fertile fronds of a Fern, in fact the compound pollen-sacs, or synangia as they are technically called, almost exactly agree with the spore-sacs of a particular family of Ferns—the Marattiaceae, a limited group, now mainly tropical, which was probably more prominent in the later Palaeozoic times than at present. The scaly hairs, or ramenta, which clothe every part of the plant, are also like those of Ferns.

It is not likely that the characters in which the Bennettiteae resemble the Ferns came to them directly from ancestors belonging to that class; an extensive group of Seed-plants, the Pteridospermeae, existed in Palaeozoic times and bear evident marks of affinity with the Fern phylum. The fern-like characters so remarkably persistent in the highly organised Cycadophyta of the Mesozoic were in all likelihood derived through the Pteridosperms, plants which show an unmistakable approach to the cycadophytic type.

The family Bennettiteae thus presents an extraordinary association of characters, exhibiting, side by side, features which belong to the Angiosperms, the Gymnosperms and the Ferns.

ii. ORIGIN OF SEED-PLANTS.

The general relation of the gymnospermous Seed-plants to the Higher Cryptogamia was cleared up, independently of fossil evidence, by the brilliant researches of Hofmeister, dating from the middle of the past century. (W. Hofmeister, "On the Germination, Development and Fructification of the Higher Cryptogamia", Ray Society, London, 1862. The original German treatise appeared in 1851.) He showed that "the embryo-sac of the Coniferae may be looked upon as a spore remaining enclosed in its sporangium; the prothallium which it forms does not come to the light." (Ibid. page 438.) He thus determined the homologies on the female side. Recognising, as some previous observers had already done, that the microspores of those Cryptogams in which two kinds of spore are developed, are equivalent to the pollen-grains of the higher plants, he further pointed out that fertilisation "in the Rhizocarpeae and Selaginellae takes place by free spermatozoa, and in the Coniferae by a pollen-tube, in the interior of which spermatozoa are probably formed"—a remarkable instance of prescience, for though spermatozoids have not been found in the Conifers proper, they were demonstrated in the allied groups Cycadaceae and Ginkgo, in 1896, by the Japanese botanists Ikeno and Hirase. A new link was thus established between the Gymnosperms and the Cryptogams.

It remained uncertain, however, from which line of Cryptogams the gymnospermous Seed-plants had sprung. The great point of morphological comparison was the presence of two kinds of spore, and this was known to occur in the recent Lycopods and Water-ferns (Rhizocarpeae) and was also found in fossil representatives of the third phylum, that of the

Horsetails. As a matter of fact all the three great Cryptogamic classes have found champions to maintain their claim to the ancestry of the Seed-plants, and in every case fossil evidence was called in. For a long time the Lycopods were the favourites, while the Ferns found the least support. The writer remembers, however, in the year 1881, hearing the late Prof. Sachs maintain, in a lecture to his class, that the descent of the Cycads could be traced, not merely from Ferns, but from a definite family of Ferns, the Marattiaceae, a view which, though in a somewhat crude form, anticipated more modern ideas.

Williamson appears to have been the first to recognise the presence, in the Carboniferous flora, of plants combining the characters of Ferns and Cycads. (See especially his "Organisation of the Fossil Plants of the Coal-Measures", Part XIII. "Phil. Trans. Royal Soc." 1887 B. page 299.) This conclusion was first reached in the case of the genera Heterangium and Lyginodendron, plants, which with a wholly fern-like habit, were found to unite an anatomical structure holding the balance between that of Ferns and Cycads, Heterangium inclining more to the former and Lyginodendron to the latter. Later researches placed Williamson's original suggestion on a firmer basis, and clearly proved the intermediate nature of these genera, and of a number of others, so far as their vegetative organs were concerned. This stage in our knowledge was marked by the institution of the class Cycadofilices by Potonie in 1897.

Nothing, however, was known of the organs of reproduction of the Cycadofilices, until F.W. Oliver, in 1903, identified a fossil seed, Lagenostoma Lomaxi, as belonging to Lyginodendron, the identification depending, in the first instance, on the recognition of an identical form of gland, of very characteristic structure, on the vegetative organs of Lyginodendron and on the cupule enveloping the seed. This evidence was supported by the discovery of a close anatomical agreement in other respects, as well as by constant association between the seed and the plant. (F.W. Oliver and D.H. Scott, "On the Structure of the Palaeozoic Seed, Lagenostoma Lomaxi, etc." "Phil. Trans. Royal Soc." Vol. 197 B. 1904.) The structure of the seed of Lyginodendron, proved to be of the same general type as that of the Cycads, as shown especially by the presence of a pollen-chamber or special cavity for the reception of the pollen-grains, an organ only known in the Cycads and Ginkgo among recent plants.

Within a few months after the discovery of the seed of Lyginodendron, Kidston found the large, nut-like seed of a Neuropteris, another fern-like Carboniferous plant, in actual connection with the pinnules of the frond, and since then seeds have been observed on the frond in species of Aneimites and Pecopteris, and a vast body of evidence, direct or indirect, has accumulated, showing that a large proportion of the Palaeozoic plants formerly classed as Ferns were in reality reproduced by seeds of the same type as those of recent Cycadaceae. (A summary of the evidence will be found in the writer's article "On the present position of Palaeozoic Botany", "Progressus Rei Botanicae", 1907, page 139, and "Studies in Fossil Botany", Vol. II. (2nd edition) London, 1909.) At the same time, the anatomical structure, where it is open to investigation, confirms the suggestion given by the habit, and shows that these early seed-bearing plants had a real affinity with Ferns. This conclusion received strong corroboration when Kidston, in 1905, discovered the male organs of Lyginodendron, and showed that they were identical with a fructification of the genus Crossotheca, hitherto regarded as belonging to Marattiaceous Ferns. (Kidston, "On the Microsporangia of the Pteridospermeae, etc." "Phil. Trans. Royal Soc." Vol. 198, B. 1906.)

The general conclusion which follows from the various observations alluded to, is that in Palaeozoic times there was a great body of plants (including, as it appears, a large majority of the fossils previously regarded as Ferns) which had attained the rank of Spermophyta, bearing seeds of a Cycadean type on fronds scarcely differing from the vegetative foliage, and in other respects, namely anatomy, habit and the structure of the pollen-bearing organs, retaining many of the characters of Ferns. From this extensive class of plants, to which the name Pteridospermeae has been given, it can scarcely be doubted that the abundant Cycadophyta, of the succeeding Mesozoic period, were derived. This conclusion is of far-reaching significance, for we have already found reason to think that the Angiosperms themselves sprang, in later times, from the Cycadophytic stock; it thus appears that the Fern-phylum, taken in a broad sense, ultimately represents the source from which the main line of descent of the Phanerogams took its rise.

It must further be borne in mind that in the Palaeozoic period there existed another group of seed-bearing plants, the Cordaiteae, far more advanced than the Pteridospermeae, and in many respects approaching the Coniferae, which themselves begin to appear in the latest Palaeozoic rocks. The Cordaiteae, while wholly different in habit from the contemporary fern-like Seed-plants, show unmistakable signs of a common origin with them. Not only is there a whole series of forms connecting the anatomical structure of the Cordaiteae with that of the Lyginodendreae among Pteridosperms, but a still more important point is that the seeds of the Cordaiteae, which have long been known, are of the same Cycadean type as those of the Pteridosperms, so that it is not always possible, as yet, to discriminate between the seeds of the two groups. These facts indicate that the same fern-like stock which gave rise to the Cycadophyta and through them, as appears probable, to the Angiosperms, was also the source of the Cordaiteae, which in their turn show manifest affinity with some at least of the Coniferae. Unless the latter are an artificial group, a view which does not commend itself to the writer, it would appear probable that the Gymnosperms generally, as well as the Angiosperms, were derived from an ancient race of Cryptogams, most nearly related to the Ferns. (Some botanists, however, believe that the Coniferae, or some of them, are probably more nearly related to the Lycopods. See Seward and Ford, "The Araucarieae, Recent and Extinct", "Phil. Trans. Royal Soc." Vol. 198 B. 1906.)

It may be mentioned here that the small gymnospermous group Gnetales (including the extraordinary West African plant Welwitschia) which were formerly regarded by some authorities as akin to the Equisetales, have recently been referred, on better grounds, to a common origin with the Angiosperms, from the Mesozoic Cycadophyta.

The tendency, therefore, of modern work on the palaeontological record of the Seed-plants has been to exalt the importance of the Fern-phylum, which, on present evidence, appears to be that from which the great majority, possibly the whole, of the Spermophyta have been derived.

One word of caution, however, is necessary. The Seed-plants are of enormous antiquity; both the Pteridosperms and the more highly organised family Cordaiteae, go back as far in geological history (namely to the Devonian) as the Ferns themselves or any other Vascular Cryptogams. It must therefore be understood that in speaking of the derivation of the Spermophyta from the Fern-phylum, we refer to that phylum at a very early stage, probably earlier than the most ancient period to which our record of land-plants extends. The affinity between the oldest Seed-plants and the Ferns, in the widest sense, seems established, but the common stock from which they actually arose is still unknown; though no doubt nearer to the Ferns than to any other group, it must have differed widely from the Ferns as we now know them, or perhaps even from any which the fossil record has yet revealed to us.

iii. THE ORIGIN OF THE HIGHER CRYPTOGAMIA.

The Sub-kingdom of the higher Spore-plants, the Cryptogamia possessing a vascular system, was more prominent in early geological periods than at present. It is true that the dominance of the Pteridophyta in Palaeozoic times has been much exaggerated owing to the assumption that everything which looked like a Fern really was a Fern. But, allowing for the fact, now established, that most of the Palaeozoic fern-like plants were already Spermophyta, there remains a vast mass of Cryptogamic forms of that period, and the familiar statement that they formed the main constituent of the Coal-forests still holds good. The three classes, Ferns (Filicales), Horsetails (Equisetales) and Club-mosses (Lycopodiales), under which we now group the Vascular Cryptogams, all extend back in geological history as far as we have any record of the flora of the land; in the Palaeozoic, however, a fourth class, the Sphenophyllales, was present.

As regards the early history of the Ferns, which are of special interest from their relation to the Seed-plants, it is impossible to speak quite positively, owing to the difficulty of discriminating between true fossil Ferns and the Pteridosperms which so closely simulated them. The difficulty especially affects the question of the position of Marattiaceous Ferns in the Palaeozoic Floras. This family, now so restricted, was until recently believed to have been one of the most important groups of Palaeozoic plants, especially during later Carboniferous and Permian times. Evidence both from anatomy and from sporangial characters appeared to establish this conclusion. Of late, however, doubts have arisen, owing to the discovery that some supposed members of the Marattiaceae bore seeds, and that a form of fructification previously referred to that family (Crossotheca) was really the pollen-bearing apparatus of a Pteridosperm (Lyginodendron). The question presents much difficulty; though it seems certain that our ideas of the extent of the family in Palaeozoic times will have to be restricted, there is still a decided balance of evidence in favour of the view that a considerable body of Marattiaceous Ferns actually existed. The plants in question were of large size (often arborescent) and highly organised—they represent, in fact, one of the highest developments of the Fern-stock, rather than a primitive type of the class.

There was, however, in the Palaeozoic period, a considerable group of comparatively simple Ferns (for which Arber has proposed the collective name Primofilices); the best known of these are referred to the family Botryopterideae, consisting of plants of small or moderate dimensions, with, on the whole, a simple anatomical structure, in certain cases actually simpler than that of any recent Ferns. On the other hand the sporangia of these plants were usually borne on special fertile fronds, a mark of rather high differentiation. This group goes back to the Devonian and includes some of the earliest types of Fern with which we are acquainted. It is probable that the Primofilices (though not the particular family Botryopterideae) represent the stock from which the various families of modern Ferns, already developed in the Mesozoic period, may have sprung.

None of the early Ferns show any clear approach to other classes of Vascular Cryptogams; so far as the fossil record affords any evidence, Ferns have always been plants with relatively large and usually compound leaves. There is no indication of their derivation from a microphyllous ancestry, though, as we shall see, there is some slight evidence for the converse hypothesis. Whatever the origin of the Ferns may have been it is hidden in the older rocks.

It has, however, been held that certain other Cryptogamic phyla had a common origin with the Ferns. The Equisetales are at present a well-defined group; even in the rich Palaeozoic floras the habit, anatomy and reproductive characters usually render the members of this class unmistakable, in spite of the great development and stature which they then attained. It is interesting, however, to find that in the oldest known representatives of the Equisetales the leaves were highly developed and dichotomously divided, thus differing greatly from the mere scale-leaves of the recent Horsetails, or even from the simple linear leaves of the later Calamites. The early members of the class, in their forked leaves, and in anatomical characters, show an approximation to the Sphenophyllales, which are chiefly represented by the large genus Sphenophyllum, ranging through the Palaeozoic from the Middle Devonian onwards. These were plants with rather slender, ribbed stems, bearing whorls of wedge-shaped or deeply forked leaves, six being the typical number in each whorl. From their weak habit it has been conjectured, with much probability, that they may have been climbing plants, like the scrambling Bedstraws of our hedgerows. The anatomy of the stem is simple and root-like; the cones are remarkable for the fact that each scale or sporophyll is a double structure, consisting of a lower, usually sterile lobe and one or more upper lobes bearing the sporangia; in one species both parts of the sporophyll were fertile. Sphenophyllum was evidently much specialised; the only other known genus is based on an isolated cone, Cheirostrobus, of Lower Carboniferous age, with an extraordinarily complex structure. In this genus especially, but also in the entire group, there is an evident relation to the Equisetales; hence it is of great interest that Nathorst has described, from the Devonian of Bear Island in the Arctic regions, a new genus Pseudobornia, consisting of large plants, remarkable for their highly compound leaves which, when found detached, were taken for the fronds of a Fern. The whorled arrangement of the leaves, and the habit of the plant, suggest affinities either with the Equisetales or the Sphenophyllales; Nathorst makes the genus the type of a new class, the Pseudoborniales. (A.G. Nathorst, "Zur Oberdevonischen Flora der Baren-Insel", "Kongl. Svenska Vetenskaps-Akademiens Handlingar" Bd. 36, No. 3, Stockholm, 1902.)

The available data, though still very fragmentary, certainly suggest that both Equisetales and Sphenophyllales may have sprung from a common stock having certain fern-like characters. On the other hand the Sphenophylls, and especially the peculiar genus Cheirostrobus, have in their anatomy a good deal in common with the Lycopods, and of late years they have been regarded as the derivatives of a stock common to that class and the Equisetales. At any rate the characters of the Sphenophyllales and of the new group Pseudoborniales suggest the existence, at a very early period, of a synthetic race of plants, combining the characters of various phyla of the Vascular Cryptogams. It may further be mentioned that the Psilotaceae, an isolated epiphytic family hitherto referred to the Lycopods, have been regarded by several recent authors as the last survivors of the Sphenophyllales, which they resemble both in their anatomy and in the position of their sporangia.

The Lycopods, so far as their early history is known, are remarkable rather for their high development in Palaeozoic times than for any indications of a more primitive ancestry. In the recent Flora, two of the four living genera (Excluding Psilotaceae.) (Selaginella and Isoetes) have spores of two kinds, while the other two (Lycopodium and Phylloglossum) are homosporous. Curiously enough, no certain instance of a homosporous Palaeozoic Lycopod has yet been discovered, though well-preserved fructifications are numerous. Wherever the facts have been definitely ascertained, we find two kinds of spore, differentiated quite as sharply as in any living members of the group. Some of the Palaeozoic Lycopods, in fact, went further, and produced bodies of the nature of seeds, some of which were actually regarded, for many years, as the seeds of Gymnosperms. This specially advanced form of fructification goes back at least as far as the Lower Carboniferous, while the oldest known genus of Lycopods, Bothrodendron, which is found in the Devonian, though not seed-bearing, was typically heterosporous, if we may judge from the Coal-measure species. No doubt homosporous Lycopods existed, but the great prevalence of the higher mode of reproduction in days which to us appear ancient, shows how long a course of evolution must have already been passed through before the oldest known members of the group came into being. The other characters of the Palaeozoic Lycopods tell the same tale; most of them attained the stature of trees, with a corresponding elaboration of anatomical structure, and even the herbaceous forms show no special simplicity. It appears from recent work that herbaceous Lycopods, indistinguishable from our recent Selaginellas, already existed in the time of the Coal-measures, while one herbaceous form (Miadesmia) is known to have borne seeds.

The utmost that can be said for primitiveness of character in Palaeozoic Lycopods is that the anatomy of the stem, in its primary ground-plan, as distinguished from its secondary growth, was simpler than that of most Lycopodiums and Selaginellas at the present day. There are also some peculiarities in the underground organs (Stigmaria) which suggest the possibility of a somewhat imperfect differentiation between root and stem, but precisely parallel difficulties are met with in the case of the living Selaginellas, and in some degree in species of Lycopodium.

In spite of their high development in past ages the Lycopods, recent and fossil, constitute, on the whole, a homogeneous group, and there is little at present to connect them with other phyla. Anatomically some relation to the Sphenophylls is indicated, and perhaps the recent Psilotaceae give some support to this connection, for while their nearest alliance appears to be with the Sphenophylls, they approach the Lycopods in anatomy, habit, and mode of branching.

The typically microphyllous character of the Lycopods, and the simple relation between sporangium and sporophyll which obtains throughout the class, have led various botanists to regard them as the most primitive phylum of the Vascular Cryptogams. There is nothing in the fossil record to disprove this view, but neither is there anything to support it, for this class so far as we know is no more ancient than the megaphyllous Cryptogams, and its earliest representatives show no special simplicity. If the indications of affinity with Sphenophylls are of any value the Lycopods are open to suspicion of reduction from a megaphyllous ancestry, but there is no direct palaeontological evidence for such a history.

The general conclusions to which we are led by a consideration of the fossil record of the Vascular Cryptogams are still very hypothetical, but may be provisionally stated as follows:

The Ferns go back to the earliest known period. In Mesozoic times practically all the existing families had appeared; in the Palaeozoic the class was less extensive than formerly believed, a majority of the supposed Ferns of that age having proved to be seed-bearing plants. The oldest authentic representatives of the Ferns were megaphyllous plants, broadly speaking, of the same type as those of later epochs, though differing much in detail. As far back as the record extends they show no sign of becoming merged with other phyla in any synthetic group.

The Equisetales likewise have a long history, and manifestly attained their greatest development in Palaeozoic times. Their oldest forms show an approach to the extinct class Sphenophyllales, which connects them to some extent, by anatomical characters, with the Lycopods. At the same time the oldest Equisetales show a somewhat megaphyllous character, which was more marked in the Devonian Pseudoborniales. Some remote affinity with the Ferns (which has also been upheld on other grounds) may thus be indicated. It is possible that in the Sphenophyllales we may have the much-modified representatives of a very ancient synthetic group.

The Lycopods likewise attained their maximum in the Palaeozoic, and show, on the whole, a greater elaboration of structure in their early forms than at any later period, while at the same time maintaining a considerable degree of uniformity in morphological characters throughout their history. The Sphenophyllales are the only other class with which they show any relation; if such a connection existed, the common point of origin must lie exceedingly far back.

The fossil record, as at present known, cannot, in the nature of things, throw any direct light on what is perhaps the most disputed question in the morphology of plants—the origin of the alternating generations of the higher Cryptogams and the Spermophyta. At the earliest period to which terrestrial plants have been traced back all the groups of Vascular Cryptogams were in a highly advanced stage of evolution, while innumerable Seed-plants—presumably the descendants of Cryptogamic ancestors—were already flourishing. On the other hand we know practically nothing of Palaeozoic Bryophyta, and the evidence even for their existence at that period cannot be termed conclusive. While there are thus no palaeontological grounds for the hypothesis that the Vascular plants came of a Bryophytic stock, the question of their actual origin remains unsolved.

III. NATURAL SELECTION.

Hitherto we have considered the palaeontological record of plants in relation to Evolution. The question remains, whether the record throws any light on the theory of which Darwin and Wallace were the authors—that of Natural Selection. The subject is clearly one which must be investigated by other methods than those of the palaeontologist; still there are certain important points involved, on which the palaeontological record appears to bear.

One of these points is the supposed distinction between morphological and adaptive characters, on which Nageli, in particular, laid so much stress. The question is a difficult one; it was discussed by Darwin ("Origin of Species" (6th edition), pages 170-176.), who, while showing that the apparent distinction is in part to be explained by our imperfect knowledge of function, recognised the existence of important morphological characters which are not adaptations. The following passage expresses his conclusion. "Thus, as I am inclined to believe, morphological differences, which we consider as important—such as the arrangement of the leaves, the divisions of the flower or of the ovarium, the position of the ovules, etc.—first appeared in many cases as fluctuating variations, which sooner or later became constant through the nature of the organism and of the surrounding conditions, as well as through the inter-crossing of distinct individuals, but not through natural selection; for as these morphological characters do not affect the welfare of the species, any slight deviations in them could not have been governed or accumulated through this latter agency." (Ibid. page 176.)

This is a sufficiently liberal concession; Nageli, however, went much further when he said: "I do not know among plants a morphological modification which can be explained on utilitarian principles." (See "More Letters", Vol. II. page 375 (footnote).) If this were true the field of Natural Selection would be so seriously restricted, as to leave the theory only a very limited importance.

It can be shown, as the writer believes, that many typical "morphological characters," on which the distinction between great classes of plants is based, were adaptive in origin, and even that their constancy is due to their functional importance. Only one or two cases will be mentioned, where the fossil evidence affects the question.

The pollen-tube is one of the most important morphological characters of the Spermophyta as now existing—in fact the name Siphonogama is used by Engler in his classification, as expressing a peculiarly constant character of the Seed-plants. Yet the pollen-tube is a manifest adaptation, following on the adoption of the seed-habit, and serving first to bring the spermatozoids with greater precision to their goal, and ultimately to relieve them of the necessity for independent movement. The pollen-tube is constant because it has proved to be indispensable.

In the Palaeozoic Seed-plants there are a number of instances in which the pollen-grains, contained in the pollen-chamber of a seed, are so beautifully preserved that the presence of a group of cells within the grain can be demonstrated; sometimes we can even see how the cell-walls broke down to emit the sperms, and quite lately it is said that the sperms themselves have been recognised. (F.W. Oliver, "On Physostoma elegans, an archaic type of seed from the Palaeozoic Rocks", "Annals of Botany", January, 1909. See also the earlier papers there cited.) In no case, however, is there as yet any satisfactory evidence for the formation of a pollen-tube; it is probable that in these early Seed-plants the pollen-grains remained at about the evolutionary level of the microspores in Pilularia or Selaginella, and discharged their spermatozoids directly, leaving them to find their own way to the female cells. It thus appears that there were once Spermophyta without pollen-tubes. The pollen-tube method ultimately prevailed, becoming a constant "morphological character," for no other reason than because, under the new conditions, it provided a more perfect mechanism for the accomplishment of the act of fertilisation. We have still, in the Cycads and Ginkgo, the transitional case, where the tube remains short, serves mainly as an anchor and water-reservoir, but yet is able, by its slight growth, to give the spermatozoids a "lift" in the right direction. In other Seed-plants the sperms are mere passengers, carried all the way by the pollen-tube; this fact has alone rendered the Angiospermous method of fertilisation through a stigma possible.

We may next take the seed itself—the very type of a morphological character. Our fossil record does not go far enough back to tell us the origin of the seed in the Cycadophyta and Pteridosperms (the main line of its development) but some interesting sidelights may be obtained from the Lycopod phylum. In two Palaeozoic genera, as we have seen, seed-like organs are known to have been developed, resembling true seeds in the presence of an integument and of a single functional embryo-sac, as well as in some other points. We will call these organs "seeds" for the sake of shortness. In one genus (Lepidocarpon) the seeds were borne on a cone indistinguishable from that of the ordinary cryptogamic Lepidodendreae, the typical Lycopods of the period, while the seed itself retained much of the detailed structure of the sporangium of that family. In the second genus, Miadesmia, the seed-bearing plant was herbaceous, and much like a recent Selaginella. (See Margaret

139

Benson, "Miadesmia membranacea, a new Palaeozoic Lycopod with a seed-like structure", "Phil. Trans. Royal Soc. Vol." 199, B. 1908.) The seeds of the two genera are differently constructed, and evidently had an independent origin. Here, then, we have seeds arising casually, as it were, at different points among plants which otherwise retain all the characters of their cryptogamic fellows; the seed is not yet a morphological character of importance. To suppose that in these isolated cases the seed sprang into being in obedience to a Law of Advance ("Vervollkommungsprincip"), from which other contemporary Lycopods were exempt, involves us in unnecessary mysticism. On the other hand it is not difficult to see how these seeds may have arisen, as adaptive structures, under the influence of Natural Selection. The seed-like structure afforded protection to the prothallus, and may have enabled the embryo to be launched on the world in greater security. There was further, as we may suppose, a gain in certainty of fertilisation. As the writer has pointed out elsewhere, the chances against the necessary association of the small male with the large female spores must have been enormously great when the cones were borne high up on tall trees. The same difficulty may have existed in the case of the herbaceous Miadesmia, if, as Miss Benson conjectures, it was an epiphyte. One way of solving the problem was for pollination to take place while the megaspore was still on the parent plant, and this is just what the formation of an ovule or seed was likely to secure.

The seeds of the Pteridosperms, unlike those of the Lycopod stock, have not yet been found in statu nascendi—in all known cases they were already highly developed organs and far removed from the cryptogamic sporangium. But in two respects we find that these seeds, or some of them, had not yet realised their possibilities. In the seed of Lyginodendron and other cases the micropyle, or orifice of the integument, was not the passage through which the pollen entered; the open neck of the pollen-chamber protruded through the micropyle and itself received the pollen. We have met with an analogous case, at a more advanced stage of evolution, in the Bennettiteae, where the wall of the gynaecium, though otherwise closed, did not provide a stigma to catch the pollen, but allowed the micropyles of the ovules to protrude and receive the pollen in the old gymnospermous fashion. The integument in the one case and the pistil in the other had not yet assumed all the functions to which the organ ultimately became adapted. Again, no Palaeozoic seed has yet been found to contain an embryo, though the preservation is often good enough for it to have been recognised if present. It is probable that the nursing of the embryo had not yet come to be one of the functions of the seed, and that the whole embryonic development was relegated to the germination stage.

In these two points, the reception of the pollen by the micropyle and the nursing of the embryo, it appears that many Palaeozoic seeds were imperfect, as compared with the typical seeds of later times. As evolution went on, one function was superadded on another, and it appears impossible to resist the conclusion that the whole differentiation of the seed was a process of adaptation, and consequently governed by Natural Selection, just as much as the specialisation of the rostellum in an Orchid, or of the pappus in a Composite.

Did space allow, other examples might be added. We may venture to maintain that the glimpses which the fossil record allows us into early stages in the evolution of organs now of high systematic importance, by no means justify the belief in any essential distinction between morphological and adaptive characters.

Another point, closely connected with Darwin's theory, on which the fossil history of plants has been supposed to have some bearing, is the question of Mutation, as opposed to indefinite variation. Arber and Parkin, in their interesting memoir on the Origin of Angiosperms, have suggested calling in Mutation to explain the apparently sudden transition from the cycadean to the angiospermous type of foliage, in late Mesozoic times, though they express themselves with much caution, and point out "a distinct danger that Mutation may become the last resort of the phylogenetically destitute"!

The distinguished French palaeobotanists, Grand'Eury (C. Grand'Eury, "Sur les mutations de quelques Plantes fossiles du Terrain houiller". "Comptes Rendus", CXLII. page 25, 1906.) and Zeiller (R. Zeiller "Les Vegetaux fossiles et leurs Enchainements", "Revue du

Mois", III. February, 1907.), are of opinion, to quote the words of the latter writer, that the facts of fossil Botany are in agreement with the sudden appearance of new forms, differing by marked characters from those that have given them birth; he adds that these results give more amplitude to this idea of Mutation, extending it to groups of a higher order, and even revealing the existence of discontinuous series between the successive terms of which we yet recognise bonds of filiation. (Loc. cit. page 23.)

If Zeiller's opinion should be confirmed, it would no doubt be a serious blow to the Darwinian theory. As Darwin said: "Under a scientific point of view, and as leading to further investigation, but little advantage is gained by believing that new forms are suddenly developed in an inexplicable manner from old and widely different forms, over the old belief in the creation of species from the dust of the earth." ("Origin of Species", page 424.)

It most however be pointed out, that such mutations as Zeiller, and to some extent Arber and Parkin, appear to have in view, bridging the gulf between different Orders and Classes, bear no relation to any mutations which have been actually observed, such as the comparatively small changes, of sub-specific value, described by De Vries in the type-case of Oenothera Lamarckiana. The results of palaeobotanical research have undoubtedly tended to fill up gaps in the Natural System of plants—that many such gaps still persist is not surprising; their presence may well serve as an incentive to further research but does not, as it seems to the writer, justify the assumption of changes in the past, wholly without analogy among living organisms.

As regards the succession of species, there are no greater authorities than Grand'Eury and Zeiller, and great weight must be attached to their opinion that the evidence from continuous deposits favours a somewhat sudden change from one specific form to another. At the same time it will be well to bear in mind that the subject of the "absence of numerous intermediate varieties in any single formation" was fully discussed by Darwin. ("Origin of Species", pages 275-282, and page 312.); the explanation which he gave may go a long way to account for the facts which recent writers have regarded as favouring the theory of saltatory mutation.

The rapid sketch given in the present essay can do no more than call attention to a few salient points, in which the palaeontological records of plants has an evident bearing on the Darwinian theory. At the present day the whole subject of palaeobotany is a study in evolution, and derives its chief inspiration from the ideas of Darwin and Wallace. In return it contributes something to the verification of their teaching; the recent progress of the subject, in spite of the immense difficulties which still remain, has added fresh force to Darwin's statement that "the great leading facts in palaeontology agree admirably with the theory of descent with modification through variation and natural selection." (Ibid. page 313.)

XIII. THE INFLUENCE OF ENVIRONMENT ON THE FORMS OF PLANTS. By Georg Klebs, PH.D.

Professor of Botany in the University of Heidelberg.

The dependence of plants on their environment became the object of scientific research when the phenomena of life were first investigated and physiology took its place as a special branch of science. This occurred in the course of the eighteenth century as the result of the pioneer work of Hales, Duhamel, Ingenhousz, Senebier and others. In the nineteenth century, particularly in the second half, physiology experienced an unprecedented development in that it began to concern itself with the experimental study of nutrition and growth, and with the phenomena associated with stimulus and movement; on the other hand, physiology neglected phenomena connected with the production of form, a department of knowledge which was the province of morphology, a purely descriptive science. It was in the middle of the last century that the growth of comparative morphology and the study of phases of development reached their highest point.

The forms of plants appeared to be the expression of their inscrutable inner nature; the stages passed through in the development of the individual were regarded as the outcome of purely internal and hidden laws. The feasibility of experimental inquiry seemed therefore remote. Meanwhile, the recognition of the great importance of such a causal morphology emerged from the researches of the physiologists of that time, more especially from those of Hofmeister (Hofmeister, "Allgemeine Morphologie", Leipzig, 1868, page 579.), and afterwards from the work of Sachs. (Sachs, "Stoff und Form der Pflanzenorgane", Vol. I. 1880; Vol. II. 1882. "Gesammelte Abhandlungen uber Pflanzen-Physiologie", II. Leipzig, 1893.) Hofmeister, in speaking of this line of inquiry, described it as "the most pressing and immediate aim of the investigator to discover to what extent external forces acting on the organism are of importance in determining its form." This advance was the outcome of the influence of that potent force in biology which was created by Darwin's "Origin of Species" (1859).

The significance of the splendid conception of the transformation of species was first recognised and discussed by Lamarck (1809); as an explanation of transformation he at once seized upon the idea—an intelligible view—that the external world is the determining factor. Lamarck (Lamarck, "Philosophie zoologique", pages 223-227. Paris, 1809.) endeavoured, more especially, to demonstrate from the behaviour of plants that changes in environment induce change in form which eventually leads to the production of new species. In the case of animals, Lamarck adopted the teleological view that alterations in the environment first lead to alterations in the needs of the organisms, which, as the result of a kind of conscious effort of will, induce useful modifications and even the development of new organs. His work has not exercised any influence on the progress of science: Darwin himself confessed in regard to Lamarck's work—"I got not a fact or idea from it." ("Life and Letters", Vol. II. page 215.)

On a mass of incomparably richer and more essential data Darwin based his view of the descent of organisms and gained for it general acceptance; as an explanation of modification he elaborated the ingeniously conceived selection theory. The question of special interest in this connection, namely what is the importance of the influence of the environment, Darwin always answered with some hesitation and caution, indeed with a certain amount of indecision.

The fundamental principle underlying his theory is that of general variability as a whole, the nature and extent of which, especially in cultivated organisms, are fully dealt with in his well-known book. (Darwin, "The variation of Animals and Plants under domestication", 2 vols., edition 1, 1868; edition 2, 1875; popular edition 1905.) In regard to the question as to the cause of variability Darwin adopts a consistently mechanical view. He says: "These several considerations alone render it probable that variability of every kind is directly or indirectly caused by changed conditions of life. Or, to put the case under another point of view, if it were possible to expose all the individuals of a species during many generations to absolutely uniform conditions of life, there would be no variability." ("The variation of Animals and Plants" (2nd edition), Vol. II. page 242.) Darwin did not draw further conclusions from this general principle.

Variations produced in organisms by the environment are distinguished by Darwin as "the definite" and "the indefinite." (Ibid. II. page 260. See also "Origin of Species" (6th edition), page 6.) The first occur "when all or nearly all the offspring of an individual exposed to certain conditions during several generations are modified in the same manner." Indefinite variation is much more general and a more important factor in the production of new species; as a result of this, single individuals are distinguished from one another by "slight" differences, first in one then in another character. There may also occur, though this is very rare, more marked modifications, "variations which seem to us in our ignorance to arise spontaneously." ("Origin of Species" (6th edition), page 421.) The selection theory demands the further postulate that such changes, "whether extremely slight or strongly marked," are inherited. Darwin was no nearer to an experimental proof of this assumption than to the discovery of the actual cause of variability. It was not until the later years of his life that Darwin was occupied with the "perplexing problem... what causes almost every cultivated plant to vary" ("Life and Letters", Vol. III. page 342.): he began to make experiments on the influence of the soil, but these were soon given up.

In the course of the violent controversy which was the outcome of Darwin's work the fundamental principles of his teaching were not advanced by any decisive observations. Among the supporters and opponents, Nageli (Nageli, "Theorie der Abstammungslehre", Munich, 1884; cf. Chapter III.) was one of the few who sought to obtain proofs by experimental methods. His extensive cultural experiments with alpine Hieracia led him to form the opinion that the changes which are induced by an alteration in the food-supply, in climate or in habitat, are not inherited and are therefore of no importance from the point of view of the production of species. And yet Nageli did attribute an important influence to the external world; he believed that adaptations of plants arise as reactions to continuous stimuli, which supply a need and are therefore useful. These opinions, which recall the teleological aspect of Lamarckism, are entirely unsupported by proof. While other far-reaching attempts at an explanation of the theory of descent were formulated both in Nageli's time and afterwards, some in support of, others in opposition to Darwin, the necessity of investigating, from different standpoints, the underlying causes, variability and heredity, was more and more realised. To this category belong the statistical investigations undertaken by Quetelet and Galton, the researches into hybridisation, to which an impetus was given by the re-discovery of the Mendelian law of segregation, as also by the culture experiments on mutating species following the work of de Vries, and lastly the consideration of the question how far variation and heredity are governed by external influences. These latter problems, which are concerned in general with the causes of form-production and form-modification, may be treated in a short summary which falls under two heads, one having reference to the conditions of form-production in single species, the other being concerned with the conditions governing the transformation of species.

I. THE INFLUENCE OF EXTERNAL CONDITIONS ON FORM-PRODUCTION IN SINGLE SPECIES.

The members of plants, which we express by the terms stem, leaf, flower, etc. are capable of modification within certain limits; since Lamarck's time this power of modification has been brought more or less into relation with the environment. We are concerned not only with the question of experimental demonstration of this relationship, but, more generally, with an examination of the origin of forms, the sequences of stages in development that are governed by recognisable causes. We have to consider the general problem; to study the conditions of all typical as well as of atypic forms, in other words, to found a physiology of form.

If we survey the endless variety of plant-forms and consider the highly complex and still little known processes in the interior of cells, and if we remember that the whole of this branch of investigation came into existence only a few decades ago, we are able to grasp the fact that a satisfactory explanation of the factors determining form cannot be discovered all at once. The goal is still far away. We are not concerned now with the controversial question, whether, on the whole, the fundamental processes in the development of form can be recognised by physiological means. A belief in the possibility of this can in any case do no harm. What we may and must attempt is this—to discover points of attack on one side or another, which may enable us by means of experimental methods to come into closer touch with these elusive and difficult problems. While we are forced to admit that there is at present much that is insoluble there remains an inexhaustible supply of problems capable of solution.

The object of our investigations is the species; but as regards the question, what is a species, science of to-day takes up a position different from that of Darwin. For him it was the Linnean species which illustrates variation: we now know, thanks to the work of Jordan, de Bary, and particularly to that of de Vries (de Vries, "Die Mutationstheorie", Leipzig, 1901, Vol. I. page 33.), that the Linnean species consists of a large or small number of entities, elementary species. In experimental investigation it is essential that observations be made on a pure species, or, as Johannsen (Johannsen, "Ueber Erblichkeit in Populationen und reinen Linien", Jena, 1903.) says, on a pure "line." What has long been recognised as necessary in the investigation of fungi, bacteria and algae must also be insisted on in the case of flowering plants; we must start with a single individual which is reproduced vegetatively or by strict self-fertilisation. In dioecious plants we must aim at the reproduction of brothers and sisters.

We may at the outset take it for granted that a pure species remains the same under similar external conditions; it varies as these vary. IT IS CHARACTERISTIC OF A SPECIES THAT IT ALWAYS EXHIBITS A CONSTANT RELATION TO A PARTICULAR ENVIRONMENT. In the case of two different species, e.g. the hay and anthrax bacilli or two varieties of Campanula with blue and white flowers respectively, a similar environment produces a constant difference. The cause of this is a mystery.

According to the modern standpoint, the living cell is a complex chemico-physical system which is regarded as a dynamical system of equilibrium, a conception suggested by Herbert Spencer and which has acquired a constantly increasing importance in the light of modern developments in physical chemistry. The various chemical compounds, proteids, carbohydrates, fats, the whole series of different ferments, etc. occur in the cell in a definite physical arrangement. The two systems of two species must as a matter of fact possess a constant difference, which it is necessary to define by a special term. We say, therefore, that the SPECIFIC STRUCTURE is different.

By way of illustrating this provisionally, we may assume that the proteids of the two species possess a constant chemical difference. This conception of specific structure is specially important in its bearing on a further treatment of the subject. In the original cell, eventually also in every cell of a plant, the characters which afterwards become apparent must exist somewhere; they are integral parts of the capabilities or potentialities of specific structure. Thus not only the characters which are exhibited under ordinary conditions in nature, but also many others which become apparent only under special conditions (In this connection I leave out of account, as before, the idea of material carriers of heredity which since the publication of Darwin's Pangenesis hypothesis has been frequently suggested. See my remarks in "Variationen der Bluten", "Pringsheim's Jahrb. Wiss. Bot." 1905, page 298; also Detto, "Biol. Centralbl." 1907, page 81, "Die Erklarbarkeit der Ontogenese durch materielle Anlagen".), are to be included as such potentialities in cells; the conception of specific structure includes the WHOLE OF THE POTENTIALITIES OF A SPECIES; specific structure comprises that which we must always assume without being able to explain it.

A relatively simple substance, such as oxalate of lime, is known under a great number of different crystalline forms belonging to different systems (Compare Kohl's work on "Anatomisch-phys. Untersuchungen uber Kalksalze", etc. Marburg, 1889.); these may occur as single crystals, concretions or as concentric sphaerites. The power to assume this variety of form is in some way inherent in the molecular structure, though we cannot, even in this case, explain the necessary connection between structure and crystalline form. These potentialities can only become operative under the influence of external conditions; their stimulation into activity depends on the degree of concentration of the various solutions, on the nature of the particular calcium salt, on the acid or alkaline reactions. Broadly speaking, the plant cell behaves in a similar way. The manifestation of each form, which is inherent as a potentiality in the specific structure, is ultimately to be referred to external conditions.

An insight into this connection is, however, rendered exceedingly difficult, often quite impossible, because the environment never directly calls into action the potentialities. Its influence is exerted on what we may call the inner world of the organism, the importance of which increases with the degree of differentiation. The production of form in every plant depends upon processes in the interior of the cells, and the nature of these determines which among the possible characters is to be brought to light. In no single case are we acquainted with the internal process responsible for the production of a particular form. All possible factors may play a part, such as osmotic pressure, permeability of the protoplasm, the degree of concentration of the various chemical substances, etc.; all these factors should be included in the category of INTERNAL CONDITIONS. This inner world appears the more hidden from our ken because it is always represented by a certain definite state, whether we are dealing with a single cell or with a small group of cells. These have been produced from pre-existing cells and they in turn from others; the problem is constantly pushed back through a succession of generations until it becomes identified with that of the origin of species.

A way, however, is opened for investigation; experience teaches us that this inner world is not a constant factor: on the contrary, it appears to be very variable. The dependence of VARIABLE INTERNAL on VARIABLE EXTERNAL conditions gives us the key with which research may open the door. In the lower plants this dependence is at once apparent, each cell is directly subject to external influences. In the higher plants with their different organs, these influences were transmitted to cells in course of development along exceedingly complex lines. In the case of the growing-point of a bud, which is capable of producing a complete plant, direct influences play a much less important part than those exerted through other organs, particularly through the roots and leaves, which are essential in nutrition. These correlations, as we may call them, are of the greatest importance as aids to an understanding of form-production. When a bud is produced on a particular part of a plant, it undergoes definite internal modifications induced by the influence of other organs, the activity of which is governed by the environment, and as the result of this it develops along a certain direction; it may, for example, become a flower. The particular direction of development is determined before the rudiment is differentiated and is exerted so strongly that further development ensues without interruption, even though the external conditions vary considerably and exert a positively inimical influence: this produces the impression that development proceeds entirely independently of the outer world. The widespread belief that such independence exists is very premature and at all events unproven.

The state of the young rudiment is the outcome of previous influences of the external world communicated through other organs. Experiments show that in certain cases, if the efficiency of roots and leaves as organs concerned with nutrition is interfered with, the production of flowers is affected, and their characters, which are normally very constant, undergo far-reaching modifications. To find the right moment at which to make the necessary alteration in the environment is indeed difficult and in many cases not yet possible. This is especially the case with fertilised eggs, which in a higher degree than buds have acquired, through parental influences, an apparently fixed internal organisation, and this seems to have pre-determined their development. It is, however, highly probable that it will be possible, by influencing the parents, to alter the internal organisation and to switch off development on to other lines.

Having made these general observations I will now cite a few of the many facts at our disposal, in order to illustrate the methods and aim of the experimental methods of research. As a matter of convenience I will deal separately with modification of development and with modification of single organs.

I. EFFECT OF ENVIRONMENT UPON THE COURSE OF DEVELOPMENT.

Every plant, whether an alga or a flowering plant passes, under natural conditions, through a series of developmental stages characteristic of each species, and these consist in a regular sequence of definite forms. It is impossible to form an opinion from mere observation and description as to what inner changes are essential for the production of the several forms. We must endeavour to influence the inner factors by known external conditions in such a way that the individual stages in development are separately controlled and the order of their sequence determined at will by experimental treatment. Such control over the course of development may be gained with special certainty in the case of the lower organisms.

With these it is practicable to control the principal conditions of cultivation and to vary them in various ways. By this means it has been demonstrated that each developmental stage depends upon special external conditions, and in cases where our knowledge is sufficient, a particular stage may be obtained at will. In the Green Algae (See Klebs, "Die Bedingung der Fortpflanzung... ", Jena, 1896; also "Jahrb. fur Wiss. Bot." 1898 and 1900; "Probleme der Entwickelung, III." "Biol. Centralbl." 1904, page 452.), as in the case of Fungi, we may classify the stages of development into purely vegetative growth (growth, cell-division, branching), asexual reproduction (formation of zoospores, conidia) and sexual processes (formation of male and female sexual organs). By modifying the external conditions it is possible to induce algae or fungi (Vaucheria, Saprolegnia) to grow continuously for several years or, in the course of a few days, to die after an enormous production of asexual or sexual cells. In some instances even an almost complete stoppage of growth may be caused, reproductive cells being scarcely formed before the organism is again compelled to resort to reproduction. Thus the sequence of the different stages in development can be modified as we may desire.

The result of a more thorough investigation of the determining conditions appears to produce at first sight a confused impression of all sorts of possibilities. Even closely allied species exhibit differences in regard to the connection between their development and external conditions. It is especially noteworthy that the same form in development may be produced as the result of very different alterations in the environment. At the same time we can undoubtedly detect a certain unity in the multiplicity of the individual phenomena.

If we compare the factors essential for the different stages in development, we see that the question always resolves itself into one of modification of similar conditions common to all life-processes. We should rather have inferred that there exist specific external stimuli for each developmental stage, for instance, certain chemical agencies. Experiments hitherto made support the conclusion that QUANTITATIVE alterations in the general conditions of life produce different types of development. An alga or a fungus grows so long as all the conditions of nutrition remain at a certain optimum for growth. In order to bring about asexual reproduction, e.g. the formation of zoospores, it is sometimes necessary to increase the degree of intensity of external factors; sometimes, on the other hand, these must be reduced in intensity. In the case of many algae a decrease in light-intensity or in the amount of salts in the culture solution, or in the temperature, induces asexual reproduction, while in others, on the contrary, an increase in regard to each of these factors is required to produce the same result. This holds good for the quantitative variations which induce sexual reproduction in algae. The controlling factor is found to be a reduction in the supply of nutritive salts and the exposure of the plants to prolonged illumination or, better still, an increase in the intensity of the light, the efficiency of illumination depending on the consequent formation of organic substances such as carbohydrates.

The quantitative alterations of external conditions may be spoken of as releasing stimuli. They produce, in the complex equilibrium of the cell, quantitative modifications in the arrangement and distribution of mass, by means of which other chemical processes are at once set in motion, and finally a new condition of equilibrium is attained. But the commonly expressed view that the environment can as a rule act only as a releasing agent is incorrect, because it overlooks an essential point. The power of a cell to receive stimuli is only acquired as the result of previous nutrition, which has produced a definite condition of concentration of different substances. Quantities are in this case the determining factors. The distribution of quantities is especially important in the sexual reproduction of algae, for which a vigorous production of the materials formed during carbon-assimilation appears to be essential.

In the Flowering plants, on the other hand, for reasons already mentioned, the whole problem is more complicated. Investigations on changes in the course of development of fertilised eggs have hitherto been unsuccessful; the difficulty of influencing egg-cells deeply immersed in tissue constitutes a serious obstacle. Other parts of plants are, however, convenient objects of experiment; e.g. the growing apices of buds which serve as cuttings for reproductive purposes, or buds on tubers, runners, rhizomes, etc. A growing apex consists of cells capable of division in which, as in egg-cells, a complete series of latent possibilities of development is embodied. Which of these possibilities becomes effective depends upon the action of the outer world transmitted by organs concerned with nutrition.

Of the different stages which a flowering plant passes through in the course of its development we will deal only with one in order to show that, in spite of its great complexity, the problem is, in essentials, equally open to attack in the higher plants and in the simplest organisms. The most important stage in the life of a flowering plant is the transition from purely vegetative growth to sexual reproduction—that is, the production of flowers. In certain cases it can be demonstrated that there is no internal cause, dependent simply on the specific structure, which compels a plant to produce its flowers after a definite period of vegetative growth. (Klebs, "Willkurliche Entwickelungsanderungen", Jena 1903; see also "Probleme der Entwickelung", I. II. "Centralbl." 1904.)

One extreme case, that of exceptionally early flowering, has been observed in nature and more often in cultivation. A number of plants under certain conditions are able to flower soon after germination. (Cf. numerous records of this kind by Diels, "Jugendformen und Bluten", Berlin, 1906.) This shortening of the period of development is exhibited in the most striking form in trees, as in the oak (Mobius, "Beitrage zur Lehre von der Fortpflanzung", Jena, 1897, page 89.), flowering seedlings of which have been observed from one to three years old, whereas normally the tree does not flower until it is sixty or eighty years old.

Another extreme case is represented by prolonged vegetative growth leading to the complete suppression of flower-production. This result may be obtained with several plants, such as Glechoma, the sugar beet, Digitalis, and others, if they are kept during the winter in a warm, damp atmosphere, and in rich soil; in the following spring or summer they fail to flower. (Klebs, "Willkurliche Aenderungen", etc. Jena, 1903, page 130.) Theoretically, however, experiments are of greater importance in which the production of flowers is inhibited by very favourable conditions of nutrition (Klebs, "Ueber kunstliche Metamorphosen", Stuttgart, 1906, page 115) ("Abh. Naturf. Ges. Halle", XXV.) occurring at the normal flowering period. Even in the case of plants of Sempervivum several years old, which, as is shown by control experiments on precisely similar plants, are on the point of flowering, flowering is rendered impossible if they are forced to very vigorous growth by an abundant supply of water and salts in the spring. Flowering, however, occurs, if such plants are cultivated in relatively dry sandy soil and in the presence of strong light. Careful researches into the conditions of growth have led, in the cases Sempervivum, to the following results: (1) With a strong light and vigorous carbon-assimilation a considerably increased supply of water and nutritive salts produces active vegetative growth. (2) With a vigorous carbon-assimilation in strong light, and a decrease in the supply of water and salts active flower-production is induced. (3) If an average supply of water and salts is given both processes are possible; the intensity of carbon-assimilation determines which of the two is manifested. A diminution in the production of organic substances, particularly of carbohydrates, induces vegetative growth. This can be effected by culture in feeble light or in light deprived of the yellow-red rays: on the other hand, flower-production follows an increase in light-intensity. These results are essentially in agreement with well-known observations on cultivated plants, according to which, the application of much moisture, after a plentiful supply of manure composed of inorganic salts, hinders the flower-production of many vegetables, while a decrease in the supply of water and salts favours flowering.

ii. INFLUENCE OF THE ENVIRONMENT ON THE FORM OF SINGLE ORGANS. (A considerable number of observations bearing on this question are given by Goebel in his "Experimentelle Morphologie der Pflanzen", Leipzig, 1908. It is not possible to deal here with the alteration in anatomical structure; cf. Kuster, "Pathologische Pflanzenanatomie", Jena, 1903.)

If we look closely into the development of a flowering plant, we notice that in a given species differently formed organs occur in definite positions. In a potato plant colourless runners are formed from the base of the main stem which grow underground and produce tubers at their tips: from a higher level foliage shoots arise nearer the apex. External appearances suggest that both the place of origin and the form of these organs were predetermined in the egg-cell or in the tuber. But it was shown experimentally by the well-known investigator Knight (Knight, "Selection from the Physiological and Horticultural Papers", London, 1841.) that tubers may be developed on the aerial stem in place of foliage shoots. These observations were considerably extended by Vochting. (Vochting, "Ueber die Bildung der Knollen", Cassel, 1887; see also "Bot. Zeit." 1902, 87.) In one kind of potato, germinating tubers were induced to form foliage shoots under the influence of a higher temperature; at a lower temperature they formed tuber-bearing shoots. Many other examples of the conversion of foliage-shoots into runners and rhizomes, or vice versa, have been described by Goebel and others. As in the asexual reproduction of algae quantitative alteration in the amount of moisture, light, temperature, etc. determines whether this or that form of shoot is produced. If the primordia of these organs are exposed to altered conditions of nutrition at a sufficiently early stage a complete substitution of one organ for another is effected. If the rudiment has reached a certain stage in development before it is exposed to these influences, extraordinary intermediate forms are obtained, bearing the characters of both organs.

The study of regeneration following injury is of greater importance as regards the problem of the development and place of origin of organs. (Reference may be made to the full summary of results given by Goebel in his "Experimentelle Morphologie", Leipzig and Berlin, 1908, Section IV.) Only in relatively very rare cases is there a complete re-formation of the injured organ itself, as e.g. in the growing-apex. Much more commonly injury leads to the development of complementary formations, it may be the rejuvenescence of a hitherto dormant rudiment, or it may be the formation of such ab initio. In all organs, stems, roots, leaves, as well as inflorescences, this kind of regeneration, which occurs in a great variety of ways according to the species, may be observed on detached pieces of the plant. Cases are also known, such, for example, as the leaves of many plants which readily form roots but not shoots, where a complete regeneration does not occur.

The widely spread power of reacting to wounding affords a very valuable means of inducing a fresh development of buds and roots on places where they do not occur in normal circumstances. Injury creates special conditions, but little is known as yet in regard to alterations directly produced in this way. Where the injury consists in the separation of an organ from its normal connections, the factors concerned are more comprehensible. A detached leaf, e.g., is at once cut off from a supply of water and salts, and is deprived of the means of getting rid of organic substances which it produces; the result is a considerable alteration in the degree of concentration. No experimental investigation on these lines has yet been made. Our ignorance has often led to the view that we are dealing with a force whose specific quality is the restitution of the parts lost by operation; the proof, therefore, that in certain cases a similar production of new roots or buds may be induced without previous injury and simply by a change in external conditions assumes an importance. (Klebs, "Willkurliche Entwickelung", page 100; also, "Probleme der Entwickelung", "Biol. Centralbl." 1904, page 610.)

A specially striking phenomenon of regeneration, exhibited also by uninjured plants, is afforded by polarity, which was discovered by Vochting. (See the classic work of Vochting, "Ueber Organbildung im Pflanzenreich", I. Bonn, 1888; also "Bot. Zeit." 1906, page 101; cf. Goebel, "Experimentelle Morphologie", Leipzig and Berlin, 1908, Section V, Polaritat.) It is found, for example, that roots are formed from the base of a detached piece of stem and shoots from the apex. Within the limits of this essay it is impossible to go into this difficult question; it is, however, important from the point of view of our general survey to emphasise the fact that the physiological distinctions between base and apex of pieces of stem are only of a quantitative kind, that is, they consist in the inhibition of certain phenomena or in favouring them. As a matter of fact roots may be produced from the apices of willows and cuttings of other plants; the distinction is thus obliterated under the influence of environment. The fixed polarity of cuttings from full grown stems cannot be destroyed; it is the expression of previous development. Vochting speaks of polarity as a fixed inherited character. This is an unconvincing conclusion, as nothing can be deduced from our present knowledge as to the causes which led up to polarity. We know that the fertilised egg, like the embryo, is fixed at one end by which it hangs freely in the embryo-sac and afterwards in the endosperm. From the first, therefore, the two ends have different natures, and these are revealed in the differentiation into root-apex and stem-apex. A definite direction in the flow of food-substances is correlated with this arrangement, and this eventually leads to a polarity in the tissues. This view requires experimental proof, which in the case of the egg-cells of flowering plants hardly appears possible; but it derives considerable support from the fact that in herbaceous plants, e.g. Sempervivum (Klebs, "Variationen der Bluten", "Jahrb. Wiss. Bot." 1905, page 260.), rosettes or flower-shoots are formed in response to external conditions at the base, in the middle, or at the apex of the stem, so that polarity as it occurs under normal conditions cannot be the result of unalterable hereditary factors. On the other hand, the lower plants should furnish decisive evidence on this question, and the experiments of Stahl, Winkler, Kniep, and others indicate the right method of attacking the problem.

The relation of leaf-form to environment has often been investigated and is well known. The leaves of bog and water plants (Cf.Goebel, loc. cit. chapter II.; also Gluck, "Untersuchungen uber Wasser- und Sumpfgewachse", Jena, Vols. I.-II. 1905-06.) afford the most striking examples of modifications: according as they are grown in water, moist or dry air, the form of the species characteristic of the particular habitat is produced, since the stems are also modified. To the same group of phenomena belongs the modification of the forms of leaves and stems in plants on transplantation from the plains to the mountains (Bonnier, "Recherches sur l'Anatomie experimentale des Vegetaux", Corbeil, 1895.) or vice versa. Such variations are by no means isolated examples. All plants exhibit a definite alteration in form as the result of prolonged cultivation in moist or dry air, in strong or feeble light, or in darkness, or in salt solutions of different composition and strength.

Every individual which is exposed to definite combinations of external factors exhibits eventually the same type of modification. This is the type of variation which Darwin termed "definite." It is easy to realise that indefinite or fluctuating variations belong essentially to the same class of phenomena; both are reactions to changes in environment. In the production of individual variations two different influences undoubtedly cooperate. One set of variations is caused by different external conditions, during the production, either of sexual cells or of vegetative primordia; another set is the result of varying external conditions during the development of the embryo into an adult plant. The two sets of influences cannot as yet be sharply differentiated. If, for purposes of vegetative reproduction, we select pieces of the same parent-plant of a pure species, the second type of variation predominates. Individual fluctuations depend essentially in such cases on small variations in environment during development.

These relations must be borne in mind if we wish to understand the results of statistical methods. Since the work of Quetelet, Galton, and others the statistical examination of individual differences in animals and plants has become a special science, which is primarily based on the consideration that the application of the theory of probability renders possible mathematical statement and control of the results. The facts show that any character, size of leaf, length of stem, the number of members in a flower, etc. do not vary haphazard but in a very regular manner. In most cases it is found that there is a value which occurs most commonly, the average or medium value, from which the larger and smaller deviations, the so-called plus and minus variations fall away in a continuous series and end in a limiting value. In the simpler cases a falling off occurs equally on both sides of the curve; the curve constructed from such data agrees very closely with the Gaussian curve of error. In more complicated cases irregular curves of different kinds are obtained which may be calculated on certain suppositions.

The regular fluctuations about a mean according to the rule of probability is often attributed to some law underlying variability. (de Vries, "Mutationstheorie", Vol. I. page 35, Leipzig, 1901.) But there is no such law which compels a plant to vary in a particular manner. Every experimental investigation shows, as we have already remarked, that the fluctuation of characters depends on fluctuation in the external factors. The applicability of the method of probability follows from the fact that the numerous individuals of a species are influenced by a limited number of variable conditions. (Klebs, "Willkurl. Ent." Jena, 1903, page 141.) As each of these conditions includes within certain limits all possible values and exhibits all possible combinations, it follows that, according to the rules of probability, there must be a mean value, about which the larger and smaller deviations are distributed. Any character will be found to have the mean value which corresponds with that combination of determining factors which occurs most frequently. Deviations towards plus and minus values will be correspondingly produced by rarer conditions.

A conclusion of fundamental importance may be drawn from this conception, which is, to a certain extent, supported by experimental investigation. (Klebs, "Studien uber Variation", "Arch. fur Entw." 1907.) There is no normal curve for a particular CHARACTER, there is only a curve for the varying combinations of conditions occurring in nature or under cultivation. Under other conditions entirely different curves may be obtained with other variants as a mean value. If, for example, under ordinary conditions the number 10 is the most frequent variant for the stamens of Sedum spectabile, in special circumstances (red light) this is replaced by the number 5. The more accurately we know the conditions for a particular form or number, and are able to reproduce it by experiment, the nearer we are to achieving our aim of rendering a particular variation impossible or of making it dominant.

In addition to the individual variations of a species, more pronounced fluctuations occur relatively rarely and sporadically which are spoken of as "single variations," or if specially striking as abnormalities or monstrosities. These forms have long attracted the attention of morphologists; a large number of observations of this kind are given in the handbooks of Masters (Masters, "Vegetable Teratology", London, 1869.) and Penzig (Penzig, "Pflanzen-Teratologie", Vols I. and II. Genua, 1890-94.) These variations, which used to be regarded as curiosities, have now assumed considerable importance in connection with the causes of form-development. They also possess special interest in relation to the question of heredity, a subject which does not at present concern us, as such deviations from normal development undoubtedly arise as individual variations induced by the influence of environment.

Abnormal developments of all kinds in stems, leaves, and flowers, may be produced by parasites, insects, or fungi. They may also be induced by injury, as Blaringhem (Blaringhem, "Mutation et traumatismes", Paris, 1907.) has more particularly demonstrated, which, by cutting away the leading shoots of branches in an early stage of development, caused fasciation, torsion, anomalous flowers, etc. The experiments of Blaringhem point to the probability that disturbances in the conditions of food-supply consequent on injury are the cause of the production of monstrosities. This is certainly the case in my experiments with species of Sempervivum (Klebs, "Kunstliche Metamorphosen", Stuttgart, 1906.); individuals, which at first formed normal flowers, produced a great variety of abnormalities as the result of changes in nutrition, we may call to mind the fact that the formation of inflorescences occurs normally when a vigorous production of organic compounds, such as starch, sugar, etc. follows a diminution in the supply of mineral salts. On the other hand, the development of inflorescences is entirely suppressed if, at a suitable moment before the actual foundations have been laid, water and mineral salts are supplied to the roots. If, during the week when the inflorescence has just been laid down and is growing very slowly, the supply of water and salts is increased, the internal conditions of the cells are essentially changed. At a later stage, after the elongation of the inflorescence, rosettes of leaves are produced instead of flowers, and structures intermediate between the two kinds of organs; a number of peculiar plant-forms are thus obtained (Cf. Lotsy, "Vorlesungen uber Deszendenztheorien", Vol. II. pl. 3, Jena, 1908.) Abnormalities in the greatest variety are produced in flowers by varying the time at which the stimulus is applied, and by the cooperation of other factors such as temperature, darkness, etc. In number and arrangement the several floral members vary within wide limits; sepals, petals, stamens, and carpels are altered in form and colour, a transformation of stamens to carpels and from carpels to stamens occurs in varying degrees. The majority of the deviations observed had not previously been seen either under natural conditions or in cultivation; they were first brought to light through the influence of external factors.

Such transformations of flowers become apparent at a time, which is separated by about two months from the period at which the particular cause began to act. There is, therefore, no close connection between the appearance of the modifications and the external conditions which prevail at the moment. When we are ignorant of the causes which are operative so long before the results are seen, we gain the impression that such variations as occur are spontaneous or autonomous expressions of the inner nature of the plant. It is much more likely that, as in Sempervivum, they were originally produced by an external stimulus which had previously reached the sexual cells or the young embryo. In any case abnormalities of this kind appear to be of a special type as compared with ordinary fluctuating variations. Darwin pointed out this difference; Bateson (Bateson, "Materials for the study of Variation", London, 1894, page 5.) has attempted to make the distinction sharper, at the same time emphasising its importance in heredity.

Bateson applies the term CONTINUOUS to small variations connected with one another by transitional stages, while those which are more striking and characterised from the first by a certain completeness, he names DISCONTINUOUS. He drew attention to a great difficulty which stands in the way of Lamarck's hypothesis, as also of Darwin's view. "According to both theories, specific diversity of form is consequent upon diversity of environment, and diversity of environment is thus the ultimate measure of diversity of specific form. Here then we meet the difficulty that diverse environments often shade into each other insensibly and form a continuous series, whereas the Specific Forms of life which are subject to them on the whole form a Discontinuous Series." This difficulty is, however, not of fundamental importance as well authenticated facts have been adduced showing that by alteration of the environment discontinuous variations, such as alterations in the number and form of members of a flower, may be produced. We can as yet no more explain how this happens than we can explain the existence of continuous variations. We can only assert that both kinds of variation arise in response to quantitative alterations in external conditions. The question as to which kind of variation is produced depends on the greater or less degree of alteration; it is correlated with the state of the particular cells at the moment.

In this short sketch it is only possible to deal superficially with a small part of the subject. It has been clearly shown that in view of the general dependence of development on the factors of the environment a number of problems are ready for experimental treatment. One must, however, not forget that the science of the physiology of form has not progressed beyond its initial stages. Just now our first duty is to demonstrate the dependence on external factors in as many forms of plants as possible, in order to obtain a more thorough control of all the different plant-forms. The problem is not only to produce at will (and independently of their normal mode of life) forms which occur in nature, but also to stimulate into operation potentialities which necessarily lie dormant under the conditions which prevail in nature. The constitution of a species is much richer in possibilities of development than would appear to be the case under normal conditions. It remains for man to stimulate into activity all the potentialities.

But the control of plant-form is only a preliminary step—the foundation stones on which to erect a coherent scientific structure. We must discover what are the internal processes in the cell produced by external factors, which as a necessary consequence result in the appearance of a definite form. We are here brought into contact with the most obscure problem of life. Progress can only be made pari passu with progress in physics and chemistry, and with the growth of our knowledge of nutrition, growth, etc.

Let us take one of the simplest cases—an alteration in form. A cylindrical cell of the alga Stigeoclonium assumes, as Livingstone (Livingstone, "On the nature of the stimulus which causes the change of form, etc." "Botanical Gazette", XXX. 1900; also XXXII. 1901.) has shown, a spherical form when the osmotic pressure of the culture fluid is increased; or a spore of Mucor, which, in a sugar solution grows into a branched filament, in the presence of a small quantity of acid (hydrogen ions) becomes a comparatively large sphere. (Ritter, "Ueber Kugelhefe, etc." "Ber. bot. Gesell." Berlin, XXV. page 255, 1907.) In both cases there has undoubtedly been an alteration in the osmotic pressure of the cell-sap, but this does not suffice to explain the alteration in form, since the unknown alterations, which are induced in the protoplasm, must in their turn influence the cell-membrane. In the case of the very much more complex alterations in form, such as we encounter in the course of development of plants, there do not appear to be any clues which lead us to a deeper insight into the phenomena. Nevertheless we continue the attempt, seeking with the help of any available hypothesis for points of attack, which may enable us to acquire a more complete mastery of physiological methods. To quote a single example; I may put the question, what internal changes produce a transition from vegetative growth to sexual reproduction?

The facts, which are as clearly established from the lower as for the higher plants, teach us that quantitative alteration in the environment produces such a transition. This suggests the conclusion that quantitative internal changes in the cells, and with them disturbances in the degree of concentration, are induced, through which the chemical reactions are led in the direction of sexual reproduction. An increase in the production of organic substances in the presence of light, chiefly of the carbohydrates, with a simultaneous decrease in the amount of inorganic salts and water, are the cause of the disturbance and at the same time of the alteration in the direction of development. Possibly indeed mineral salts as such are not in question, but only in the form of other organic combinations, particularly proteid material, so that we are concerned with an alteration in the relation of the carbohydrates and proteids. The difficulties of such researches are very great because the methods are not yet sufficiently exact to demonstrate the frequently small quantitative differences in chemical composition. Questions relating to the enzymes, which are of the greatest importance in all these life-processes, are especially complicated. In any case it is the necessary result of such an hypothesis that we must employ chemical methods of investigation in dealing with problems connected with the physiology of form.

II. INFLUENCE OF ENVIRONMENT ON THE TRANSFORMATION OF SPECIES.

The study of the physiology of form-development in a pure species has already yielded results and makes slow but sure progress. The physiology of the possibility of the transformation of one species into another is based, as yet, rather on pious hope than on accomplished fact. From the first it appeared to be hopeless to investigate physiologically the origin of Linnean species and at the same time that of the natural system, an aim which Darwin had before him in his enduring work. The historical sequence of events, of which an organism is the expression, can only be treated hypothetically with the help of facts supplied by comparative morphology, the history of development, geographical distribution, and palaeontology. (See Lotsy, "Vorlesungen" (Jena, I. 1906, II. 1908), for summary of the facts.) A glance at the controversy which is going on today in regard to different hypotheses shows that the same material may lead different investigators to form entirely different opinions. Our ultimate aim is to find a solution of the problem as to the cause of the origin of species. Indeed such attempts are now being made: they are justified by the fact that under cultivation new and permanent strains are produced; the fundamental importance of this was first grasped by Darwin. New points of view in regard to these lines of inquiry have been adopted by H. de Vries who has succeeded in obtaining from Oenothera Lamarckiana a number of constant "elementary" species. Even if it is demonstrated that he was simply dealing with the complex splitting up of a hybrid (Bateson, "Reports to the Evolution Committee of the Royal Society", London, 1902; cf. also Lotsy, "Vorlesungen", Vol. I. page 234.), the facts adduced in no sense lose their very great value.

We must look at the problem in its simplest form; we find it in every case where a new race differs essentially from the original type in a single character only; for example, in the colour of the flowers or in the petalody of the stamens (doubling of flowers). In this connection we must keep in view the fact that every visible character in a plant is the resultant of the cooperation of specific structure, with its various potentialities, and the influence of the environment. We know, that in a pure species all characters vary, that a blue-flowering Campanula or a red Sempervivum can be converted by experiment into white-flowering forms, that a transformation of stamens into petals may be caused by fungi or by the influence of changed conditions of nutrition, or that plants in dry and poor soil become dwarfed. But so far as the experiments justify a conclusion, it would appear that such alterations are not inherited by the offspring. Like all other variations they appear only so long as special conditions prevail in the surroundings.

It has been shown that the case is quite different as regards the white-flowering, double or dwarf races, because these retain their characters when cultivated under practically identical conditions, and side by side with the blue, single-flowering or tall races. The problem may therefore be stated thus: how can a character, which appears in the one case only under the strictly limited conditions of the experiment, in other cases become apparent under the very much wider conditions of ordinary cultivation? If a character appears, in these circumstances, in the case of all individuals, we then speak of constant races. In such simple cases the essential point is not the creation of a new character but rather an ALTERATION OF THIS CHARACTER IN ACCORDANCE WITH THE ENVIRONMENT. In the examples mentioned the modified character in the simple varieties (or a number of characters in elementary species) appears more or less suddenly and is constant in the above sense. The result is what de Vries has termed a Mutation. In this connection we must bear in mind the fact that no difference, recognisable externally, need exist between individual variation and mutation. Even the most minute quantitative difference between two plants may be of specific value if it is preserved under similar external conditions during many successive generations. We do not know how this happens. We may state the problem in other terms; by saying that the specific structure must be altered. It is possible, to some extent, to explain this sudden alteration, if we regard it as a chemical alteration of structure either in the specific qualities of the proteids or of the unknown carriers of life. In the case of many organic compounds their morphological characters (the physical condition, crystalline form, etc.) are at once changed by alteration of atomic relations or by incorporation of new radicals. (For instance ethylchloride (C_2H_5Cl) is a gas at 21 deg C., ethylenechloride ($C_2H_4Cl_2$) a fluid boiling at 84 deg C., beta trichlorethane ($C_2H_3Cl_3$) a fluid boiling at 113 deg C., perchlorethane (C_2Cl_6) a crystalline substance. Klebs, ("Willkurliche Entwickelungsanderungen" page 158.) Much more important, however, would be an answer to the question, whether an individual variation can be converted experimentally into an inherited character—a mutation in de Vries's sense.

In all circumstances we may recognise as a guiding principle the assumption adopted by Lamarck, Darwin, and many others, that the inheritance of any one character, or in more general terms, the transformation of one species into another, is, in the last instance, to be referred to a change in the environment. From a causal-mechanical point of view it is not a priori conceivable that one species can ever become changed into another so long as external conditions remain constant. The inner structure of a species must be essentially altered by external influences. Two methods of experimental research may be adopted, the effect of crossing distinct species and, secondly, the effect of definite factors of the environment.

The subject of hybridisation is dealt with in another part of this essay. It is enough to refer here to the most important fact, that as the result of combinations of characters of different species new and constant forms are produced. Further, Tschermack, Bateson and others have demonstrated the possibility that hitherto unknown inheritable characters may be produced by hybridisation.

The other method of producing constant races by the influence of special external conditions has often been employed. The sporeless races of Bacteria and Yeasts (Cf. Detto,

"Die Theorie der direkten Anpassung... ", pages 98 et seq., Jena, 1904; see also Lotsy, "Vorlesungen", II. pages 636 et seq., where other similar cases are described.) are well known, in which an internal alteration of the cells is induced by the influence of poison or higher temperature, so that the power of producing spores even under normal conditions appears to be lost. A similar state of things is found in some races which under certain definite conditions lose their colour or their virulence. Among the phanerogams the investigations of Schubler on cereals afford parallel cases, in which the influence of a northern climate produces individuals which ripen their seeds early; these seeds produce plants which seed early in southern countries. Analogous results were obtained by Cieslar in his experiments; seeds of conifers from the Alps when planted in the plains produced plants of slow growth and small diameter.

All these observations are of considerable interest theoretically; they show that the action of environment certainly induces such internal changes, and that these are transmitted to the next generation. But as regards the main question, whether constant races may be obtained by this means, the experiments cannot as yet supply a definite answer. In phanerogams, the influence very soon dies out in succeeding generations; in the case of bacteria, in which it is only a question of the loss of a character it is relatively easy for this to reappear. It is not impossible, that in all such cases there is a material hanging-on of certain internal conditions, in consequence of which the modification of the character persists for a time in the descendants, although the original external conditions are no longer present.

Thus a slow dying-out of the effect of a stimulus was seen in my experiments on Veronica chamaedrys. (Klebs, "Kunstliche Metamorphosen", Stuttgart, 1906, page 132.) During the cultivation of an artificially modified inflorescence I obtained a race showing modifications in different directions, among which twisting was especially conspicuous. This plant, however, does not behave as the twisted race of Dipsacus isolated by de Vries (de Vries, "Mutationstheorie", Vol. II. Leipzig, 1903, page 573.), which produced each year a definite percentage of twisted individuals. In the vegetative reproduction of this Veronica the torsion appeared in the first, also in the second and third year, but with diminishing intensity. In spite of good cultivation this character has apparently now disappeared; it disappeared still more quickly in seedlings. In another character of the same Veronica chamaedrys the influence of the environment was stronger. The transformation of the inflorescences to foliage-shoots formed the starting-point; it occurred only under narrowly defined conditions, namely on cultivation as a cutting in moist air and on removal of all other leaf-buds. In the majority (7/10) of the plants obtained from the transformed shoots, the modification appeared in the following year without any interference. Of the three plants which were under observation several years the first lost the character in a short time, while the two others still retain it, after vegetative propagation, in varying degrees. The same character occurs also in some of the seedlings; but anything approaching a constant race has not been produced.

Another means of producing new races has been attempted by Blaringhem. (Blaringhem, "Mutation et Traumatisme", Paris, 1907.) On removing at an early stage the main shoots of different plants he observed various abnormalities in the newly formed basal shoots. From the seeds of such plants he obtained races, a large percentage of which exhibited these abnormalities. Starting from a male Maize plant with a fasciated inflorescence, on which a proportion of the flowers had become male, a new race was bred in which hermaphrodite flowers were frequently produced. In the same way Blaringhem obtained, among other similar results, a race of barley with branched ears. These races, however, behaved in essentials like those which have been demonstrated by de Vries to be inconstant, e.g. Trifolium pratense quinquefolium and others. The abnormality appears in a proportion of the individuals and only under very special conditions. It must be remembered too that Blaringhem worked with old cultivated plants, which from the first had been disposed to split into a great variety of races. It is possible, but difficult to prove, that injury contributed to this result.

A third method has been adopted by MacDougal (MacDougal, "Heredity and Origin of species", "Monist", 1906; "Report of department of botanical research", "Fifth Year-book of the Carnegie Institution of Washington", page 119, 1907.) who injected strong (10 percent) sugar solution or weak solutions of calcium nitrate and zinc sulphate into young carpels of different plants. From the seeds of a plant of Raimannia odorata the carpels of which had been thus treated he obtained several plants distinguished from the parent-forms by the absence of hairs and by distinct forms of leaves. Further examination showed that he had here to do with a new elementary species. MacDougal also obtained a more or less distinct mutant of Oenothera biennis. We cannot as yet form an opinion as to how far the effect is due to the wound or to the injection of fluid as such, or to its chemical properties. This, however, is not so essential as to decide whether the mutant stands in any relation to the influence of external factors. It is at any rate very important that this kind of investigation should be carried further.

If it could be shown that new and inherited races were obtained by MacDougal's method, it would be safe to conclude that the same end might be gained by altering the conditions of the food-stuff conducted to the sexual cells. New races or elementary species, however, arise without wounding or injection. This at once raises the much discussed question, how far garden-cultivation has led to the creation of new races? Contrary to the opinion expressed by Darwin and others, de Vries ("Mutationstheorie", Vol. I. pages 412 et seq.) tried to show that garden-races have been produced only from spontaneous types which occur in a wild state or from sub-races, which the breeder has accidentally discovered but not originated. In a small number of cases only has de Vries adduced definite proof. On the other side we have the work of Korschinsky (Korschinsky, "Heterogenesis und Evolution", "Flora", 1901.) which shows that whole series of garden-races have made their appearance only after years of cultivation. In the majority of races we are entirely ignorant of their origin.

It is, however, a fact that if a plant is removed from natural conditions into cultivation, a well-marked variation occurs. The well-known plant-breeder L. de Vilmorin (L. de Vilmorin, "Notices sur l'amelioration des plantes", Paris, 1886, page 36.), speaking from his own experience, states that a plant is induced to "affoler," that is to exhibit all possible variations from which the breeder may make a further selection only after cultivation for several generations. The effect of cultivation was particularly striking in Veronica chamaedrys (Klebs, "Kunstliche Metamorphosen", Stuttgart, 1906, page 152.) which, in spite of its wide distribution in nature, varies very little. After a few years of cultivation this "good" and constant species becomes highly variable. The specimens on which the experiments were made were three modified inflorescence cuttings, the parent-plants of which certainly exhibited no striking abnormalities. In a short time many hitherto latent potentialities became apparent, so that characters, never previously observed, or at least very rarely, were exhibited, such as scattered leaf-arrangement, torsion, terminal or branched inflorescences, the conversion of the inflorescence into foliage-shoots, every conceivable alteration in the colour of flowers, the assumption of a green colour by parts of the flowers, the proliferation of flowers.

All this points to some disturbance in the species resulting from methods of cultivation. It has, however, not yet been possible to produce constant races with any one of these modified characters. But variations appeared among the seedlings, some of which, e.g. yellow variegation, were not inheritable, while others have proved constant. This holds good, so far as we know at present, for a small rose-coloured form which is to be reckoned as a mutation. Thus the prospect of producing new races by cultivation appears to be full of promise.

So long as the view is held that good nourishment, i.e. a plentiful supply of water and salts, constitutes the essential characteristic of garden-cultivation, we can hardly conceive that new mutations can be thus produced. But perhaps the view here put forward in regard to the production of form throws new light on this puzzling problem.

Good manuring is in the highest degree favourable to vegetative growth, but is in no way equally favourable to the formation of flowers. The constantly repeated expression, good or favourable nourishment, is not only vague but misleading, because circumstances favourable to growth differ from those which promote reproduction; for the production of every form there are certain favourable conditions of nourishment, which may be defined for each species. Experience shows that, within definite and often very wide limits, it does not depend upon the ABSOLUTE AMOUNT of the various food substances, but upon their respective degrees of concentration. As we have already stated, the production of flowers follows a relative increase in the amount of carbohydrates formed in the presence of light, as compared with the inorganic salts on which the formation of albuminous substances depends. (Klebs, "Kunstliche Metamorphosen", page 117.) The various modifications of flowers are due to the fact that a relatively too strong solution of salts is supplied to the rudiments of these organs. As a general rule every plant form depends upon a certain relation between the different chemical substances in the cells and is modified by an alteration of that relation.

During long cultivation under conditions which vary in very different degrees, such as moisture, the amount of salts, light intensity, temperature, oxygen, it is possible that sudden and special disturbances in the relations of the cell substances have a directive influence on the inner organisation of the sexual cells, so that not only inconstant but also constant varieties will be formed.

Definite proof in support of this view has not yet been furnished, and we must admit that the question as to the cause of heredity remains, fundamentally, as far from solution as it was in Darwin's time. As the result of the work of many investigators, particularly de Vries, the problem is constantly becoming clearer and more definite. The penetration into this most difficult and therefore most interesting problem of life and the creation by experiment of new races or elementary species are no longer beyond the region of possibility.

XIV. EXPERIMENTAL STUDY OF THE INFLUENCE OF ENVIRONMENT ON ANIMALS. By Jacques Loeb, M.D. Professor of Physiology in the University of California.

I. INTRODUCTORY REMARKS.

What the biologist calls the natural environment of an animal is from a physical point of view a rather rigid combination of definite forces. It is obvious that by a purposeful and systematic variation of these and by the application of other forces in the laboratory, results must be obtainable which do not appear in the natural environment. This is the reasoning underlying the modern development of the study of the effects of environment upon animal life. It was perhaps not the least important of Darwin's services to science that the boldness of his conceptions gave to the experimental biologist courage to enter upon the attempt of controlling at will the life-phenomena of animals, and of bringing about effects which cannot be expected in Nature.

The systematic physico-chemical analysis of the effect of outside forces upon the form and reactions of animals is also our only means of unravelling the mechanism of heredity beyond the scope of the Mendelian law. The manner in which a germ-cell can force upon the adult certain characters will not be understood until we succeed in varying and controlling hereditary characteristics; and this can only be accomplished on the basis of a systematic study of the effects of chemical and physical forces upon living matter.

Owing to limitation of space this sketch is necessarily very incomplete, and it must not be inferred that studies which are not mentioned here were considered to be of minor importance. All the writer could hope to do was to bring together a few instances of the experimental analysis of the effect of environment, which indicate the nature and extent of our control over life-phenomena and which also have some relation to the work of Darwin. In the selection of these instances preference is given to those problems which are not too technical for the general reader.

The forces, the influence of which we shall discuss, are in succession chemical agencies, temperature, light, and gravitation. We shall also treat separately the effect of these forces upon form and instinctive reactions.

II. THE EFFECTS OF CHEMICAL AGENCIES.

(a) HETEROGENEOUS HYBRIDISATION.

It was held until recently that hybridisation is not possible except between closely related species and that even among these a successful hybridisation cannot always be counted upon. This view was well supported by experience. It is, for instance, well known that the majority of marine animals lay their unfertilised eggs in the ocean and that the males shed their sperm also into the sea-water. The numerical excess of the spermatozoa over the ova in the sea-water is the only guarantee that the eggs are fertilised, for the spermatozoa are carried to the eggs by chance and are not attracted by the latter. This statement is the result of numerous experiments by various authors, and is contrary to common belief. As a rule all or the majority of individuals of a species in a given region spawn on the same day, and when this occurs the sea-water constitutes a veritable suspension of sperm. It has been shown by experiment that in fresh sea-water the sperm may live and retain its fertilising power for several days. It is thus unavoidable that at certain periods more than one kind of spermatozoon is suspended in the sea-water and it is a matter of surprise that the most heterogeneous hybridisations do not constantly occur. The reason for this becomes obvious if we bring together mature eggs and equally mature and active sperm of a different family. When this is done no egg is, as a rule, fertilised. The eggs of a sea-urchin can be fertilised by sperm of their own species, or, though in smaller numbers, by the sperm of other species of sea-urchins, but not by the sperm of other groups of echinoderms, e.g. starfish, brittle-stars, holothurians or crinoids, and still less by the sperm of more distant groups of animals. The consensus of opinion seemed to be that the spermatozoon must enter the egg through a narrow opening or canal, the so-called micropyle, and that the micropyle allowed only the spermatozoa of the same or of a closely related species to enter the egg.

It seemed to the writer that the cause of this limitation of hybridisation might be of another kind and that by a change in the constitution of the sea-water it might be possible to bring about heterogenous hybridisations, which in normal sea-water are impossible. This assumption proved correct. Sea-water has a faintly alkaline reaction (in terms of the physical chemist its concentration of hydroxyl ions is about (10 to the power minus six)N at Pacific Grove, California, and about (10 to the power minus 5)N at Woods Hole, Massachusetts). If we slightly raise the alkalinity of the sea-water by adding to it a small but definite quantity of sodium hydroxide or some other alkali, the eggs of the sea-urchin can be fertilised with the sperm of widely different groups of animals, possibly with the sperm of any marine animal which sheds it into the ocean. In 1903 it was shown that if we add from about 0.5 to 0.8 cubic centimetre N/10 sodium hydroxide to 50 cubic centimetres of sea-water, the eggs of Strongylocentrotus purpuratus (a sea-urchin which is found on the coast of California) can be fertilised in large quantities by the sperm of various kinds of starfish, brittle-stars and holothurians; while in normal sea-water or with less sodium hydroxide not a single egg of the same female could be fertilised with the starfish sperm which proved effective in the hyper-alkaline sea-water. The sperm of the various forms of starfish was not equally effective for these hybridisations; the sperm of Asterias ochracea and A. capitata gave the best results, since it was possible to fertilise 50 per cent or more of the sea-urchin eggs, while the sperm of Pycnopodia and Asterina fertilised only 2 per cent of the same eggs.

157

Godlewski used the same method for the hybridisation of the sea-urchin eggs with the sperm of a crinoid (Antedon rosacea). Kupelwieser afterwards obtained results which seemed to indicate the possibility of fertilising the eggs of Strongylocentrotus with the sperm of a mollusc (Mytilus.) Recently, the writer succeeded in fertilising the eggs of Strongylocentrotus franciscanus with the sperm of a mollusc—Chlorostoma. This result could only be obtained in sea-water the alkalinity of which had been increased (through the addition of 0.8 cubic centimetre N/10 sodium hydroxide to 50 cubic centimetres of sea-water). We thus see that by increasing the alkalinity of the sea-water it is possible to effect heterogeneous hybridisations which are at present impossible in the natural environment of these animals.

It is, however, conceivable that in former periods of the earth's history such heterogeneous hybridisations were possible. It is known that in solutions like sea-water the degree of alkalinity must increase when the amount of carbon-dioxide in the atmosphere is diminished. If it be true, as Arrhenius assumes, that the Ice age was caused or preceded by a diminution in the amount of carbon-dioxide in the air, such a diminution must also have resulted in an increase of the alkalinity of the sea-water, and one result of such an increase must have been to render possible heterogeneous hybridisations in the ocean which in the present state of alkalinity are practically excluded.

But granted that such hybridisations were possible, would they have influenced the character of the fauna? In other words, are the hybrids between sea-urchin and starfish, or better still, between sea-urchin and mollusc, capable of development, and if so, what is their character? The first experiment made it appear doubtful whether these heterogeneous hybrids could live. The sea-urchin eggs which were fertilised in the laboratory by the spermatozoa of the starfish, as a rule, died earlier than those of the pure breeds. But more recent results indicate that this was due merely to deficiencies in the technique of the earlier experiments. The writer has recently obtained hybrid larvae between the sea-urchin egg and the sperm of a mollusc (Chlorostoma) which, in the laboratory, developed as well and lived as long as the pure breeds of the sea-urchin, and there was nothing to indicate any difference in the vitality of the two breeds.

So far as the question of heredity is concerned, all the experiments on heterogeneous hybridisation of the egg of the sea-urchin with the sperm of starfish, brittle-stars, crinoids and molluscs, have led to the same result, namely, that the larvae have purely maternal characteristics and differ in no way from the pure breed of the form from which the egg is taken. By way of illustration it may be said that the larvae of the sea-urchin reach on the third day or earlier (according to species and temperature) the so-called pluteus stage, in which they possess a typical skeleton; while neither the larvae of the starfish nor those of the mollusc form a skeleton at the corresponding stage. It was, therefore, a matter of some interest to find out whether or not the larvae produced by the fertilisation of the sea-urchin egg with the sperm of starfish or mollusc would form the normal and typical pluteus skeleton. This was invariably the case in the experiments of Godlewski, Kupelwieser, Hagedoorn, and the writer. These hybrid larvae were exclusively maternal in character.

It might be argued that in the case of heterogeneous hybridisation the sperm-nucleus does not fuse with the egg-nucleus, and that, therefore, the spermatozoon cannot transmit its hereditary substances to the larvae. But these objections are refuted by Godlewski's experiments, in which he showed definitely that if the egg of the sea-urchin is fertilised with the sperm of a crinoid the fusion of the egg-nucleus and sperm-nucleus takes place in the normal way. It remains for further experiments to decide what the character of the adult hybrids would be.

(b). ARTIFICIAL PARTHENOGENESIS.

Possibly in no other field of Biology has our ability to control life-phenomena by outside conditions been proved to such an extent as in the domain of fertilisation. The reader knows that the eggs of the overwhelming majority of animals cannot develop unless a spermatozoon enters them. In this case a living agency is the cause of development and the problem arises whether it is possible to accomplish the same result through the application

158

of well-known physico-chemical agencies. This is, indeed, true, and during the last ten years living larvae have been produced by chemical agencies from the unfertilised eggs of sea-urchins, starfish, holothurians and a number of annelids and molluscs; in fact this holds true in regard to the eggs of practically all forms of animals with which such experiments have been tried long enough. In each form the method of procedure is somewhat different and a long series of experiments is often required before the successful method is found.

The facts of Artificial Parthenogenesis, as the chemical fertilisation of the egg is called, have, perhaps, some bearing on the problem of evolution. If we wish to form a mental image of the process of evolution we have to reckon with the possibility that parthenogenetic propagation may have preceded sexual reproduction. This suggests also the possibility that at that period outside forces may have supplied the conditions for the development of the egg which at present the spermatozoon has to supply. For this, if for no other reason, a brief consideration of the means of artificial parthenogenesis may be of interest to the student of evolution.

It seemed necessary in these experiments to imitate as completely as possible by chemical agencies the effects of the spermatozoon upon the egg. When a spermatozoon enters the egg of a sea-urchin or certain starfish or annelids, the immediate effect is a characteristic change of the surface of the egg, namely the formation of the so-called membrane of fertilisation. The writer found that we can produce this membrane in the unfertilised egg by certain acids, especially the monobasic acids of the fatty series, e.g. formic, acetic, propionic, butyric, etc. Carbon-dioxide is also very efficient in this direction. It was also found that the higher acids are more efficient than the lower ones, and it is possible that the spermatozoon induces membrane-formation by carrying into the egg a higher fatty acid, namely oleic acid or one of its salts or esters.

The physico-chemical process which underlies the formation of the membrane seems to be the cause of the development of the egg. In all cases in which the unfertilised egg has been treated in such a way as to cause it to form a membrane it begins to develop. For the eggs of certain animals membrane-formation is all that is required to induce a complete development of the unfertilised egg, e.g. in the starfish and certain annelids. For the eggs of other animals a second treatment is necessary, presumably to overcome some of the injurious effects of acid treatment. Thus the unfertilised eggs of the sea-urchin Strongylocentrotus purpuratus of the Californian coast begin to develop when membrane-formation has been induced by treatment with a fatty acid, e.g. butyric acid; but the development soon ceases and the eggs perish in the early stages of segmentation, or after the first nuclear division. But if we treat the same eggs, after membrane-formation, for from 35 to 55 minutes (at 15 deg C.) with sea-water the concentration (osmotic pressure) of which has been raised through the addition of a definite amount of some salt or sugar, the eggs will segment and develop normally, when transferred back to normal sea-water. If care is taken, practically all the eggs can be caused to develop into plutei, the majority of which may be perfectly normal and may live as long as larvae produced from eggs fertilised with sperm.

It is obvious that the sea-urchin egg is injured in the process of membrane-formation and that the subsequent treatment with a hypertonic solution only acts as a remedy. The nature of this injury became clear when it was discovered that all the agencies which cause haemolysis, i.e. the destruction of the red blood corpuscles, also cause membrane-formation in unfertilised eggs, e.g. fatty acids or ether, alcohols or chloroform, etc., or saponin, solanin, digitalin, bile salts and alkali. It thus happens that the phenomena of artificial parthenogenesis are linked together with the phenomena of haemolysis which at present play so important a role in the study of immunity. The difference between cytolysis (or haemolysis) and fertilisation seems to be this, that the latter is caused by a superficial or slight cytolysis of the egg, while if the cytolytic agencies have time to act on the whole egg the latter is completely destroyed. If we put unfertilised eggs of a sea-urchin into sea-water which contains a trace of saponin we notice that, after a few minutes, all the eggs form the typical membrane of fertilisation. If the eggs are then taken out of the saponin solution, freed from all traces of saponin by repeated washing in normal sea-water, and transferred to the hypertonic sea-water for from 35 to 55 minutes, they develop into larvae. If, however, they are left in the sea-water containing the saponin they undergo, a few minutes after membrane-formation, the disintegration known in pathology as CYTOLYSIS. Membrane-formation is, therefore, caused by a superficial or incomplete cytolysis. The writer believes that the subsequent treatment of the egg with hypertonic sea-water is needed only to overcome the destructive effects of this partial cytolysis. The full reasons for this belief cannot be given in a short essay.

Many pathologists assume that haemolysis or cytolysis is due to a liquefaction of certain fatty or fat-like compounds, the so-called lipoids, in the cell. If this view is correct, it would be necessary to ascribe the fertilisation of the egg to the same process.

The analogy between haemolysis and fertilisation throws, possibly, some light on a curious observation. It is well known that the blood corpuscles, as a rule, undergo cytolysis if injected into the blood of an animal which belongs to a different family. The writer found last year that the blood of mammals, e.g. the rabbit, pig, and cattle, causes the egg of Strongylocentrotus to form a typical fertilisation-membrane. If such eggs are afterwards treated for a short period with hypertonic sea-water they develop into normal larvae (plutei). Some substance contained in the blood causes, presumably, a superficial cytolysis of the egg and thus starts its development.

We can also cause the development of the sea-urchin egg without membrane-formation. The early experiments of the writer were done in this way and many experimenters still use such methods. It is probable that in this case the mechanism of fertilisation is essentially the same as in the case where the membrane-formation is brought about, with this difference only, that the cytolytic effect is less when no fertilisation-membrane is formed. This inference is corroborated by observations on the fertilisation of the sea-urchin egg with ox blood. It very frequently happens that not all of the eggs form membranes in this process. Those eggs which form membranes begin to develop, but perish if they are not treated with hypertonic sea-water. Some of the other eggs, however, which do not form membranes, develop directly into normal larvae without any treatment with hypertonic sea-water, provided they are exposed to the blood for only a few minutes. Presumably some blood enters the eggs and causes the cytolytic effects in a less degree than is necessary for membrane-formation, but in a sufficient degree to cause their development. The slightness of the cytolytic effect allows the egg to develop without treatment with hypertonic sea-water.

Since the entrance of the spermatozoon causes that degree of cytolysis which leads to membrane-formation, it is probable that, in addition to the cytolytic or membrane-forming substance (presumably a higher fatty acid), it carries another substance into the egg which counteracts the deleterious cytolytic effects underlying membrane-formation.

The question may be raised whether the larvae produced by artificial parthenogenesis can reach the mature stage. This question may be answered in the affirmative, since Delage has succeeded in raising several parthenogenetic sea-urchin larvae beyond the metamorphosis into the adult stage and since in all the experiments made by the writer the parthenogenetic plutei lived as long as the plutei produced from fertilised eggs.

(c). ON THE PRODUCTION OF TWINS FROM ONE EGG THROUGH A CHANGE IN THE CHEMICAL CONSTITUTION OF THE SEA-WATER.

The reader is probably familiar with the fact that there exist two different types of human twins. In the one type the twins differ as much as two children of the same parents born at different periods; they may or may not have the same sex. In the second type the twins have invariably the same sex and resemble each other most closely. Twins of the latter type are produced from the same egg, while twins of the former type are produced from two different eggs.

The experiments of Driesch and others have taught us that twins originate from one egg in this manner, namely, that the first two cells into which the egg divides after fertilisation become separated from each other. This separation can be brought about by a change in the chemical constitution of the sea-water. Herbst observed that if the fertilised eggs of the sea-urchin are put into sea-water which is freed from calcium, the cells into which the egg divides have a tendency to fall apart. Driesch afterwards noticed that eggs of the sea-urchin treated with sea-water which is free from lime have a tendency to give rise to twins. The writer has recently found that twins can be produced not only by the absence of lime, but also through the absence of sodium or of potassium; in other words, through the absence of one or two of the three important metals in the sea-water. There is, however, a second condition, namely, that the solution used for the production of twins must have a neutral or at least not an alkaline reaction.

The procedure for the production of twins in the sea-urchin egg consists simply in this:—the eggs are fertilised as usual in normal sea-water and then, after repeated washing in a neutral solution of sodium chloride (of the concentration of the sea-water), are placed in a neutral mixture of potassium chloride and calcium chloride, or of sodium chloride and potassium chloride, or of sodium chloride and calcium chloride, or of sodium chloride and magnesium chloride. The eggs must remain in this solution until half an hour or an hour after they have reached the two-cell stage. They are then transferred into normal sea-water and allowed to develop. From 50 to 90 per cent of the eggs of Strongylocentrotus purpuratus treated in this manner may develop into twins. These twins may remain separate or grow partially together and form double monsters, or heal together so completely that only slight or even no imperfections indicate that the individual started its career as a pair of twins. It is also possible to control the tendency of such twins to grow together by a change in the constitution of the sea-water. If we use as a twin-producing solution a mixture of sodium, magnesium and potassium chlorides (in the proportion in which these salts exist in the sea-water) the tendency of the twins to grow together is much more pronounced than if we use simply a mixture of sodium chloride and magnesium chloride.

The mechanism of the origin of twins, as the result of altering the composition of the sea-water, is revealed by observation of the first segmentation of the egg in these solutions. This cell-division is modified in a way which leads to a separation of the first two cells. If the egg is afterwards transferred back into normal sea-water, each of these two cells develops into an independent embryo. Since normal sea-water contains all three metals, sodium, calcium, and potassium, and since it has besides an alkaline reaction, we perceive the reason why twins are not normally produced from one egg. These experiments suggest the possibility of a chemical cause for the origin of twins from one egg or of double monstrosities in mammals. If, for some reason, the liquids which surround the human egg a short time before and after the first cell-division are slightly acid, and at the same time lacking in one of the three important metals, the conditions for the separation of the first two cells and the formation of identical twins are provided.

In conclusion it may be pointed out that the reverse result, namely, the fusion of normally double organs, can also be brought about experimentally through a change in the chemical constitution of the sea-water. Stockard succeeded in causing the eyes of fish embryos (Fundulus heteroclitus) to fuse into a single cyclopean eye through the addition of magnesium chloride to the sea-water. When he added about 6 grams of magnesium chloride to 100 cubic centimetres of sea-water and placed the fertilised eggs in the mixture, about 50 per cent of the eggs gave rise to one-eyed embryos. "When the embryos were studied the one-eyed condition was found to result from the union or fusion of the 'anlagen' of the two eyes. Cases were observed which showed various degrees in this fusion; it appeared as though the optic vessels were formed too far forward and ventral, so that their antero-ventro-median surfaces fused. This produces one large optic cup, which in all cases gives more or less evidence of its double nature." (Stockard, "Archiv f. Entwickelungsmechanik", Vol. 23, page 249, 1907.)

We have confined ourselves to a discussion of rather simple effects of the change in the constitution of the sea-water upon development. It is a priori obvious, however, that an unlimited number of pathological variations might be produced by a variation in the concentration and constitution of the sea-water, and experience confirms this statement. As an example we may mention the abnormalities observed by Herbst in the development of sea-urchins through the addition of lithium to sea-water. It is, however, as yet impossible to connect in a rational way the effects produced in this and similar cases with the cause which produced them; and it is also impossible to define in a simple way the character of the change produced.

III. THE INFLUENCE OF TEMPERATURE.

(a) THE INFLUENCE OF TEMPERATURE UPON THE DENSITY OF PELAGIC ORGANISMS AND THE DURATION OF LIFE.

It has often been noticed by explorers who have had a chance to compare the faunas in different climates that in polar seas such species as thrive at all in those regions occur, as a rule, in much greater density than they do in the moderate or warmer regions of the ocean. This refers to those members of the fauna which live at or near the surface, since they alone lend themselves to a statistical comparison. In his account of the Valdivia expedition, Chun (Chun, "Aus den Tiefen des Weltmeeres", page 225, Jena, 1903.) calls especial attention to this quantitative difference in the surface fauna and flora of different regions. "In the icy water of the Antarctic, the temperature of which is below 0 deg C., we find an astonishingly rich animal and plant life. The same condition with which we are familiar in the Arctic seas is repeated here, namely, that the quantity of plankton material exceeds that of the temperate and warm seas." And again, in regard to the pelagic fauna in the region of the Kerguelen Islands, he states: "The ocean is alive with transparent jelly fish, Ctenophores (Bolina and Callianira) and of Siphonophore colonies of the genus Agalma."

The paradoxical character of this general observation lies in the fact that a low temperature retards development, and hence should be expected to have the opposite effect from that mentioned by Chun. Recent investigations have led to the result that life-phenomena are affected by temperature in the same sense as the velocity of chemical reactions. In the case of the latter van't Hoff had shown that a decrease in temperature by 10 degrees reduces their velocity to one half or less, and the same has been found for the influence of temperature on the velocity of physiological processes. Thus Snyder and T.B. Robertson found that the rate of heartbeat in the tortoise and in Daphnia is reduced to about one-half if the temperature is lowered 10 deg C., and Maxwell, Keith Lucas, and Snyder found the same influence of temperature for the rate with which an impulse travels in the nerve. Peter observed that the rate of development in a sea-urchin's egg is reduced to less than one-half if the temperature (within certain limits) is reduced by 10 degrees. The same effect of temperature upon the rate of development holds for the egg of the frog, as Cohen and Peter calculated from the experiments of O. Hertwig. The writer found the same temperature-coefficient for the rate of maturation of the egg of a mollusc (Lottia).

All these facts prove that the velocity of development of animal life in Arctic regions, where the temperature is near the freezing point of water, must be from two to three times smaller than in regions where the temperature of the ocean is about 10 deg C. and from four to nine times smaller than in seas the temperature of which is about 20 deg C. It is, therefore, exactly the reverse of what we should expect when authors state that the density of organisms at or near the surface of the ocean in polar regions is greater than in more temperate regions.

The writer believes that this paradox finds its explanation in experiments which he has recently made on the influence of temperature on the duration of life of cold-blooded marine animals. The experiments were made on the fertilised and unfertilised eggs of the sea-urchin, and yielded the result that for the lowering of temperature by 1 deg C. the duration of life was about doubled. Lowering the temperature by 10 degrees therefore prolongs the life of the organism 2 to the power 10, i.e. over a thousand times, and a lowering by 20 degrees prolongs it about one million times. Since this prolongation of life is far in excess of the retardation of development through a lowering of temperature, it is obvious that, in spite of the retardation of development in Arctic seas, animal life must be denser there than in temperate or tropical seas. The excessive increase of the duration of life at the poles will necessitate the simultaneous existence of more successive generations of the same species in these regions than in the temperate or tropical regions.

The writer is inclined to believe that these results have some bearing upon a problem which plays an important role in theories of evolution, namely, the cause of natural death. It has been stated that the processes of differentiation and development lead also to the natural death of the individual. If we express this in chemical terms it means that the chemical processes which underlie development also determine natural death. Physical chemistry has taught us to identify two chemical processes even if only certain of their features are known. One of these means of identification is the temperature coefficient. When two chemical processes are identical, their velocity must be reduced by the same amount if the temperature is lowered to the same extent. The temperature coefficient for the duration of life of cold-blooded organisms seems, however, to differ enormously from the temperature coefficient for their rate of development. For a difference in temperature of 10 deg C. the duration of life is altered five hundred times as much as the rate of development; and, for a change of 20 deg C., it is altered more than a hundred thousand times as much. From this we may conclude that, at least for the sea-urchin eggs and embryo, the chemical processes which determine natural death are certainly not identical with the processes which underlie their development. T.B. Robertson has also arrived at the conclusion, for quite different reasons, that the process of senile decay is essentially different from that of growth and development.

(b) CHANGES IN THE COLOUR OF BUTTERFLIES PRODUCED THROUGH THE INFLUENCE OF TEMPERATURE.

The experiments of Dorfmeister, Weismann, Merrifield, Standfuss, and Fischer, on seasonal dimorphism and the aberration of colour in butterflies have so often been discussed in biological literature that a short reference to them will suffice. By seasonal dimorphism is meant the fact that species may appear at different seasons of the year in a somewhat different form or colour. Vanessa prorsa is the summer form, Vanessa levana the winter form of the same species. By keeping the pupae of Vanessa prorsa several weeks at a temperature of from 0 deg to 1 deg Weismann succeeded in obtaining from the summer chrysalids specimens which resembled the winter variety, Vanessa levana.

If we wish to get a clear understanding of the causes of variation in the colour and pattern of butterflies, we must direct our attention to the experiments of Fischer, who worked with more extreme temperatures than his predecessors, and found that almost identical aberrations of colour could be produced by both extremely high and extremely low temperatures. This can be clearly seen from the following tabulated results of his observations. At the head of each column the reader finds the temperature to which Fischer submitted the pupae, and in the vertical column below are found the varieties that were produced. In the vertical column A are given the normal forms:

(Temperatures in deg C.)

0 to -20	0 to +10	A. (Normal forms)	+35 to +37	+36 to +41	+42 to +46
ichnusoides	*polaris*	*urticae*	*ichnusa*	*polaris*	*ichnusoides*
(nigrita)			*(nigrita)*	*antigone*	*fischeri io*
-	*fischeri*	*antigone*	*(iokaste)*		*(iokaste)*
testudo	*dixeyi*	*polychloros erythromelas*	*dixeyi*	*testudo hygiaea*	*artemis*
antiopa	*epione*	*artemis*	*hygiaea*	*elymi wiskotti*	*cardui* -
wiskotti	*elymi*	*klymene*	*merrifieldi atalanta*	-	*merrifieldi klymene*
weismanni	*porima*	*prorsa*	-	*porima*	*weismanni*

The reader will notice that the aberrations produced at a very low temperature (from 0 to -20 deg C.) are absolutely identical with the aberrations produced by exposing the pupae to extremely high temperatures (42 to 46 deg C.). Moreover the aberrations produced by a moderately low temperature (from 0 to 10 deg C.) are identical with the aberrations produced by a moderately high temperature (36 to 41 deg C.)

From these observations Fischer concludes that it is erroneous to speak of a specific effect of high and of low temperatures, but that there must be a common cause for the aberration found at the high as well as at the low temperature limits. This cause he seems to find in the inhibiting effects of extreme temperatures upon development.

If we try to analyse such results as Fischer's from a physico-chemical point of view, we must realise that what we call life consists of a series of chemical reactions, which are connected in a catenary way; inasmuch as one reaction or group of reactions (a) (e.g. hydrolyses) causes or furnishes the material for a second reaction or group of reactions (b) (e.g. oxydations). We know that the temperature coefficient for physiological processes varies slightly at various parts of the scale; as a rule it is higher near 0 and lower near 30 deg. But we know also that the temperature coefficients do not vary equally from the various physiological processes. It is, therefore, to be expected that the temperature coefficients for the group of reactions of the type (a) will not be identical through the whole scale with the temperature coefficients for the reactions of the type (b). If therefore a certain substance is formed at the normal temperature of the animal in such quantities as are needed for the catenary reaction (b), it is not to be expected that this same perfect balance will be maintained for extremely high or extremely low temperatures; it is more probable that one group of reactions will exceed the other and thus produce aberrant chemical effects, which may underlie the colour aberrations observed by Fischer and other experimenters.

It is important to notice that Fischer was also able to produce aberrations through the application of narcotics. Wolfgang Ostwald has produced experimentally, through variation of temperature, dimorphism of form in Daphnia. Lack of space precludes an account of these important experiments, as of so many others.

IV. THE EFFECTS OF LIGHT.

At the present day nobody seriously questions the statement that the action of light upon organisms is primarily one of a chemical character. While this chemical action is of the utmost importance for organisms, the nutrition of which depends upon the action of chlorophyll, it becomes of less importance for organisms devoid of chlorophyll. Nevertheless, we find animals in which the formation of organs by regeneration is not possible unless they are exposed to light. An observation made by the writer on the regeneration of polyps in a hydroid, Eudendrium racemosum, at Woods Hole, may be mentioned as an instance of this. If the stem of this hydroid, which is usually covered with polyps, is put into an aquarium the polyps soon fall off. If the stems are kept in an aquarium where light strikes them during the day, a regeneration of numerous polyps takes place in a few days. If, however, the stems of Eudendrium are kept permanently in the dark, no polyps are formed even after an interval of some weeks; but they are formed in a few days after the same stems have been transferred from the dark to the light. Diffused daylight suffices for this effect. Goldfarb, who repeated these experiments, states that an exposure of comparatively short duration is sufficient for this effect, it is possible that the light favours the formation of substances which are a prerequisite for the origin of polyps and their growth.

Of much greater significance than this observation are the facts which show that a large number of animals assume, to some extent, the colour of the ground on which they are placed. Pouchet found through experiments upon crustaceans and fish that this influence of the ground on the colour of animals is produced through the medium of the eyes. If the eyes are removed or the animals made blind in another way these phenomena cease. The second general fact found by Pouchet was that the variation in the colour of the animal is brought about through an action of the nerves on the pigment-cells of the skin; the nerve-action being induced through the agency of the eye.

The mechanism and the conditions for the change in colouration were made clear through the beautiful investigations of Keeble and Gamble, on the colour-change in crustaceans. According to these authors the pigment-cells can, as a rule, be considered as consisting of a central body from which a system of more or less complicated ramifications or processes spreads out in all directions. As a rule, the centre of the cell contains one or more different pigments which under the influence of nerves can spread out separately or together into the ramifications. These phenomena of spreading and retraction of the pigments into or from the ramifications of the pigment-cells form on the whole the basis for the colour changes under the influence of environment. Thus Keeble and Gamble observed that Macromysis flexuosa appears transparent and colourless or grey on sandy ground. On a dark ground their colour becomes darker. These animals have two pigments in their chromatophores, a brown pigment and a whitish or yellow pigment; the former is much more plentiful than the latter. When the animal appears transparent all the pigment is contained in the centre of the cells, while the ramifications are free from pigment. When the animal appears brown both pigments are spread out into the ramifications. In the condition of maximal spreading the animals appear black.

This is a comparatively simple case. Much more complicated conditions were found by Keeble and Gamble in other crustaceans, e.g. in Hippolyte cranchii, but the influence of the surroundings upon the colouration of this form was also satisfactorily analysed by these authors.

While many animals show transitory changes in colour under the influence of their surroundings, in a few cases permanent changes can be produced. The best examples of this are those which were observed by Poulton in the chrysalids of various butterflies, especially the small tortoise-shell. These experiments are so well known that a short reference to them will suffice. Poulton (Poulton, E.B., "Colours of Animals" (The International Scientific Series), London, 1890, page 121.) found that in gilt or white surroundings the pupae became light coloured and there was often an immense development of the golden spots, "so that in many cases the whole surface of the pupae glittered with an apparent metallic lustre. So remarkable was the appearance that a physicist to whom I showed the chrysalids, suggested that I had played a trick and had covered them with goldleaf." When black surroundings were used "the pupae were as a rule extremely dark, with only the smallest trace, and often no trace at all, of the golden spots which are so conspicuous in the lighter form." The susceptibility of the animal to this influence of its surroundings was found to be greatest during a definite period when the caterpillar undergoes the metamorphosis into the chrysalis stage. As far as the writer is aware, no physico-chemical explanation, except possibly Wiener's suggestion of colour-photography by mechanical colour adaptation, has ever been offered for the results of the type of those observed by Poulton.

V. EFFECTS OF GRAVITATION.

(a) EXPERIMENTS ON THE EGG OF THE FROG.

Gravitation can only indirectly affect life-phenomena; namely, when we have in a cell two different non-miscible liquids (or a liquid and a solid) of different specific gravity, so that a change in the position of the cell or the organ may give results which can be traced to a change in the position of the two substances. This is very nicely illustrated by the frog's egg, which has two layers of very viscous protoplasm one of which is black and one white. The dark one occupies normally the upper position in the egg and may therefore be assumed to possess a smaller specific gravity than the white substance. When the egg is turned with the white pole upwards a tendency of the white protoplasm to flow down again manifests itself. It is, however, possible to prevent or retard this rotation of the highly viscous protoplasm, by compressing the eggs between horizontal glass plates. Such compression experiments may lead to rather interesting results, as O. Schultze first pointed out. Pflueger had already shown that the first plane of division in a fertilised frog's egg is vertical and Roux established the fact that the first plane of division is identical with the plane of symmetry of the later embryo. Schultze found that if the frog's egg is turned upside down at the time of its first division and kept in this abnormal position, through compression between two glass plates for about 20 hours, a small number of eggs may give rise to twins. It is possible, in this case, that the tendency of the black part of the egg to rotate upwards along the surface of the egg leads to a separation of its first cells, such a separation leading to the formation of twins.

T.H. Morgan made an interesting additional observation. He destroyed one half of the egg after the first segmentation and found that the half which remained alive gave rise to only one half of an embryo, thus confirming an older observation of Roux. When, however, Morgan put the egg upside down after the destruction of one of the first two cells, and

165

compressed the eggs between two glass plates, the surviving half of the egg gave rise to a perfect embryo of half size (and not to a half embryo of normal size as before.) Obviously in this case the tendency of the protoplasm to flow back to its normal position was partially successful and led to a partial or complete separation of the living from the dead half; whereby the former was enabled to form a whole embryo, which, of course, possessed only half the size of an embryo originating from a whole egg.

(b) EXPERIMENTS ON HYDROIDS.

A striking influence of gravitation can be observed in a hydroid, Antennularia antennina, from the bay of Naples. This hydroid consists of a long straight main stem which grows vertically upwards and which has at regular intervals very fine and short bristle-like lateral branches, on the upper side of which the polyps grow. The main stem is negatively geotropic, i.e. its apex continues to grow vertically upwards when we put it obliquely into the aquarium, while the roots grow vertically downwards. The writer observed that when the stem is put horizontally into the water the short lateral branches on the lower side give rise to an altogether different kind of organ, namely, to roots, and these roots grow indefinitely in length and attach themselves to solid bodies; while if the stem had remained in its normal position no further growth would have occurred in the lateral branches. From the upper side of the horizontal stem new stems grow out, mostly directly from the original stem, occasionally also from the short lateral branches. It is thus possible to force upon this hydroid an arrangement of organs which is altogether different from the hereditary arrangement. The writer had called the change in the hereditary arrangement of organs or the transformation of organs by external forces HETEROMORPHOSIS. We cannot now go any further into this subject, which should, however, prove of interest in relation to the problem of heredity.

If it is correct to apply inferences drawn from the observation on the frog's egg to the behaviour of Antennularia, one might conclude that the cells of Antennularia also contain non-miscible substances of different specific gravity, and that wherever the specifically lighter substance comes in contact with the sea-water (or gets near the surface of the cell) the growth of a stem is favoured; while contact with the sea-water of the specifically heavier of the substances, will favour the formation of roots.

VI. THE EXPERIMENTAL CONTROL OF ANIMAL INSTINCTS.

(a) EXPERIMENTS ON THE MECHANISM OF HELIOTROPIC REACTIONS IN ANIMALS.

Since the instinctive reactions of animals are as hereditary as their morphological character, a discussion of experiments on the physico-chemical character of the instinctive reactions of animals should not be entirely omitted from this sketch. It is obvious that such experiments must begin with the simplest type of instincts, if they are expected to lead to any results; and it is also obvious that only such animals must be selected for this purpose, the reactions of which are not complicated by associative memory, or, as it may preferably be termed, associative hysteresis.

The simplest type of instincts is represented by the purposeful motions of animals to or from a source of energy, e.g. light; and it is with some of these that we intend to deal here. When we expose winged aphides (after they have flown away from the plant), or young caterpillars of Porthesia chrysorrhoea (when they are aroused from their winter sleep) or marine or freshwater copepods and many other animals, to diffused daylight falling in from a window, we notice a tendency among these animals to move towards the source of light. If the animals are naturally sensitive, or if they are rendered sensitive through the agencies which we shall mention later, and if the light is strong enough, they move towards the source of light in as straight a line as the imperfections and peculiarities of their locomotor apparatus will permit. It is also obvious that we are here dealing with a forced reaction in which the animals have no more choice in the direction of their motion than have the iron filings in their arrangement in a magnetic field. This can be proved very nicely in the case of starving caterpillars of Porthesia. The writer put such caterpillars into a glass tube the axis of which was at right angles to the plane of the window: the caterpillars went to the window side of the tube and remained there, even if leaves of their food-plant were put into the tube directly behind them. Under such conditions the animals actually died from starvation, the light preventing them from turning to the food, which they eagerly ate when the light allowed them to do so. One cannot say that these animals, which we call positively helioptropic, are attracted by the light, since it can be shown that they go towards the source of the light even if in so doing they move from places of a higher to places of a lower degree of illumination.

The writer has advanced the following theory of these instinctive reactions. Animals of the type of those mentioned are automatically orientated by the light in such a way that symmetrical elements of their retina (or skin) are struck by the rays of light at the same angle. In this case the intensity of light is the same for both retinae or symmetrical parts of the skin.

This automatic orientation is determined by two factors, first a peculiar photo-sensitiveness of the retina (or skin), and second a peculiar nervous connection between the retina and the muscular apparatus. In symmetrically built heliotropic animals in which the symmetrical muscles participate equally in locomotion, the symmetrical muscles work with equal energy as long as the photo-chemical processes in both eyes are identical. If, however, one eye is struck by stronger light than the other, the symmetrical muscles will work unequally and in positively heliotropic animals those muscles will work with greater energy which bring the plane of symmetry back into the direction of the rays of light and the head towards the source of light. As soon as both eyes are struck by the rays of light at the same angle, there is no more reason for the animal to deviate from this direction and it will move in a straight line. All this holds good on the supposition that the animals are exposed to only one source of light and are very sensitive to light.

Additional proof for the correctness of this theory was furnished through the experiments of G.H. Parker and S.J. Holmes. The former worked on a butterfly, Vanessa antiope, the latter on other arthropods. All the animals were in a marked degree positively heliotropic. These authors found that if one cornea is blackened in such an animal, it moves continually in a circle when it is exposed to a source of light, and in these motions the eye which is not covered with paint is directed towards the centre of the circle. The animal behaves, therefore, as if the darkened eye were in the shade.

(b) THE PRODUCTION OF POSITIVE HELIOTROPISM BY ACIDS AND OTHER MEANS AND THE PERIODIC DEPTH-MIGRATIONS OF PELAGIC ANIMALS.

When we observe a dense mass of copepods collected from a freshwater pond, we notice that some have a tendency to go to the light while others go in the opposite direction and many, if not the majority, are indifferent to light. It is an easy matter to make the negatively heliotropic or the indifferent copepods almost instantly positively heliotropic by adding a small but definite amount of carbon-dioxide in the form of carbonated water to the water in which the animals are contained. If the animals are contained in 50 cubic centimetres of water it suffices to add from three to six cubic centimetres of carbonated water to make all the copepods energetically positively heliotropic. This heliotropism lasts about half an hour (probably until all the carbon-dioxide has again diffused into the air.) Similar results may be obtained with any other acid.

The same experiments may be made with another freshwater crustacean, namely Daphnia, with this difference, however, that it is as a rule necessary to lower the temperature of the water also. If the water containing the Daphniae is cooled and at the same time carbon-dioxide added, the animals which were before indifferent to light now become most strikingly positively heliotropic. Marine copepods can be made positively heliotropic by the lowering of the temperature alone, or by a sudden increase in the concentration of the sea-water.

These data have a bearing upon the depth-migrations of pelagic animals, as was pointed out years ago by Theo. T. Groom and the writer. It is well known that many animals living near the surface of the ocean or freshwater lakes, have a tendency to migrate upwards towards evening and downwards in the morning and during the day. These periodic motions are determined to a large extent, if not exclusively, by the heliotropism of these animals. Since the consumption of carbon-dioxide by the green plants ceases towards evening, the tension of this gas in the water must rise and this must have the effect of inducing positive heliotropism or increasing its intensity. At the same time the temperature of the water near the surface is lowered and this also increases the positive heliotropism in the organisms.

The faint light from the sky is sufficient to cause animals which are in a high degree positively heliotropic to move vertically upwards towards the light, as experiments with such pelagic animals, e.g. copepods, have shown. When, in the morning, the absorption of carbon-dioxide by the green algae begins again and the temperature of the water rises, the animals lose their positive heliotropism, and slowly sink down or become negatively heliotropic and migrate actively downwards.

These experiments have also a bearing upon the problem of the inheritance of instincts. The character which is transmitted in this case is not the tendency to migrate periodically upwards and downwards, but the positive heliotropism. The tendency to migrate is the outcome of the fact that periodically varying external conditions induce a periodic change in the sense and intensity of the heliotropism of these animals. It is of course immaterial for the result, whether the carbon-dioxide or any other acid diffuse into the animal from the outside or whether they are produced inside in the tissue cells of the animals. Davenport and Cannon found that Daphniae, which at the beginning of the experiment, react sluggishly to light react much more quickly after they have been made to go to the light a few times. The writer is inclined to attribute this result to the effect of acids, e.g. carbon-dioxide, produced in the animals themselves in consequence of their motion. A similar effect of the acids was shown by A.D. Waller in the case of the response of nerve to stimuli.

The writer observed many years ago that winged male and female ants are positively helioptropic and that their heliotropic sensitiveness increases and reaches its maximum towards the period of nuptial flight. Since the workers show no heliotropism it looks as if an internal secretion from the sexual glands were the cause of their heliotropic sensitiveness. V. Kellogg has observed that bees also become intensely positively heliotropic at the period of their wedding flight, in fact so much so that by letting light fall into the observation hive from above, the bees are prevented from leaving the hive through the exit at the lower end.

We notice also the reverse phenomenon, namely, that chemical changes produced in the animal destroy its heliotropism. The caterpillars of Porthesia chrysorrhoea are very strongly positively heliotropic when they are first aroused from their winter sleep. This heliotropic sensitiveness lasts only as long as they are not fed. If they are kept permanently without food they remain permanently positively heliotropic until they die from starvation. It is to be inferred that as soon as these animals take up food, a substance or substances are formed in their bodies which diminish or annihilate their heliotropic sensitiveness.

The heliotropism of animals is identical with the heliotropism of plants. The writer has shown that the experiments on the effect of acids on the heliotropism of copepods can be repeated with the same result in Volvox. It is therefore erroneous to try to explain these heliotropic reactions of animals on the basis of peculiarities (e.g. vision) which are not found in plants.

We may briefly discuss the question of the transmission through the sex cells of such instincts as are based upon heliotropism. This problem reduces itself simply to that of the method whereby the gametes transmit heliotropism to the larvae or to the adult. The writer has expressed the idea that all that is necessary for this transmission is the presence in the eyes (or in the skin) of the animal of a photo-sensitive substance. For the transmission of this the gametes need not contain anything more than a catalyser or ferment for the synthesis of the photo-sensitive substance in the body of the animal. What has been said in regard to animal heliotropism might, if space permitted, be extended, mutatis mutandis, to geotropism and stereotropism.

(c) THE TROPIC REACTIONS OF CERTAIN TISSUE-CELLS AND THE MORPHOGENETIC EFFECTS OF THESE REACTIONS.

Since plant-cells show heliotropic reactions identical with those of animals, it is not surprising that certain tissue-cells also show reactions which belong to the class of tropisms. These reactions of tissue-cells are of special interest by reason of their bearing upon the inheritance of morphological characters. An example of this is found in the tiger-like marking of the yolk-sac of the embryo of Fundulus and in the marking of the young fish itself. The writer found that the former is entirely, and the latter at least in part, due to the creeping of the chromatophores upon the blood-vessels. The chromatophores are at first scattered irregularly over the yolk-sac and show their characteristic ramifications. There is at that time no definite relation between blood-vessels and chromatophores. As soon as a ramification of a chromatophore comes in contact with a blood-vessel the whole mass of the chromatophore creeps gradually on the blood-vessel and forms a complete sheath around the vessel, until finally all the chromatophores form a sheath around the vessels and no more pigment cells are found in the meshes between the vessels. Nobody who has not actually watched the process of the creeping of the chromatophores upon the blood-vessels would anticipate that the tiger-like colouration of the yolk-sac in the later stages of the development was brought about in this way. Similar facts can be observed in regard to the first marking of the embryo itself. The writer is inclined to believe that we are here dealing with a case of chemotropism, and that the oxygen of the blood may be the cause of the spreading of the chromatophores around the blood-vessels. Certain observations seem to indicate the possibility that in the adult the chromatophores have, in some forms at least, a more rigid structure and are prevented from acting in the way indicated. It seems to the writer that such observations as those made on Fundulus might simplify the problem of the hereditary transmission of certain markings.

Driesch has found that a tropism underlies the arrangement of the skeleton in the pluteus larvae of the sea-urchin. The position of this skeleton is predetermined by the arrangement of the mesenchyme cells, and Driesch has shown that these cells migrate actively to the place of their destination, possibly led there under the influence of certain chemical substances. When Driesch scattered these cells mechanically before their migration, they nevertheless reached their destination.

In the developing eggs of insects the nuclei, together with some cytoplasm, migrate to the periphery of the egg. Herbst pointed out that this might be a case of chemotropism, caused by the oxygen surrounding the egg. The writer has expressed the opinion that the formation of the blastula may be caused generally by a tropic reaction of the blastomeres, the latter being forced by an outside influence to creep to the surface of the egg.

These examples may suffice to indicate that the arrangement of definite groups of cells and the morphological effects resulting therefrom may be determined by forces lying outside the cells. Since these forces are ubiquitous and constant it appears as if we were dealing exclusively with the influence of a gamete; while in reality all that it is necessary for the gamete to transmit is a certain form of irritability.

(d) FACTORS WHICH DETERMINE PLACE AND TIME FOR THE DEPOSITION OF EGGS.

For the preservation of species the instinct of animals to lay their eggs in places in which the young larvae find their food and can develop is of paramount importance. A simple example of this instinct is the fact that the common fly lays its eggs on putrid material which serves as food for the young larvae. When a piece of meat and of fat of the same animal are placed side by side, the fly will deposit its eggs upon the meat on which the larvae can grow, and not upon the fat, on which they would starve. Here we are dealing with the effect of a volatile nitrogenous substance which reflexly causes the peristaltic motions for the laying of the egg in the female fly.

Kammerer has investigated the conditions for the laying of eggs in two forms of salamanders, e.g. Salamandra atra and S. maculosa. In both forms the eggs are fertilised in the body and begin to develop in the uterus. Since there is room only for a few larvae in the uterus, a large number of eggs perish and this number is the greater the longer the period of gestation. It thus happens that when the animals retain their eggs a long time, very few young ones are born; and these are in a rather advanced stage of development, owing to the long time which elapsed since they were fertilised. When the animal lays its eggs comparatively soon after copulation, many eggs (from 12 to 72) are produced and the larvae are of course in an early stage of development. In the early stage the larvae possess gills and can therefore live in water, while in later stages they have no gills and breathe through their lungs. Kammerer showed that both forms of Salamandra can be induced to lay their eggs early or late, according to the physical conditions surrounding them. If they are kept in water or in proximity to water and in a moist atmosphere they have a tendency to lay their eggs earlier and a comparatively high temperature enhances the tendency to shorten the period of gestation. If the salamanders are kept in comparative dryness they show a tendency to lay their eggs rather late and a low temperature enhances this tendency.

Since Salamandra atra is found in rather dry alpine regions with a relatively low temperature and Salamandra maculosa in lower regions with plenty of water and a higher temperature, the fact that S. atra bears young which are already developed and beyond the stage of aquatic life, while S. maculosa bears young ones in an earlier stage, has been termed adaptation. Kammerer's experiments, however, show that we are dealing with the direct effects of definite outside forces. While we may speak of adaptation when all or some of the variables which determine a reaction are unknown, it is obviously in the interest of further scientific progress to connect cause and effect directly whenever our knowledge allows us to do so.

VII. CONCLUDING REMARKS.

The discovery of De Vries, that new species may arise by mutation and the wide if not universal applicability of Mendel's Law to phenomena of heredity, as shown especially by Bateson and his pupils, must, for the time being, if not permanently, serve as a basis for theories of evolution. These discoveries place before the experimental biologist the definite task of producing mutations by physico-chemical means. It is true that certain authors claim to have succeeded in this, but the writer wishes to apologise to these authors for his inability to convince himself of the validity of their claims at the present moment. He thinks that only continued breeding of these apparent mutants through several generations can afford convincing evidence that we are here dealing with mutants rather than with merely pathological variations.

What was said in regard to the production of new species by physico-chemical means may be repeated with still more justification in regard to the second problem of transformation, namely the making of living from inanimate matter. The purely morphological imitations of bacteria or cells which physicists have now and then proclaimed as artificially produced living beings; or the plays on words by which, e.g. the regeneration of broken crystals and the regeneration of lost limbs by a crustacean were declared identical, will not appeal to the biologist. We know that growth and development in animals and plants are determined by definite although complicated series of catenary chemical reactions, which result in the synthesis of a DEFINITE compound or group of compounds, namely, NUCLEINS.

The nucleins have the peculiarity of acting as ferments or enzymes for their own synthesis. Thus a given type of nucleus will continue to synthesise other nuclein of its own kind. This determines the continuity of a species; since each species has, probably, its own specific nuclein or nuclear material. But it also shows us that whoever claims to have succeeded in making living matter from inanimate will have to prove that he has succeeded in producing nuclein material which acts as a ferment for its own synthesis and thus reproduces itself. Nobody has thus far succeeded in this, although nothing warrants us in taking it for granted that this task is beyond the power of science.

XV. THE VALUE OF COLOUR IN THE STRUGGLE FOR LIFE. By E.B. Poulton.

Hope Professor of Zoology in the
University of Oxford.

INTRODUCTION.

The following pages have been written almost entirely from the historical stand-point. Their principal object has been to give some account of the impressions produced on the mind of Darwin and his great compeer Wallace by various difficult problems suggested by the colours of living nature. In order to render the brief summary of Darwin's thoughts and opinions on the subject in any way complete, it was found necessary to say again much that has often been said before. No attempt has been made to display as a whole the vast contribution of Wallace; but certain of its features are incidentally revealed in passages quoted from Darwin's letters. It is assumed that the reader is familiar with the well-known theories of Protective Resemblance, Warning Colours, and Mimicry both Batesian and Mullerian. It would have been superfluous to explain these on the present occasion; for a far more detailed account than could have been attempted in these pages has recently appeared. (Poulton, "Essays on Evolution" Oxford, 1908, pages 293-382.) Among the older records I have made a point of bringing together the principal observations scattered through the note-books and collections of W.J. Burchell. These have never hitherto found a place in any memoir dealing with the significance of the colours of animals.

INCIDENTAL COLOURS.

Darwin fully recognised that the colours of living beings are not necessarily of value as colours, but that they may be an incidental result of chemical or physical structure. Thus he wrote to T. Meehan, Oct. 9, 1874: "I am glad that you are attending to the colours of dioecious flowers; but it is well to remember that their colours may be as unimportant to them as those of a gall, or, indeed, as the colour of an amethyst or ruby is to these gems." ("More Letters of Charles Darwin", Vol. I. pages 354, 355. See also the admirable account of incidental colours in "Descent of Man" (2nd edition), 1874, pages 261, 262.)

Incidental colours remain as available assets of the organism ready to be turned to account by natural selection. It is a probable speculation that all pigmentary colours were originally incidental; but now and for immense periods of time the visible tints of animals have been modified and arranged so as to assist in the struggle with other organisms or in courtship. The dominant colouring of plants, on the other hand, is an essential element in the paramount physiological activity of chlorophyll. In exceptional instances, however, the shapes and visible colours of plants may be modified in order to promote concealment.

TELEOLOGY AND ADAPTATION.

In the department of Biology which forms the subject of this essay, the adaptation of means to an end is probably more evident than in any other; and it is therefore of interest to compare, in a brief introductory section, the older with the newer teleological views.

The distinctive feature of Natural Selection as contrasted with other attempts to explain the process of Evolution is the part played by the struggle for existence. All naturalists in all ages must have known something of the operations of "Nature red in tooth and claw"; but it was left for this great theory to suggest that vast extermination is a necessary condition of progress, and even of maintaining the ground already gained.

Realising that fitness is the outcome of this fierce struggle, thus turned to account for the first time, we are sometimes led to associate the recognition of adaptation itself too exclusively with Natural Selection. Adaptation had been studied with the warmest enthusiasm nearly forty years before this great theory was given to the scientific world, and it is difficult now to realise the impetus which the works of Paley gave to the study of Natural History. That they did inspire the naturalists of the early part of the last century is clearly shown in the following passages.

In the year 1824 the Ashmolean Museum at Oxford was intrusted to the care of J.S. Duncan of New College. He was succeeded in this office by his brother, P.B. Duncan, of the same College, author of a History of the Museum, which shows very clearly the influence of Paley upon the study of nature, and the dominant position given to his teachings: "Happily at this time (1824) a taste for the study of natural history had been excited in the University by Dr Paley's very interesting work on Natural Theology, and the very popular lectures of Dr Kidd on Comparative Anatomy, and Dr Buckland on Geology." In the arrangement of the contents of the Museum the illustration of Paley's work was given the foremost place by J.S. Duncan: "The first division proposes to familiarize the eye to those relations of all natural objects which form the basis of argument in Dr Paley's Natural Theology; to induce a mental habit of associating the view of natural phenomena with the conviction that they are the media of Divine manifestation; and by such association to give proper dignity to every branch of natural science." (From "History and Arrangement of the Ashmolean Museum" by P.B. Duncan: see pages vi, vii of "A Catalogue of the Ashmolean Museum", Oxford, 1836.)

The great naturalist, W.J. Burchell, in his classical work shows the same recognition of adaptation in nature at a still earlier date. Upon the subject of collections he wrote ("Travels in the Interior of Southern Africa", London, Vol. I. 1822, page 505. The references to Burchell's observations in the present essay are adapted from the author's article in "Report of the British and South African Associations", 1905, Vol. III. pages 57-110.): "It must not be supposed that these charms (the pleasures of Nature) are produced by the mere discovery of new objects: it is the harmony with which they have been adapted by the Creator to each other, and to the situations in which they are found, which delights the observer in countries where Art has not yet introduced her discords." The remainder of the passage is so admirable that I venture to quote it: "To him who is satisfied with amassing collections of curious objects, simply for the pleasure of possessing them, such objects can afford, at best, but a childish gratification, faint and fleeting; while he who extends his view beyond the narrow field of nomenclature, beholds a boundless expanse, the exploring of which is worthy of the philosopher, and of the best talents of a reasonable being."

On September 14, 1811, Burchell was at Zand Valley (Vlei), or Sand Pool, a few miles south-west of the site of Prieska, on the Orange River. Here he found a Mesembryanthemum (M. turbiniforme, now M. truncatum) and also a "Gryllus" (Acridian), closely resembling the pebbles with which their locality was strewn. He says of both of these, "The intention of Nature, in these instances, seems to have been the same as when she gave to the Chameleon the power of accommodating its color, in a certain degree, to that of the object nearest to it, in order to compensate for the deficiency of its locomotive powers. By their form and colour, this insect may pass unobserved by those birds, which otherwise would soon extirpate a species so little able to elude its pursuers, and this juicy little Mesembryanthemum may generally escape the notice of cattle and wild animals." (Loc. cit. pages 310, 311. See Sir William Thiselton-Dyer "Morphological Notes", XI.; "Protective Adaptations", I.; "Annals of Botany", Vol. XX. page 124. In plates VII., VIII. and IX. accompanying this article the author represents the species observed by Burchell, together with others in which analogous adaptations exist. He writes: "Burchell was clearly on the track on which Darwin reached the goal. But the time had not come for emancipation from the old teleology. This, however, in no respect detracts from the merit or value of his work. For, as Huxley has pointed out ("Life and Letters of Thomas Henry Huxley", London, 1900, I. page 457), the facts of the old teleology are immediately transferable to Darwinism, which simply supplies them with a natural in place of a supernatural explanation.") Burchell here

172

seems to miss, at least in part, the meaning of the relationship between the quiescence of the Acridian and its cryptic colouring. Quiescence is an essential element in the protective resemblance to a stone—probably even more indispensable than the details of the form and colouring. Although Burchell appears to overlook this point he fully recognised the community between protection by concealment and more aggressive modes of defence; for, in the passage of which a part is quoted above, he specially refers to some earlier remarks on page 226 of his Vol. I. We here find that even when the oxen were resting by the Juk rivier (Yoke river), on July 19, 1811, Burchell observed "Geranium spinosum, with a fleshy stem and large white flowers...; and a succulent species of Pelargonium... so defended by the old panicles, grown to hard woody thorns, that no cattle could browze upon it." He goes on to say, "In this arid country, where every juicy vegetable would soon be eaten up by the wild animals, the Great Creating Power, with all-provident wisdom, has given to such plants either an acrid or poisonous juice, or sharp thorns, to preserve the species from annihilation... " All these modes of defence, especially adapted to a desert environment, have since been generally recognised, and it is very interesting to place beside Burchell's statement the following passage from a letter written by Darwin, Aug. 7, 1868, to G.H. Lewes; "That Natural Selection would tend to produce the most formidable thorns will be admitted by every one who has observed the distribution in South America and Africa (vide Livingstone) of thorn-bearing plants, for they always appear where the bushes grow isolated and are exposed to the attacks of mammals. Even in England it has been noticed that all spine-bearing and sting-bearing plants are palatable to quadrupeds, when the thorns are crushed." ("More Letters", I. page 308.)

ADAPTATION AND NATURAL SELECTION.

I have preferred to show the influence of the older teleology upon Natural History by quotations from a single great and insufficiently appreciated naturalist. It might have been seen equally well in the pages of Kirby and Spence and those of many other writers. If the older naturalists who thought and spoke with Burchell of "the intention of Nature" and the adaptation of beings "to each other, and to the situations in which they are found," could have conceived the possibility of evolution, they must have been led, as Darwin was, by the same considerations to Natural Selection. This was impossible for them, because the philosophy which they followed contemplated the phenomena of adaptation as part of a static immutable system. Darwin, convinced that the system is dynamic and mutable, was prevented by these very phenomena from accepting anything short of the crowning interpretation offered by Natural Selection. ("I had always been much struck by such adaptations (e.g. woodpecker and tree-frog for climbing, seeds for dispersal), and until these could be explained it seemed to me almost useless to endeavour to prove by indirect evidence that species have been modified." "Autobiography" in "Life and Letters of Charles Darwin", Vol. I. page 82. The same thought is repeated again and again in Darwin's letters to his friends. It is forcibly urged in the Introduction to the "Origin" (1859), page 3.) And the birth of Darwin's unalterable conviction that adaptation is of dominant importance in the organic world,—a conviction confirmed and ever again confirmed by his experience as a naturalist—may probably be traced to the influence of the great theologian. Thus Darwin, speaking of his Undergraduate days, tells us in his "Autobiography" that the logic of Paley's "Evidences of Christianity" and "Moral Philosophy" gave him as much delight as did Euclid.

"The careful study of these works, without attempting to learn any part by rote, was the only part of the academical course which, as I then felt and as I still believe, was of the least use to me in the education of my mind. I did not at that time trouble myself about Paley's premises; and taking these on trust, I was charmed and convinced by the long line of argumentation." ("Life and Letters", I. page 47.)

When Darwin came to write the "Origin" he quoted in relation to Natural Selection one of Paley's conclusions. "No organ will be formed, as Paley has remarked, for the purpose of causing pain or for doing an injury to its possessor." ("Origin of Species" (1st edition) 1859, page 201.)

The study of adaptation always had for Darwin, as it has for many, a peculiar charm. His words, written Nov. 28, 1880, to Sir W. Thiselton-Dyer, are by no means inapplicable to-day: "Many of the Germans are very contemptuous about making out use of organs; but they may sneer the souls out of their bodies, and I for one shall think it the most interesting part of natural history." ("More Letters" II. page 428.)

PROTECTIVE AND AGGRESSIVE RESEMBLANCE: PROCRYPTIC AND ANTICRYPTIC COLOURING.

Colouring for the purpose of concealment is sometimes included under the head Mimicry, a classification adopted by H.W. Bates in his classical paper. Such an arrangement is inconvenient, and I have followed Wallace in keeping the two categories distinct.

The visible colours of animals are far more commonly adapted for Protective Resemblance than for any other purpose. The concealment of animals by their colours, shapes and attitudes, must have been well known from the period at which human beings first began to take an intelligent interest in Nature. An interesting early record is that of Samuel Felton, who (Dec. 2, 1763) figured and gave some account of an Acridian (Phyllotettix) from Jamaica. Of this insect he says "THE THORAX is like a leaf that is raised perpendicularly from the body." ("Phil. Trans. Roy. Soc." Vol. LIV. Tab. VI. page 55.)

Both Protective and Aggressive Resemblances were appreciated and clearly explained by Erasmus Darwin in 1794: "The colours of many animals seem adapted to their purposes of concealing themselves either to avoid danger, or to spring upon their prey." ("Zoonomia", Vol. I. page 509, London, 1794.)

Protective Resemblance of a very marked and beautiful kind is found in certain plants, inhabitants of desert areas. Examples observed by Burchell almost exactly a hundred years ago have already been mentioned. In addition to the resemblance to stones Burchell observed, although he did not publish the fact, a South African plant concealed by its likeness to the dung of birds. (Sir William Thiselton-Dyer has suggested the same method of concealment ("Annals of Botany", Vol. XX. page 123). Referring to Anacampseros papyracea, figured on plate IX., the author says of its adaptive resemblance: "At the risk of suggesting one perhaps somewhat far-fetched, I must confess that the aspect of the plant always calls to my mind the dejecta of some bird, and the more so owing to the whitening of the branches towards the tips" (loc. cit. page 126). The student of insects, who is so familiar with this very form of protective resemblance in larvae, and even perfect insects, will not be inclined to consider the suggestion far-fetched.) The observation is recorded in one of the manuscript journals kept by the great explorer during his journey. I owe the opportunity of studying it to the kindness of Mr Francis A. Burchell of the Rhodes University College, Grahamstown. The following account is given under the date July 5, 1812, when Burchell was at the Makkwarin River, about half-way between the Kuruman River and Litakun the old capital of the Bachapins (Bechuanas): "I found a curious little Crassula (not in flower) so snow white, that I should never has (have) distinguished it from the white limestones... It was an inch high and a little branchy,... and was at first mistaken for the dung of birds of the passerine order. I have often had occasion to remark that in stony place(s) there grow many small succulent plants and abound insects (chiefly Grylli) which have exactly the same colour as the ground and must for ever escape observation unless a person sit on the ground and observe very attentively."

The cryptic resemblances of animals impressed Darwin and Wallace in very different degrees, probably in part due to the fact that Wallace's tropical experiences were so largely derived from the insect world, in part to the importance assigned by Darwin to Sexual Selection "a subject which had always greatly interested me," as he says in his "Autobiography", ("Life and Letters", Vol. I. page 94.) There is no reference to Cryptic Resemblance in Darwin's section of the Joint Essay, although he gives an excellent short account of Sexual Selection (see page 295). Wallace's section on the other hand contains the following statement: "Even the peculiar colours of many animals, especially insects, so closely resembling the soil or the leaves or the trunks on which they habitually reside, are explained on the same principle; for though in the course of ages varieties of many tints may have occurred, YET THOSE RACES HAVING COLOURS BEST ADAPTED TO CONCEALMENT FROM THEIR ENEMIES WOULD INEVITABLY SURVIVE THE LONGEST." ("Journ. Proc. Linn. Soc." Vol. III. 1859, page 61. The italics are Wallace's.)

It would occupy too much space to attempt any discussion of the difference between the views of these two naturalists, but it is clear that Darwin, although fully believing in the efficiency of protective resemblance and replying to St George Mivart's contention that Natural Selection was incompetent to produce it ("Origin" (6th edition) London, 1872, pages 181, 182; see also page 66.), never entirely agreed with Wallace's estimate of its importance. Thus the following extract from a letter to Sir Joseph Hooker, May 21, 1868, refers to Wallace: "I find I must (and I always distrust myself when I differ from him) separate rather widely from him all about birds' nests and protection; he is riding that hobby to death." ("More Letters", I. page 304.) It is clear from the account given in "The Descent of Man", (London, 1874, pages 452-458. See also "Life and Letters", III. pages 123-125, and "More Letters", II. pages 59-63, 72-74, 76-78, 84-90, 92, 93.), that the divergence was due to the fact that Darwin ascribed more importance to Sexual Selection than did Wallace, and Wallace more importance to Protective Resemblance than Darwin. Thus Darwin wrote to Wallace, Oct. 12 and 13, 1867: "By the way, I cannot but think that you push protection too far in some cases, as with the stripes on the tiger." ("More Letters", I. page 283.) Here too Darwin was preferring the explanation offered by Sexual Selection ("Descent of Man" (2nd edition) 1874, pages 545, 546.), a preference which, considering the relation of the colouring of the lion and tiger to their respective environments, few naturalists will be found to share. It is also shown that Darwin contemplated the possibility of cryptic colours such as those of Patagonian animals being due to sexual selection influenced by the aspect of surrounding nature.

Nearly a year later Darwin in his letter of May 5, 1868?, expressed his agreement with Wallace's views: "Expect that I should put sexual selection as an equal, or perhaps as even a more important agent in giving colour than Natural Selection for protection." ("More Letters", II. pages 77, 78.) The conclusion expressed in the above quoted passage is opposed by the extraordinary development of Protective Resemblance in the immature stages of animals, especially insects.

It must not be supposed, however, that Darwin ascribed an unimportant role to Cryptic Resemblances, and as observations accumulated he came to recognise their efficiency in fresh groups of the animal kingdom. Thus he wrote to Wallace, May 5, 1867: "Haeckel has recently well shown that the transparency and absence of colour in the lower oceanic animals, belonging to the most different classes, may be well accounted for on the principle of protection." ("More Letters", II. page 62. See also "Descent of Man", page 261.) Darwin also admitted the justice of Professor E.S. Morse's contention that the shells of molluscs are often adaptively coloured. ("More Letters", II. page 95.) But he looked upon cryptic colouring and also mimicry as more especially Wallace's departments, and sent to him and to Professor Meldola observations and notes bearing upon these subjects. Thus the following letter given to me by Dr A.R. Wallace and now, by kind permission, published for the first time, accompanied a photograph of the chrysalis of Papilio sarpedon choredon, Feld., suspended from a leaf of its food-plant:

July 9th, Down, Beckenham, Kent.

My Dear Wallace,

Dr G. Krefft has sent me the enclosed from Sydney. A nurseryman saw a caterpillar feeding on a plant and covered the whole up, but when he searched for the cocoon (pupa), was long before he could find it, so good was its imitation in colour and form to the leaf to which it was attached. I hope that the world goes well with you. Do not trouble yourself by acknowledging this.

Ever yours

Ch. Darwin.

Another deeply interesting letter of Darwin's bearing upon protective resemblance, has only recently been shown to me by my friend Professor E.B. Wilson, the great American Cytologist. With his kind consent and that of Mr Francis Darwin, this letter, written four months before Darwin's death on April 19, 1882, is reproduced here (The letter is addressed: "Edmund B. Wilson, Esq., Assistant in Biology, John Hopkins University, Baltimore Md, U. States."):

December 21, 1881.

Dear Sir,

I thank you much for having taken so much trouble in describing fully your interesting and curious case of mimickry.

I am in the habit of looking through many scientific Journals, and though my memory is now not nearly so good as it was, I feel pretty sure that no such case as yours has been described (amongst the nudibranch) molluscs. You perhaps know the case of a fish allied to Hippocampus, (described some years ago by Dr Gunther in "Proc. Zoolog. Socy.") which clings by its tail to sea-weeds, and is covered with waving filaments so as itself to look like a piece of the same sea-weed. The parallelism between your and Dr Gunther's case makes both of them the more interesting; considering how far a fish and a mollusc stand apart. It would be difficult for anyone to explain such cases by the direct action of the environment.—I am glad that you intend to make further observations on this mollusc, and I hope that you will give a figure and if possible a coloured figure.

With all good wishes from an old brother naturalist,

I remain, Dear Sir,

Yours faithfully,

Charles Darwin.

Professor E.B. Wilson has kindly given the following account of the circumstances under which he had written to Darwin: "The case to which Darwin's letter refers is that of the nudibranch mollusc Scyllaea, which lives on the floating Sargassum and shows a really astonishing resemblance to the plant, having leaf-shaped processes very closely similar to the fronds of the sea-weed both in shape and in colour. The concealment of the animal may be judged from the fact that we found the animal quite by accident on a piece of Sargassum that had been in a glass jar in the laboratory for some time and had been closely examined in the search for hydroids and the like without disclosing the presence upon it of two large specimens of the Scyllaea (the animal, as I recall it, is about two inches long). It was first detected by its movements alone, by someone (I think a casual visitor to the laboratory) who was looking closely at the Sargassum and exclaimed 'Why, the sea-weed is moving its leaves'! We found the example in the summer of 1880 or 1881 at Beaufort, N.C., where the Johns Hopkins laboratory was located for the time being. It must have been seen by many others, before or since.

"I wrote and sent to Darwin a short description of the case at the suggestion of Brooks, with whom I was at the time a student. I was, of course, entirely unknown to Darwin (or to anyone else) and to me the principal interest of Darwin's letter is the evidence that it gives of his extraordinary kindness and friendliness towards an obscure youngster who had of course absolutely no claim upon his time or attention. The little incident made an indelible impression upon my memory and taught me a lesson that was worth learning."

VARIABLE PROTECTIVE RESEMBLANCE.

The wonderful power of rapid colour adjustment possessed by the cuttle-fish was observed by Darwin in 1832 at St Jago, Cape de Verd Islands, the first place visited during the voyage of the "Beagle". From Rio he wrote to Henslow, giving the following account of his observations, May 18, 1832: "I took several specimens of an Octopus which possessed a most marvellous power of changing its colours, equalling any chameleon, and evidently accommodating the changes to the colour of the ground which it passed over. Yellowish green, dark brown, and red, were the prevailing colours; this fact appears to be new, as far as

I can find out." ("Life and Letters", I. pages 235, 236. See also Darwin's "Journal of Researches", 1876, pages 6-8, where a far more detailed account is given together with a reference to "Encycl. of Anat. and Physiol.")

Darwin was well aware of the power of individual colour adjustment, now known to be possessed by large numbers of lepidopterous pupae and larvae. An excellent example was brought to his notice by C.V. Riley ("More Letters" II, pages 385, 386.), while the most striking of the early results obtained with the pupae of butterflies—those of Mrs M.E. Barber upon Papilio nireus—was communicated by him to the Entomological Society of London. ("Trans. Ent. Soc. Lond." 1874, page 519. See also "More Letters", II. page 403.)

It is also necessary to direct attention to C.W. Beebe's ("Zoologica: N.Y. Zool. Soc." Vol. I. No. 1, Sept. 25, 1907: "Geographic variation in birds with especial reference to the effects of humidity".) recent discovery that the pigmentation of the plumage of certain birds is increased by confinement in a superhumid atmosphere. In Scardafella inca, on which the most complete series of experiments was made, the changes took place only at the moults, whether normal and annual or artificially induced at shorter periods. There was a corresponding increase in the choroidal pigment of the eye. At a certain advanced stage of feather pigmentation a brilliant iridescent bronze or green tint made its appearance on those areas where iridescence most often occurs in allied genera. Thus in birds no less than in insects, characters previously regarded as of taxonomic value, can be evoked or withheld by the forces of the environment.

WARNING OR APOSEMATIC COLOURS.

From Darwin's description of the colours and habits it is evident that he observed, in 1833, an excellent example of warning colouring in a little South American toad (Phryniscus nigricans). He described it in a letter to Henslow, written from Monte Video, Nov. 24, 1832: "As for one little toad, I hope it may be new, that it may be christened 'diabolicus.' Milton must allude to this very individual when he talks of 'squat like a toad'; its colours are by Werner ("Nomenclature of Colours", 1821) ink black, vermilion red and buff orange." ("More Letters", I. page 12.) In the "Journal of Researches" (1876, page 97.) its colours are described as follows: "If we imagine, first, that it had been steeped in the blackest ink, and then, when dry, allowed to crawl over a board, freshly painted with the brightest vermilion, so as to colour the soles of its feet and parts of its stomach, a good idea of its appearance will be gained." "Instead of being nocturnal in its habits, as other toads are, and living in damp obscure recesses, it crawls during the heat of the day about the dry sand-hillocks and arid plains,... " The appearance and habits recall T. Belt's well-known description of the conspicuous little Nicaraguan frog which he found to be distasteful to a duck. ("The Naturalist in Nicaragua" (2nd edition) London, 1888, page 321.)

The recognition of the Warning Colours of caterpillars is due in the first instance to Darwin, who, reflecting on Sexual Selection, was puzzled by the splendid colours of sexually immature organisms. He applied to Wallace "who has an innate genius for solving difficulties." ("Descent of Man", page 325. On this and the following page an excellent account of the discovery will be found, as well as in Wallace's "Natural Selection", London, 1875, pages 117-122.) Darwin's original letter exists ("Life and Letters", III. pages 93, 94.), and in it we are told that he had taken the advice given by Bates: "You had better ask Wallace." After some consideration Wallace replied that he believed the colours of conspicuous caterpillars and perfect insects were a warning of distastefulness and that such forms would be refused by birds. Darwin's reply ("Life and Letters", III. pages 94, 95.) is extremely interesting both for its enthusiasm at the brilliancy of the hypothesis and its caution in acceptance without full confirmation:

"Bates was quite right; you are the man to apply to in a difficulty. I never heard anything more ingenious than your suggestion, and I hope you may be able to prove it true. That is a splendid fact about the white moths (A single white moth which was rejected by young turkeys, while other moths were greedily devoured: "Natural Selection", 1875, page 78.); it warms one's very blood to see a theory thus almost proved to be true."

Two years later the hypothesis was proved to hold for caterpillars of many kinds by J. Jenner Weir and A.G. Butler, whose observations have since been abundantly confirmed by many naturalists. Darwin wrote to Weir, May 13, 1869: "Your verification of Wallace's suggestion seems to me to amount to quite a discovery." ("More Letters", II. page 71 (footnote).)

RECOGNITION OR EPISEMATIC CHARACTERS.

This principle does not appear to have been in any way foreseen by Darwin, although he draws special attention to several elements of pattern which would now be interpreted by many naturalists as epismes. He believed that the markings in question interfered with the cryptic effect, and came to the conclusion that, even when common to both sexes, they "are the result of sexual selection primarily applied to the male." ("Descent of Man", page 544.) The most familiar of all recognition characters was carefully explained by him, although here too explained as an ornamental feature now equally transmitted to both sexes: "The hare on her form is a familiar instance of concealment through colour; yet this principle partly fails in a closely-allied species, the rabbit, for when running to its burrow, it is made conspicuous to the sportsman, and no doubt to all beasts of prey, by its upturned white tail." ("Descent of Man", page 542.)

The analogous episematic use of the bright colours of flowers to attract insects for effecting cross-fertilisation and of fruits to attract vertebrates for effecting dispersal is very clearly explained in the "Origin". (Edition 1872, page 161. For a good example of Darwin's caution in dealing with exceptions see the allusion to brightly coloured fruit in "More Letters", II. page 348.)

It is not, at this point, necessary to treat sematic characters at any greater length. They will form the subject of a large part of the following section, where the models of Batesian (Pseudaposematic) mimicry are considered as well as the Mullerian (Synaposematic) combinations of Warning Colours.

MIMICRY,—BATESIAN OR PSEUDAPOSEMATIC, MULLERIAN OR SYNAPOSEMATIC.

The existence of superficial resemblances between animals of various degrees of affinity must have been observed for hundreds of years. Among the early examples, the best known to me have been found in the manuscript note-books and collections of W.J. Burchell, the great traveller in Africa (1810-15) and Brazil (1825-30). The most interesting of his records on this subject are brought together in the following paragraphs.

Conspicuous among well-defended insects are the dark steely or iridescent greenish blue fossorial wasps or sand-wasps, Sphex and the allied genera. Many Longicorn beetles mimic these in colour, slender shape of body and limbs, rapid movements, and the readiness with which they take to flight. On Dec. 21, 1812, Burchell captured one such beetle (Promeces viridis) at Kosi Fountain on the journey from the source of the Kuruman River to Klaarwater. It is correctly placed among the Longicorns in his catalogue, but opposite to its number is the comment "Sphex! totus purpureus."

In our own country the black-and-yellow colouring of many stinging insects, especially the ordinary wasps, affords perhaps the commonest model for mimicry. It is reproduced with more or less accuracy on moths, flies and beetles. Among the latter it is again a Longicorn which offers one of the best-known, although by no means one of the most perfect, examples. The appearance of the well-known "wasp-beetle" (Clytus arietis) in the living state is sufficiently suggestive to prevent the great majority of people from touching it. In Burchell's Brazilian collection there is a nearly allied species (Neoclytus curvatus) which appears to be somewhat less wasp-like than the British beetle. The specimen bears the number "1188," and the date March 27, 1827, when Burchell was collecting in the neighbourhood of San Paulo. Turning to the corresponding number in the Brazilian note-book we find this record: "It runs rapidly like an ichneumon or wasp, of which it has the appearance."

178

The formidable, well-defended ants are as freely mimicked by other insects as the sand-wasps, ordinary wasps and bees. Thus on February 17, 1901, Guy A.K. Marshall captured, near Salisbury, Mashonaland, three similar species of ants (Hymenoptera) with a bug (Hemiptera) and a Locustid (Orthoptera), the two latter mimicking the former. All the insects, seven in number, were caught on a single plant, a small bushy vetch. ("Trans. Ent. Soc. Lond." 1902, page 535, plate XIX. figs. 53-59.)

This is an interesting recent example from South Africa, and large numbers of others might be added—the observations of many naturalists in many lands; but nearly all of them known since that general awakening of interest in the subject which was inspired by the great hypotheses of H.W. Bates and Fritz Muller. We find, however, that Burchell had more than once recorded the mimetic resemblance to ants. An extremely ant-like bug (the larva of a species of Alydus) in his Brazilian collection is labelled "1141," with the date December 8, 1826, when Burchell was at the Rio das Pedras, Cubatao, near Santos. In the note-book the record is as follows: "1141 Cimex. I collected this for a Formica."

Some of the chief mimics of ants are the active little hunting spiders belonging to the family Attidae. Examples have been brought forward during many recent years, especially by my friends Dr and Mrs Peckham, of Milwaukee, the great authorities on this group of Araneae. Here too we find an observation of the mimetic resemblance recorded by Burchell, and one which adds in the most interesting manner to our knowledge of the subject. A fragment, all that is now left, of an Attid spider, captured on June 30, 1828, at Goyaz, Brazil, bears the following note, in this case on the specimen and not in the note-book: "Black... runs and seems like an ant with large extended jaws." My friend Mr R.I. Pocock, to whom I have submitted the specimen, tells me that it is not one of the group of species hitherto regarded as ant-like, and he adds, "It is most interesting that Burchell should have noticed the resemblance to an ant in its movements. This suggests that the perfect imitation in shape, as well as in movement, seen in many species was started in forms of an appropriate size and colour by the mimicry of movement alone." Up to the present time Burchell is the only naturalist who has observed an example which still exhibits this ancestral stage in the evolution of mimetic likeness.

Following the teachings of his day, Burchell was driven to believe that it was part of the fixed and inexorable scheme of things that these strange superficial resemblances existed. Thus, when he found other examples of Hemipterous mimics, including one (Luteva macrophthalma) with "exactly the manners of a Mantis," he added the sentence, "In the genus Cimex (Linn.) are to be found the outward resemblances of insects of many other genera and orders" (February 15, 1829). Of another Brazilian bug, which is not to be found in his collection, and cannot therefore be precisely identified, he wrote: "Cimex... Nature seems to have intended it to imitate a Sphex, both in colour and the rapid palpitating and movement of the antennae" (November 15, 1826). At the same time it is impossible not to feel the conviction that Burchell felt the advantage of a likeness to stinging insects and to aggressive ants, just as he recognised the benefits conferred on desert plants by spines and by concealment. Such an interpretation of mimicry was perfectly consistent with the theological doctrines of his day. (See Kirby and Spence, "An Introduction to Entomology" (1st edition), London, Vol. II. 1817, page 223.)

The last note I have selected from Burchell's manuscript refers to one of the chief mimics of the highly protected Lycid beetles. The whole assemblage of African insects with a Lycoid colouring forms a most important combination and one which has an interesting bearing upon the theories of Bates and Fritz Muller. This most wonderful set of mimetic forms, described in 1902 by Guy A.K. Marshall, is composed of flower-haunting beetles belonging to the family Lycidae, and the heterogeneous group of varied insects which mimic their conspicuous and simple scheme of colouring. The Lycid beetles, forming the centre or "models" of the whole company, are orange-brown in front for about two-thirds of the exposed surface, black behind for the remaining third. They are undoubtedly protected by qualities which make them excessively unpalatable to the bulk of insect-eating animals. Some experimental proof of this has been obtained by Mr Guy Marshall. What are the forms which surround them? According to the hypothesis of Bates they would be, at any rate mainly, palatable hard-pressed insects which only hold their own in the struggle for life by a fraudulent imitation of the trade-mark of the successful and powerful Lycidae. According to Fritz Muller's hypothesis we should expect that the mimickers would be highly protected, successful and abundant species, which (metaphorically speaking) have found it to their advantage to possess an advertisement, a danger-signal, in common with each other, and in common with the beetles in the centre of the group.

How far does the constitution of this wonderful combination—the largest and most complicated as yet known in all the world—convey to us the idea of mimicry working along the lines supposed by Bates or those suggested by Muller? Figures 1 to 52 of Mr Marshall's coloured plate ("Trans. Ent. Soc. Lond." 1902, plate XVIII. See also page 517, where the group is analysed.) represent a set of forty-two or forty-three species or forms of insects captured in Mashonaland, and all except two in the neighbourhood of Salisbury. The combination includes six species of Lycidae; nine beetles of five groups all specially protected by nauseous qualities, Telephoridae, Melyridae, Phytophaga, Lagriidae, Cantharidae; six Longicorn beetles; one Coprid beetle; eight stinging Hymenoptera; three or four parasitic Hymenoptera (Braconidae, a group much mimicked and shown by some experiments to be distasteful); five bugs (Hemiptera, a largely unpalatable group); three moths (Arctiidae and Zygaenidae, distasteful families); one fly. In fact the whole combination, except perhaps one Phytophagous, one Coprid and the Longicorn beetles, and the fly, fall under the hypothesis of Muller and not under that of Bates. And it is very doubtful whether these exceptions will be sustained: indeed the suspicion of unpalatability already besets the Longicorns and is always on the heels,—I should say the hind tarsi—of a Phytophagous beetle.

This most remarkable group which illustrates so well the problem of mimicry and the alternative hypotheses proposed for its solution, was, as I have said, first described in 1902. Among the most perfect of the mimetic resemblances in it is that between the Longicorn beetle, Amphidesmus analis, and the Lycidae. It was with the utmost astonishment and pleasure that I found this very resemblance had almost certainly been observed by Burchell. A specimen of the Amphidesmus exists in his collection and it bears "651." Turning to the same number in the African Catalogue we find that the beetle is correctly placed among the Longicorns, that it was captured at Uitenhage on Nov. 18, 1813, and that it was found associated with Lycid beetles in flowers ("consocians cum Lycis 78-87 in floribus"). Looking up Nos. 78-87 in the collection and catalogue, three species of Lycidae are found, all captured on Nov. 18, 1813, at Uitenhage. Burchell recognised the wide difference in affinity, shown by the distance between the respective numbers; for his catalogue is arranged to represent relationships. He observed, what students of mimicry are only just beginning to note and record, the coincidence between model and mimic in time and space and in habits. We are justified in concluding that he observed the close superficial likeness although he does not in this case expressly allude to it.

One of the most interesting among the early observations of superficial resemblance between forms remote in the scale of classification was made by Darwin himself, as described in the following passage from his letter to Henslow, written from Monte Video, Aug. 15, 1832: "Amongst the lower animals nothing has so much interested me as finding two species of elegantly coloured true Planaria inhabiting the dewy forest! The false relation they bear to snails is the most extraordinary thing of the kind I have ever seen." ("More Letters", I. page 9.)

Many years later, in 1867, he wrote to Fritz Muller suggesting that the resemblance of a soberly coloured British Planarian to a slug might be due to mimicry. ("Life and Letters", III. page 71.)

The most interesting copy of Bates's classical memoir on Mimicry ("Contributions to an Insect Fauna of the Amazon Valley". "Trans. Linn. Soc." Vol. XXIII. 1862, page 495.), read before the Linnean Society in 1861, is that given by him to the man who has done most to support and extend the theory. My kind friend has given that copy to me; it bears the inscription:

"Mr A.R. Wallace from his old travelling companion the Author."

Only a year and a half after the publication of the "Origin", we find that Darwin wrote to Bates on the subject which was to provide such striking evidence of the truth of Natural Selection: "I am glad to hear that you have specially attended to 'mimetic' analogies—a most curious subject; I hope you publish on it. I have for a long time wished to know whether what Dr Collingwood asserts is true—that the most striking cases generally occur between insects inhabiting the same country." (The letter is dated April 4, 1861. "More Letters", I. page 183.)

The next letter, written about six months later, reveals the remarkable fact that the illustrious naturalist who had anticipated Edward Forbes in the explanation of arctic forms on alpine heights ("I was forestalled in only one important point, which my vanity has always made me regret, namely, the explanation by means of the Glacial period of the presence of the same species of plants and of some few animals on distant mountain summits and in the arctic regions. This view pleased me so much that I wrote it out in extenso, and I believe that it was read by Hooker some years before E. Forbes published his celebrated memoir on the subject. In the very few points in which we differed, I still think that I was in the right. I have never, of course, alluded in print to my having independently worked out this view." "Autobiography, Life and Letters", I. page 88.), had also anticipated H.W. Bates in the theory of Mimicry: "What a capital paper yours will be on mimetic resemblances! You will make quite a new subject of it. I had thought of such cases as a difficulty; and once, when corresponding with Dr Collingwood, I thought of your explanation; but I drove it from my mind, for I felt that I had not knowledge to judge one way or the other." (The letter is dated Sept. 25, 1861: "More Letters", I. page 197.)

Bates read his paper before the Linnean Society, Nov. 21, 1861, and Darwin's impressions on hearing it were conveyed in a letter to the author dated Dec. 3: "Under a general point of view, I am quite convinced (Hooker and Huxley took the same view some months ago) that a philosophic view of nature can solely be driven into naturalists by treating special subjects as you have done. Under a special point of view, I think you have solved one of the most perplexing problems which could be given to solve." ("Life and Letters", II. page 378.) The memoir appeared in the following year, and after reading it Darwin wrote as follows, Nov. 20, 1862: "... In my opinion it is one of the most remarkable and admirable papers I ever read in my life... I am rejoiced that I passed over the whole subject in the "Origin", for I should have made a precious mess of it. You have most clearly stated and solved a wonderful problem... Your paper is too good to be largely appreciated by the mob of naturalists without souls; but, rely on it, that it will have LASTING value, and I cordially congratulate you on your first great work. You will find, I should think, that Wallace will fully appreciate it." ("Life and Letters", II. pages 391-393.) Four days later, Nov. 24, Darwin wrote to Hooker on the same subject: "I have now finished his paper...' it seems to me admirable. To my mind the act of segregation of varieties into species was never so plainly brought forward, and there are heaps of capital miscellaneous observations." ("More Letters", I. page 214.)

Darwin was here referring to the tendency of similar varieties of the same species to pair together, and on Nov. 25 he wrote to Bates asking for fuller information on this subject. ("More Letters", I. page 215. See also parts of Darwin's letter to Bates in "Life and Letters", II. page 392.) If Bates's opinion were well founded, sexual selection would bear a most important part in the establishment of such species. (See Poulton, "Essays on Evolution", 1908, pages 65, 85-88.) It must be admitted, however, that the evidence is as yet quite insufficient to establish this conclusion. It is interesting to observe how Darwin at once fixed on the part of Bates's memoir which seemed to bear upon sexual selection. A review of Bates's theory of Mimicry was contributed by Darwin to the "Natural History Review" (New Ser. Vol. III. 1863, page 219.) and an account of it is to be found in the "Origin" (Edition 1872, pages 375-378.) and in "The Descent of Man". (Edition 1874, pages 323-325.)

Darwin continually writes of the value of hypothesis as the inspiration of inquiry. We find an example in his letter to Bates, Nov. 22, 1860: "I have an old belief that a good observer really means a good theorist, and I fully expect to find your observations most valuable." ("More Letters", I. page 176.) Darwin's letter refers to many problems upon which Bates had theorised and observed, but as regards Mimicry itself the hypothesis was thought out after the return of the letter from the Amazons, when he no longer had the opportunity of testing it by the observation of living Nature. It is by no means improbable that, had he been able to apply this test, Bates would have recognised that his division of butterfly resemblances into two classes,—one due to the theory of mimicry, the other to the influence of local conditions,—could not be sustained.

Fritz Muller's contributions to the problem of Mimicry were all made in S.E. Brazil, and numbers of them were communicated, with other observations on natural history, to Darwin, and by him sent to Professor R. Meldola who published many of the facts. Darwin's letters to Meldola (Poulton, "Charles Darwin and the theory of Natural Selection", London, 1896, pages 199-218.) contain abundant proofs of his interest in Muller's work upon Mimicry. One deeply interesting letter (Loc. cit. pages 201, 202.) dated Jan. 23, 1872, proves that Fritz Muller before he originated the theory of Common Warning Colours (Synaposematic Resemblance or Mullerian Mimicry), which will ever be associated with his name, had conceived the idea of the production of mimetic likeness by sexual selection.

Darwin's letter to Meldola shows that he was by no means inclined to dismiss the suggestion as worthless, although he considered it daring. "You will also see in this letter a strange speculation, which I should not dare to publish, about the appreciation of certain colours being developed in those species which frequently behold other forms similarly ornamented. I do not feel at all sure that this view is as incredible as it may at first appear. Similar ideas have passed through my mind when considering the dull colours of all the organisms which inhabit dull-coloured regions, such as Patagonia and the Galapagos Is." A little later, on April 5, he wrote to Professor August Weismann on the same subject: "It may be suspected that even the habit of viewing differently coloured surrounding objects would influence their taste, and Fritz Muller even goes so far as to believe that the sight of gaudy butterflies might influence the taste of distinct species." ("Life and Letters", III. page 157.)

This remarkable suggestion affords interesting evidence that F. Muller was not satisfied with the sufficiency of Bates's theory. Nor is this surprising when we think of the numbers of abundant conspicuous butterflies which he saw exhibiting mimetic likenesses. The common instances in his locality, and indeed everywhere in tropical America, were anything but the hard-pressed struggling forms assumed by the theory of Bates. They belonged to the groups which were themselves mimicked by other butterflies. Fritz Muller's suggestion also shows that he did not accept Bates's alternative explanation of a superficial likeness between models themselves, based on some unknown influence of local physico-chemical forces. At the same time Muller's own suggestion was subject to this apparently fatal objection, that the sexual selection he invoked would tend to produce resemblances in the males rather than the females, while it is well known that when the sexes differ the females are almost invariably more perfectly mimetic than the males and in a high proportion of cases are mimetic while the males are non-mimetic.

The difficulty was met several years later by Fritz Muller's well-known theory, published in 1879 ("Kosmos", May 1879, page 100.), and immediately translated by Meldola and brought before the Entomological Society. ("Proc. Ent. Soc. Lond." 1879, page xx.) Darwin's letter to Meldola dated June 6, 1879, shows "that the first introduction of this new and most suggestive hypothesis into this country was due to the direct influence of Darwin himself, who brought it before the notice of the one man who was likely to appreciate it at its true value and to find the means for its presentation to English naturalists." ("Charles Darwin and the Theory of Natural Selection", page 214.) Of the hypothesis itself Darwin wrote "F. Muller's view of the mutual protection was quite new to me." (Ibid. page 213.) The hypothesis of Mullerian mimicry was at first strongly opposed. Bates himself could never make up his mind to accept it. As the Fellows were walking out of the meeting at which

Professor Meldola explained the hypothesis, an eminent entomologist, now deceased, was heard to say to Bates: "It's a case of save me from my friends!" The new ideas encountered and still encounter to a great extent the difficulty that the theory of Bates had so completely penetrated the literature of natural history. The present writer has observed that naturalists who have not thoroughly absorbed the older hypothesis are usually far more impressed by the newer one than are those whose allegiance has already been rendered. The acceptance of Natural Selection itself was at first hindered by similar causes, as Darwin clearly recognised: "If you argue about the non-acceptance of Natural Selection, it seems to me a very striking fact that the Newtonian theory of gravitation, which seems to every one now so certain and plain, was rejected by a man so extraordinarily able as Leibnitz. The truth will not penetrate a preoccupied mind." (To Sir J. Hooker, July 28, 1868, "More Letters", I. page 305. See also the letter to A.R. Wallace, April 30, 1868, in "More Letters" II. page 77, lines 6-8 from top.)

There are many naturalists, especially students of insects, who appear to entertain an inveterate hostility to any theory of mimicry. Some of them are eager investigators in the fascinating field of geographical distribution, so essential for the study of Mimicry itself. The changes of pattern undergone by a species of Erebia as we follow it over different parts of the mountain ranges of Europe is indeed a most interesting inquiry, but not more so than the differences between e.g. the Acraea johnstoni of S.E. Rhodesia and of Kilimanjaro. A naturalist who is interested by the Erebia should be equally interested by the Acraea; and so he would be if the student of mimicry did not also record that the characteristics which distinguish the northern from the southern individuals of the African species correspond with the presence, in the north but not in the south, of certain entirely different butterflies. That this additional information should so greatly weaken, in certain minds, the appeal of a favourite study, is a psychological problem of no little interest. This curious antagonism is I believe confined to a few students of insects. Those naturalists who, standing rather farther off, are able to see the bearings of the subject more clearly, will usually admit the general support yielded by an ever-growing mass of observations to the theories of Mimicry propounded by H.W. Bates and Fritz Muller. In like manner natural selection itself was in the early days often best understood and most readily accepted by those who were not naturalists. Thus Darwin wrote to D.T. Ansted, Oct. 27, 1860: "I am often in despair in making the generality of NATURALISTS even comprehend me. Intelligent men who are not naturalists and have not a bigoted idea of the term species, show more clearness of mind." ("More Letters", I. page 175.)

Even before the "Origin" appeared Darwin anticipated the first results upon the mind of naturalists. He wrote to Asa Gray, Dec. 21, 1859: "I have made up my mind to be well abused; but I think it of importance that my notions should be read by intelligent men, accustomed to scientific argument, though NOT naturalists. It may seem absurd, but I think such men will drag after them those naturalists who have too firmly fixed in their heads that a species is an entity." ("Life and Letters" II. page 245.)

Mimicry was not only one of the first great departments of zoological knowledge to be studied under the inspiration of natural Selection, it is still and will always remain one of the most interesting and important of subjects in relation to this theory as well as to evolution. In mimicry we investigate the effect of environment in its simplest form: we trace the effects of the pattern of a single species upon that of another far removed from it in the scale of classification. When there is reason to believe that the model is an invader from another region and has only recently become an element in the environment of the species native to its second home, the problem gains a special interest and fascination. Although we are chiefly dealing with the fleeting and changeable element of colour we expect to find and we do find evidence of a comparatively rapid evolution. The invasion of a fresh model is for certain species an unusually sudden change in the forces of the environment and in some instances we have grounds for the belief that the mimetic response has not been long delayed.

MIMICRY AND SEX.

Ever since Wallace's classical memoir on mimicry in the Malayan Swallowtail butterflies, those naturalists who have written on the subject have followed his interpretation of the marked prevalence of mimetic resemblance in the female sex as compared with the male. They have believed with Wallace that the greater dangers of the female, with slower flight and often alighting for oviposition, have been in part met by the high development of this special mode of protection. The fact cannot be doubted. It is extremely common for a non-mimetic male to be accompanied by a beautifully mimetic female and often by two or three different forms of female, each mimicking a different model. The male of a polymorphic mimetic female is, in fact, usually non-mimetic (e.g. Papilio dardanus = merope), or if a mimic (e.g. the Nymphaline genus Euripus), resembles a very different model. On the other hand a non-mimetic female accompanied by a mimetic male is excessively rare. An example is afforded by the Oriental Nymphaline, Cethosia, in which the males of some species are rough mimics of the brown Danaines. In some of the orb-weaving spiders the males mimic ants, while the much larger females are non-mimetic. When both sexes mimic, it is very common in butterflies and is also known in moths, for the females to be better and often far better mimics than the males.

Although still believing that Wallace's hypothesis in large part accounts for the facts briefly summarised above, the present writer has recently been led to doubt whether it offers a complete explanation. Mimicry in the male, even though less beneficial to the species than mimicry in the female, would still surely be advantageous. Why then is it so often entirely restricted to the female? While the attempt to find an answer to this question was haunting me, I re-read a letter written by Darwin to Wallace, April 15, 1868, containing the following sentences: "When female butterflies are more brilliant than their males you believe that they have in most cases, or in all cases, been rendered brilliant so as to mimic some other species, and thus escape danger. But can you account for the males not having been rendered equally brilliant and equally protected? Although it may be most for the welfare of the species that the female should be protected, yet it would be some advantage, certainly no disadvantage, for the unfortunate male to enjoy an equal immunity from danger. For my part, I should say that the female alone had happened to vary in the right manner, and that the beneficial variations had been transmitted to the same sex alone. Believing in this, I can see no improbability (but from analogy of domestic animals a strong probability) that variations leading to beauty must often have occurred in the males alone, and been transmitted to that sex alone. Thus I should account in many cases for the greater beauty of the male over the female, without the need of the protective principle." ("More Letters", II. pages 73, 74. On the same subject—"the gay-coloured females of Pieris" (Perrhybris (Mylothris) pyrrha of Brazil), Darwin wrote to Wallace, May 5, 1868, as follows: "I believe I quite follow you in believing that the colours are wholly due to mimicry; and I further believe that the male is not brilliant from not having received through inheritance colour from the female, and from not himself having varied; in short, that he has not been influenced by selection." It should be noted that the male of this species does exhibit a mimetic pattern on the under surface. "More Letters" II. page 78.)

The consideration of the facts of mimicry thus led Darwin to the conclusion that the female happens to vary in the right manner more commonly than the male, while the secondary sexual characters of males supported the conviction "that from some unknown cause such characters (viz. new characters arising in one sex and transmitted to it alone) apparently appear oftener in the male than in the female." (Letter from Darwin to Wallace, May 5, 1867, "More Letters", II. Page 61.)

Comparing these conflicting arguments we are led to believe that the first is the stronger. Mimicry in the male would be no disadvantage but an advantage, and when it appears would be and is taken advantage of by selection. The secondary sexual characters of males would be no advantage but a disadvantage to females, and, as Wallace thinks, are withheld from this sex by selection. It is indeed possible that mimicry has been hindered and often prevented from passing to the males by sexual selection. We know that Darwin was much impressed ("Descent of Man", page 325.) by Thomas Belt's daring and brilliant suggestion that the white patches which exist, although ordinarily concealed, on the wings of mimetic males of certain Pierinae (Dismorphia), have been preserved by preferential mating. He supposed this result to have been brought about by the females exhibiting a deep-seated preference for males that displayed the chief ancestral colour, inherited from periods before any mimetic pattern had been evolved in the species. But it has always appeared to me that Belt's deeply interesting suggestion requires much solid evidence and repeated confirmation before it can be accepted as a valid interpretation of the facts. In the present state of our knowledge, at any rate of insects and especially of Lepidoptera, it is probable that the female is more apt to vary than the male and that an important element in the interpretation of prevalent female mimicry is provided by this fact.

In order adequately to discuss the question of mimicry and sex it would be necessary to analyse the whole of the facts, so far as they are known in butterflies. On the present occasion it is only possible to state the inferences which have been drawn from general impressions,—inferences which it is believed will be sustained by future inquiry.

(1) Mimicry may occasionally arise in one sex because the differences which distinguish it from the other sex happen to be such as to afford a starting-point for the resemblance. Here the male is at no disadvantage as compared with the female, and the rarity of mimicry in the male alone (e.g. Cethosia) is evidence that the great predominance of female mimicry is not to be thus explained.

(2) The tendency of the female to dimorphism and polymorphism has been of great importance in determining this predominance. Thus if the female appear in two different forms and the male in only one it will be twice as probable that she will happen to possess a sufficient foundation for the evolution of mimicry.

(3) The appearance of melanic or partially melanic forms in the female has been of very great service, providing as it does a change of ground-colour. Thus the mimicry of the black generally red-marked American "Aristolochia swallowtails" (Pharmacophagus) by the females of Papilio swallowtails was probably begun in this way.

(4) It is probably incorrect to assume with Haase that mimicry always arose in the female and was later acquired by the male. Both sexes of the third section of swallowtails (Cosmodesmus) mimic Pharmacophagus in America, far more perfectly than do the females of Papilio. But this is not due to Cosmodesmus presenting us with a later stage of history begun in Papilio; for in Africa Cosmodesmus is still mimetic (of Danainae) in both sexes although the resemblances attained are imperfect, while many African species of Papilio have non-mimetic males with beautifully mimetic females. The explanation is probably to be sought in the fact that the females of Papilio are more variable and more often tend to become dimorphic than those of Cosmodesmus, while the latter group has more often happened to possess a sufficient foundation for the origin of the resemblance in patterns which, from the start, were common to male and female.

(5) In very variable species with sexes alike, mimicry can be rapidly evolved in both sexes out of very small beginnings. Thus the reddish marks which are common in many individuals of Limenitis arthemis were almost certainly the starting-point for the evolution of the beautifully mimetic L. archippus. Nevertheless in such cases, although there is no reason to suspect any greater variability, the female is commonly a somewhat better mimic than the male and often a very much better mimic. Wallace's principle seems here to supply the obvious interpretation.

(6) When the difference between the patterns of the model and presumed ancestor of the mimic is very great, the female is often alone mimetic; when the difference is comparatively small, both sexes are commonly mimetic. The Nymphaline genus Hypolimnas is a good example. In Hypolimnas itself the females mimic Danainae with patterns very different from those preserved by the non-mimetic males: in the sub-genus Euralia, both sexes resemble the black and white Ethiopian Danaines with patterns not very dissimilar from that which we infer to have existed in the non-mimetic ancestor.

(7) Although a melanic form or other large variation may be of the utmost importance in facilitating the start of a mimetic likeness, it is impossible to explain the evolution of any

185

detailed resemblance in this manner. And even the large colour variation itself may well be the expression of a minute and "continuous" change in the chemical and physical constitution of pigments.

SEXUAL SELECTION (EPIGAMIC CHARACTERS).

We do not know the date at which the idea of Sexual Selection arose in Darwin's mind, but it was probably not many years after the sudden flash of insight which, in October 1838, gave to him the theory of Natural Selection. An excellent account of Sexual Selection occupies the concluding paragraph of Part I. of Darwin's Section of the Joint Essay on Natural Selection, read July 1st, 1858, before the Linnean Society. ("Journ. Proc. Linn. Soc." Vol. III. 1859, page 50.) The principles are so clearly and sufficiently stated in these brief sentences that it is appropriate to quote the whole: "Besides this natural means of selection, by which those individuals are preserved, whether in their egg, or larval, or mature state, which are best adapted to the place they fill in nature, there is a second agency at work in most unisexual animals, tending to produce the same effect, namely, the struggle of the males for the females. These struggles are generally decided by the law of battle, but in the case of birds, apparently, by the charms of their song, by their beauty or their power of courtship, as in the dancing rock-thrush of Guiana. The most vigorous and healthy males, implying perfect adaptation, must generally gain the victory in their contests. This kind of selection, however, is less rigorous than the other; it does not require the death of the less successful, but gives to them fewer descendants. The struggle falls, moreover, at a time of year when food is generally abundant, and perhaps the effect chiefly produced would be the modification of the secondary sexual characters, which are not related to the power of obtaining food, or to defence from enemies, but to fighting with or rivalling other males. The result of this struggle amongst the males may be compared in some respects to that produced by those agriculturists who pay less attention to the careful selection of all their young animals, and more to the occasional use of a choice mate."

A full exposition of Sexual Selection appeared in the "The Descent of Man" in 1871, and in the greatly augmented second edition, in 1874. It has been remarked that the two subjects, "The Descent of Man and Selection in Relation to Sex", seem to fuse somewhat imperfectly into the single work of which they form the title. The reason for their association is clearly shown in a letter to Wallace, dated May 28, 1864: "... I suspect that a sort of sexual selection has been the most powerful means of changing the races of man." ("More Letters", II. page 33.)

Darwin, as we know from his Autobiography ("Life and Letters", I. page 94.), was always greatly interested in this hypothesis, and it has been shown in the preceding pages that he was inclined to look favourably upon it as an interpretation of many appearances usually explained by Natural Selection. Hence Sexual Selection, incidentally discussed in other sections of the present essay, need not be considered at any length, in the section specially allotted to it.

Although so interested in the subject and notwithstanding his conviction that the hypothesis was sound, Darwin was quite aware that it was probably the most vulnerable part of the "Origin". Thus he wrote to H.W. Bates, April 4, 1861: "If I had to cut up myself in a review I would have (worried?) and quizzed sexual selection; therefore, though I am fully convinced that it is largely true, you may imagine how pleased I am at what you say on your belief." ("More Letters", I. page 183.)

The existence of sound-producing organs in the males of insects was, Darwin considered, the strongest evidence in favour of the operation of sexual selection in this group. ("Life and Letters", III. pages 94, 138.) Such a conclusion has received strong support in recent years by the numerous careful observations of Dr F.A. Dixey ("Proc. Ent. Soc. Lond." 1904, page lvi; 1905, pages xxxvii, liv; 1906, page ii.) and Dr G.B. Longstaff ("Proc. Ent. Soc. Lond." 1905, page xxxv; "Trans. Ent. Soc. Lond." 1905, page 136; 1908, page 607.) on the scents of male butterflies. The experience of these naturalists abundantly confirms and extends the account given by Fritz Muller ("Jen. Zeit." Vol. XI. 1877, page 99; "Trans. Ent. Soc. Lond." 1878, page 211.) of the scents of certain Brazilian butterflies. It is a remarkable fact that the apparently epigamic scents of male butterflies should be pleasing to man while the apparently aposematic scents in both sexes of species with warning colours should be displeasing to him. But the former is far more surprising than the latter. It is not perhaps astonishing that a scent which is ex hypothesi unpleasant to an insect-eating Vertebrate should be displeasing to the human sense; but it is certainly wonderful that an odour which is ex hypothesi agreeable to a female butterfly should also be agreeable to man.

Entirely new light upon the seasonal appearance of epigamic characters is shed by the recent researches of C.W. Beebe ("The American Naturalist", Vol. XLII. No. 493, Jan. 1908, page 34.), who caused the scarlet tanager (Piranga erythromelas) and the bobolink (Dolichonyx oryzivorus) to retain their breeding plumage through the whole year by means of fattening food, dim illumination, and reduced activity. Gradual restoration to the light and the addition of meal-worms to the diet invariably brought back the spring song, even in the middle of winter. A sudden alteration of temperature, either higher or lower, caused the birds nearly to stop feeding, and one tanager lost weight rapidly and in two weeks moulted into the olive-green winter plumage. After a year, and at the beginning of the normal breeding season, "individual tanagers and bobolinks were gradually brought under normal conditions and activities," and in every case moulted from nuptial plumage to nuptial plumage. "The dull colours of the winter season had been skipped." The author justly claims to have established "that the sequence of plumage in these birds is not in any way predestined through inheritance..., but that it may be interrupted by certain factors in the environmental complex."

XVI. GEOGRAPHICAL DISTRIBUTION OF PLANTS. By Sir William Thiselton-Dyer, K.C.M.G., C.I.E. Sc.D., F.R.S.

The publication of "The Origin of Species" placed the study of Botanical Geography on an entirely new basis. It is only necessary to study the monumental "Geographie Botanique raisonnee" of Alphonse De Candolle, published four years earlier (1855), to realise how profound and far-reaching was the change. After a masterly and exhaustive discussion of all available data De Candolle in his final conclusions could only arrive at a deadlock. It is sufficient to quote a few sentences:—

"L'opinion de Lamarck est aujourd'hui abandonee par tous les naturalistes qui ont etudie sagement les modifications possibles des etres organises...

"Et si l'on s'ecarte des exagerations de Lamarck, si l'on suppose un premier type de chaque genre, de chaque famille tout au moins, on se trouve encore a l'egard de l'origine de ces types en presence de la grande question de la creation.

"Le seul parti a prendre est donc d'envisager les etres organises comme existant depuis certaines epoques, avec leurs qualites particulieres." (Vol. II. page 1107.)

Reviewing the position fourteen years afterwards, Bentham remarked:—"These views, generally received by the great majority of naturalists at the time De Candolle wrote, and still maintained by a few, must, if adhered to, check all further enquiry into any connection of facts with causes," and he added, "there is little doubt but that if De Candolle were to revise his work, he would follow the example of so many other eminent naturalists, and... insist that the present geographical distribution of plants was in most instances a derivative one, altered from a very different former distribution." ("Pres. Addr." (1869) "Proc. Linn. Soc." 1868-69, page lxviii.)

Writing to Asa Gray in 1856, Darwin gave a brief preliminary account of his ideas as to the origin of species, and said that geographical distribution must be one of the tests of their validity. ("Life and Letters", II. page 78.) What is of supreme interest is that it was also their starting-point. He tells us:—"When I visited, during the voyage of H.M.S. "Beagle", the Galapagos Archipelago,... I fancied myself brought near to the very act of creation. I often asked myself how these many peculiar animals and plants had been produced: the simplest answer seemed to be that the inhabitants of the several islands had descended from each other, undergoing modification in the course of their descent." ("The Variation of Animals and Plants" (2nd edition), 1890, I. pages 9, 10.) We need not be surprised then, that in writing in 1845 to Sir Joseph Hooker, he speaks of "that grand subject, that almost keystone of the laws of creation, Geographical Distribution." ("Life and Letters", I. page 336.)

Yet De Candolle was, as Bentham saw, unconsciously feeling his way, like Lyell, towards evolution, without being able to grasp it. They both strove to explain phenomena by means of agencies which they saw actually at work. If De Candolle gave up the ultimate problem as insoluble:—"La creation ou premiere formation des etres organises echappe, par sa nature et par son anciennete, a nos moyens d'observation" (Loc. cit. page 1106.), he steadily endeavoured to minimise its scope. At least half of his great work is devoted to the researches by which he extricated himself from a belief in species having had a multiple origin, the view which had been held by successive naturalists from Gmelin to Agassiz. To account for the obvious fact that species constantly occupy disseverbed areas, De Candolle made a minute study of their means of transport. This was found to dispose of the vast majority of cases, and the remainder he accounted for by geographical change. (Loc. cit. page 1116.)

But Darwin strenuously objected to invoking geographical change as a solution of every difficulty. He had apparently long satisfied himself as to the "permanence of continents and great oceans." Dana, he tells us "was, I believe, the first man who maintained" this ("Life and Letters", III. page 247. Dana says:—"The continents and oceans had their general outline or form defined in earliest time," "Manual of Geology", revised edition. Philadelphia, 1869, page 732. I have no access to an earlier edition.), but he had himself probably arrived at it independently. Modern physical research tends to confirm it. The earth's centre of gravity, as pointed out by Pratt from the existence of the Pacific Ocean, does not coincide with its centre of figure, and it has been conjectured that the Pacific Ocean dates its origin from the separation of the moon from the earth.

The conjecture appears to be unnecessary. Love shows that "the force that keeps the Pacific Ocean on one side of the earth is gravity, directed more towards the centre of gravity than the centre of the figure." ("Report of the 77th Meeting of the British Association" (Leicester, 1907), London, 1908, page 431.) I can only summarise the conclusions of a technical but masterly discussion. "The broad general features of the distribution of continent and ocean can be regarded as the consequences of simple causes of a dynamical character," and finally, "As regards the contour of the great ocean basins, we seem to be justified in saying that the earth is approximately an oblate spheroid, but more nearly an ellipsoid with three unequal axes, having its surface furrowed according to the formula for a certain spherical harmonic of the third degree" (Ibid. page 436.), and he shows that this furrowed surface must be produced "if the density is greater in one hemispheroid than in the other, so that the position of the centre of gravity is eccentric." (Ibid. page 431.) Such a modelling of the earth's surface can only be referred to a primitive period of plasticity. If the furrows account for the great ocean basins, the disposition of the continents seems equally to follow. Sir George Darwin has pointed out that they necessarily "arise from a supposed primitive viscosity or plasticity of the earth's mass. For during this course of evolution the earth's mass must have suffered a screwing motion, so that the polar regions have travelled a little from west to east relatively to the equator. This affords a possible explanation of the north and south trend of our great continents." ("Encycl. Brit." (9th edition), Vol. XXIII. "Tides", page 379.)

It would be trespassing on the province of the geologist to pursue the subject at any length. But as Wallace ("Island Life" (2nd edition), 1895, page 103.), who has admirably vindicated Darwin's position, points out, the "question of the permanence of our continents... lies at the root of all our inquiries into the great changes of the earth and its inhabitants." But he proceeds: "The very same evidence which has been adduced to prove the GENERAL stability and permanence of our continental areas also goes to prove that they have been subjected to wonderful and repeated changes in DETAIL." (Loc. cit. page 101.) Darwin of course would have admitted this, for with a happy expression he insisted to Lyell (1856) that "the skeletons, at least, of our continents are ancient." ("More Letters", II. page 135.) It is impossible not to admire the courage and tenacity with which he carried on the conflict single-handed. But he failed to convince Lyell. For we still find him maintaining in the last edition of the "Principles": "Continents therefore, although permanent for whole geological epochs, shift their positions entirely in the course of ages." (Lyell's "Principles of Geology" (11th edition), London, 1872, I. page 258.)

Evidence, however, steadily accumulates in Darwin's support. His position still remains inexpugnable that it is not permissible to invoke geographical change to explain difficulties in distribution without valid geological and physical support. Writing to Mellard Reade, who in 1878 had said, "While believing that the ocean-depths are of enormous age, it is impossible to reject other evidences that they have once been land," he pointed out "the statement from the 'Challenger' that all sediment is deposited within one or two hundred miles from the shores." ("More Letters", II. page 146.) The following year Sir Archibald Geikie ("Geographical Evolution", "Proc. R. Geogr. Soc." 1879, page 427.) informed the Royal Geographical Society that "No part of the results obtained by the 'Challenger' expedition has a profounder interest for geologists and geographers than the proof which they furnish that the floor of the ocean basins has no real analogy among the sedimentary formations which form most of the framework of the land."

Nor has Darwin's earlier argument ever been upset. "The fact which I pointed out many years ago, that all oceanic islands are volcanic (except St Paul's, and now that is viewed by some as the nucleus of an ancient volcano), seem to me a strong argument that no continent ever occupied the great oceans." ("More Letters", II. page 146.)

Dr Guppy, who devoted several years to geological and botanical investigations in the Pacific, found himself forced to similar conclusions. "It may be at once observed," he says, "that my belief in the general principle that islands have always been islands has not been shaken," and he entirely rejects "the hypothesis of a Pacific continent." He comes back, in full view of the problems on the spot, to the position from which, as has been seen, Darwin started: "If the distribution of a particular group of plants or animals does not seem to accord with the present arrangement of the land, it is by far the safest plan, even after exhausting all likely modes of explanation, not to invoke the intervention of geographical changes; and I scarcely think that our knowledge of any one group of organisms is ever sufficiently precise to justify a recourse to hypothetical alterations in the present relations of land and sea." ("Observations of a Naturalist in the Pacific between 1896 and 1899", London, 1903, I. page 380.) Wallace clinches the matter when he finds "almost the whole of the vast areas of the Atlantic, Pacific, Indian, and Southern Oceans, without a solitary relic of the great islands or continents supposed to have sunk beneath their waves." ("Island Life", page 105.)

Writing to Wallace (1876), Darwin warmly approves the former's "protest against sinking imaginary continents in a quite reckless manner, as was stated by Forbes, followed, alas, by Hooker, and caricatured by Wollaston and (Andrew) Murray." ("Life and Letters", III. page 230.) The transport question thus became of enormously enhanced importance. We need not be surprised then at his writing to Lyell in 1856:—"I cannot avoid thinking that Forbes's 'Atlantis' was an ill-service to science, as checking a close study of means of dissemination" (Ibid. II. page 78.), and Darwin spared no pains to extend our knowledge of them. He implores Hooker, ten years later, to "admit how little is known on the subject," and summarises with some satisfaction what he had himself achieved:—"Remember how recently you and others thought that salt water would soon kill seeds... Remember that no one knew that seeds would remain for many hours in the crops of birds and retain their vitality; that fish eat seeds, and that when the fish are devoured by birds the seeds can germinate, etc. Remember that every year many birds are blown to Madeira and to the Bermudas. Remember that dust is blown 1000 miles across the Atlantic." ("More Letters", I. page 483.)

It has always been the fashion to minimise Darwin's conclusions, and these have not escaped objection. The advocatus diaboli has a useful function in science. But in attacking Darwin his brief is generally found to be founded on a slender basis of facts. Thus Winge and Knud Andersen have examined many thousands of migratory birds and found "that their crops and stomachs were always empty. They never observed any seeds adhering to the feathers, beaks or feet of the birds." (R.F. Scharff, "European Animals", page 64, London, 1907.) The most considerable investigation of the problem of Plant Dispersal since Darwin is that of Guppy. He gives a striking illustration of how easily an observer may be led into error by relying on negative evidence.

"When Ekstam published, in 1895, the results of his observations on the plants of Nova Zembla, he observed that he possessed no data to show whether swimming and wading birds fed on berries; and he attached all importance to dispersal by winds. On subsequently visiting Spitzbergen he must have been at first inclined, therefore, to the opinion of Nathorst, who, having found only a solitary species of bird (a snow-sparrow) in that region, naturally concluded that birds had been of no importance as agents in the plant-stocking. However, Ekstam's opportunities were greater, and he tells us that in the craws of six specimens of Lagopus hyperboreus shot in Spitzbergen in August he found represented almost 25 per cent. of the usual phanerogamic flora of that region in the form of fruits, seeds, bulbils, flower-buds, leaf-buds, etc... "

"The result of Ekstam's observations in Spitzbergen was to lead him to attach a very considerable importance in plant dispersal to the agency of birds; and when in explanation of the Scandinavian elements in the Spitzbergen flora he had to choose between a former land connection and the agency of birds, he preferred the bird." (Guppy, op. cit. II. pages 511, 512.)

Darwin objected to "continental extensions" on geological grounds, but he also objected to Lyell that they do not "account for all the phenomena of distribution on islands" ("Life and Letters", II. page 77.), such for example as the absence of Acacias and Banksias in New Zealand. He agreed with De Candolle that "it is poor work putting together the merely POSSIBLE means of distribution." But he also agreed with him that they were the only practicable door of escape from multiple origins. If they would not work then "every one who believes in single centres will have to admit continental extensions" (Ibid. II. page 82.), and that he regarded as a mere counsel of despair:—"to make continents, as easily as a cook does pancakes." (Ibid. II. page 74.)

The question of multiple origins however presented itself in another shape where the solution was much more difficult. The problem, as stated by Darwin, is this:—"The identity of many plants and animals, on mountain-summits, separated from each other by hundreds of miles of lowlands... without the apparent possibility of their having migrated from one point to the other." He continues, "even as long ago as 1747, such facts led Gmelin to conclude that the same species must have been independently created at several distinct points; and we might have remained in this same belief, had not Agassiz and others called vivid attention to the Glacial period, which affords... a simple explanation of the facts." ("Origin of Species" (6th edition) page 330.)

The "simple explanation" was substantially given by E. Forbes in 1846. It is scarcely too much to say that it belongs to the same class of fertile and far-reaching ideas as "natural selection" itself. It is an extraordinary instance, if one were wanted at all, of Darwin's magnanimity and intense modesty that though he had arrived at the theory himself, he acquiesced in Forbes receiving the well-merited credit. "I have never," he says, "of course alluded in print to my having independently worked out this view." But he would have been more than human if he had not added:—"I was forestalled in... one important point, which my vanity has always made me regret." ("Life and Letters", I. page 88.)

Darwin, however, by applying the theory to trans-tropical migration, went far beyond Forbes. The first enunciation to this is apparently contained in a letter to Asa Gray in 1858. The whole is too long to quote, but the pith is contained in one paragraph. "There is a considerable body of geological evidence that during the Glacial epoch the whole world was colder; I inferred that,... from erratic boulder phenomena carefully observed by me on both the east and west coast of South America. Now I am so bold as to believe that at the height of the Glacial epoch, AND WHEN ALL TROPICAL PRODUCTIONS MUST HAVE BEEN CONSIDERABLY DISTRESSED, several temperate forms slowly travelled into the heart of the Tropics, and even reached the southern hemisphere; and some few southern forms penetrated in a reverse direction northward." ("Life and Letters", II. page 136.) Here again it is clear that though he credits Agassiz with having called vivid attention to the Glacial period, he had himself much earlier grasped the idea of periods of refrigeration.

Putting aside the fact, which has only been made known to us since Darwin's death, that he had anticipated Forbes, it is clear that he gave the theory a generality of which the latter had no conception. This is pointed out by Hooker in his classical paper "On the Distribution of Arctic Plants" (1860). "The theory of a southern migration of northern types being due to the cold epochs preceding and during the glacial, originated, I believe, with the late Edward Forbes; the extended one, of the trans-tropical migration, is Mr Darwin's." ("Linn. Trans." XXIII. page 253. The attempt appears to have been made to claim for Heer priority in what I may term for short the arctic-alpine theory (Scharff, "European Animals", page 128). I find no suggestion of his having hit upon it in his correspondence with Darwin or Hooker. Nor am I aware of any reference to his having done so in his later publications. I am indebted to his biographer, Professor Schroter, of Zurich, for an examination of his earlier papers with an equally negative result.) Assuming that local races have derived from a common ancestor, Hooker's great paper placed the fact of the migration on an impregnable basis. And, as he pointed out, Darwin has shown that "such an explanation meets the difficulty of accounting for the restriction of so many American and Asiatic arctic types to their own peculiar longitudinal zones, and for what is a far greater difficulty, the representation of the same arctic genera by most closely allied species in different longitudes."

The facts of botanical geography were vital to Darwin's argument. He had to show that they admitted of explanation without assuming multiple origins for species, which would be fatal to the theory of Descent. He had therefore to strengthen and extend De Candolle's work as to means of transport. He refused to supplement them by hypothetical geographical changes for which there was no independent evidence: this was simply to attempt to explain ignotum per ignotius. He found a real and, as it has turned out, a far-reaching solution in climatic change due to cosmical causes which compelled the migration of species as a condition of their existence. The logical force of the argument consists in dispensing with any violent assumption, and in showing that the principle of descent is adequate to explain the ascertained facts.

It does not, I think, detract from the merit of Darwin's conclusions that the tendency of modern research has been to show that the effects of the Glacial period were less simple, more localised and less general than he perhaps supposed. He admitted that "equatorial refrigeration... must have been small." ("More Letters", I. page 177.) It may prove possible to dispense with it altogether. One cannot but regret that as he wrote to Bates:—"the sketch in the 'Origin' gives a very meagre account of my fuller MS. essay on this subject." (Loc. cit.) Wallace fully accepted "the effect of the Glacial epoch in bringing about the present distribution of Alpine and Arctic plants in the NORTHERN HEMISPHERE," but rejected "the lowering of the temperature of the tropical regions during the Glacial period" in order to account for their presence in the SOUTHERN hemisphere. ("More Letters", II. page 25 (footnote 1).) The divergence however does not lie very deep. Wallace attaches more importance to ordinary means of transport. "If plants can pass in considerable numbers and variety over wide seas and oceans, it must be yet more easy for them to traverse continuous areas of land, wherever mountain-chains offer suitable stations." ("Island Life" (2nd edition), London, 1895, page 512.) And he argues that such periodical changes of climate, of which the Glacial period may be taken as a type, would facilitate if not stimulate the process. (Loc. cit. page 518.)

It is interesting to remark that Darwin drew from the facts of plant distribution one of his most ingenious arguments in support of this theory. (See "More Letters", I. page 424.) He tells us, "I was led to anticipate that the species of the larger genera in each country would oftener present varieties, than the species of the smaller genera." ("Origin", page 44.) He argues "where, if we may use the expression, the manufactory of species has been active, we ought generally to find the manufactory still in action." (Ibid. page 45.) This proved to be the case. But the labour imposed upon him in the study was immense. He tabulated local floras "belting the whole northern hemisphere" ("More Letters", I. page 107.), besides voluminous works such as De Candolle's "Prodromus". The results scarcely fill a couple of pages. This is a good illustration of the enormous pains which he took to base any statement on a secure foundation of evidence, and for this the world, till the publication of his letters, could not do him justice. He was a great admirer of Herbert Spencer, whose "prodigality of original thought" astonished him. "But," he says, "the reflection constantly recurred to me that each suggestion, to be of real value to service, would require years of work." (Ibid. II. page 235.)

At last the ground was cleared and we are led to the final conclusion. "If the difficulties be not insuperable in admitting that in the long course of time all the individuals of the same species belonging to the same genus, have proceeded from some one source; then all the grand leading facts of geographical distribution are explicable on the theory of migration, together with subsequent modification and the multiplication of new forms." ("Origin", page 360.) In this single sentence Darwin has stated a theory which, as his son F. Darwin has said with justice, has "revolutionized botanical geography." ("The Botanical Work of Darwin", "Ann. Bot." 1899, page xi.) It explains how physical barriers separate and form botanical regions; how allied species become concentrated in the same areas; how, under similar physical conditions, plants may be essentially dissimilar, showing that descent and not the surroundings is the controlling factor; how insular floras have acquired their peculiarities; in short how the most various and apparently uncorrelated problems fall easily and inevitably into line.

The argument from plant distribution was in fact irresistible. A proof, if one were wanted, was the immediate conversion of what Hooker called "the stern keen intellect" ("More Letters", I. page 134.) of Bentham, by general consent the leading botanical systematist at the time. It is a striking historical fact that a paper of his own had been set down for reading at the Linnean Society on the same day as Darwin's, but had to give way. In this he advocated the fixity of species. He withdrew it after hearing Darwin's. We can hardly realise now the momentous effect on the scientific thought of the day of the announcement of the new theory. Years afterwards (1882) Bentham, notwithstanding his habitual restraint, could not write of it without emotion. "I was forced, however reluctantly, to give up my long-cherished convictions, the results of much labour and study." The revelation came without preparation. Darwin, he wrote, "never made any communications to me in relation to his views and labours." But, he adds, "I... fully adopted his theories and conclusions, notwithstanding the severe pain and disappointment they at first occasioned me." ("Life and Letters", II. page 294.) Scientific history can have few incidents more worthy. I do not know what is most striking in the story, the pathos or the moral dignity of Bentham's attitude.

Darwin necessarily restricted himself in the "Origin" to establishing the general principles which would account for the facts of distribution, as a part of his larger argument, without attempting to illustrate them in particular cases. This he appears to have contemplated doing in a separate work. But writing to Hooker in 1868 he said:—"I shall to the day of my death keep up my full interest in Geographical Distribution, but I doubt whether I shall ever have strength to come in any fuller detail than in the "Origin" to this grand subject." ("More Letters", II. page 7.) This must be always a matter for regret. But we may gather some indication of his later speculations from the letters, the careful publication of which by F. Darwin has rendered a service to science, the value of which it is difficult to exaggerate. They admit us to the workshop, where we see a great theory, as it were, in the making. The later ideas that they contain were not it is true public property at the time. But they were communicated to the leading biologists of the day and indirectly have had a large influence.

If Darwin laid the foundation, the present fabric of Botanical Geography must be credited to Hooker. It was a happy partnership. The far-seeing, generalising power of the one was supplied with data and checked in conclusions by the vast detailed knowledge of the other. It may be permitted to quote Darwin's generous acknowledgment when writing the "Origin":—"I never did pick any one's pocket, but whilst writing my present chapter I keep on feeling (even when differing most from you) just as if I were stealing from you, so much do I owe to your writings and conversation, so much more than mere acknowledgements show." ("Life and Letters", II. page 148 (footnote).) Fourteen years before he had written to Hooker: "I know I shall live to see you the first authority in Europe on... Geographical Distribution." (Ibid. I. page 336.) We owe it to Hooker that no one now undertakes the flora of a country without indicating the range of the species it contains. Bentham tells us: "After De Candolle, independently of the great works of Darwin... the first important addition to the science of geographical botany was that made by Hooker in his "Introductory Essay to the Flora of Tasmania", which, though contemporaneous only with the "Origin of Species", was drawn up with a general knowledge of his friend's observations and views." (Pres. Addr. (1869), "Proc. Linn. Soc." 1868-69, page lxxiv.) It cannot be doubted that this and the great memoir on the "Distribution of Arctic Plants" were only less epoch-making than the "Origin" itself, and must have supplied a powerful support to the general theory of organic evolution.

Darwin always asserted his "entire ignorance of Botany." ("More Letters", I. page 400.) But this was only part of his constant half-humorous self-depreciation. He had been a pupil of Henslow, and it is evident that he had a good working knowledge of systematic botany. He could find his way about in the literature and always cites the names of plants with scrupulous accuracy. It was because he felt the want of such a work for his own researches that he urged the preparation of the "Index Kewensis", and undertook to defray the expense. It has been thought singular that he should have been elected a "correspondant" of the Academie des Sciences in the section of Botany, but it is not surprising that his work in Geographical Botany made the botanists anxious to claim him. His heart went with them. "It has always pleased me," he tells us, "to exalt plants in the scale of organised beings." ("Life and Letters", I. page 98.) And he declares that he finds "any proposition more easily tested in botanical works (Ibid. II. page 99.) than in zoological."

In the "Introductory Essay" Hooker dwelt on the "continuous current of vegetation from Scandinavia to Tasmania" ("Introductory Essay to the Flora of Tasmania", London, 1859. Reprinted from the "Botany of the Antarctic Expedition", Part III., "Flora of Tasmania", Vol I. page ciii.), but finds little evidence of one in the reverse direction. "In the New World, Arctic, Scandinavian, and North American genera and species are continuously extended from the north to the south temperate and even Antarctic zones; but scarcely one Antarctic species, or even genus advances north beyond the Gulf of Mexico" (page civ.). Hooker considered that this negatived "the idea that the Southern and Northern Floras have had common origin within comparatively modern geological epochs." (Loc. cit.) This is no doubt a correct conclusion. But it is difficult to explain on Darwin's view alone, of alternating cold in the two hemispheres, the preponderant migration from the north to the south. He suggests, therefore, that it "is due to the greater extent of land in the north and to the northern forms... having... been advanced through natural selection and competition to a higher stage of perfection or dominating power." ("Origin of Species" (6th edition), page 340; cf. also "Life and Letters", II. page 142.) The present state of the Flora of New Zealand affords a striking illustration of the correctness of this view. It is poor in species, numbering only some 1400, of which three-fourths are endemic. They seem however quite unable to resist the invasion of new comers and already 600 species of foreign origin have succeeded in establishing themselves.

If we accept the general configuration of the earth's surface as permanent a continuous and progressive dispersal of species from the centre to the circumference, i.e. southwards, seems inevitable. If an observer were placed above a point in St George's Channel from which one half of the globe was visible he would see the greatest possible quantity of land spread out in a sort of stellate figure. The maritime supremacy of the English race has perhaps flowed from the central position of its home. That such a disposition would facilitate a centrifugal migration of land organisms is at any rate obvious, and fluctuating conditions of climate operating from the pole would supply an effective means of propulsion. As these became more rigorous animals at any rate would move southwards to escape them. It would be equally the case with plants if no insuperable obstacle interposed. This implies a mobility in plants, notwithstanding what we know of means of transport which is at first sight paradoxical. Bentham has stated this in a striking way: "Fixed and immovable as is the individual plant, there is no class in which the race is endowed with greater facilities for the widest dispersion... Plants cast away their offspring in a dormant state, ready to be carried to any distance by those external agencies which we may deem fortuitous, but without which many a race might perish from the exhaustion of the limited spot of soil in which it is rooted." (Pres. Addr.(1869), "Proc. Linn. Soc." 1868-69, pages lxvi, lxvii.)

I have quoted this passage from Bentham because it emphasises a point which Darwin for his purpose did not find it necessary to dwell upon, though he no doubt assumed it. Dispersal to a distance is, so to speak, an accidental incident in the life of a species. Lepidium Draba, a native of South-eastern Europe, owes its prevalence in the Isle of Thanet to the disastrous Walcheren expedition; the straw-stuffing of the mattresses of the fever-stricken soldiers who were landed there was used by a farmer for manure. Sir Joseph Hooker ("Royal Institution Lecture", April 12, 1878.) tells us that landing on Lord Auckland's Island, which was uninhabited, "the first evidence I met with of its having been previously visited by man was the English chickweed; and this I traced to a mound that marked the grave of a British sailor, and that was covered with the plant, doubtless the offspring of seed that had adhered to the spade or mattock with which the grave had been dug."

Some migration from the spot where the individuals of a species have germinated is an essential provision against extinction. Their descendants otherwise would be liable to suppression by more vigorous competitors. But they would eventually be extinguished inevitably, as pointed out by Bentham, by the exhaustion of at any rate some one necessary constituent of the soil. Gilbert showed by actual analysis that the production of a "fairy ring" is simply due to the using up by the fungi of the available nitrogen in the enclosed area which continually enlarges as they seek a fresh supply on the outside margin. Anyone who cultivates a garden can easily verify the fact that every plant has some adaptation for varying degrees of seed-dispersal. It cannot be doubted that slow but persistent terrestrial migration has played an enormous part in bringing about existing plant-distribution, or that climatic changes would intensify the effect because they would force the abandonment of a former area and the occupation of a new one. We are compelled to admit that as an incident of the Glacial period a whole flora may have moved down and up a mountain side, while only some of its constituent species would be able to take advantage of means of long-distance transport.

I have dwelt on the importance of what I may call short-distance dispersal as a necessary condition of plant life, because I think it suggests the solution of a difficulty which leads Guppy to a conclusion with which I am unable to agree. But the work which he has done taken as a whole appears to me so admirable that I do so with the utmost respect. He points out, as Bentham had already done, that long-distance dispersal is fortuitous. And being so it cannot have been provided for by previous adaptation. He says (Guppy, op. cit. II. page 99.): "It is not conceivable that an organism can be adapted to conditions outside its environment." To this we must agree; but, it may be asked, do the general means of plant dispersal violate so obvious a principle? He proceeds: "The great variety of the modes of dispersal of seeds is in itself an indication that the dispersing agencies avail themselves in a hap-hazard fashion of characters and capacities that have been developed in other connections." (Loc. cit. page 102.) "Their utility in these respects is an accident in the plant's life." (Loc. cit. page 100.) He attributes this utility to a "determining agency," an influence which constantly reappears in various shapes in the literature of Evolution and is ultra-scientific in the sense that it bars the way to the search for material causes. He goes so far as to doubt whether fleshy fruits are an adaptation for the dispersal of their contained seeds. (Loc. cit. page 102.) Writing as I am from a hillside which is covered by hawthorn bushes sown by birds, I confess I can feel little doubt on the subject myself. The essential fact which Guppy brings out is that long-distance unlike short-distance dispersal is not universal and purposeful, but selective and in that sense accidental. But it is not difficult to see how under favouring conditions one must merge into the other.

Guppy has raised one novel point which can only be briefly referred to but which is of extreme interest. There are grounds for thinking that flowers and insects have mutually reacted upon one another in their evolution. Guppy suggests that something of the same kind may be true of birds. I must content myself with the quotation of a single sentence. "With the secular drying of the globe and the consequent differentiation of climate is to be connected the suspension to a great extent of the agency of birds as plant dispersers in later ages, not only in the Pacific Islands but all over the tropics. The changes of climate, birds and plants have gone on together, the range of the bird being controlled by the climate, and the distribution of the plant being largely dependent on the bird." (Loc.cit. II. page 221.)

Darwin was clearly prepared to go further than Hooker in accounting for the southern flora by dispersion from the north. Thus he says: "We must, I suppose, admit that every yard of land has been successively covered with a beech-forest between the Caucasus and Japan." ("More Letters", II. page 9.) Hooker accounted for the dissevered condition of the southern flora by geographical change, but this Darwin could not admit. He suggested to Hooker that the Australian and Cape floras might have had a point of connection through Abyssinia (Ibid. I. page 447.), an idea which was promptly snuffed out. Similarly he remarked to Bentham (1869): "I suppose you think that the Restiaceae, Proteaceae, etc., etc. once extended over the whole world, leaving fragments in the south." (Ibid. I. page 380.) Eventually he conjectured "that there must have been a Tertiary Antarctic continent, from which various forms radiated to the southern extremities of our present continents." ("Life and Letters", III. page 231.) But characteristically he could not admit any land connections and trusted to "floating ice for transporting seed." ("More Letters", I. page 116.) I am far from saying that this theory is not deserving of serious attention, though there seems to be no positive evidence to support it, and it immediately raises the difficulty how did such a continent come to be stocked?

We must, however, agree with Hooker that the common origin of the northern and southern floras must be referred to a remote past. That Darwin had this in his mind at the time of the publication of the "Origin" is clear from a letter to Hooker. "The view which I should have looked at as perhaps most probable (though it hardly differs from yours) is that the whole world during the Secondary ages was inhabited by marsupials, araucarias (Mem.— Fossil wood of this nature in South America), Banksia, etc.; and that these were supplanted and exterminated in the greater area of the north, but were left alive in the south." (Ibid. I. page 453.) Remembering that Araucaria, unlike Banksia, belongs to the earlier Jurassic not to the angiospermous flora, this view is a germinal idea of the widest generality.

The extraordinary congestion in species of the peninsulas of the Old World points to the long-continued action of a migration southwards. Each is in fact a cul-de-sac into which they have poured and from which there is no escape. On the other hand the high degree of specialisation in the southern floras and the little power the species possess of holding their own in competition or in adaptation to new conditions point to long-continued isolation. "An island... will prevent free immigration and competition, hence a greater number of ancient forms will survive." (Ibid. I. page 481.) But variability is itself subject to variation.

The nemesis of a high degree of protected specialisation is the loss of adaptability. (See Lyell, "The Geological Evidences of the Antiquity of Man", London, 1863, page 446.) It is probable that many elements of the southern flora are doomed: there is, for example, reason to think that the singular Stapelieae of S. Africa are a disappearing group. The tree Lobelias which linger in the mountains of Central Africa, in Tropical America and in the Sandwich Islands have the aspect of extreme antiquity. I may add a further striking illustration from Professor Seward: "The tall, graceful fronds of Matonia pectinata, forming miniature forests on the slopes of Mount Ophir and other districts in the Malay Peninsula in association with Dipteris conjugata and Dipteris lobbiana, represent a phase of Mesozoic life which survives 'Like a dim picture of the drowned past.'" ("Report of the 73rd Meeting of the British Assoc." (Southport, 1903), London, 1904, page 844.)

The Matonineae are ferns with an unusually complex vascular system and were abundant "in the northern hemisphere during the earlier part of the Mesozoic era."

It was fortunate for science that Wallace took up the task which his colleague had abandoned. Writing to him on the publication of his "Geographical Distribution of Animals" Darwin said: "I feel sure that you have laid a broad and safe foundation for all future work on Distribution. How interesting it will be to see hereafter plants treated in strict relation to your views." ("More Letters", II. page 12.) This hope was fulfilled in "Island Life". I may quote a passage from it which admirably summarises the contrast between the northern and the southern floras.

"Instead of the enormous northern area, in which highly organised and dominant groups of plants have been developed gifted with great colonising and aggressive powers, we have in the south three comparatively small and detached areas, in which rich floras have been developed with SPECIAL adaptations to soil, climate, and organic environment, but comparatively impotent and inferior beyond their own domain." (Wallace, "Island Life", pages 527, 528.)

It will be noticed that in the summary I have attempted to give of the history of the subject, efforts have been concentrated on bringing into relation the temperate floras of the northern and southern hemispheres, but no account has been taken of the rich tropical vegetation which belts the world and little to account for the original starting-point of existing vegetation generally. It must be remembered on the one hand that our detailed knowledge of the floras of the tropics is still very incomplete and far inferior to that of temperate regions; on the other hand palaeontological discoveries have put the problem in an entirely new light. Well might Darwin, writing to Heer in 1875, say: "Many as have been the wonderful discoveries in Geology during the last half-century, I think none have exceeded in interest your results with respect to the plants which formerly existed in the arctic regions." ("More Letters", II. page 240.)

As early as 1848 Debey had described from the Upper Cretaceous rocks of Aix-la-Chapelle Flowering plants of as high a degree of development as those now existing. The fact was commented upon by Hooker ("Introd. Essay to the Flora of Tasmania", page xx.), but its full significance seems to have been scarcely appreciated. For it implied not merely that their evolution must have taken place but the foundations of existing distribution must have been laid in a preceding age. We now know from the discoveries of the last fifty years that the remains of the Neocomian flora occur over an area extending through 30 deg of latitude. The conclusion is irresistible that within this was its centre of distribution and probably of origin.

Darwin was immensely impressed with the outburst on the world of a fully fledged angiospermous vegetation. He warmly approved the brilliant theory of Saporta that this happened "as soon (as) flower-frequenting insects were developed and favoured intercrossing." ("More Letters", II. page 21.) Writing to him in 1877 he says: "Your idea that dicotyledonous plants were not developed in force until sucking insects had been evolved seems to me a splendid one. I am surprised that the idea never occurred to me, but this is always the case when one first hears a new and simple explanation of some mysterious phenomenon." ("Life and Letters", III. page 285. Substantially the same idea had occurred

196

earlier to F.W.A. Miquel. Remarking that "sucking insects (Haustellata)... perform in nature the important duty of maintaining the existence of the vegetable kingdom, at least as far as the higher orders are concerned," he points our that "the appearance in great numbers of haustellate insects occurs at and after the Cretaceous epoch, when the plants with pollen and closed carpels (Angiosperms) are found, and acquire little by little the preponderance in the vegetable kingdom." "Archives Neerlandaises", III. (1868). English translation in "Journ. of Bot." 1869, page 101.)

Even with this help the abruptness still remains an almost insoluble problem, though a forecast of floral structure is now recognised in some Jurassic and Lower Cretaceous plants. But the gap between this and the structural complexity and diversity of angiosperms is enormous. Darwin thought that the evolution might have been accomplished during a period of prolonged isolation. Writing to Hooker (1881) he says: "Nothing is more extraordinary in the history of the Vegetable Kingdom, as it seems to me, than the APPARENTLY very sudden or abrupt development of the higher plants. I have sometimes speculated whether there did not exist somewhere during long ages an extremely isolated continent, perhaps near the South Pole." ("Life and Letters", III. page 248.)

The present trend of evidence is, however, all in favour of a northern origin for flowering plants, and we can only appeal to the imperfection of the geological record as a last resource to extricate us from the difficulty of tracing the process. But Darwin's instinct that at some time or other the southern hemisphere had played an important part in the evolution of the vegetable kingdom did not mislead him. Nothing probably would have given him greater satisfaction than the masterly summary in which Seward has brought together the evidence for the origin of the Glossopteris flora in Gondwana land.

"A vast continental area, of which remnants are preserved in Australia, South Africa and South America... A tract of enormous extent occupying an area, part of which has since given place to a southern ocean, while detached masses persist as portions of more modern continents, which have enabled us to read in their fossil plants and ice-scratched boulders the records of a lost continent, in which the Mesozoic vegetation of the northern continent had its birth." ("Encycl. Brit." (10th edition 1902), Vol. XXXI. ("Palaeobotany; Mesozoic"), page 422.) Darwin would probably have demurred on physical grounds to the extent of the continent, and preferred to account for the transoceanic distribution of its flora by the same means which must have accomplished it on land.

It must in fairness be added that Guppy's later views give some support to the conjectural existence of the "lost continent." "The distribution of the genus Dammara" (Agathis) led him to modify his earlier conclusions. He tells us:—"In my volume on the geology of Vanua Levu it was shown that the Tertiary period was an age of submergence in the Western Pacific, and a disbelief in any previous continental condition was expressed. My later view is more in accordance with that of Wichmann, who, on geological grounds, contended that the islands of the Western Pacific were in a continental condition during the Palaeozoic and Mesozoic periods, and that their submergence and subsequent emergence took place in Tertiary times." (Guppy, op. cit. II. page 304.)

The weight of the geological evidence I am unable to scrutinise. But though I must admit the possibility of some unconscious bias in my own mind on the subject, I am impressed with the fact that the known distribution of the Glossopteris flora in the southern hemisphere is precisely paralleled by that of Proteaceae and Restiaceae in it at the present time. It is not unreasonable to suppose that both phenomena, so similar, may admit of the same explanation. I confess it would not surprise me if fresh discoveries in the distribution of the Glossopteris flora were to point to the possibility of its also having migrated southwards from a centre of origin in the northern hemisphere.

Darwin, however, remained sceptical "about the travelling of plants from the north EXCEPT DURING THE TERTIARY PERIOD." But he added, "such speculations seem to me hardly scientific, seeing how little we know of the old floras." ("Life and Letters", III. page 247.) That in later geological times the south has been the grave of the weakened offspring of the aggressive north can hardly be doubted. But if we look to the Glossopteris flora for the ancestry of Angiosperms during the Secondary period, Darwin's prevision might be justified, though he has given us no clue as to how he arrived at it.

It may be true that technically Darwin was not a botanist. But in two pages of the "Origin" he has given us a masterly explanation of "the relationship, with very little identity, between the productions of North America and Europe." (Pages 333, 334.) He showed that this could be accounted for by their migration southwards from a common area, and he told Wallace that he "doubted much whether the now called Palaearctic and Neartic regions ought to be separated." ("Life and Letters", III. page 230.) Catkin-bearing deciduous trees had long been seen to justify Darwin's doubt: oaks, chestnuts, beeches, hazels, hornbeams, birches, alders, willows and poplars are common both to the Old and New World. Newton found that the separate regions could not be sustained for birds, and he is now usually followed in uniting them as the Holartic. One feels inclined to say in reading the two pages, as Lord Kelvin did to a correspondent who asked for some further development of one of his papers, It is all there. We have only to apply the principle to previous geological ages to understand why the flora of the Southern United States preserves a Cretaceous facies. Applying it still further we can understand why, when the northern hemisphere gradually cooled through the Tertiary period, the plants of the Eocene "suggest a comparison of the climate and forests with those of the Malay Archipelago and Tropical America." (Clement Reid, "Encycl. Brit." (10th edition), Vol. XXXI. ("Palaeobotany; Tertiary"), page 435.) Writing to Asa Gray in 1856 with respect to the United States flora, Darwin said that "nothing has surprised me more than the greater generic and specific affinity with East Asia than with West America." ("More Letters", I. page 434.) The recent discoveries of a Tulip tree and a Sassafras in China afford fresh illustrations. A few years later Asa Gray found the explanation in both areas being centres of preservation of the Cretaceous flora from a common origin. It is interesting to note that the paper in which this was enunciated at once established his reputation.

In Europe the latitudinal range of the great mountain chains gave the Miocene flora no chance of escape during the Glacial period, and the Mediterranean appears to have equally intercepted the flow of alpine plants to the Atlas. (John Ball in Appendix G, page 438, in "Journal of a Tour in Morocco and the Great Atlas", J.D. Hooker and J. Ball, London, 1878.) In Southern Europe the myrtle, the laurel, the fig and the dwarf-palm are the sole representatives of as many great tropical families. Another great tropical family, the Gesneraceae has left single representatives from the Pyrenees to the Balkans; and in the former a diminutive yam still lingers. These are only illustrations of the evidence which constantly accumulates and which finds no rational explanation except that which Darwin has given to it.

The theory of southward migration is the key to the interpretation of the geographical distribution of plants. It derived enormous support from the researches of Heer and has now become an accepted commonplace. Saporta in 1888 described the vegetable kingdom as "emigrant pour suivre une direction determinee et marcher du nord au sud, a la recherche de regions et de stations plus favorables, mieux appropriees aux adaptations acquises, a meme que la temperature terrestre perd ses conditions premieres." ("Origine Paleontologique des arbres", Paris, 1888, page 28.) If, as is so often the case, the theory now seems to be a priori inevitable, the historian of science will not omit to record that the first germ sprang from the brain of Darwin.

In attempting this sketch of Darwin's influence on Geographical Distribution, I have found it impossible to treat it from an external point of view. His interest in it was unflagging; all I could say became necessarily a record of that interest and could not be detached from it. He was in more or less intimate touch with everyone who was working at it. In reading the letters we move amongst great names. With an extraordinary charm of persuasive correspondence he was constantly suggesting, criticising and stimulating. It is hardly an exaggeration to say that from the quiet of his study at Down he was founding and directing a wide-world school.

POSTSCRIPTUM.

Since this essay was put in type Dr Ernst's striking account of the "New Flora of the Volcanic Island of Krakatau" (Cambridge, 1909.) has reached me. All botanists must feel a debt of gratitude to Prof. Seward for his admirable translation of a memoir which in its original form is practically unprocurable and to the liberality of the Cambridge University Press for its publication. In the preceding pages I have traced the laborious research by which the methods of Plant Dispersal were established by Darwin. In the island of Krakatau nature has supplied a crucial experiment which, if it had occurred earlier, would have at once secured conviction of their efficiency. A quarter of a century ago every trace of organic life in the island was "destroyed and buried under a thick covering of glowing stones." Now, it is "again covered with a mantle of green, the growth being in places so luxuriant that it is necessary to cut one's way laboriously through the vegetation." (Op. cit. page 4.) Ernst traces minutely how this has been brought about by the combined action of wind, birds and sea currents, as means of transport. The process will continue, and he concludes:—"At last after a long interval the vegetation on the desolated island will again acquire that wealth of variety and luxuriance which we see in the fullest development which Nature has reached in the primaeval forest in the tropics." (Op. cit. page 72.) The possibility of such a result revealed itself to the insight of Darwin with little encouragement or support from contemporary opinion.

One of the most remarkable facts established by Ernst is that this has not been accomplished by the transport of seeds alone. "Tree stems and branches played an important part in the colonisation of Krakatau by plants and animals. Large piles of floating trees, stems, branches and bamboos are met with everywhere on the beach above high-water mark and often carried a considerable distance inland. Some of the animals on the island, such as the fat Iguana (Varanus salvator) which suns itself in the beds of streams, may have travelled on floating wood, possibly also the ancestors of the numerous ants, but certainly plants." (Op. cit. page 56.) Darwin actually had a prevision of this. Writing to Hooker he says:— "Would it not be a prodigy if an unstocked island did not in the course of ages receive colonists from coasts whence the currents flow, trees are drifted and birds are driven by gales?" ("More Letters", I. page 483.) And ten years earlier:—"I must believe in the... whole plant or branch being washed into the sea; with floods and slips and earthquakes; this must continually be happening." ("Life and Letters", II. pages 56, 57.) If we give to "continually" a cosmic measure, can the fact be doubted? All this, in the light of our present knowledge, is too obvious to us to admit of discussion. But it seems to me nothing less than pathetic to see how in the teeth of the obsession as to continental extension, Darwin fought single-handed for what we now know to be the truth.

Guppy's heart failed him when he had to deal with the isolated case of Agathis which alone seemed inexplicable by known means of transport. But when we remember that it is a relic of the pre-Angiospermous flora, and is of Araucarian ancestry, it cannot be said that the impossibility, in so prolonged a history, of the bodily transference of cone-bearing branches or even of trees, compels us as a last resort to fall back on continental extension to account for its existing distribution.

When Darwin was in the Galapagos Archipelago, he tells us that he fancied himself "brought near to the very act of creation." He saw how new species might arise from a common stock. Krakatau shows us an earlier stage and how by simple agencies, continually at work, that stock might be supplied. It also shows us how the mixed and casual elements of a new colony enter into competition for the ground and become mutually adjusted. The study of Plant Distribution from a Darwinian standpoint has opened up a new field of research in Ecology. The means of transport supply the materials for a flora, but their ultimate fate depends on their equipment for the "struggle for existence." The whole subject can no longer be regarded as a mere statistical inquiry which has seemed doubtless to many of somewhat arid interest. The fate of every element of the earth's vegetation has sooner or later depended on its ability to travel and to hold its own under new conditions. And the means by which it has secured success is an each case a biological problem which demands and will reward the most attentive study. This is the lesson which Darwin has bequeathed to us. It is summed up in the concluding paragraph of the "Origin" ("Origin of Species" (6th edition), page 429.):—"It is interesting to contemplate a tangled bank, clothed with many plants of many kinds, with birds singing on the bushes, with various insects flitting about, and with worms crawling through the damp earth, and to reflect that these elaborately constructed forms, so different from each other, and dependent upon each other in so complex a manner, have all been produced by laws acting around us."

XVII. GEOGRAPHICAL DISTRIBUTION OF ANIMALS. By Hans Gadow, M.A., Ph.D., F.R.S.

Strickland Curator and Lecturer on Zoology in the University of Cambridge.

The first general ideas about geographical distribution may be found in some of the brilliant speculations contained in Buffon's "Histoire Naturelle". The first special treatise on the subject was however written in 1777 by E.A.W. Zimmermann, Professor of Natural Science at Brunswick, whose large volume, "Specimen Zoologiae Geographicae Quadrupedum"..., deals in a statistical way with the mammals; important features of the large accompanying map of the world are the ranges of mountains and the names of hundreds of genera indicating their geographical range. In a second work he laid special stress on domesticated animals with reference to the spreading of the various races of Mankind.

In the following year appeared the "Philosophia Entomologica" by J.C. Fabricius, who was the first to divide the world into eight regions. In 1803 G.R. Treviranus ("Biologie oder Philosophie der lebenden Natur", Vol. II. Gottingen, 1803.) devoted a long chapter of his great work on "Biologie" to a philosophical and coherent treatment of the distribution of the whole animal kingdom. Remarkable progress was made in 1810 by F. Tiedemann ("Anatomie und Naturgeschichte der Vogel". Heidelberg, 1810.) of Heidelberg. Few, if any, of the many subsequent Ornithologists seem to have appreciated, or known of, the ingenious way in which Tiedemann marshalled his statistics in order to arrive at general conclusions. There are, for instance, long lists of birds arranged in accordance with their occurrence in one or more continents: by correlating the distribution of the birds with their food he concludes "that the countries of the East Indian flora have no vegetable feeders in common with America," and "that it is probably due to the great peculiarity of the African flora that Africa has few phytophagous kinds in common with other countries, whilst zoophagous birds have a far more independent, often cosmopolitan, distribution." There are also remarkable chapters on the influence of environment, distribution, and migration, upon the structure of the Birds! In short, this anatomist dealt with some of the fundamental causes of distribution.

Whilst Tiedemann restricted himself to Birds, A. Desmoulins in 1822 wrote a short but most suggestive paper on the Vertebrata, omitting the birds; he combated the view recently proposed by the entomologist Latreille that temperature was the main factor in distribution. Some of his ten main conclusions show a peculiar mixture of evolutionary ideas coupled with the conception of the stability of species: whilst each species must have started from but one creative centre, there may be several "analogous centres of creation" so far as genera and families are concerned. Countries with different faunas, but lying within the same climatic zones, are proof of the effective and permanent existence of barriers preventing an exchange between the original creative centres.

The first book dealing with the "geography and classification" of the whole animal kingdom was written by W. Swainson ("A Treatise on the Geography and Classification of

Animals", Lardner's "Cabinet Cyclopaedia" London, 1835.) in 1835. He saw in the five races of Man the clue to the mapping of the world into as many "true zoological divisions," and he reconciled the five continents with his mystical quinary circles.

Lyell's "Principles of Geology" should have marked a new epoch, since in his "Elements" he treats of the past history of the globe and the distribution of animals in time, and in his "Principles" of their distribution in space in connection with the actual changes undergone by the surface of the world. But as the sub-title of his great work "Modern changes of the Earth and its inhabitants" indicates, he restricted himself to comparatively minor changes, and, emphatically believing in the permanency of the great oceans, his numerous and careful interpretations of the effect of the geological changes upon the dispersal of animals did after all advance the problem but little.

Hitherto the marine faunas had been neglected. This was remedied by E. Forbes, who established nine homozoic zones, based mainly on the study of the mollusca, the determining factors being to a great extent the isotherms of the sea, whilst the 25 provinces were given by the configuration of the land. He was followed by J.D. Dana, who, taking principally the Crustacea as a basis, and as leading factors the mean temperatures of the coldest and of the warmest months, established five latitudinal zones. By using these as divisors into an American, Afro-European, Oriental, Arctic and Antarctic realm, most of which were limited by an eastern and western land-boundary, he arrived at about threescore provinces.

In 1853 appeared L.K. Schmarda's ("Die geographische Verbreitung der Thiere", Wien, 1853.) two volumes, embracing the whole subject. Various centres of creation being, according to him, still traceable, he formed the hypothesis that these centres were originally islands, which later became enlarged and joined together to form the great continents, so that the original faunas could overlap and mix whilst still remaining pure at their respective centres. After devoting many chapters to the possible physical causes and modes of dispersal, he divided the land into 21 realms which he shortly characterises, e.g. Australia as the only country inhabited by marsupials, monotremes and meliphagous birds. Ten main marine divisions were diagnosed in a similar way. Although some of these realms were not badly selected from the point of view of being applicable to more than one class of animals, they were obviously too numerous for general purposes, and this drawback was overcome, in 1857, by P.L. Sclater. ("On the general Geographical Distribution of the members of the class Aves", "Proc. Linn. Soc." (Zoology II. 1858, pages 130-145.)) Starting with the idea, that "each species must have been created within and over the geographical area, which it now occupies," he concluded "that the most natural primary ontological divisions of the Earth's surface" were those six regions, which since their adoption by Wallace in his epoch-making work, have become classical. Broadly speaking, these six regions are equivalent to the great masses of land; they are convenient terms for geographical facts, especially since the Palaearctic region expresses the unity of Europe with the bulk of Asia. Sclater further brigaded the regions of the Old World as Palaeogaea and the two Americas as Neogaea, a fundamental mistake, justifiable to a certain extent only since he based his regions mainly upon the present distribution of the Passerine birds.

Unfortunately these six regions are not of equal value. The Indian countries and the Ethiopian region (Africa south of the Sahara) are obviously nothing but the tropical, southern continuations or appendages of one greater complex. Further, the great eastern mass of land is so intimately connected with North America that this continent has much more in common with Europe and Asia than with South America. Therefore, instead of dividing the world longitudinally as Sclater had done, Huxley, in 1868 ("On the classification and distribution of the Alectoromorphae and Heteromorphae", "Proc. Zool. Soc." 1868, page 294.), gave weighty reasons for dividing it transversely. Accordingly he established two primary divisions, Arctogaea or the North world in a wider sense, comprising Sclater's Indian, African, Palaearctic and Neartic regions; and Notogaea, the Southern world, which he divided into (1) Austro-Columbia (an unfortunate substitute for the neotropical region), (2) Australasia, and (3) New Zealand, the number of big regions thus being reduced to three but for the separation of New Zealand upon rather negative characters. Sclater was the first to accept these four great regions and showed, in 1874 ("The geographical distribution of Mammals", "Manchester Science Lectures", 1874.), that they were well borne out by the present distribution of the Mammals.

Although applicable to various other groups of animals, for instance to the tailless Amphibia and to Birds (Huxley himself had been led to found his two fundamental divisions on the distribution of the Gallinaceous birds), the combination of South America with Australia was gradually found to be too sweeping a measure. The obvious and satisfactory solution was provided by W.T. Blanford (Anniversary address (Geological Society, 1889), "Proc. Geol. Soc." 1889-90, page 67; "Quart. Journ." XLVI 1890.), who in 1890 recognised three main divisions, namely Australian, South American, and the rest, for which the already existing terms (although used partly in a new sense, as proposed by an anonymous writer in "Natural Science", III. page 289) "Notogaea," "Neogaea" and "Arctogaea" have been gladly accepted by a number of English writers.

After this historical survey of the search for larger and largest or fundamental centres of animal creation, which resulted in the mapping of the world into zoological regions and realms of after all doubtful value, we have to return to the year 1858. The eleventh and twelfth chapters of "The Origin of Species" (1859), dealing with "Geographical Distribution," are based upon a great amount of observation, experiment and reading. As Darwin's main problem was the origin of species, nature's way of making species by gradual changes from others previously existing, he had to dispose of the view, held universally, of the independent creation of each species and at the same time to insist upon a single centre of creation for each species; and in order to emphasise his main point, the theory of descent, he had to disallow convergent, or as they were then called, analogous forms. To appreciate the difficulty of his position we have to take the standpoint of fifty years ago, when the immutability of the species was an axiom and each was supposed to have been created within or over the geographical area which it now occupies. If he once admitted that a species could arise from many individuals instead of from one pair, there was no way of shutting the door against the possibility that these individuals may have been so numerous that they occupied a very large district, even so large that it had become as discontinuous as the distribution of many a species actually is. Such a concession would at once be taken as an admission of multiple, independent, origin instead of descent in Darwin's sense.

For the so-called multiple, independently repeated creation of species as an explanation of their very wide and often quite discontinuous distribution, he substituted colonisation from the nearest and readiest source together with subsequent modification and better adaptation to their new home.

He was the first seriously to call attention to the many accidental means, "which more properly should be called occasional means of distribution," especially to oceanic islands. His specific, even individual, centres of creation made migrations all the more necessary, but their extent was sadly baulked by the prevailing dogma of the permanency of the oceans. Any number of small changes ("many islands having existed as halting places, of which not a wreck now remains" ("The Origin of Species" (1st edition), page 396.).) were conceded freely, but few, if any, great enough to permit migration of truly terrestrial creatures. The only means of getting across the gaps was by the principle of the "flotsam and jetsam," a theory which Darwin took over from Lyell and further elaborated so as to make it applicable to many kinds of plants and animals, but sadly deficient, often grotesque, in the case of most terrestrial creatures.

Another very fertile source was Darwin's strong insistence upon the great influence which the last glacial epoch must have had upon the distribution of animals and plants. Why was the migration of northern creatures southwards of far-reaching and most significant importance? More northerners have established themselves in southern lands than vice versa, because there is such a great mass of land in the north and greater continents imply greater intensity of selection. "The productions of real islands have everywhere largely yielded to continental forms." (Ibid. page 380.)... "The Alpine forms have almost everywhere largely yielded to the more dominant forms generated in the larger areas and more efficient workshops of the North."

Let us now pass in rapid survey the influence of the publication of "The Origin of Species" upon the study of Geographical Distribution in its wider sense.

Hitherto the following thought ran through the minds of most writers: Wherever we examine two or more widely separated countries their respective faunas are very different, but where two faunas can come into contact with each other, they intermingle. Consequently these faunas represent centres of creation, whence the component creatures have spread peripherally so far as existing boundaries allowed them to do so. This is of course the fundamental idea of "regions." There is not one of the numerous writers who considered the possibility that these intermediate belts might represent not a mixture of species but transitional forms, the result of changes undergone by the most peripheral migrants in adaptation to their new surroundings. The usual standpoint was also that of Pucheran ("Note sur l'equateur zoologique", "Rev. et Mag. de Zoologie", 1855; also several other papers, ibid. 1865, 1866, and 1867.) in 1855. But what a change within the next ten years! Pucheran explains the agreement in coloration between the desert and its fauna as "une harmonie post-etablie"; the Sahara, formerly a marine basin, was peopled by immigrants from the neighbouring countries, and these new animals adapted themselves to the new environment. He also discusses, among other similar questions, the Isthmus of Panama with regard to its having once been a strait. From the same author may be quoted the following passage as a strong proof of the new influence: "By the radiation of the contemporaneous faunas, each from one centre, whence as the various parts of the world successively were formed and became habitable, they spread and became modified according to the local physical conditions."

The "multiple" origin of each species as advocated by Sclater and Murray, although giving the species a broader basis, suffered from the same difficulties. There was only one alternative to the old orthodox view of independent creation, namely the bold acceptance of land-connections to an extent for which geological and palaeontological science was not yet ripe. Those who shrank from either view, gave up the problem as mysterious and beyond the human intellect. This was the expressed opinion of men like Swainson, Lyell and Humboldt. Only Darwin had the courage to say that the problem was not insoluble. If we admit "that in the long course of time the individuals of the same species, and likewise of allied species, have proceeded from some one source; then I think all the grand leading facts of geographical distribution are explicable on the theory of migration... together with subsequent modification and the multiplication of new forms." We can thus understand how it is that in some countries the inhabitants "are linked to the extinct beings which formerly inhabited the same continent." We can see why two areas, having nearly the same physical conditions, should often be inhabited by very different forms of life,... and "we can see why in two areas, however distant from each other, there should be a correlation, in the presence of identical species... and of distinct but representative species." ("The Origin of Species" (1st edition), pages 408, 409.)

Darwin's reluctance to assume great geological changes, such as a land-connection of Europe with North America, is easily explained by the fact that he restricted himself to the distribution of the present and comparatively recent species. "I do not believe that it will ever be proved that within the recent period continents which are now quite separate, have been continuously, or almost continuously, united with each other, and with the many existing oceanic islands." (Ibid. page 357.) Again, "believing... that our continents have long remained in nearly the same relative position, though subjected to large, but partial oscillations of level," that means to say within the period of existing species, or "within the recent period." (Ibid. page. 370.) The difficulty was to a great extent one of his own making. Whilst almost everybody else believed in the immutability of the species, which implies an enormous age, logically since the dawn of creation, to him the actually existing species as the latest results of evolution, were necessarily something very new, so young that only the very latest of the geological epochs could have affected them. It has since come to our knowledge that a great number of terrestrial "recent" species, even those of the higher classes of Vertebrates, date much farther back than had been thought possible. Many of them reach well into the Miocene, a time since which the world seems to have assumed the main outlines of the present continents.

In the year 1866 appeared A. Murray's work on the "Geographical Distribution of Mammals", a book which has perhaps received less recognition than it deserves. His treatment of the general introductory questions marks a considerable advance of our problem, although, and partly because, he did not entirely agree with Darwin's views as laid down in the first edition of "The Origin of Species", which after all was the great impulse given to Murray's work. Like Forbes he did not shrink from assuming enormous changes in the configuration of the continents and oceans because the theory of descent, with its necessary postulate of great migrations, required them. He stated, for instance, "that a Miocene Atlantis sufficiently explains the common distribution of animals and plants in Europe and America up to the glacial epoch." And next he considers how, and by what changes, the rehabilitation and distribution of these lands themselves were effected subsequent to that period. Further, he deserves credit for having cleared up a misunderstanding of the idea of specific centres of creation. Whilst for instance Schmarda assumed without hesitation that the same species, if occurring at places separated by great distances, or apparently insurmountable barriers, had been there created independently (multiple centres), Lyell and Darwin held that each species had only one single centre, and with this view most of us agree, but their starting point was to them represented by one individual, or rather one single pair. According to Murray, on the other hand, this centre of a species is formed by all the individuals of a species, all of which equally undergo those changes which new conditions may impose upon them. In this respect a new species has a multiple origin, but this in a sense very different from that which was upheld by L. Agassiz. As Murray himself puts it: "To my multiple origin, communication and direct derivation is essential. The species is compounded of many influences brought together through many individuals, and distilled by Nature into one species; and, being once established it may roam and spread wherever it finds the conditions of life not materially different from those of its original centre." (Murray, "The Geographical Distribution of Mammals", page 14. London, 1866.) This declaration fairly agrees with more modern views, and it must be borne in mind that the application of the single-centre principle to the genera, families and larger groups in the search for descent inevitably leads to one creative centre for the whole animal kingdom, a condition as unwarrantable as the myth of Adam and Eve being the first representatives of Mankind.

It looks as if it had required almost ten years for "The Origin of Species" to show its full effect, since the year 1868 marks the publication of Haeckel's "Naturliche Schoepfungsgeschichte" in addition to other great works. The terms "Oecology" (the relation of organisms to their environment) and "Chorology" (their distribution in space) had been given us in his "Generelle Morphologie" in 1866. The fourteenth chapter of the "History of Creation" is devoted to the distribution of organisms, their chorology, with the emphatic assertion that "not until Darwin can chorology be spoken of as a separate science, since he supplied the acting causes for the elucidation of the hitherto accumulated mass of facts." A map (a "hypothetical sketch") shows the monophyletic origin and the routes of distribution of Man.

Natural Selection may be all-mighty, all-sufficient, but it requires time, so much that the countless aeons required for the evolution of the present fauna were soon felt to be one of the most serious drawbacks of the theory. Therefore every help to ease and shorten this process should have been welcomed. In 1868 M. Wagner (The first to formulate clearly the fundamental idea of a theory of migration and its importance in the origin of new species was L. von Buch, who in his "Physikalische Beschreibung der Canarischen Inseln", written in 1825, wrote as follows: "Upon the continents the individuals of the genera by spreading far, form, through differences of the locality, food and soil, varieties which finally become constant as new species, since owing to the distances they could never be crossed with other varieties and thus be brought back to the main type. Next they may again, perhaps upon different roads, return to the old home where they find the old type likewise changed, both having become so different that they can interbreed no longer. Not so upon islands, where the individuals shut up in narrow valleys or within narrow districts, can always meet one another and thereby destroy every new attempt towards the fixing of a new variety." Clearly von Buch explains here why island types remain fixed, and why these types themselves have become so different from their continental congeners.—Actually von Buch is aware of a most important point, the difference in the process of development which exists between a new species b, which is the result of an ancestral species a having itself changed into b and thereby vanished itself, and a new species c which arose through separation out of the same ancestral a, which itself persists as such unaltered. Von Buch's prophetic view seems to have escaped Lyell's and even Wagner's notice.) came to the rescue with his "Darwin'sche Theorie und das Migrations-Gesetz der Organismen". (Leipzig, 1868.) He shows that migration, i.e. change of locality, implies new environmental conditions (never mind whether these be new stimuli to variation, or only acting as their selectors or censors), and moreover secures separation from the original stock and thus eliminates or lessens the reactionary dangers of panmixia. Darwin accepted Wagner's theory as "advantageous." Through the heated polemics of the more ardent selectionists Wagner's theory came to grow into an alternative instead of a help to the theory of selectional evolution. Separation is now rightly considered a most important factor by modern students of geographical distribution.

For the same year, 1868, we have to mention Huxley, whose Arctogaea and Notogaea are nothing less than the reconstructed main masses of land of the Mesozoic period. Beyond doubt the configuration of land at that remote period has left recognisable traces in the present continents, but whether they can account for the distribution of such a much later group as the Gallinaceous birds is more than questionable. In any case he took for his text a large natural group of birds, cosmopolitan as a whole, but with a striking distribution. The Peristeropodes, or pigeon-footed division, are restricted to the Australian and Neotropical regions, in distinction to the Alectoropodes (with the hallux inserted at a level above the front toes) which inhabit the whole of the Arctogaea, only a few members having spread into the South World. Further, as Asia alone has its Pheasants and allies, so is Africa characterised by its Guinea-fowls and relations, America has the Turkey as an endemic genus, and the Grouse tribe in a wider sense has its centre in the holarctic region: a splendid object lesson of descent, world-wide spreading and subsequent differentiation. Huxley, by the way, was the first—at least in private talk—to state that it will be for the morphologist, the well-trained anatomist, to give the casting vote in questions of geographical distribution, since he alone can determine whether we have to deal with homologous, or analogous, convergent, representative forms.

It seems late to introduce Wallace's name in 1876, the year of the publication of his standard work. ("The Geographical Distribution of Animals", 2 vols. London, 1876.) We cannot do better than quote the author's own words, expressing the hope that his "book should bear a similar relation to the eleventh and twelfth chapters of the "Origin of Species" as Darwin's "Animals and Plants under Domestication" does to the first chapter of that work," and to add that he has amply succeeded. Pleading for a few primary centres he accepts Sclater's six regions and does not follow Huxley's courageous changes which Sclater himself had accepted in 1874. Holding the view of the permanence of the oceans he accounts for the colonisation of outlying islands by further elaborating the views of Lyell and Darwin, especially in his fascinating "Island Life", with remarkable chapters on the Ice Age, Climate and Time and other fundamental factors. His method of arriving at the degree of relationship of the faunas of the various regions is eminently statistical. Long lists of genera determine by their numbers the affinity and hence the source of colonisation. In order to make sure of his material he performed the laborious task of evolving a new classification of the host of Passerine birds. This statistical method has been followed by many authors, who, relying more upon quantity than quality, have obscured the fact that the key to the present distribution lies in the past changes of the earth's surface. However, with Wallace begins the modern study of the geographical distribution of animals and the sudden interest taken in this subject by an ever widening circle of enthusiasts far beyond the professional brotherhood.

A considerable literature has since grown up, almost bewildering in its range, diversity of aims and style of procedure. It is a chaos, with many paths leading into the maze, but as yet very few take us to a position commanding a view of the whole intricate terrain with its impenetrable tangle and pitfalls.

One line of research, not initiated but greatly influenced by Wallace's works, became so prominent as to almost constitute a period which may be characterised as that of the search by specialists for either the justification or the amending of his regions. As class after class of animals was brought up to reveal the secret of the true regions, some authors saw in their different results nothing but the faultiness of previously established regions; others looked upon eventual agreements as their final corroboration, especially when for instance such diverse groups as mammals and scorpions could, with some ingenuity, be made to harmonise. But the obvious result of all these efforts was the growing knowledge that almost every class seemed to follow principles of its own. The regions tallied neither in extent nor in numbers, although most of them gravitated more and more towards three centres, namely Australia, South America and the rest of the world. Still zoologists persisted in the search, and the various modes and capabilities of dispersal of the respective groups were thought sufficient explanation of the divergent results in trying to bring the mapping of the world under one scheme.

Contemporary literature is full of devices for the mechanical dispersal of animals. Marine currents, warm and cold, were favoured all the more since they showed the probable original homes of the creatures in question. If these could not stand sea-water, they floated upon logs or icebergs, or they were blown across by storms; fishes were lifted over barriers by waterspouts, and there is on record even an hypothetical land tortoise, full of eggs, which colonised an oceanic island after a perilous sea voyage upon a tree trunk. Accidents will happen, and beyond doubt many freaks of discontinuous distribution have to be accounted for by some such means. But whilst sufficient for the scanty settlers of true oceanic islands, they cannot be held seriously to account for the rich fauna of a large continent, over which palaeontology shows us that the immigrants have passed like waves. It should also be borne in mind that there is a great difference between flotsam and jetsam. A current is an extension of the same medium and the animals in it may suffer no change during even a long voyage, since they may be brought from one littoral to another where they will still be in the same or but slightly altered environment. But the jetsam is in the position of a passenger who has been carried off by the wrong train. Almost every year some American land birds arrive at our western coasts and none of them have gained a permanent footing although such visits must have taken place since prehistoric times. It was therefore argued that only those groups of animals should be used for locating and defining regions which were absolutely bound to the soil. This method likewise gave results not reconcilable with each other, even when the distribution of fossils was taken into account, but it pointed to the absolute necessity of searching for former land-connections regardless of their extent and the present depths to which they may have sunk.

That the key to the present distribution lies in the past had been felt long ago, but at last it was appreciated that the various classes of animals and plants have appeared in successive geological epochs and also at many places remote from each other. The key to the distribution of any group lies in the configuration of land and water of that epoch in which it made its first appearance. Although this sounds like a platitude, it has frequently been ignored. If, for argument's sake, Amphibia were evolved somewhere upon the great southern land-mass of Carboniferous times (supposed by some to have stretched from South America across Africa to Australia), the distribution of this developing class must have proceeded upon lines altogether different from that of the mammals which dated perhaps from lower Triassic times, when the old south continental belt was already broken up. The broad lines of this distribution could never coincide with that of the other, older class, no matter whether the original mammalian centre was in the Afro-Indian, Australian, or Brazilian portion. If all the various groups of animals had come into existence at the same time and at the same place, then it would be possible, with sufficient geological data, to construct a map showing the generalised results applicable to the whole animal kingdom. But the premises are wrong. Whatever regions we may seek to establish applicable to all classes, we are necessarily mixing up several principles, namely geological, historical, i.e. evolutionary, with present day statistical facts. We might as well attempt one compound picture representing a chick's growth into an adult bird and a child's growth into manhood.

In short there are no general regions, not even for each class separately, unless this class be one which is confined to a comparatively short geological period. Most of the great classes have far too long a history and have evolved many successive main groups. Let us take the mammals. Marsupials live now in Australia and in both Americas, because they already existed in Mesozoic times; Ungulata existed at one time or other all over the world except in Australia, because they are post-Cretaceous; Insectivores, although as old as any Placentalia, are cosmopolitan excepting South America and Australia; Stags and Bears, as examples of comparatively recent Arctogaeans, are found everywhere with the exception of Ethiopia and Australia. Each of these groups teaches a valuable historical lesson, but when these are combined into the establishment of a few mammalian "realms," they mean nothing but statistical majorities. If there is one at all, Australia is such a realm backed against the rest of the world, but as certainly it is not a mammalian creative centre!

Well then, if the idea of generally applicable regions is a mare's nest, as was the search for the Holy Grail, what is the object of the study of geographical distribution? It is nothing less than the history of the evolution of life in space and time in the widest sense. The attempt to account for the present distribution of any group of organisms involves the aid of every branch of science. It bids fair to become a history of the world. It started in a mild, statistical way, restricting itself to the present fauna and flora and to the present configuration of land and water. Next came Oceanography concerned with the depths of the seas, their currents and temperatures; then enquiries into climatic changes, culminating in irreconcilable astronomical hypotheses as to glacial epochs; theories about changes of the level of the seas, mainly from the point of view of the physicist and astronomer. Then came more and more to the front the importance of the geological record, hand in hand with the palaeontological data and the search for the natural affinities, the genetic system of the organisms. Now and then it almost seems as if the biologists had done their share by supplying the problems and that the physicists and geologists would settle them, but in reality it is not so. The biologists not only set the problems, they alone can check the offered solutions. The mere fact of palms having flourished in Miocene Spitzbergen led to an hypothetical shifting of the axis of the world rather than to the assumption, by way of explanation, that the palms themselves might have changed their nature. One of the most valuable aids in geological research, often the only means for reconstructing the face of the earth in by-gone periods, is afforded by fossils, but only the morphologist can pronounce as to their trustworthiness as witnesses, because of the danger of mistaking analogous for homologous forms. This difficulty applies equally to living groups, and it is so important that a few instances may not be amiss.

There is undeniable similarity between the faunas of Madagascar and South America. This was supported by the Centetidae and Dendrobatidae, two entire "families," as also by other facts. The value of the Insectivores, Solenodon in Cuba, Centetes in Madagascar, has been much lessened by their recognition as an extremely ancient group and as a case of convergence, but if they are no longer put into the same family, this amendment is really to a great extent due to their widely discontinuous distribution. The only systematic difference of the Dendrobatidae from the Ranidae is the absence of teeth, morphologically a very unimportant character, and it is now agreed, on the strength of their distribution, that these little arboreal, conspicuously coloured frogs, Dendrobates in South America, Mantella in Madagascar, do not form a natural group, although a third genus, Cardioglossa in West Africa, seems also to belong to them. If these creatures lived all on the same continent, we should unhesitatingly look upon them as forming a well-defined, natural little group. On the other hand the Aglossa, with their three very divergent genera, namely Pipa in South

207

America, Xenopus and Hymenochirus in Africa, are so well characterised as one ancient group that we use their distribution unhesitatingly as a hint of a former connection between the two continents. We are indeed arguing in vicious circles. The Ratitae as such are absolutely worthless since they are a most heterogeneous assembly, and there are untold groups, of the artificiality of which many a zoo-geographer had not the slightest suspicion when he took his statistical material, the genera and families, from some systematic catalogues or similar lists. A lamentable instance is that of certain flightless Rails, recently extinct or sub-fossil, on the isalnds of Mauritius, Rodriguez and Chatham. Being flightless they have been used in support of a former huge Antarctic continent, instead of ruling them out of court as Rails which, each in its island, have lost the power of flight, a process which must have taken place so recently that it is difficult, upon morphological grounds, to justify their separation into Aphanapteryx in Mauritius, Erythromachus in Rodriguez and Diaphorapteryx on Chatham Island. Morphologically they may well form but one genus, since they have sprung from the same stock and have developed upon the same lines; they are therefore monogenetic: but since we know that they have become what they are independently of each other (now unlike any other Rails), they are polygenetic and therefore could not form one genus in the old Darwinian sense. Further, they are not a case of convergence, since their ancestry is not divergent but leads into the same stratum.

THE RECONSTRUCTION OF THE GEOGRAPHY OF SUCCESSIVE EPOCHS.

A promising method is the study by the specialist of a large, widely distributed group of animals from an evolutionary point of view. Good examples of this method are afforded by A.E. Ortmann's ("The geographical distribution of Freshwater Decapods and its bearing upon ancient geography", "Proc. Amer. Phil. Soc." Vol. 41, 1902.) exhaustive paper and by A.W. Grabau's "Phylogeny of Fusus and its Allies" ("Smithsonian Misc. Coll." 44, 1904.) After many important groups of animals have been treated in this way—as yet sparingly attempted—the results as to hypothetical land-connections etc. are sure to be corrective and supplementary, and their problems will be solved, since they are not imaginary.

The same problems are attacked, in the reverse way, by starting with the whole fauna of a country and thence, so to speak, letting the research radiate. Some groups will be considered as autochthonous, others as immigrants, and the directions followed by them will be inquired into; the search may lead far and in various directions, and by comparison of results, by making compound maps, certain routes will assume definite shape, and if they lead across straits and seas they are warrants to search for land-connections in the past. (A fair sample of this method is C.H. Eigenmann's "The Freshwater Fishes of South and Middle America", "Popular Science Monthly", Vol. 68, 1906.) There are now not a few maps purporting to show the outlines of land and water at various epochs. Many of these attempts do not tally with each other, owing to the lamentable deficiencies of geological and fossil data, but the bolder the hypothetical outlines are drawn, the better, and this is preferable to the insertion of bays and similar detail which give such maps a fallacious look of certainty where none exists. Moreover it must be borne in mind that, when we draw a broad continental belt across an ocean, this belt need never have existed in its entirety at any one time. The features of dispersal, intended to be explained by it, would be accomplished just as well by an unknown number of islands which have joined into larger complexes while elsewhere they subsided again: like pontoon-bridges which may be opened anywhere, or like a series of superimposed dissolving views of land and sea-scapes. Hence the reconstructed maps of Europe, the only continent tolerably known, show a considerable number of islands in puzzling changes, while elsewhere, e.g. in Asia, we have to be satisfied with sweeping generalisations.

At present about half-a-dozen big connections are engaging our attention, leaving as comparatively settled the extent and the duration of such minor "bridges" as that between Africa and Madagascar, Tasmania and Australia, the Antilles and Central America, Europe and North Africa. (Not a few of those who are fascinated by, and satisfied with, the statistical aspect of distribution still have a strong dislike to the use of "bridges" if these lead over deep seas, and they get over present discontinuous occurrences by a former "universal

208

or sub-universal distribution" of their groups.) This is indeed an easy method of cutting the knot, but in reality they shunt the question only a stage or two back, never troubling to explain how their groups managed to attain to that sub-universal range; or do they still suppose that the whole world was originally one paradise where everything lived side by side, until sin and strife and glacial epochs left nothing but scattered survivors?

The permanence of the great ocean-basins had become a dogma since it was found that a universal elevation of the land to the extent of 100 fathoms would produce but little changes, and when it was shown that even the 1000 fathom-line followed the great masses of land rather closely, and still leaving the great basins (although transgression of the sea to the same extent would change the map of the world beyond recognition), by general consent one mile was allowed as the utmost speculative limit of subsidence. Naturally two or three miles, the average depth of the oceans, seems enormous, and yet such a difference in level is as nothing in comparison with the size of the Earth. On a clay model globe ten feet in diameter an ocean bed three miles deep would scarcely be detected, and the highest mountains would be smaller than the unavoidable grains in the glazed surface of our model. There are but few countries which have not be submerged at some time or other.

CONNECTION OF SOUTH EASTERN ASIA WITH AUSTRALIA. Neumayr's Sino-Australian continent during mid-Mesozoic times was probably a much changing Archipelago, with final separations subsequent to the Cretaceous period. Henceforth Australasia was left to its own fate, but for a possible connection with the antarctic continent.

AFRICA, MADAGASCAR, INDIA. The "Lemuria" of Sclater and Haeckel cannot have been more than a broad bridge in Jurassic times; whether it was ever available for the Lemurs themselves must depend upon the time of its duration, the more recent the better, but it is difficult to show that it lasted into the Miocene.

AFRICA AND SOUTH AMERICA. Since the opposite coasts show an entire absence of marine fossils and deposits during the Mesozoic period, whilst further north and south such are known to exist and are mostly identical on either side, Neumayr suggested the existence of a great Afro-Son American mass of land during the Jurassic epoch. Such land is almost a necessity and is supported by many facts; it would easily explain the distribution of numerous groups of terrestrial creatures. Moreover to the north of this hypothetical land, somewhere across from the Antilles and Guiana to North Africa and South Western Europe, existed an almost identical fauna of Corals and Molluscs, indicating either a coast-line or a series of islands interrupted by shallow seas, just as one would expect if, and when, a Brazil-Ethiopian mass of land were breaking up. Lastly from Central America to the Mediterranean stretches one of the Tertiary tectonic lines of the geologists. Here also the great question is how long this continent lasted. Apparently the South Atlantic began to encroach from the south so that by the later Cretaceous epoch the land was reduced to a comparatively narrow Brazil-West Africa, remnants of which persisted certainly into the early Tertiary, until the South Atlantic joined across the equator with the Atlantic portion of the "Thetys," leaving what remained of South America isolated from the rest of the world.

ANTARCTIC CONNECTIONS. Patagonia and Argentina seem to have joined Antartica during the Cretaceous epoch, and this South Georgian bridge had broken down again by mid-Tertiary times when South America became consolidated. The Antarctic continent, presuming that it existed, seems also to have been joined, by way of Tasmania, with Australia, also during the Cretaceous epoch, and it is assumed that the great Australia-Antarctic-Patagonian land was severed first to the south of Tasmania and then at the South Georgian bridge. No connection, and this is important, is indicated between Antarctica and either Africa or Madagascar.

So far we have followed what may be called the vicissitudes of the great Permo-Carboniferous Gondwana land in its fullest imaginary extent, an enormous equatorial and south temperate belt from South America to Africa, South India and Australia, which seems to have provided the foundation of the present Southern continents, two of which temporarily joined Antarctica, of which however we know nothing except that it exists now.

Let us next consider the Arctic and periarctic lands. Unfortunately very little is known about the region within the arctic circle. If it was all land, or more likely great changing archipelagoes, faunistic exchange between North America, Europe and Siberia would present no difficulties, but there is one connection which engages much attention, namely a land where now lies the North temperate and Northern part of the Atlantic ocean. How far south did it ever extend and what is the latest date of a direct practicable communication, say from North Western Europe to Greenland? Connections, perhaps often interrupted, e.g. between Greenland and Labrador, at another time between Greenland and Scandinavia, seem to have existed at least since the Permo-Carboniferous epoch. If they existed also in late Cretaceous and in Tertiary times, they would of course easily explain exchanges which we know to have repeatedly taken place between America and Europe, but they are not proved thereby, since most of these exchanges can almost as easily have occurred across the polar regions, and others still more easily by repeated junction of Siberia with Alaska.

Let us now describe a hypothetical case based on the supposition of connecting bridges. Not to work in a circle, we select an important group which has not served as a basis for the reconstruction of bridges; and it must be a group which we feel justified in assuming to be old enough to have availed itself of ancient land-connections.

The occurrence of one species of Peripatus in the whole of Australia, Tasmania and New Zealand (the latter being joined to Australia by way of New Britain in Cretaceous times but not later) puts the genus back into this epoch, no unsatisfactory assumption to the morphologist. The apparent absence of Peripatus in Madagascar indicates that it did not come from the east into Africa, that it was neither Afro-Indian, nor Afro-Australian; nor can it have started in South America. We therefore assume as its creative centre Australia or Malaya in the Cretaceous epoch, whence its occurrence in Sumatra, Malay Peninsula, New Britain, New Zealand and Australia is easily explained. Then extension across Antarctica to Patagonia and Chile, whence it could spread into the rest of South America as this became consolidated in early Tertiary times. For getting to the Antilles and into Mexico it would have to wait until the Miocene, but long before that time it could arrive in Africa, there surviving as a Congolese and a Cape species. This story is unsupported by a single fossil. Peripatus may have been "sub-universal" all over greater Gondwana land in Carboniferous times, and then its absence from Madagascar would be difficult to explain, but the migrations suggested above amount to little considering that the distance from Tasmania to South America could be covered in far less time than that represented by the whole of the Eocene epoch alone.

There is yet another field, essentially the domain of geographical distribution, the cultivation of which promises fair to throw much light upon Nature's way of making species. This is the study of the organisms with regard to their environment. Instead of revealing pedigrees or of showing how and when the creatures got to a certain locality, it investigates how they behaved to meet the ever changing conditions of their habitats. There is a facies, characteristic of, and often peculiar to, the fauna of tropical moist forests, another of deserts, of high mountains, of underground life and so forth; these same facies are stamped upon whole associations of animals and plants, although these may be—and in widely separated countries generally are—drawn from totally different families of their respective orders. It does not go to the root of the matter to say that these facies have been brought about by the extermination of all the others which did not happen to fit into their particular environment. One might almost say that tropical moist forests must have arboreal frogs and that these are made out of whatever suitable material happened to be available; in Australia and South America Hylidae, in Africa Ranidae, since there Hylas are absent. The deserts must have lizards capable of standing the glare, the great changes of temperature, of running over or burrowing into the loose sand. When as in America Iguanids are available, some of these are thus modified, while in Africa and Asia the Agamids are drawn upon. Both in the Damara and in the Transcaspian deserts, a Gecko has been turned into a runner upon sand!

We cannot assume that at various epochs deserts, and at others moist forests were continuous all over the world. The different facies and associations were developed at various times and places. Are we to suppose that, wherever tropical forests came into existence, amongst the stock of humivagous lizards were always some which presented those nascent variations which made them keep step with the similarly nascent forests, the overwhelming rest being eliminated? This principle would imply that the same stratum of lizards always had variations ready to fit any changed environment, forests and deserts, rocks and swamps. The study of Ecology indicates a different procedure, a great, almost boundless plasticity of the organism, not in the sense of an exuberant moulding force, but of a readiness to be moulded, and of this the "variations" are the visible outcome. In most cases identical facies are produced by heterogeneous convergences and these may seem to be but superficial, affecting only what some authors are pleased to call the physiological characters; but environment presumably affects first those parts by which the organism comes into contact with it most directly, and if the internal structures remain unchanged, it is not because these are less easily modified but because they are not directly affected. When they are affected, they too change deeply enough.

That the plasticity should react so quickly—indeed this very quickness seems to have initiated our mistaking the variations called forth for something performed—and to the point, is itself the outcome of the long training which protoplasm has undergone since its creation.

In Nature's workshop he does not succeed who has ready an arsenal of tools for every conceivable emergency, but he who can make a tool at the spur of the moment. The ordeal of the practical test is Charles Darwin's glorious conception of Natural Selection.

XVIII. DARWIN AND GEOLOGY. By
J.W. Judd, C.B., LL.D., F.R.S.

(Mr Francis Darwin has related how his father occasionally came up from Down to spend a few days with his brother Erasmus in London, and, after his brother's death, with his daughter, Mrs Litchfield. On these occasions, it was his habit to arrange meetings with Huxley, to talk over zoological questions, with Hooker, to discuss botanical problems, and with Lyell to hold conversations on geology. After the death of Lyell, Darwin, knowing my close intimacy with his friend during his later years, used to ask me to meet him when he came to town, and "talk geology." The "talks" took place sometimes at Jermyn Street Museum, at other times in the Royal College of Science, South Kensington; but more frequently, after having lunch with him, at his brother's or his daughter's house. On several occasions, however, I had the pleasure of visiting him at Down. In the postscript of a letter (of April 15, 1880) arranging one of these visits, he writes: "Since poor, dear Lyell's death, I rarely have the pleasure of geological talk with anyone.")

In one of the very interesting conversations which I had with Charles Darwin during the last seven years of his life, he asked me in a very pointed manner if I were able to recall the circumstances, accidental or otherwise, which had led me to devote myself to geological studies. He informed me that he was making similar inquiries of other friends, and I gathered from what he said that he contemplated at that time a study of the causes producing SCIENTIFIC BIAS in individual minds. I have no means of knowing how far this project ever assumed anything like concrete form, but certain it is that Darwin himself often indulged in the processes of mental introspection and analysis; and he has thus fortunately left us—in his fragments of autobiography and in his correspondence—the materials from which may be reconstructed a fairly complete history of his own mental development.

211

There are two perfectly distinct inquiries which we have to undertake in connection with the development of Darwin's ideas on the subject of evolution:

FIRST. How, when, and under what conditions was Darwin led to a conviction that species were not immutable, but were derived from pre-existing forms?

SECONDLY. By what lines of reasoning and research was he brought to regard "natural selection" as a vera causa in the process of evolution?

It is the first of these inquiries which specially interests the geologist; though geology undoubtedly played a part—and by no means an insignificant part—in respect to the second inquiry.

When, indeed, the history comes to be written of that great revolution of thought in the nineteenth century, by which the doctrine of evolution, from being the dream of poets and visionaries, gradually grew to be the accepted creed of naturalists, the paramount influence exerted by the infant science of geology—and especially that resulting from the publication of Lyell's epoch-making work, the "Principles of Geology"—cannot fail to be regarded as one of the leading factors. Herbert Spencer in his "Autobiography" bears testimony to the effect produced on his mind by the recently published "Principles", when, at the age of twenty, he had already begun to speculate on the subject of evolution (Herbert Spencer's "Autobiography", London, 1904, Vol. I. pages 175-177.); and Alfred Russel Wallace is scarcely less emphatic concerning the part played by Lyell's teaching in his scientific education. (See "My Life; a record of Events and Opinions", London, 1905, Vol. I. page 355, etc. Also his review of Lyell's "Principles" in "Quarterly Review" (Vol. 126), 1869, pages 359-394. See also "The Darwin-Wallace Celebration by the Linnean Society" (1909), page 118.) Huxley wrote in 1887 "I owe more than I can tell to the careful study of the "Principles of Geology" in my young days." ("Science and Pseudo Science"; "Collected Essays", London, 1902, Vol. V. page 101.) As for Charles Darwin, he never tired—either in his published writings, his private correspondence or his most intimate conversations—of ascribing the awakening of his enthusiasm and the direction of his energies towards the elucidation of the problem of development to the "Principles of Geology" and the personal influence of its author. Huxley has well expressed what the author of the "Origin of Species" so constantly insisted upon, in the statements "Darwin's greatest work is the outcome of the unflinching application to Biology of the leading idea and the method applied in the "Principles" to Geology ("Proc. Roy. Soc." Vol. XLIV. (1888), page viii.; "Collected Essays" II. page 268, 1902.), and "Lyell, for others, as for myself, was the chief agent in smoothing the road for Darwin." ("Life and Letters of Charles Darwin" II. page 190.)

We propose therefore to consider, first, what Darwin owed to geology and its cultivators, and in the second place how he was able in the end so fully to pay a great debt which he never failed to acknowledge. Thanks to the invaluable materials contained in the "Life and Letters of Charles Darwin" (3 vols.) published by Mr Francis Darwin in 1887; and to "More Letters of Charles Darwin" (2 vols.) issued by the same author, in conjunction with Professor A.C. Seward, in 1903, we are permitted to follow the various movements in Darwin's mind, and are able to record the story almost entirely in his own words. (The first of these works is indicated in the following pages by the letters "L.L."; the second by "M.L.")

From the point of view of the geologist, Darwin's life naturally divides itself into four periods. In the first, covering twenty-two years, various influences were at work militating, now for and now against, his adoption of a geological career; in the second period—the five memorable years of the voyage of the "Beagle"—the ardent sportsman with some natural-history tastes, gradually became the most enthusiastic and enlightened of geologists; in the third period, lasting ten years, the valuable geological recruit devoted nearly all his energies and time to geological study and discussion and to preparing for publication the numerous observations made by him during the voyage; the fourth period, which covers the latter half of his life, found Darwin gradually drawn more and more from geological to biological studies, though always retaining the deepest interest in the progress and fortunes of his "old love." But geologists gladly recognise the fact that Darwin immeasurably better served their science by this biological work, than he could possibly have done by confining himself to purely geological questions.

From his earliest childhood, Darwin was a collector, though up to the time when, at eight years of age, he went to a preparatory school, seals, franks and similar trifles appear to have been the only objects of his quest. But a stone, which one of his schoolfellows at that time gave to him, seems to have attracted his attention and set him seeking for pebbles and minerals; as the result of this newly acquired taste, he says (writing in 1838) "I distinctly recollect the desire I had of being able to know something about every pebble in front of the hall door—it was my earliest and only geological aspiration at that time." ("M.L." I. page 3.) He further suspects that while at Mr Case's school "I do not remember any mental pursuits except those of collecting stones," etc... "I was born a naturalist." ("M.L." I. page 4.)

The court-yard in front of the hall door at the Mount House, Darwin's birthplace and the home of his childhood, is surrounded by beds or rockeries on which lie a number of pebbles. Some of these pebbles (in quite recent times as I am informed) have been collected to form a "cobbled" space in front of the gate in the outer wall, which fronts the hall door; and a similar "cobbled area," there is reason to believe, may have existed in Darwin's childhood before the door itself. The pebbles, which were obtained from a neighbouring gravel-pit, being derived from the glacial drift, exhibit very striking differences in colour and form. It was probably this circumstance which awakened in the child his love of observation and speculation. It is certainly remarkable that "aspirations" of the kind should have arisen in the mind of a child of 9 or 10!

When he went to Shrewsbury School, he relates "I continued collecting minerals with much zeal, but quite unscientifically,—all that I cared about was a new-NAMED mineral, and I hardly attempted to classify them." ("L.L." I. page 34.)

There has stood from very early times in Darwin's native town of Shrewsbury, a very notable boulder which has probably marked a boundary and is known as the "Bell-stone"—giving its name to a house and street. Darwin tells us in his "Autobiography" that while he was at Shrewsbury School at the age of 13 or 14 "an old Mr Cotton in Shropshire, who knew a good deal about rocks" pointed out to me "... the 'bell-stone'; he told me that there was no rock of the same kind nearer than Cumberland or Scotland, and he solemnly assured me that the world would come to an end before anyone would be able to explain how this stone came where it now lay"! Darwin adds "This produced a deep impression on me, and I meditated over this wonderful stone." ("L.L." I. page 41.)

The "bell-stone" has now, owing to the necessities of building, been removed a short distance from its original site, and is carefully preserved within the walls of a bank. It is a block of irregular shape 3 feet long and 2 feet wide, and about 1 foot thick, weighing probably not less than one-third of a ton. By the courtesy of the directors of the National Provincial Bank of England, I have been able to make a minute examination of it, and Professors Bonney and Watts, with Mr Harker and Mr Fearnsides have given me their valuable assistance. The rock is a much altered andesite and was probably derived from the Arenig district in North Wales, or possibly from a point nearer the Welsh Border. (I am greatly indebted to the Managers of the Bank at Shrewsbury for kind assistance in the examination of this interesting memorial: and Mr H.T. Beddoes, the Curator of the Shrewsbury Museum, has given me some archaeological information concerning the stone. Mr Richard Cotton was a good local naturalist, a Fellow both of the Geological and Linnean Societies; and to the officers of these societies I am indebted for information concerning him. He died in 1839, and although he does not appear to have published any scientific papers, he did far more for science by influencing the career of the school boy!) It was of course brought to where Shrewsbury now stands by the agency of a glacier—as Darwin afterwards learnt.

We can well believe from the perusal of these reminiscences that, at this time, Darwin's mind was, as he himself says, "prepared for a philosophical treatment of the subject" of Geology. ("L.L." I. page 41.) When at the age of 16, however, he was entered as a medical student at Edinburgh University, he not only did not get any encouragement of his scientific tastes, but was positively repelled by the ordinary instruction given there. Dr Hope's lectures on Chemistry, it is true, interested the boy, who with his brother Erasmus had made a laboratory in the toolhouse, and was nicknamed "Gas" by his schoolfellows, while undergoing solemn and public reprimand from Dr Butler at Shrewsbury School for thus wasting his time. ("L.L." I. page 35.) But most of the other Edinburgh lectures were "intolerably dull," "as dull as the professors" themselves, "something fearful to remember." In after life the memory of these lectures was like a nightmare to him. He speaks in 1840 of Jameson's lectures as something "I... for my sins experienced!" ("L.L." I. page 340.) Darwin especially signalises these lectures on Geology and Zoology, which he attended in his second year, as being worst of all "incredibly dull. The sole effect they produced on me was the determination never so long as I lived to read a book on Geology, or in any way to study the science!" ("L.L." I. page 41.)

The misfortune was that Edinburgh at that time had become the cockpit in which the barren conflict between "Neptunism" and "Plutonism" was being waged with blind fury and theological bitterness. Jameson and his pupils, on the one hand, and the friends and disciples of Hutton, on the other, went to the wildest extremes in opposing each other's peculiar tenets. Darwin tells us that he actually heard Jameson "in a field lecture at Salisbury Craigs, discoursing on a trap-dyke, with amygdaloidal margins and the strata indurated on each side, with volcanic rocks all around us, say that it was a fissure filled with sediment from above, adding with a sneer that there were men who maintained that it had been injected from beneath in a molten condition." ("L.L." I. pages 41-42.) "When I think of this lecture," added Darwin, "I do not wonder that I determined never to attend to Geology." (This was written in 1876 and Darwin had in the summer of 1839 revisited and carefully studied the locality ("L.L." I. page 290.) It is probable that most of Jameson's teaching was of the same controversial and unilluminating character as this field-lecture at Salisbury Craigs.

There can be no doubt that, while at Edinburgh, Darwin must have become acquainted with the doctrines of the Huttonian School. Though so young, he mixed freely with the scientific society of the city, Macgillivray, Grant, Leonard Horner, Coldstream, Ainsworth and others being among his acquaintances, while he attended and even read papers at the local scientific societies. It is to be feared, however, that what Darwin would hear most of, as characteristic of the Huttonian teaching, would be assertions that chalk-flints were intrusions of molten silica, that fossil wood and other petrifactions had been impregnated with fused materials, that heat—but never water—was always the agent by which the induration and crystallisation of rock-materials (even siliceous conglomerate, limestone and rock-salt) had been effected! These extravagant "anti-Wernerian" views the young student might well regard as not one whit less absurd and repellant than the doctrine of the "aqueous precipitation" of basalt. There is no evidence that Darwin, even if he ever heard of them, was in any way impressed, in his early career, by the suggestive passages in Hutton and Playfair, to which Lyell afterwards called attention, and which foreshadowed the main principles of Uniformitarianism.

As a matter of fact, I believe that the influence of Hutton and Playfair in the development of a philosophical theory of geology has been very greatly exaggerated by later writers on the subject. Just as Wells and Matthew anticipated the views of Darwin on Natural Selection, but without producing any real influence on the course of biological thought, so Hutton and Playfair adumbrated doctrines which only became the basis of vivifying theory in the hands of Lyell. Alfred Russel Wallace has very justly remarked that when Lyell wrote the "Principles of Geology", "the doctrines of Hutton and Playfair, so much in advance of their age, seemed to be utterly forgotten." ("Quarterly Review", Vol. CXXVI. (1869), page 363.) In proof of this it is only necessary to point to the works of the great masters of English geology, who preceded Lyell, in which the works of Hutton and his followers are scarcely ever mentioned. This is true even of the "Researches in Theoretical Geology" and the other works of the sagacious De la Beche. (Of the strength and persistence of the prejudice felt against Lyell's views by his contemporaries, I had a striking illustration some little time after Lyell's death. One of the old geologists who in the early years of the century had done really good work in connection with the Geological Society expressed a hope that I was not "one of those who had been carried away by poor Lyell's fads." My surprise was indeed great when further conversation showed me that the whole of the "Principles" were included in the "fads"!) Darwin himself possessed a copy of Playfair's "Illustrations of the Huttonian Theory", and occasionally quotes it; but I have met with only one reference to Hutton, and that a somewhat enigmatical one, in all Darwin's writings. In a letter to Lyell in 1841, when his mind was much exercised concerning glacial questions, he says "What a grand new feature all this ice work is in Geology! How old Hutton would have stared!" ("M.L." II. page 149.)

As a consequence of the influences brought to bear on his mind during his two years' residence in Edinburgh, Darwin, who had entered that University with strong geological aspirations, left it and proceeded to Cambridge with a pronounced distaste for the whole subject. The result of this was that, during his career as an under-graduate, he neglected all the opportunities for geological study. During that important period of life, when he was between eighteen and twenty years of age, Darwin spent his time in riding, shooting and beetle-hunting, pursuits which were undoubtedly an admirable preparation for his future work as an explorer; but in none of his letters of this period does he even mention geology. He says, however, "I was so sickened with lectures at Edinburgh that I did not even attend Sedgwick's eloquent and interesting lectures." ("L.L." I. page 48.)

It was only after passing his examination, and when he went up to spend two extra terms at Cambridge, that geology again began to attract his attention. The reading of Sir John Herschel's "Introduction to the Study of Natural Philosophy", and of Humboldt's "Personal Narrative", a copy of which last had been given to him by his good friend and mentor Henslow, roused his dormant enthusiasm for science, and awakened in his mind a passionate desire for travel. And it was from Henslow, whom he had accompanied in his excursions, but without imbibing any marked taste, at that time, for botany, that the advice came to think of and to "begin the study of geology." ("L.L." I. page 56.) This was in 1831, and in the summer vacation of that year we find him back again at Shrewsbury "working like a tiger" at geology and endeavouring to make a map and section of Shropshire—work which he says was not "as easy as I expected." ("L.L." I. page 189.) No better field for geological studies could possibly be found than Darwin's native county.

Writing to Henslow at this time, and referring to a form of the instrument devised by his friend, Darwin says: "I am very glad to say I think the clinometer will answer admirably. I put all the tables in my bedroom at every conceivable angle and direction. I will venture to say that I have measured them as accurately as any geologist going could do." But he adds: "I have been working at so many things that I have not got on much with geology. I suspect the first expedition I take, clinometer and hammer in hand, will send me back very little wiser and a good deal more puzzled than when I started." ("L.L." I. page 189.) Valuable aid was, however, at hand, for at this time Sedgwick, to whom Darwin had been introduced by the ever-helpful Henslow, was making one of his expeditions into Wales, and consented to accept the young student as his companion during the geological tour. ("L.L." I. page 56.) We find Darwin looking forward to this privilege with the keenest interest. ("L.L." I. page 189.)

When at the beginning of August (1831), Sedgwick arrived at his father's house in Shrewsbury, where he spent a night, Darwin began to receive his first and only instruction as a field-geologist. The journey they took together led them through Llangollen, Conway, Bangor, and Capel Curig, at which latter place they parted after spending many hours in examining the rocks at Cwm Idwal with extreme care, seeking for fossils but without success. Sedgwick's mode of instruction was admirable—he from time to time sent the pupil off on a line parallel to his own, "telling me to bring back specimens of the rocks and to mark the stratification on a map." ("L.L." I. page 57.) On his return to Shrewsbury, Darwin wrote to Henslow, "My trip with Sedgwick answered most perfectly," ("L.L." I. page 195.), and in the following year he wrote again from South America to the same friend, "Tell Professor Sedgwick he does not know how much I am indebted to him for the Welsh expedition; it has given me an interest in Geology which I would not give up for any consideration. I do not think I ever spent a more delightful three weeks than pounding the north-west mountains." ("L.L." I. pages 237-8.)

It would be a mistake, however, to suppose that at this time Darwin had acquired anything like the affection for geological study, which he afterwards developed. After parting with Sedgwick, he walked in a straight line by compass and map across the mountains to Barmouth to visit a reading party there, but taking care to return to Shropshire before September 1st, in order to be ready for the shooting. For as he candidly tells us, "I should have thought myself mad to give up the first days of partridge-shooting for geology or any other science!" ("L.L." I. page 58.)

Any regret we may be disposed to feel that Darwin did not use his opportunities at Edinburgh and Cambridge to obtain systematic and practical instruction in mineralogy and geology, will be mitigated, however, when we reflect on the danger which he would run of being indoctrinated with the crude "catastrophic" views of geology, which were at that time prevalent in all the centres of learning.

Writing to Henslow in the summer of 1831, Darwin says "As yet I have only indulged in hypotheses, but they are such powerful ones that I suppose, if they were put into action but for one day, the world would come to an end." ("L.L." I. page 189.)

May we not read in this passage an indication that the self-taught geologist had, even at this early stage, begun to feel a distrust for the prevalent catastrophism, and that his mind was becoming a field in which the seeds which Lyell was afterwards to sow would "fall on good ground"?

The second period of Darwin's geological career—the five years spent by him on board the "Beagle"—was the one in which by far the most important stage in his mental development was accomplished. He left England a healthy, vigorous and enthusiastic collector; he returned five years later with unique experiences, the germs of great ideas, and a knowledge which placed him at once in the foremost ranks of the geologists of that day. Huxley has well said that "Darwin found on board the "Beagle" that which neither the pedagogues of Shrewsbury, nor the professoriate of Edinburgh, nor the tutors of Cambridge had managed to give him." ("Proc. Roy. Soc." Vol. XLIV. (1888), page IX.) Darwin himself wrote, referring to the date at which the voyage was expected to begin: "My second life will then commence, and it shall be as a birthday for the rest of my life." ("L.L." I. page 214.); and looking back on the voyage after forty years, he wrote; "The voyage of the 'Beagle' has been by far the most important event in my life, and has determined my whole career;... I have always felt that I owe to the voyage the first real training or education of my mind; I was led to attend closely to several branches of natural history, and thus my powers of observation were improved, though they were always fairly developed." ("L.L." I. page 61.)

Referring to these general studies in natural history, however, Darwin adds a very significant remark: "The investigation of the geology of the places visited was far more important, as reasoning here comes into play. On first examining a new district nothing can appear more hopeless than the chaos of rocks; but by recording the stratification and nature of the rocks and fossils at many points, always reasoning and predicting what will be found elsewhere, light soon begins to dawn on the district, and the structure of the whole becomes more or less intelligible." ("L.L." I. page 62.)

The famous voyage began amid doubts, discouragements and disappointments. Fearful of heart-disease, sad at parting from home and friends, depressed by sea-sickness, the young explorer, after being twice driven back by baffling winds, reached the great object of his ambition, the island of Teneriffe, only to find that, owing to quarantine regulations, landing was out of the question.

But soon this inauspicious opening of the voyage was forgotten. Henslow had advised his pupil to take with him the first volume of Lyell's "Principles of Geology", then just published—but cautioned him (as nearly all the leaders in geological science at that day would certainly have done) "on no account to accept the views therein advocated." ("L.L." I. page 73.) It is probable that the days of waiting, discomfort and sea-sickness at the beginning of the voyage were relieved by the reading of this volume. For he says that when he landed, three weeks after setting sail from Plymouth, in St Jago, the largest of the Cape de Verde Islands, the volume had already been "studied attentively; and the book was of the highest service to me in many ways... " His first original geological work, he declares, "showed me clearly the wonderful superiority of Lyell's manner of treating geology, compared with that of any other author, whose works I had with me or ever afterwards read." ("L.L." I. page 62.)

At St Jago Darwin first experienced the joy of making new discoveries, and his delight was unbounded. Writing to his father he says, "Geologising in a volcanic country is most delightful; besides the interest attached to itself, it leads you into most beautiful and retired spots." ("L.L." I. page 228.) To Henslow he wrote of St Jago: "Here we spent three most delightful weeks... St Jago is singularly barren, and produces few plants or insects, so that my hammer was my usual companion, and in its company most delightful hours I spent." "The geology was pre-eminently interesting, and I believe quite new; there are some facts on a large scale of upraised coast (which is an excellent epoch for all the volcanic rocks to date from), that would interest Mr Lyell." ("L.L." I. page 235.) After more than forty years the memory of this, his first geological work, seems as fresh as ever, and he wrote in 1876, "The geology of St Jago is very striking, yet simple: a stream of lava formerly flowed over the bed of the sea, formed of triturated recent shells and corals, which it has baked into a hard white rock. Since then the whole island has been upheaved. But the line of white rock revealed to me a new and important fact, namely, that there had been afterwards subsidence round the craters, which had since been in action, and had poured forth lava." ("L.L." I. page 65.)

It was at this time, probably, that Darwin made his first attempt at drawing a sketch-map and section to illustrate the observations he had made (see his "Volcanic Islands", pages 1 and 9). His first important geological discovery, that of the subsidence of strata around volcanic vents (which has since been confirmed by Mr Heaphy in New Zealand and other authors) awakened an intense enthusiasm, and he writes: "It then first dawned on me that I might perhaps write a book on the geology of the various countries visited, and this made me thrill with delight. That was a memorable hour to me, and how distinctly I can call to mind the low cliff of lava beneath which I rested, with the sun glaring hot, a few strange desert plants growing near, and with living corals in the tidal pools at my feet." ("L.L." I. page 66.)

But it was when the "Beagle", after touching at St Paul's rock and Tristan d'Acunha (for a sufficient time only to collect specimens), reached the shores of South America, that Darwin's real work began; and he was able, while the marine surveys were in progress, to make many extensive journeys on land. His letters at this time show that geology had become his chief delight, and such exclamations as "Geology carries the day," "I find in Geology a never failing interest," etc. abound in his correspondence.

Darwin's time was divided between the study of the great deposits of red mud—the Pampean formation—with its interesting fossil bones and shells affording proofs of slow and constant movements of the land, and the underlying masses of metamorphic and plutonic rocks. Writing to Henslow in March, 1834, he says: "I am quite charmed with Geology, but, like the wise animal between two bundles of hay, I do not know which to like best; the old crystalline groups of rocks, or the softer and fossiliferous beds. When puzzling about stratification, etc., I feel inclined to cry 'a fig for your big oysters, and your bigger megatheriums.' But then when digging out some fine bones, I wonder how any man can tire his arms with hammering granite." ("L.L." I. page 249.) We are told by Darwin that he loved to reason about and attempt to predict the nature of the rocks in each new district before he arrived at it.

This love of guessing as to the geology of a district he was about to visit is amusingly expressed by him in a letter (of May, 1832) to his cousin and old college-friend, Fox. After alluding to the beetles he had been collecting—a taste his friend had in common with himself—he writes of geology that "It is like the pleasure of gambling. Speculating on first arriving, what the rocks may be, I often mentally cry out 3 to 1 tertiary against primitive; but the latter have hitherto won all the bets." ("L.L." I. page 233.)

Not the least important of the educational results of the voyage to Darwin was the acquirement by him of those habits of industry and method which enabled him in after life to accomplish so much—in spite of constant failures of health. From the outset, he daily undertook and resolutely accomplished, in spite of sea-sickness and other distractions, four important tasks. In the first place he regularly wrote up the pages of his Journal, in which, paying great attention to literary style and composition, he recorded only matters that would be of general interest, such as remarks on scenery and vegetation, on the peculiarities and habits of animals, and on the characters, avocations and political institutions of the various races of men with whom he was brought in contact. It was the freshness of these observations that gave his "Narrative" so much charm. Only in those cases in which his ideas had become fully crystallised, did he attempt to deal with scientific matters in this journal. His second task was to write in voluminous note-books facts concerning animals and plants, collected on sea or land, which could not be well made out from specimens preserved in spirit; but he tells us that, owing to want of skill in dissecting and drawing, much of the time spent in this work was entirely thrown away, "a great pile of MS. which I made during the voyage has proved almost useless." ("L.L." I. page 62.) Huxley confirmed this judgment on his biological work, declaring that "all his zeal and industry resulted, for the most part, in a vast accumulation of useless manuscript." ("Proc. Roy. Soc." Vol. XLIV. (1888), page IX.) Darwin's third task was of a very different character and of infinitely greater value. It consisted in writing notes of his journeys on land—the notes being devoted to the geology of the districts visited by him. These formed the basis, not only of a number of geological papers published on his return, but also of the three important volumes forming "The Geology of the voyage of the 'Beagle'". On July 24th, 1834, when little more than half of the voyage had been completed, Darwin wrote to Henslow, "My notes are becoming bulky. I have about 600 small quarto pages full; about half of this is Geology." ("M.L." I. page 14.) The last, and certainly not the least important of all his duties, consisted in numbering, cataloguing, and packing his specimens for despatch to Henslow, who had undertaken the care of them. In his letters he often expresses the greatest solicitude lest the value of these specimens should be impaired by the removal of the numbers corresponding to his manuscript lists. Science owes much to Henslow's patient care of the collections sent to him by Darwin. The latter wrote in Henslow's biography, "During the five years' voyage, he regularly corresponded with me and guided my efforts; he received, opened, and took care of all the specimens sent home in many large boxes." ("Life of Henslow", by L. Jenyns (Blomefield), London, 1862, page 53.)

Darwin's geological specimens are now very appropriately lodged for the most part in the Sedgwick Museum, Cambridge, his original Catalogue with subsequent annotations being preserved with them. From an examination of these catalogues and specimens we are able to form a fair notion of the work done by Darwin in his little cabin in the "Beagle", in the intervals between his land journeys.

Besides writing up his notes, it is evident that he was able to accomplish a considerable amount of study of his specimens, before they were packed up for despatch to Henslow. Besides hand-magnifiers and a microscope, Darwin had an equipment for blowpipe-analysis, a contact-goniometer and magnet; and these were in constant use by him. His small library of reference (now included in the Collection of books placed by Mr F. Darwin in the Botany School at Cambridge ("Catalogue of the Library of Charles Darwin now in the Botany School, Cambridge". Compiled by H.W. Rutherford; with an introduction by Francis Darwin. Cambridge, 1908.)) appears to have been admirably selected, and in all probability contained (in addition to a good many works relating to South America) a fair number of excellent books of reference. Among those relating to mineralogy, he possessed the manuals of Phillips, Alexander Brongniart, Beudant, von Kobell and Jameson: all the "Cristallographie" of Brochant de Villers and, for blowpipe work, Dr Children's translation of the book of Berzelius on the subject. In addition to these, he had Henry's "Experimental Chemistry" and Ure's "Dictionary" (of Chemistry). A work, he evidently often employed, was P. Syme's book on "Werner's Nomenclature of Colours"; while, for Petrology, he used Macculloch's "Geological Classification of Rocks". How diligently and well he employed his instruments and books is shown by the valuable observations recorded in the annotated Catalogues drawn up on board ship.

These catalogues have on the right-hand pages numbers and descriptions of the specimens, and on the opposite pages notes on the specimens—the result of experiments made at the time and written in a very small hand. Of the subsequently made pencil notes, I shall have to speak later. (I am greatly indebted to my friend Mr A. Harker, F.R.S., for his assistance in examining these specimens and catalogues. He has also arranged the specimens in the Sedgwick Museum, so as to make reference to them easy. The specimens from Ascension and a few others are however in the Museum at Jermyn Street.)

It is a question of great interest to determine the period and the occasion of Darwin's first awakening to the great problem of the transmutation of species. He tells us himself that his grandfather's "Zoonomia" had been read by him "but without producing any effect," and that his friend Grant's rhapsodies on Lamarck and his views on evolution only gave rise to "astonishment." ("L.L." I. page 38.)

Huxley, who had probably never seen the privately printed volume of letters to Henslow, expressed the opinion that Darwin could not have perceived the important bearing of his discovery of bones in the Pampean Formation, until they had been studied in England, and their analogies pronounced upon by competent comparative anatomists. And this seemed to be confirmed by Darwin's own entry in his pocket-book for 1837, "In July opened first notebook on Transmutation of Species. Had been greatly struck from about the month of previous March on character of South American fossils... " ("L.L." I. page 276.)

The second volume of Lyell's "Principles of Geology" was published in January, 1832, and Darwin's copy (like that of the other two volumes, in a sadly dilapidated condition from constant use) has in it the inscription, "Charles Darwin, Monte Video. Nov. 1832." As everyone knows, Darwin in dedicating the second edition of his Journal of the Voyage to Lyell declared, "the chief part of whatever scientific merit this journal and the other works of the author may possess, has been derived from studying the well-known and admirable 'Principles of Geology'".

In the first chapter of this second volume of the "Principles", Lyell insists on the importance of the species question to the geologist, but goes on to point out the difficulty of accepting the only serious attempt at a transmutation theory which had up to that time appeared—that of Lamarck. In subsequent chapters he discusses the questions of the modification and variability of species, of hybridity, and of the geographical distribution of plants and animals. He then gives vivid pictures of the struggle for existence, ever going on between various species, and of the causes which lead to their extinction—not by overwhelming catastrophes, but by the silent and almost unobserved action of natural causes. This leads him to consider theories with regard to the introduction of new species, and, rejecting the fanciful notions of "centres or foci of creation," he argues strongly in favour of the view, as most reconcileable with observed facts, that "each species may have had its origin in a single pair, or individual, where an individual was sufficient, and species may have been created in succession at such times and in such places as to enable them to multiply and endure for an appointed period, and occupy an appointed space on the globe." ("Principles of Geology", Vol. II. (1st edition 1832), page 124. We now know, as has been so well pointed out by Huxley, that Lyell, as early as 1827, was prepared to accept the doctrine of the transmutation of species. In that year he wrote to Mantell, "What changes species may really undergo! How impossible will it be to distinguish and lay down a line, beyond which some of the so-called extinct species may have never passed into recent ones" (Lyell's "Life and Letters" Vol. I. page 168). To Sir John Herschel in 1836, he wrote, "In regard to the origination of new species, I am very glad to find that you think it probable that it may be carried on through the intervention of intermediate causes. I left this rather to be inferred, not thinking it worth while to offend a certain class of persons by embodying in words what would only be a speculation" (Ibid. page 467). He expressed the same views to Whewell in 1837 (Ibid. Vol. II. page 5.), and to Sedgwick (Ibid. Vol. II. page 36) to whom he says, of "the theory, that the creation of new species is going on at the present day"—"I really entertain it," but "I have studiously avoided laying the doctrine down dogmatically as capable of proof" (see Huxley in "L.L." II. pages 190-195.))

After pointing out how impossible it would be for a naturalist to prove that a newly DISCOVERED species was really newly CREATED (Mr F. Darwin has pointed out that his father (like Lyell) often used the term "Creation" in speaking of the origin of new species ("L.L." II. chapter 1.)), Lyell argued that no satisfactory evidence OF THE WAY in which these new forms were created, had as yet been discovered, but that he entertained the hope of a possible solution of the problem being found in the study of the geological record.

It is not difficult, in reading these chapters of Lyell's great work, to realise what an effect they would have on the mind of Darwin, as new facts were collected and fresh observations concerning extinct and recent forms were made in his travels. We are not surprised to find him writing home, "I am become a zealous disciple of Mr Lyell's views, as known in his admirable book. Geologising in South America, I am tempted to carry parts to a greater extent even than he does." ("L.L." I. page 263.)

Lyell's anticipation that the study of the geological record might afford a clue to the discovery of how new species originate was remarkably fulfilled, within a few months, by Darwin's discovery of fossil bones in the red Pampean mud.

It is very true that, as Huxley remarked, Darwin's knowledge of comparative anatomy must have been, at that time, slight; but that he recognised the remarkable resemblances between the extinct and existing mammals of South America is proved beyond all question by a passage in his letter to Henslow, written November 24th, 1832: "I have been very lucky with fossil bones; I have fragments of at least six distinct animals... I found a large surface of osseous polygonal plates... Immediately I saw them I thought they must belong to an enormous armadillo, living species of which genus are so abundant here," and he goes on to say that he has "the lower jaw of some large animal which, from the molar teeth, I should think belonged to the Edentata." ("M.L." I. pages 11, 12. See "Extracts of Letters addressed to Prof. Henslow by C. Darwin" (1835), page 7.)

Having found this important clue, Darwin followed it up with characteristic perseverance. In his quest for more fossil bones he was indefatigable. Mr Francis Darwin tells us, "I have often heard him speak of the despair with which he had to break off the projecting extremity of a huge, partly excavated bone, when the boat waiting for him would wait no longer." ("L.L." I. page 276 (footnote).) Writing to Haeckel in 1864, Darwin says: "I shall never forget my astonishment when I dug out a gigantic piece of armour, like that of the living armadillo." (Haeckel, "History of Creation", Vol. I. page 134, London, 1876.)

In a letter to Henslow in 1834 Darwin says: "I have just got scent of some fossil bones... what they may be I do not know, but if gold or galloping will get them they shall be mine." ("M.L." I. page 15.)

Darwin also showed his sense of the importance of the discovery of these bones by his solicitude about their safe arrival and custody. From the Falkland Isles (March, 1834), he writes to Henslow: "I have been alarmed by your expression 'cleaning all the bones' as I am afraid the printed numbers will be lost: the reason I am so anxious they should not be, is, that a part were found in a gravel with recent shells, but others in a very different bed. Now with these latter there were bones of an Agouti, a genus of animals, I believe, peculiar to America, and it would be curious to prove that some one of the genus co-existed with the Megatherium: such and many other points depend on the numbers being carefully preserved." ("Extracts from Letters etc.", pages 13-14.) In the abstract of the notes read to the Geological Society in 1835, we read: "In the gravel of Patagonia he (Darwin) also found many bones of the Megatherium and of five or six other species of quadrupeds, among which he has detected the bones of a species of Agouti. He also met with several examples of the polygonal plates, etc." ("Proc. Geol. Soc." Vol. II. pages 211-212.)

Darwin's own recollections entirely bear out the conclusion that he fully recognised, WHILE IN SOUTH AMERICA, the wonderful significance of the resemblances between the extinct and recent mammalian faunas. He wrote in his "Autobiography": "During the voyage of the 'Beagle' I had been deeply impressed by discovering in the Pampean formation great fossil animals covered with armour like that on the existing armadillos." ("L.L." I. page 82.)

The impression made on Darwin's mind by the discovery of these fossil bones, was doubtless deepened as, in his progress southward from Brazil to Patagonia, he found similar species of Edentate animals everywhere replacing one another among the living forms, while, whenever fossils occurred, they also were seen to belong to the same remarkable group of animals. (While Darwin was making these observations in South America, a similar generalisation to that at which he arrived was being reached, quite independently and almost simultaneously, with respect to the fossil and recent mammals of Australia. In the year 1831, Clift gave to Jameson a list of bones occurring in the caves and breccias of Australia, and in publishing this list the latter referred to the fact that the forms belonged to marsupials, similar to those of the existing Australian fauna. But he also stated that, as a skull had been identified (doubtless erroneously) as having belonged to a hippopotamus, other mammals than marsupials must have spread over the island in late Tertiary times. It is not necessary to point out that this paper was quite unknown to Darwin while in South America. Lyell first noticed it in the third edition of his "Principles", which was published in May, 1834 (see "Edinb. New Phil. Journ." Vol. X. (1831), pages 394-6, and Lyell's "Principles" (3rd edition), Vol. III. page 421). Darwin referred to this discovery in 1839 (see his "Journal", page 210.))

That the passage in Darwin's pocket-book for 1837 can only refer to an AWAKENING of Darwin's interest in the subject—probably resulting from a sight of the bones when they were being unpacked—I think there cannot be the smallest doubt; AND WE MAY THEREFORE CONFIDENTLY FIX UPON NOVEMBER, 1832, AS THE DATE AT WHICH DARWIN COMMENCED THAT LONG SERIES OF OBSERVATIONS AND REASONINGS WHICH EVENTUALLY CULMINATED IN THE PREPARATION OF THE "ORIGIN OF SPECIES". Equally certain is it, that it was his geological work that led Darwin into those paths of research which in the end conducted him to his great discoveries. I quite agree with the view expressed by Mr F. Darwin and Professor Seward, that Darwin, like Lyell, "thought it 'almost useless' to try to prove the truth of evolution until the cause of change was discovered" ("M.L." I. page 38.), and that possibly he may at times have vacillated in his opinions, but I believe there is evidence that, from the date mentioned, the "species question" was always more or less present in Darwin's mind. (Although we admit with Huxley that Darwin's training in comparative anatomy was very small, yet it may be remembered that he was a medical student for two years, and, if he hated the lectures, he enjoyed the society of naturalists. He had with him in the little "Beagle" library a fair number of zoological books, including works on Osteology by Cuvier, Desmarest and Lesson, as well as two French Encyclopaedias of Natural History. As a sportsman, he would obtain specimens of recent mammals in South America, and would thus have opportunities of studying their teeth and general anatomy. Keen observer, as he undoubtedly was, we need not then be surprised that he was able to make out the resemblances between the recent and fossil forms.)

It is clear that, as time went on, Darwin became more and more absorbed in his geological work. One very significant fact was that the once ardent sportsman, when he found that shooting the necessary game and zoological specimens interfered with his work with the hammer, gave up his gun to his servant. ("L.L." I. page 63.) There is clear evidence that Darwin gradually became aware how futile were his attempts to add to zoological knowledge by dissection and drawing, while he felt ever increasing satisfaction with his geological work.

The voyage fortunately extended to a much longer period (five years) than the two originally intended, but after being absent nearly three years, Darwin wrote to his sister in November, 1834, "Hurrah! hurrah! it is fixed that the 'Beagle' shall not go one mile south of Cape Tres Montes (about 200 miles south of Chiloe), and from that point to Valparaiso will be finished in about five months. We shall examine the Chonos Archipelago, entirely unknown, and the curious inland sea behind Chiloe. For me it is glorious. Cape Tres Montes is the most southern point where there is much geological interest, as there the modern beds end. The Captain then talks of crossing the Pacific; but I think we shall persuade him to finish the coast of Peru, where the climate is delightful, the country hideously sterile, but abounding with the highest interest to the geologist... I have long been grieved and most sorry at the interminable length of the voyage (though I never would have quitted it)... I could not make up my mind to return. I could not give up all the geological castles in the air I had been building up for the last two years." ("L.L." I. pages 257-58.)

In April, 1835, he wrote to another sister: "I returned a week ago from my excursion across the Andes to Mendoza. Since leaving England I have never made so successful a journey... how deeply I have enjoyed it; it was something more than enjoyment; I cannot express the delight which I felt at such a famous winding-up of all my geology in South America. I literally could hardly sleep at nights for thinking over my day's work. The scenery was so new, and so majestic; everything at an elevation of 12,000 feet bears so different an aspect from that in the lower country... To a geologist, also, there are such manifest proofs of excessive violence; the strata of the highest pinnacles are tossed about like the crust of a broken pie." ("L.L." I. pages 259-60.)

Darwin anticipated with intense pleasure his visit to the Galapagos Islands. On July 12th, 1835, he wrote to Henslow: "In a few days' time the "Beagle" will sail for the Galapagos Islands. I look forward with joy and interest to this, both as being somewhat nearer to England and for the sake of having a good look at an active volcano. Although we have seen lava in abundance, I have never yet beheld the crater." ("M.L." I. page 26.) He could little anticipate, as he wrote these lines, the important aid in the solution of the "species question" that would ever after make his visit to the Galapagos Islands so memorable. In 1832, as we have seen, the great discovery of the relations of living to extinct mammals in the same area had dawned upon his mind; in 1835 he was to find a second key for opening up the great mystery, by recognising the variations of similar types in adjoining islands among the Galapagos.

The final chapter in the second volume of the "Principles" had aroused in Darwin's mind a desire to study coral-reefs, which was gratified during his voyage across the Pacific and Indian Oceans. His theory on the subject was suggested about the end of 1834 or the beginning of 1835, as he himself tells us, before he had seen a coral-reef, and resulted from his work during two years in which he had "been incessantly attending to the effects on the shores of South America of the intermittent elevation of the land, together with denudation and the deposition of sediment." ("L.L." I. page 70.)

On arriving at the Cape of Good Hope in July, 1836, Darwin was greatly gratified by hearing that Sedgwick had spoken to his father in high terms of praise concerning the work done by him in South America. Referring to the news from home, when he reached Bahia once more, on the return voyage (August, 1836), he says: "The desert, volcanic rocks, and wild sea of Ascension... suddenly wore a pleasing aspect, and I set to work with a good-will at my old work of Geology." ("L.L." I. page 265.) Writing fifty years later, he says: "I clambered over the mountains of Ascension with a bounding step and made the volcanic rocks resound under my geological hammer!" ("L.L." I. page 66.)

That his determination was now fixed to devote his own labours to the task of working out the geological results of the voyage, and that he was prepared to leave to more practised hands the study of his biological collections, is clear from the letters he sent home at this time. From St Helena he wrote to Henslow asking that he would propose him as a Fellow of the Geological Society; and his Certificate, in Henslow's handwriting, is dated September 8th, 1836, being signed from personal knowledge by Henslow and Sedgwick. He was proposed on November 2nd and elected November 30th, being formally admitted to the Society by Lyell, who was then President, on January 4th, 1837, on which date he also read his first paper. Darwin did not become a Fellow of the Linnean Society till eighteen years later (in 1854).

An estimate of the value and importance of Darwin's geological discoveries during the voyage of the "Beagle" can best be made when considering the various memoirs and books in which the author described them. He was too cautious to allow himself to write his first impressions in his Journal, and wisely waited till he could study his specimens under better conditions and with help from others on his return. The extracts published from his correspondence with Henslow and others, while he was still abroad, showed, nevertheless, how great was the mass of observation, how suggestive and pregnant with results were the reasonings of the young geologist.

Two sets of these extracts from Darwin's letters to Henslow were printed while he was still abroad. The first of these was the series of "Geological Notes made during a survey of the East and West Coasts of South America, in the years 1832, 1833, 1834 and 1835, with an account of a transverse section of the Cordilleras of the Andes between Valparaiso and Mendoza". Professor Sedgwick, who read these notes to the Geological Society on November 18th, 1835, stated that "they were extracted from a series of letters (addressed to Professor Henslow), containing a great mass of information connected with almost every branch of natural history," and that he (Sedgwick) had made a selection of the remarks which he thought would be more especially interesting to the Geological Society. An abstract of three pages was published in the "Proceedings of the Geological Society" (Vol. II. pages 210-12.), but so unknown was the author at this time that he was described as F. Darwin, Esq., of St John's College, Cambridge! Almost simultaneously (on November 16th, 1835) a second set of extracts from these letters—this time of a general character—were read to the Philosophical Society at Cambridge, and these excited so much interest that they were privately printed in pamphlet form for circulation among the members.

Many expeditions and "scientific missions" have been despatched to various parts of the world since the return of the "Beagle" in 1836, but it is doubtful whether any, even the most richly endowed of them, has brought back such stores of new information and fresh discoveries as did that little "ten-gun brig"—certainly no cabin or laboratory was the birth-place of ideas of such fruitful character as was that narrow end of a chart-room, where the solitary naturalist could climb into his hammock and indulge in meditation.

The third and most active portion of Darwin's career as a geologist was the period which followed his return to England at the end of 1836. His immediate admission to the Geological Society, at the beginning of 1837, coincided with an important crisis in the history of geological science.

The band of enthusiasts who nearly thirty years before had inaugurated the Geological Society—weary of the fruitless conflicts between "Neptunists" and "Plutonists"—had determined to eschew theory and confine their labours to the collection of facts, their publications to the careful record of observations. Greenough, the actual founder of the Society, was an ardent Wernerian, and nearly all his fellow-workers had come, more or less directly, under the Wernerian teaching. Macculloch alone gave valuable support to the Huttonian doctrines, so far as they related to the influence of igneous activity—but the most important portion of the now celebrated "Theory of the Earth"—that dealing with the competency of existing agencies to account for changes in past geological times—was ignored by all alike. Macculloch's influence on the development of geology, which might have had far-reaching effects, was to a great extent neutralised by his peculiarities of mind and temper; and, after a stormy and troublous career, he retired from the society in 1832. In all the writings of the great pioneers in English geology, Hutton and his splendid generalisation are scarcely ever referred to. The great doctrines of Uniformitarianism, which he had foreshadowed, were completely ignored, and only his extravagances of "anti-Wernerianism" seem to have been remembered.

When between 1830 and 1832, Lyell, taking up the almost forgotten ideas of Hutton, von Hoff and Prevost, published that bold challenge to the Catastrophists—the "Principles of Geology"—he was met with the strongest opposition, not only from the outside world, which was amused by his "absurdities" and shocked by his "impiety"—but not less from his fellow-workers and friends in the Geological Society. For Lyell's numerous original observations, and his diligent collection of facts his contemporaries had nothing but admiration, and they cheerfully admitted him to the highest offices in the society, but they met his reasonings on geological theory with vehement opposition and his conclusions with coldness and contempt.

There is, indeed, a very striking parallelism between the reception of the "Principles of Geology" by Lyell's contemporaries and the manner in which the "Origin of Species" was met a quarter of a century later, as is so vividly described by Huxley. ("L.L." II. pages 179-204.) Among Lyell's fellow-geologists, two only—G. Poulett Scrope and John Herschel (Both Lyell and Darwin fully realised the value of the support of these two friends. Scrope in his appreciative reviews of the "Principles" justly pointed out what was the weakest point, the inadequate recognition of sub-aerial as compared with marine denudation. Darwin also admitted that Scrope had to a great extent forestalled him in his theory of Foliation. Herschel from the first insisted that the leading idea of the "Principles" must be applied to organic as well as to inorganic nature and must explain the appearance of new species (see Lyell's "Life and Letters", Vol. I. page 467). Darwin tells us that Herschel's "Introduction to the Study of Natural Philosophy" with Humboldt's "Personal Narrative" "stirred up in me a burning zeal" in his undergraduate days. I once heard Lyell exclaim with fervour "If ever there was a heaven-born genius it was John Herschel!")—declared themselves from the first his strong supporters. Scrope in two luminous articles in the "Quarterly Review" did for Lyell what Huxley accomplished for Darwin in his famous review in the "Times"; but Scrope unfortunately was at that time immersed in the stormy sea of politics, and devoted his great powers of exposition to the preparation of fugitive pamphlets. Herschel, like Scrope, was unable to support Lyell at the Geological Society, owing to his absence on the important astronomical mission to the Cape.

It thus came about that, in the frequent conflicts of opinion within the walls of the Geological Society, Lyell had to bear the brunt of battle for Uniformitarianism quite alone, and it is to be feared that he found himself sadly overmatched when opposed by the eloquence of Sedgwick, the sarcasm of Buckland, and the dead weight of incredulity on the part of Greenough, Conybeare, Murchison and other members of the band of pioneer workers. As time went on there is evidence that the opposition of De la Beche and Whewell somewhat relaxed; the brilliant "Paddy" Fitton (as his friends called him) was sometimes found in alliance with Lyell, but was characteristically apt to turn his weapon, as occasion served, on friend or foe alike; the amiable John Phillips "sat upon the fence." Only when a new generation arose—including Jukes, Ramsay, Forbes and Hooker—did Lyell find his teachings received with anything like favour.

We can well understand, then, how Lyell would welcome such a recruit as young Darwin—a man who had declared himself more Lyellian than Lyell, and who brought to his support facts and observations gleaned from so wide a field.

The first meeting of Lyell and Darwin was characteristic of the two men. Darwin at once explained to Lyell that, with respect to the origin of coral-reefs, he had arrived at views directly opposed to those published by "his master." To give up his own theory, cost Lyell, as he told Herschel, a "pang at first," but he was at once convinced of the immeasurable superiority of Darwin's theory. I have heard members of Lyell's family tell of the state of wild excitement and sustained enthusiasm, which lasted for days with Lyell after this interview, and his letters to Herschel, Whewell and others show his pleasure at the new light thrown upon the subject and his impatience to have the matter laid before the Geological Society.

Writing forty years afterwards, Darwin, speaking of the time of the return of the "Beagle", says: "I saw a great deal of Lyell. One of his chief characteristics was his sympathy with the work of others, and I was as much astonished as delighted at the interest which he showed when, on my return to England, I explained to him my views on coral-reefs. This encouraged me greatly, and his advice and example had much influence on me." ("L.L." I. page 68.) Darwin further states that he saw more of Lyell at this time than of any other scientific man, and at his request sent his first communication to the Geological Society. ("L.L." I. page 67.)

"Mr Lonsdale" (the able curator of the Geological Society), Darwin wrote to Henslow, "with whom I had much interesting conversation," "gave me a most cordial reception," and he adds, "If I was not much more inclined for geology than the other branches of Natural History, I am sure Mr Lyell's and Lonsdale's kindness ought to fix me. You cannot conceive anything more thoroughly good-natured than the heart-and-soul manner in which he put himself in my place and thought what would be best to do." ("L.L." I. page 275.)

Within a few days of Darwin's arrival in London we find Lyell writing to Owen as follows:

"Mrs Lyell and I expect a few friends here on Saturday next, 29th (October), to an early tea party at eight o'clock, and it will give us great pleasure if you can join it. Among others you will meet Mr Charles Darwin, whom I believe you have seen, just returned from South America, where he has laboured for zoologists as well as for hammer-bearers. I have also asked your friend Broderip." ("The Life of Richard Owen", London, 1894, Vol. I. page 102.) It would probably be on this occasion that the services of Owen were secured for the work on the fossil bones sent home by Darwin.

On November 2nd, we find Lyell introducing Darwin as his guest at the Geological Society Club; on December 14th, Lyell and Stokes proposed Darwin as a member of the Club; between that date and May 3rd of the following year, when his election to the Club took place, he was several times dining as a guest.

On January 4th, 1837, as we have already seen, Darwin was formally admitted to the Geological Society, and on the same evening he read his first paper (I have already pointed out that the notes read at the Geological Society on Nov. 18, 1835 were extracts made by Sedgwick from letters sent to Henslow, and not a paper sent home for publication by Darwin.) before the Society, "Observations of proofs of recent elevation on the coast of Chili, made during the Survey of H.M.S. "Beagle", commanded by Captain FitzRoy, R.N." By C. Darwin, F.G.S. This paper was preceded by one on the same subject by Mr A. Caldcleugh, and the reading of a letter and other communications from the Foreign Office also relating to the earthquakes in Chili.

At the meeting of the Council of the Geological Society on February 1st, Darwin was nominated as a member of the new Council, and he was elected on February 17th.

The meeting of the Geological Society on April 19th was devoted to the reading by Owen of his paper on Toxodon, perhaps the most remarkable of the fossil mammals found by Darwin in South America; and at the next meeting, on May 3rd, Darwin himself read "A Sketch of the Deposits containing extinct Mammalia in the neighbourhood of the Plata". The next following meeting, on May 17th, was devoted to Darwin's Coral-reef paper, entitled "On certain areas of elevation and subsidence in the Pacific and Indian Oceans, as deduced from the study of Coral Formations". Neither of these three early papers of Darwin were published in the Transactions of the Geological Society, but the minutes of the Council show that they were "withdrawn by the author by permission of the Council."

Darwin's activity during this session led to some rather alarming effects upon his health, and he was induced to take a holiday in Staffordshire and the Isle of Wight. He was not idle, however, for a remark of his uncle, Mr Wedgwood, led him to make those interesting observations on the work done by earthworms, that resulted in his preparing a short memoir on the subject, and this paper, "On the Formation of Mould", was read at the Society on November 1st, 1837, being the first of Darwin's papers published in full; it appeared in Vol. V. of the "Geological Transactions", pages 505-510.

During this session, Darwin attended nearly all the Council meetings, and took such an active part in the work of the Society that it is not surprising to find that he was now requested to accept the position of Secretary. After some hesitation, in which he urged his inexperience and want of knowledge of foreign languages, he consented to accept the appointment. ("L.L." I. page 285.)

At the anniversary meeting on February 16th, 1838, the Wollaston Medal was given to Owen in recognition of his services in describing the fossil mammals sent home by Darwin.

In his address, the President, Professor Whewell, dwelt at length on the great value of the papers which Darwin had laid before the Society during the preceding session.

On March 7th, Darwin read before the Society the most important perhaps of all his geological papers, "On the Connexion of certain Volcanic Phenomena in South America, and on the Formation of Mountain-Chains and Volcanoes as the effect of Continental Elevations". In this paper he boldly attacked the tenets of the Catastrophists. It is evident that Darwin at this time, taking advantage of the temporary improvement in his health, was throwing himself into the breach of Uniformitarianism with the greatest ardour. Lyell wrote to Sedgwick on April 21st, 1837, "Darwin is a glorious addition to any society of geologists, and is working hard and making way, both in his book and in our discussions." ("The Life and Letters of the Reverend Adam Sedgwick", Vol. I. page 484, Cambridge, 1890.)

We have unfortunately few records of the animated debates which took place at this time between the old and new schools of geologists. I have often heard Lyell tell how Lockhart would bring down a party of friends from the Athenaeum Club to Somerset House on Geological nights, not, as he carefully explained, that "he cared for geology, but because he liked to while the fellows fight." But it fortunately happens that a few days after this last of Darwin's great field-days, at the Geological Society, Lyell, in a friendly letter to his father-in-law, Leonard Horner, wrote a very lively account of the proceedings while his impressions were still fresh; and this gives us an excellent idea of the character of these discussions.

Neither Sedgwick nor Buckland were present on this occasion, but we can imagine how they would have chastised their two "erring pupils"—more in sorrow than in anger—had they been there. Greenough, too, was absent—possibly unwilling to countenance even by his presence such outrageous doctrines.

Darwin, after describing the great earthquakes which he had experienced in South America, and the evidence of their connection with volcanic outbursts, proceeded to show that earthquakes originated in fractures, gradually formed in the earth's crust, and were accompanied by movements of the land on either side of the fracture. In conclusion he boldly advanced the view "that continental elevations, and the action of volcanoes, are phenomena now in progress, caused by some great but slow change in the interior of the earth; and, therefore, that it might be anticipated, that the formation of mountain chains is likewise in progress: and at a rate which may be judged of by either actions, but most clearly by the growth of volcanoes." ("Proc. Geol. Soc." Vol. II. pages 654-60.)

Lyell's account ("Life, Letters and Journals of Sir Charles Lyell, Bart.", edited by his sister-in-law, Mrs Lyell, Vol. II. pages 40, 41 (Letter to Leonard Horner, 1838), 2 vols. London, 1881.) of the discussion was as follows: "In support of my heretical notions," Darwin "opened upon De la Beche, Phillips and others his whole battery of the earthquakes and volcanoes of the Andes, and argued that spaces at least a thousand miles long were simultaneously subject to earthquakes and volcanic eruptions, and that the elevation of the Pampas, Patagonia, etc., all depended on a common cause; also that the greater the contortions of strata in a mountain chain, the smaller must have been each separate and individual movement of that long series which was necessary to upheave the chain. Had they been more violent, he contended that the subterraneous fluid matter would have gushed out and overflowed, and the strata would have been blown up and annihilated. (It is interesting to compare this with what Darwin wrote to Henslow seven years earlier.) He therefore introduces a cooling of one small underground injection, and then the pumping in of other lava, or porphyry, or granite, into the previously consolidated and first-formed mass of igneous rock. (Ideas somewhat similar to this suggestion have recently been revived by Dr See ("Proc. Am. Phil. Soc." Vol. XLVII. 1908, page 262.).) When he had done his description of the reiterated strokes of his volcanic pump, De la Beche gave us a long oration about the impossibility of strata of the Alps, etc., remaining flexible for such a time as they must have done, if they were to be tilted, convoluted, or overturned by gradual small shoves. He never, however, explained his theory of original flexibility, and therefore I am as unable as ever to comprehend why flexiblility is a quality so limited in time.

226

"Phillips then got up and pronounced a panegyric upon the "Principles of Geology", and although he still differed, thought the actual cause doctrine had been so well put, that it had advanced the science and formed a date or era, and that for centuries the two opposite doctrines would divide geologists, some contending for greater pristine forces, others satisfied, like Lyell and Darwin, with the same intensity as nature now employs.

"Fitton quizzed Phillips a little for the warmth of his eulogy, saying that he (Fitton) and others, who had Mr Lyell always with them, were in the habit of admiring and quarrelling with him every day, as one might do with a sister or cousin, whom one would only kiss and embrace fervently after a long absence. This seemed to be Mr Phillips' case, coming up occasionally from the provinces. Fitton then finished this drollery by charging me with not having done justice to Hutton, who he said was for gradual elevation.

"I replied, that most of the critics had attacked me for overrating Hutton, and that Playfair understood him as I did.

"Whewell concluded by considering Hopkins' mathematical calculations, to which Darwin had often referred. He also said that we ought not to try and make out what Hutton would have taught and thought, if he had known the facts which we now know."

It may be necessary to point out, in explanation of the above narrative, that while it was perfectly clear from Hutton's rather obscure and involved writings that he advocated slow and gradual change on the earth's surface, his frequent references to violent action and earthquakes led many—including Playfair, Lyell and Whewell—to believe that he held the changes going on in the earth's interior to be of a catastrophic nature. Fitton, however, maintained that Hutton was consistently uniformitarian. Before the idea of the actual "flowing" of solid bodies under intense pressure had been grasped by geologists, De la Beche, like Playfair before him, maintained that the bending and folding of rocks must have been effected before their complete consolidation.

In concluding his account of this memorable discussion, Lyell adds: "I was much struck with the different tone in which my gradual causes was treated by all, even including De la Beche, from that which they experienced in the same room four years ago, when Buckland, De la Beche(?), Sedgwick, Whewell, and some others treated them with as much ridicule as was consistent with politeness in my presence."

This important paper was, in spite of its theoretical character, published in full in the "Transactions of the Geological Society" (Ser. 2, Vol. V. pages 601-630). It did not however appear till 1840, and possibly some changes may have been made in it during the long interval between reading and printing. During the year 1839, Darwin continued his regular attendance at the Council meetings, but there is no record of any discussions in which he may have taken part, and he contributed no papers himself to the Society. At the beginning of 1840, he was re-elected for the third time as Secretary, but the results of failing health are indicated by the circumstance that, only at one meeting early in the session, was he able to attend the Council. At the beginning of the next session (Feb. 1841) Bunbury succeeded him as Secretary, Darwin still remaining on the Council. It may be regarded as a striking indication of the esteem in which he was held by his fellow geologists, that Darwin remained on the Council for 14 consecutive years down to 1849, though his attendances were in some years very few. In 1843 and 1844 he was a Vice-president, but after his retirement at the beginning of 1850, he never again accepted re-nomination. He continued, however, to contribute papers to the Society, as we shall see, down to the end of 1862.

Although Darwin early became a member of the Geological Dining Club, it is to be feared that he scarcely found himself in a congenial atmosphere at those somewhat hilarious gatherings, where the hardy wielders of the hammer not only drank port—and plenty of it— but wound up their meal with a mixture of Scotch ale and soda water, a drink which, as reminiscent of the "field," was regarded as especially appropriate to geologists. Even after the meetings, which followed the dinners, they reassembled for suppers, at which geological dainties, like "pterodactyle pie" figured in the bill of fare, and fines of bumpers were inflicted on those who talked the "ologies."

After being present at a fair number of meetings in 1837 and 1838, Darwin's attendances at the Club fell off to two in 1839, and by 1841 he had ceased to be a member. In a letter to Lyell on Dec. 2nd, 1841, Leonard Horner wrote that the day before "At the Council, I had the satisfaction of seeing Darwin again in his place and looking well. He tried the last evening meeting, but found it too much, but I hope before the end of the season he will find himself equal to that also. I hail Darwin's recovery as a vast gain to science." Darwin's probably last attendance, this time as a guest, was in 1851, when Horner again wrote to Lyell, "Charles Darwin was at the Geological Society's Club yesterday, where he had not been for ten years—remarkably well, and grown quite stout." ("Memoirs of Leonard Horner" (privately printed), Vol. II. pages 39 and 195.)

It may be interesting to note that at the somewhat less lively dining Club—the Philosophical—in the founding of which his friends Lyell and Hooker had taken so active a part, Darwin found himself more at home, and he was a frequent attendant—in spite of his residence being at Down—from 1853 to 1864. He even made contributions on scientific questions after these dinners. In a letter to Hooker he states that he was deeply interested in the reforms of the Royal Society, which the Club was founded to promote. He says also that he had arranged to come to town every Club day "and then my head, I think, will allow me on an average to go to every other meeting. But it is grievous how often any change knocks me up." ("L.L." II. pages 42, 43.)

Of the years 1837 and 1838 Darwin himself says they were "the most active ones which I ever spent, though I was occasionally unwell, and so lost some time... I also went a little into society." ("L.L." I. pages 67, 68.) But of the four years from 1839 to 1842 he has to confess sadly "I did less scientific work, though I worked as hard as I could, than during any other equal length of time in my life. This was owing to frequently recurring unwellness, and to one long and serious illness." ("L.L." I. page 69.)

Darwin's work at the Geological Society did not by any means engage the whole of his energies, during the active years 1837 and 1838. In June of the latter year, leaving town in somewhat bad health, he found himself at Edinburgh again, and engaged in examining the Salisbury Craigs, in a very different spirit to that excited by Jameson's discourse. ("L.L." I. page 290.) Proceeding to the Highlands he then had eight days of hard work at the famous "Parallel Roads of Glen Roy", being favoured with glorious weather.

He says of the writing of the paper on the subject—the only memoir contributed by Darwin to the Royal Society, to which he had been recently elected—that it was "one of the most difficult and instructive tasks I was ever engaged on." The paper extends to 40 quarto pages and is illustrated by two plates. Though it is full of the records of careful observation and acute reasoning, yet the theory of marine beaches which he propounded was, as he candidly admitted in after years ("M.L." II page 188.), altogether wrong. The alternative lake-theory he found himself unable to accept at the time, for he could not understand how barriers could be formed at successive levels across the valleys; and until the following year, when the existence of great glaciers in the district was proved by the researches of Agassiz, Buckland and others, the difficulty appeared to him an insuperable one. Although Darwin said of this paper in after years that it "was a great failure and I am ashamed of it"—yet he retained his interest in the question ever afterwards, and he says "my error has been a good lesson to me never to trust in science to the principle of exclusion." ("M.L." II. pages 171-93.)

Although Darwin had not realised in 1838 that large parts of the British Islands had been occupied by great glaciers, he had by no means failed while in South America to recognise the importance of ice-action. His observations, as recorded in his Journal, on glaciers coming down to the sea-level, on the west coast of South America, in a latitude corresponding to a much lower one than that of the British Islands, profoundly interested geologists; and the same work contains many valuable notes on the boulders and unstratified beds in South America in which they were included.

But in 1840 Agassiz read his startling paper on the evidence of the former existence of glaciers in the British Islands, and this was followed by Buckland's memoir on the same subject. On April 14, 1841, Darwin contributed to the Geological Society his important paper "On the Distribution of Erratic Boulders and the Contemporaneous Unstratified Deposits of South America", a paper full of suggestiveness for those studying the glacial deposits of this country. It was published in the "Transactions" in 1842.

The description of traces of glacial action in North Wales, by Buckland, appears to have greatly excited the interest of Darwin. With Sedgwick he had, in 1831, worked at the stratigraphy of that district, but neither of them had noticed the very interesting surface features. ("L.L." I. page 58.) Darwin was able to make a journey to North Wales in June, 1842 (alas! it was his last effort in field-geology) and as a result he published his most able and convincing paper on the subject in the September number of the "Philosophical Magazine" for 1842. Thus the mystery of the bell-stone was at last solved and Darwin, writing many years afterwards, said "I felt the keenest delight when I first read of the action of icebergs in transporting boulders, and I gloried in the progress of Geology." ("L.L." I. page 41.) To the "Geographical Journal" he had sent in 1839 a note "On a Rock seen on an Iceberg in 16 deg S. Latitude." For the subject of ice-action, indeed, Darwin retained the greatest interest to the end of his life. ("M.L." II. pages 148-71.)

In 1846, Darwin read two papers to the Geological Society "On the dust which falls on vessels in the Atlantic, and On the Geology of the Falkland Islands"; in 1848 he contributed a note on the transport of boulders from lower to higher levels; and in 1862 another note on the thickness of the Pampean formation, as shown by recent borings at Buenos Ayres. An account of the "British Fossil Lepadidae" read in 1850, was withdrawn by him.

At the end of 1836 Darwin had settled himself in lodgings in Fitzwilliam Street, Cambridge, and devoted three months to the work of unpacking his specimens and studying his collection of rocks. The pencilled notes on the Manuscript Catalogue in the Sedgwick Museum enable us to realise his mode of work, and the diligence with which it was carried on. The letters M and H, indicate the assistance he received from time to time from Professor Miller, the crystallographer, and from his friend Henslow. Miller not only measured many of the crystals submitted to him, but evidently taught Darwin to use the reflecting goniometer himself with considerable success. The "book of measurements" in which the records were kept, appears to have been lost, but the pencilled notes in the catalogue show how thoroughly the work was done. The letter R attached to some of the numbers in the catalogue evidently refers to the fact that they were submitted to Mr Trenham Reeks (who analysed some of his specimens) at the Geological Survey quarters in Craig's Court. This was at a later date when Darwin was writing the "Volcanic Islands" and "South America".

It was about the month of March, 1837, that Darwin completed this work upon his rocks, and also the unpacking and distribution of his fossil bones and other specimens. We have seen that November, 1832, must certainly be regarded as the date when he FIRST realised the important fact that the fossil mammals of the Pampean formation were all closely related to the existing forms in South America; while October, 1835, was, as undoubtedly, the date when the study of the birds and other forms of life in the several islands of the Galapagos Islands gave him his SECOND impulse towards abandoning the prevalent view of the immutability of species. When then in his pocket-book for 1837 Darwin wrote the often quoted passage: "In July opened first note-book on Transmutation of Species. Had been greatly struck from about the month of previous March on character of South American fossils, and species on Galapagos Archipelago. These facts (especially latter), origin of all my views" ("L.L." I. page 276.), it is clear that he must refer, not to his first inception of the idea of evolution, but to the flood of recollections, the reawakening of his interest in the subject, which could not fail to result from the sight of his specimens and the reference to his notes.

Except during the summer vacation, when he was visiting his father and uncle, and with the latter making his first observations upon the work of earthworms, Darwin was busy with his arrangements for the publication of the five volumes of the "Zoology of the 'Beagle'" and in getting the necessary financial aid from the government for the preparation of the plates. He was at the same time preparing his "Journal" for publication. During the years 1837 to 1843, Darwin worked intermittently on the volumes of Zoology, all of which he edited, while he wrote introductions to those by Owen and Waterhouse and supplied notes to the others.

Although Darwin says of his Journal that the preparation of the book "was not hard work, as my MS. Journal had been written with care." Yet from the time that he settled at 36, Great Marlborough Street in March, 1837, to the following November he was occupied with this book. He tells us that the account of his scientific observations was added at this time. The work was not published till March, 1839, when it appeared as the third volume of the "Narrative of the Surveying Voyages of H.M. Ships 'Adventure' and 'Beagle' between the years 1826 and 1836". The book was probably a long time in the press, for there are no less than 20 pages of addenda in small print. Even in this, its first form, the work is remarkable for its freshness and charm, and excited a great amount of attention and interest. In addition to matters treated of in greater detail in his other works, there are many geological notes of extreme value in this volume, such as his account of lightning tubes, of the organisms found in dust, and of the obsidian bombs of Australia.

Having thus got out of hand a number of preliminary duties, Darwin was ready to set to work upon the three volumes which were designed by him to constitute "The Geology of the Voyage of the 'Beagle'". The first of these was to be on "The Structure and Distribution of Coral-reefs". He commenced the writing of the book on October 5, 1838, and the last proof was corrected on May 6, 1842. Allowing for the frequent interruptions through illness, Darwin estimated that it cost him twenty months of hard work.

Darwin has related how his theory of Coral-reefs which was begun in a more "deductive spirit" than any of his other work, for in 1834 or 1835 it "was thought out on the west coast of South America, before I had seen a true coral-reef." ("L.L." I. page 70.) The final chapter in Lyell's second volume of the "Principles" was devoted to the subject of Coral-reefs, and a theory was suggested to account for the peculiar phenomena of "atolls." Darwin at once saw the difficulty of accepting the view that the numerous and diverse atolls all represent submerged volcanic craters. His own work had for two years been devoted to the evidence of land movements over great areas in South America, and thus he was led to announce his theory of subsidence to account for barrier and encircling reefs as well as atolls.

Fortunately, during his voyage across the Pacific and Indian Oceans, in his visit to Australia and his twelve days' hard work at Keeling Island, he had opportunities for putting his theory to the test of observation.

On his return to England, Darwin appears to have been greatly surprised at the amount of interest that his new theory excited. Urged by Lyell, he read to the Geological Society a paper on the subject, as we have seen, with as little delay as possible, but this paper was "withdrawn by permission of the Council." An abstract of three pages however appeared in the "Proceedings of the Geological Society". (Vol. II. pages 552-554 (May 31, 1837).) A full account of the observations and the theory was given in the "Journal" (1839) in the 40 pages devoted to Keeling Island in particular and to Coral formations generally. ("Journal" (1st edition), pages 439-69.)

It will be readily understood what an amount of labour the book on Coral reefs cost Darwin when we reflect on the number of charts, sailing directions, narratives of voyages and other works which, with the friendly assistance of the authorities at the Admiralty, he had to consult before he could draw up his sketch of the nature and distribution of the reefs, and this was necessary before the theory, in all its important bearings, could be clearly enunciated. Very pleasing is it to read how Darwin, although arriving at a different conclusion to Lyell, shows, by quoting a very suggestive passage in the "Principles" (1st edition Vol. II. page 296.), how the latter only just missed the true solution. This passage is cited, both in the "Journal" and the volume on Coral-reefs. Lyell, as we have seen, received the new theory not merely ungrudgingly, but with the utmost enthusiasm.

In 1849 Darwin was gratified by receiving the support of Dana, after his prolonged investigation in connection with the U.S. Exploring Expedition ("M.L." II. pages 226-8.), and in 1874 he prepared a second edition of his book, in which some objections which had been raised to the theory were answered. A third edition, edited by Professor Bonney, appeared in 1880, and a fourth (a reprint of the first edition, with introduction by myself) in 1890.

Although Professor Semper, in his account of the Pelew Islands, had suggested difficulties in the acceptance of Darwin's theory, it was not till after the return of the "Challenger" expedition in 1875 that a rival theory was propounded, and somewhat heated discussions were raised as to the respective merits of the two theories. While geologists have, nearly without exception, strongly supported Darwin's views, the notes of dissent have come almost entirely from zoologists. At the height of the controversy unfounded charges of unfairness were made against Darwin's supporters and the authorities of the Geological Society, but this unpleasant subject has been disposed of, once for all, by Huxley. ("Essays upon some Controverted Questions", London, 1892, pages 314-328 and 623-625.)

Darwin's final and very characteristic utterance on the coral-reef controversy is found in a letter which he wrote to Professor Alexander Agassiz, May 5th, 1881: less than a year before his death: "If I am wrong, the sooner I am knocked on the head and annihilated so much the better. It still seems to me a marvellous thing that there should not have been much, and long-continued, subsidence in the beds of the great oceans. I wish that some doubly rich millionaire would take it into his head to have borings made in some of the Pacific and Indian atolls, and bring home cores for slicing from a depth of 500 or 600 feet." ("L.L." III. page 184.)

Though the "doubly rich millionaire" has not been forthcoming, the energy, in England, of Professor Sollas, and in New South Wales of Professor Anderson Stuart served to set on foot a project, which, aided at first by the British Association for the Advancement of Science, and afterwards taken up jointly by the Royal Society, the New South Wales Government, and the Admiralty, has led to the most definite and conclusive results.

The Committee appointed by the Royal Society to carry out the undertaking included representatives of all the views that had been put forward on the subject. The place for the experiment was, with the consent of every member of the Committee, selected by the late Admiral Sir W.J. Wharton—who was not himself an adherent of Darwin's views—and no one has ventured to suggest that his selection, the splendid atoll of Funafuti, was not a most judicious one.

By the pluck and perseverance of Professor Sollas in the preliminary expedition, and of Professor T. Edgeworth David and his pupils, in subsequent investigations of the island, the rather difficult piece of work was brought to a highly satisfactory conclusion. The New South Wales Government lent boring apparatus and workmen, and the Admiralty carried the expedition to its destination in a surveying ship which, under Captain (now Admiral) A. Mostyn Field, made the most complete survey of the atoll and its surrounding seas that has ever been undertaken in the case of a coral formation.

After some failures and many interruptions, the boring was carried to the depth of 1114 feet, and the cores obtained were sent to England. Here the examination of the materials was fortunately undertaken by a zoologist of the highest repute, Dr G.J. Hinde—who has a wide experience in the study of organisms by sections—and he was aided at all points by specialists in the British Museum of Natural History and by other naturalists. Nor were the chemical and other problems neglected.

The verdict arrived at, after this most exhaustive study of a series of cores obtained from depths twice as great as that thought necessary by Darwin, was as follows:—"The whole of the cores are found to be built up of those organisms which are seen forming coral-reefs near the surface of the ocean—many of them evidently in situ; and not the slightest indication could be detected, by chemical or microscopic means, which suggested the proximity of non-calcareous rocks, even in the lowest portions brought up."

But this was not all. Professor David succeeded in obtaining the aid of a very skilful engineer from Australia, while the Admiralty allowed Commander F.C.D. Sturdee to take a surveying ship into the lagoon for further investigations. By very ingenious methods, and with great perseverance, two borings were put down in the midst of the lagoon to the depth of nearly 200 feet. The bottom of the lagoon, at the depth of 101 1/2 feet from sea-level, was found to be covered with remains of the calcareous, green sea-weed Halimeda, mingled with many foraminifera; but at a depth of 163 feet from the surface of the lagoon the boring tools encountered great masses of coral, which were proved from the fragments brought up to belong to species that live within AT MOST 120 feet from the surface of the ocean, as admitted by all zoologists. ("The Atoll of Funafuti; Report of the Coral Reef Committee of the Royal Society", London, 1904.)

Darwin's theory, as is well known, is based on the fact that the temperature of the ocean at any considerable depth does not permit of the existence and luxuriant growth of the organisms that form the reefs. He himself estimated this limit of depth to be from 120 to 130 feet; Dana, as an extreme, 150 feet; while the recent very prolonged and successful investigations of Professor Alexander Agassiz in the Pacific and Indian Oceans lead him also to assign a limiting depth of 150 feet; the EFFECTIVE, REEF-FORMING CORALS, however, flourishing at a much smaller depth. Mr Stanley Gardiner gives for the most important reef-forming corals depths between 30 and 90 feet, while a few are found as low as 120 feet or even 180 feet.

It will thus be seen that the verdict of Funafuti is clearly and unmistakeably in favour of Darwin's theory. It is true that some zoologists find a difficulty in realising a slow sinking of parts of the ocean floor, and have suggested new and alternative explanations: but geologists generally, accepting the proofs of slow upheaval in some areas—as shown by the admirable researches of Alexander Agassiz—consider that it is absolutely necessary to admit that this elevation is balanced by subsidence in other areas. If atolls and barrier-reefs did not exist we should indeed be at a great loss to frame a theory to account for their absence.

After finishing his book on Coral-reefs, Darwin made his summer excursion to North Wales, and prepared his important memoir on the glaciers of that district: but by October (1842) we find him fairly settled at work upon the second volume of his "Geology of the 'Beagle'—Geological Observations on the Volcanic Islands, visited during the Voyage of H.M.S. 'Beagle'". The whole of the year 1843 was devoted to this work, but he tells his friend Fox that he could "manage only a couple of hours per day, and that not very regularly." ("L.L." I. page 321.) Darwin's work on the various volcanic islands examined by him had given him the most intense pleasure, but the work of writing the book by the aid of his notes and specimens he found "uphill work," especially as he feared the book would not be read, "even by geologists." (Loc. cit.)

As a matter of fact the work is full of the most interesting observations and valuable suggestions, and the three editions (or reprints) which have appeared have proved a most valuable addition to geological literature. It is not necessary to refer to the novel and often very striking discoveries described in this well-known work. The subsidence beneath volcanic vents, the enormous denudation of volcanic cones reducing them to "basal wrecks," the effects of solfataric action and the formation of various minerals in the cavities of rocks—all of these subjects find admirable illustration from his graphic descriptions. One of the most important discussions in this volume is that dealing with the "lamination" of lavas as especially well seen in the rocks of Ascension. Like Scrope, Darwin recognised the close analogy between the structure of these rocks and those of metamorphic origin—a subject which he followed out in the volume "Geological Observations on South America".

Of course in these days, since the application of the microscope to the study of rocks in thin sections, Darwin's nomenclature and descriptions of the petrological characters of the lavas appear to us somewhat crude. But it happened that the "Challenger" visited most of the volcanic islands described by Darwin, and the specimens brought home were examined by the eminent petrologist Professor Renard. Renard was so struck with the work done by Darwin, under disadvantageous conditions, that he undertook a translation of Darwin's work into French, and I cannot better indicate the manner in which the book is regarded by geologists than by quoting a passage from Renard's preface. Referring to his own work in studying the rocks brought home by the "Challenger" (Renard's descriptions of these rocks are contained in the "Challenger Reports". Mr Harker is supplementing these descriptions by a series of petrological memoirs on Darwin's specimens, the first of which appeared in the "Geological Magazine" for March, 1907.), he says:

"Je dus, en me livrant a ces recherches, suivre ligne par ligne les divers chapitres des "Observations geologiques" consacrees aux iles de l'Atlantique, oblige que j'etais de comparer d'une maniere suivie les resultats auxquels j'etais conduit avec ceux de Darwin, qui servaient de controle a mes constatations. Je ne tardai pas a eprouver une vive admiration pour ce chercheur qui, sans autre appareil que la loupe, sans autre reaction que quelques essais pyrognostiques, plus rarement quelques mesures au goniometre, parvenait a discerner la nature des agregats mineralogiques les plue complexes et les plus varies. Ce coup d'oeil qui savait embrasser de si vastes horizons, penetre ici profondement tous les details lithologiques. Avec quelle surete et quelle exactitude la structure et la composition des roches ne sont'elles pas determinees, l'origne de ces masses minerales deduite et confirmee par l'etude comparee des manifestations volcaniques d'autres regions; avec quelle science les relations entre les faits qu'il decouvre et ceux signales ailleurs par ses devanciers ne sont'elles pas etablies, et comme voici ebranlees les hypotheses regnantes, admises sans preuves, celles, par exemple, des crateres de soulevement et de la differenciation radicale des phenomenes plutoniques et volcaniques! Ce qui acheve de donner a ce livre un incomparable merite, ce sont les idees nouvelles qui s'y trouvent en germe et jetees la comme au hasard ainsi qu'un superflu d'abondance intellectuelle inepuisable." ("Observations Geologiques sur les Iles Volcaniques... ", Paris, 1902, pages vi., vii.)

While engaged in his study of banded lavas, Darwin was struck with the analogy of their structure with that of glacier ice, and a note on the subject, in the form of a letter addressed to Professor J.D. Forbes, was published in the "Proceedings of the Royal Society of Edinburgh". (Vol. II. (1844-5), pages 17, 18.)

From April, 1832, to September, 1835, Darwin had been occupied in examining the coast or making inland journeys in the interior of the South American continent. Thus while eighteen months were devoted, at the beginning and end of the voyage to the study of volcanic islands and coral-reefs, no less than three and a half years were given to South American geology. The heavy task of dealing with the notes and specimens accumulated during that long period was left by Darwin to the last. Finishing the "Volcanic Islands" on February 14th, 1844, he, in July of the same year, commenced the preparation of two important works which engaged him till near the end of the year 1846. The first was his "Geological Observations on South America", the second a recast of his "Journal", published under the short title of "A Naturalist's Voyage round the World".

The first of these works contains an immense amount of information collected by the author under great difficulties and not unfrequently at considerable risk to life and health. No sooner had Darwin landed in South America than two sets of phenomena powerfully arrested his attention. The first of these was the occurrence of great masses of red mud containing bones and shells, which afforded striking evidence that the whole continent had shared in a series of slow and gradual but often interrupted movements. The second related to the great masses of crystalline rocks which, underlying the muds, cover so great a part of the continent. Darwin, almost as soon as he landed, was struck by the circumstance that the direction, as shown by his compass, of the prominent features of these great crystalline rock-masses—their cleavage, master-joints, foliation and pegmatite veins—was the same as the orientation described by Humboldt (whose works he had so carefully studied) on the west of the same great continent.

The first five chapters of the book on South America were devoted to formations of recent date and to the evidence collected on the east and west coasts of the continent in regard to those grand earth-movements, some of which could be shown to have been accompanied by earthquake-shocks. The fossil bones, which had given him the first hint concerning the mutability of species, had by this time been studied and described by comparative anatomists, and Darwin was able to elaborate much more fully the important conclusion that the existing fauna of South America has a close analogy with that of the period immediately preceding our own.

The remaining three chapters of the book dealt with the metamorphic and plutonic rocks, and in them Darwin announced his important conclusions concerning the relations of cleavage and foliation, and on the close analogy of the latter structure with the banding found in rock-masses of igneous origin. With respect to the first of these conclusions, he received the powerful support of Daniel Sharpe, who in the years 1852 and 1854 published two papers on the structure of the Scottish Highlands, supplying striking confirmation of the correctness of Darwin's views. Although Darwin's and Sharpe's conclusions were contested by Murchison and other geologists, they are now universally accepted. In his theory concerning the origin of foliation, Darwin had been to some extent anticipated by Scrope, but he supplied many facts and illustrations leading to the gradual acceptance of a doctrine which, when first enunciated, was treated with neglect, if not with contempt.

The whole of this volume on South American geology is crowded with the records of patient observations and suggestions of the greatest value; but, as Darwin himself saw, it was a book for the working geologist and "caviare to the general." Its author, indeed, frequently expressed his sense of the "dryness" of the book; he even says "I long hesitated whether I would publish it or not," and he wrote to Leonard Horner "I am astonished that you should have had the courage to go right through my book." ("M.L." II. page 221.)

Fortunately the second book, on which Darwin was engaged at this time, was of a very different character. His "Journal", almost as he had written it on board ship, with facts and observations fresh in his mind, had been published in 1839 and attracted much attention. In 1845, he says, "I took much pains in correcting a new edition," and the work which was commenced in April, 1845, was not finished till August of that year. The volume contains a history of the voyage with "a sketch of those observations in Natural History and Geology, which I think will possess some interest for the general reader." It is not necessary to speak of the merits of this scientific classic. It became a great favourite with the general public—having passed through many editions—it was, moreover, translated into a number of different languages. Darwin was much gratified by these evidences of popularity, and naively remarks in his "Autobiography", "The success of this my first literary child tickles my vanity more than that of any of my other books" ("L.L." I. page 80.)—and this was written after the "Origin of Species" had become famous!

In Darwin's letters there are many evidences that his labours during these ten years devoted to the working out of the geological results of the voyage often made many demands on his patience and indomitable courage. Most geologists have experience of the contrast between the pleasures felt when wielding the hammer in the field, and the duller labour of plying the pen in the study. But in Darwin's case, innumerable interruptions from sickness and other causes, and the oft-deferred hope of reaching the end of his task were not the only causes operating to make the work irksome. The great project, which was destined to become the crowning achievement of his life, was now gradually assuming more definite shape, and absorbing more of his time and energies.

Nevertheless, during all this period, Darwin so far regarded his geological pursuits as his PROPER "work," that attention to other matters was always spoken of by him as "indulging in idleness." If at the end of this period the world had sustained the great misfortune of losing Darwin by death before the age of forty—and several times that event seemed only too probable—he might have been remembered only as a very able geologist of most advanced views, and a traveller who had written a scientific narrative of more than ordinary excellence!

The completion of the "Geology of the 'Beagle'" and the preparation of a revised narrative of the voyage mark the termination of that period of fifteen years of Darwin's life during which geological studies were his principal occupation. Henceforth, though his interest in geological questions remained ever keen, biological problems engaged more and more of his attention to the partial exclusion of geology.

The eight years from October, 1846, to October, 1854, were mainly devoted to the preparation of his two important monographs on the recent and fossil Cirripedia. Apart from the value of his description of the fossil forms, this work of Darwin's had an important influence on the progress of geological science. Up to that time a practice had prevailed for the student of a particular geological formation to take up the description of the plant and animal remains in it—often without having anything more than a rudimentary knowledge of the living forms corresponding to them. Darwin in his monograph gave a very admirable illustration of the enormous advantage to be gained—alike for biology and geology—by undertaking the study of the living and fossil forms of a natural group of organisms in connection with one another. Of the advantage of these eight years of work to Darwin himself, in preparing for the great task lying before him, Huxley has expressed a very strong opinion indeed. ("L.L." II. pages 247-48.)

But during these eight years of "species work," Darwin found opportunities for not a few excursions into the field of geology. He occasionally attended the Geological Society, and, as we have already seen, read several papers there during this period. His friend, Dr Hooker, then acting as botanist to the Geological Survey, was engaged in studying the Carboniferous flora, and many discussions on Palaezoic plants and on the origin of coal took place at this period. On this last subject he felt the deepest interest and told Hooker, "I shall never rest easy in Down churchyard without the problem be solved by some one before I die." ("M.L." I. pages 63, 64.)

As at all times, conversations and letters with Lyell on every branch of geological science continued with unabated vigour, and in spite of the absorbing character of the work on the Cirripedes, time was found for all. In 1849 his friend Herschel induced him to supply a chapter of forty pages on Geology to the Admiralty "Manual of Scientific Inquiry" which he was editing. This is Darwin's single contribution to books of an "educational" kind. It is remarkable for its clearness and simplicity and attention to minute details. It may be read by the student of Darwin's life with much interest, for the directions he gives to an explorer are without doubt those which he, as a self-taught geologist, proved to be serviceable during his life on the "Beagle".

On the completion of the Cirripede volumes, in 1854, Darwin was able to grapple with the immense pile of MS. notes which he had accumulated on the species question. The first sketch of 35 pages (1842), had been enlarged in 1844 into one of 230 pages ([The first draft of the "Origin" is being prepared for Press by Mr Francis Darwin and will be published by the Cambridge University Press this year (1909). A.C.S.]); but in 1856 was commenced the work (never to be completed) which was designed on a scale three or four times more extensive than that on which the "Origin of Species" was in the end written.

In drawing up those two masterly chapters of the "Origin", "On the Imperfection of the Geological Record," and "On the Geological Succession of Organic Beings", Darwin had need of all the experience and knowledge he had been gathering during thirty years, the first half of which had been almost wholly devoted to geological study. The most enlightened geologists of the day found much that was new, and still more that was startling from the manner of its presentation, in these wonderful essays. Of Darwin's own sense of the importance of the geological evidence in any presentation of his theory a striking proof will be found in a passage of the touching letter to his wife, enjoining the publication of his sketch of 1844. "In case of my sudden death," he wrote, "... the editor must be a geologist as well as a naturalist." ("L.L." II. pages 16, 17.)

In spite of the numerous and valuable palaeontological discoveries made since the publication of "The Origin of Species", the importance of the first of these two geological chapters is as great as ever. It still remains true that "Those who believe that the geological record is in any degree perfect, will at once reject the theory"—as indeed they must reject any theory of evolution. The striking passage with which Darwin concludes this chapter—in which he compares the record of the rocks to the much mutilated volumes of a human history—remains as apt an illustration as it did when first written.

And the second geological chapter, on the Succession of Organic Beings—though it has been strengthened in a thousand ways, by the discoveries concerning the pedigrees of the horse, the elephant and many other aberrant types, though new light has been thrown even on the origin of great groups like the mammals, and the gymnosperms, though not a few fresh links have been discovered in the chains of evidence, concerning the order of appearance of new forms of life—we would not wish to have re-written. Only the same line of argument could be adopted, though with innumerable fresh illustrations. Those who reject the reasonings of this chapter, neither would they be persuaded if a long and complete succession of "ancestral forms" could rise from the dead and pass in procession before them.

Among the geological discussions, which so frequently occupied Darwin's attention during the later years of his life, there was one concerning which his attitude seemed somewhat remarkable—I allude to his views on "the permanence of Continents and Ocean-basins." In a letter to Mr Mellard Reade, written at the end of 1880, he wrote: "On the whole, I lean to the side that the continents have since Cambrian times occupied approximately their present positions. But, as I have said, the question seems a difficult one, and the more it is discussed the better." ("M.L." II. page 147.) Since this was written, the important contribution to the subject by the late Dr W.T. Blanford (himself, like Darwin, a naturalist and geologist) has appeared in an address to the Geological Society in 1890; and many discoveries, like that of Dr Woolnough in Fiji, have led to considerable qualifications of the generalisation that all the islands in the great ocean are wholly of volcanic or coral origin.

I remember once expressing surprise to Darwin that, after the views which he had originated concerning the existence of areas of elevation and others of subsidence in the Pacific Ocean, and in face of the admitted difficulty of accounting for the distribution of certain terrestrial animals and plants, if the land and sea areas had been permanent in position, he still maintained that theory. Looking at me with a whimsical smile, he said: "I have seen many of my old friends make fools of themselves, by putting forward new theoretical views or revising old ones, AFTER THEY WERE SIXTY YEARS OF AGE; so, long ago, I determined that on reaching that age I would write nothing more of a speculative character."

Though Darwin's letters and conversations on geology during these later years were the chief manifestations of the interest he preserved in his "old love," as he continued to call it, yet in the sunset of that active life a gleam of the old enthusiasm for geology broke forth once more. There can be no doubt that Darwin's inability to occupy himself with field-work proved an insuperable difficulty to any attempt on his part to resume active geological research. But, as is shown by the series of charming volumes on plant-life, Darwin had found compensation in making patient and persevering experiment take the place of enterprising and exact observation; and there was one direction in which he could indulge the "old love" by employment of the new faculty.

We have seen that the earliest memoir written by Darwin, which was published in full, was a paper "On the Formation of Mould" which was read at the Geological Society on November 1st, 1837, but did not appear in the "Transactions" of the Society till 1840, where it occupied four and a half quarto pages, including some supplementary matter, obtained later, and a woodcut. This little paper was confined to observations made in his uncle's fields in Staffordshire, where burnt clay, cinders, and sand were found to be buried under a layer of black earth, evidently brought from below by earthworms, and to a recital of similar facts from Scotland obtained through the agency of Lyell. The subsequent history of Darwin's work on this question affords a striking example of the tenacity of purpose with which he continued his enquiries on any subject that interested him.

In 1842, as soon as he was settled at Down, he began a series of observations on a foot-path and in his fields, that continued with intermissions during his whole life, and he extended his enquiries from time to time to the neighbouring parks of Knole and Holwood. In 1844 we find him making a communication to the "Gardener's Chronicle" on the subject. About 1870, his attention to the question was stimulated by the circumstance that his niece (Miss L. Wedgwood) undertook to collect and weigh the worm-casts thrown up, during a whole year, on measured squares selected for the purpose, at Leith Hill Place. He also obtained information from Professor Ramsay concerning observations made by him on a pavement near his house in 1871. Darwin at this time began to realise the great importance of the action of worms to the archaeologist. At an earlier date he appears to have obtained some information concerning articles found buried on the battle-field of Shrewsbury, and the old Roman town of Uriconium, near his early home; between 1871 and 1878 Mr (afterwards Lord) Farrer carried on a series of investigations at the Roman Villa discovered on his land at Abinger; Darwin's son William examined for his father the evidence at Beaulieu Abbey, Brading, Stonehenge and other localities in the neighbourhood of his home; his sons Francis and Horace were enlisted to make similar enquiries at Chideock and Silchester; while Francis Galton contributed facts noticed in his walks in Hyde Park. By correspondence with Fritz Muller and Dr Ernst, Darwin obtained information concerning the worm-casts found in South America; from Dr Kreft those of Australia; and from Mr Scott and Dr (afterwards Sir George) King, those of India; the last-named correspondent also supplied him with much valuable information obtained in the South of Europe. Help too was obtained from the memoirs on Earthworms published by Perrier in 1874 and van Hensen in 1877, while Professor Ray Lankester supplied important facts with regard to their anatomy.

When therefore the series of interesting monographs on plant-life had been completed, Darwin set to work in bringing the information that he had gradually accumulated during forty-four years to bear on the subject of his early paper. He also utilised the skill and ingenuity he had acquired in botanical work to aid in the elucidation of many of the difficulties that presented themselves. I well remember a visit which I paid to Down at this period. At the side of the little study stood flower-pots containing earth with worms, and, without interrupting our conversation, Darwin would from time to time lift the glass plate covering a pot to watch what was going on. Occasionally, with a humorous smile, he would murmur something about a book in another room, and slip away; returning shortly, without the book but with unmistakeable signs of having visited the snuff-jar outside. After working about a year at the worms, he was able at the end of 1881 to publish the charming little book—"The Formation of Vegetable Mould through the Action of Worms, with Observations on their Habits". This was the last of his books, and its reception by reviewers and the public alike afforded the patient old worker no little gratification. Darwin's scientific career, which had begun with geological research, most appropriately ended with a return to it.

It has been impossible to sketch the origin and influence of Darwin's geological work without, at almost every step, referring to the part played by Lyell and the "Principles of Geology". Haeckel, in the chapters on Lyell and Darwin in his "History of Creation", and Huxley in his striking essay "On the Reception of the Origin of Species" ("L.L." II. pages 179-204.) have both strongly insisted on the fact that the "Origin" of Darwin was a necessary corollary to the "Principles" of Lyell.

It is true that, in an earlier essay, Huxley had spoken of the doctrine of Uniformitarianism as being, in a certain sense, opposed to that of Evolution (Huxley's Address to the Geological Society, 1869. "Collected Essays", Vol. VIII. page 305, London, 1896.); but in his later years he took up a very different and more logical position, and maintained that "Consistent uniformitarianism postulates evolution as much in the organic as in the inorganic world. The origin of a new species by other than ordinary agencies would be a vastly greater 'catastrophe' than any of those which Lyell success fully eliminated from sober geological speculation." ("L.L." II. page 190.)

Huxley's admiration for the "Principles of Geology", and his conviction of the greatness of the revolution of thought brought about by Lyell, was almost as marked as in the case of

Darwin himself. (See his Essay on "Science and Pseudo Science". "Collected Essays", Vol. V. page 90, London, 1902.) He felt, however, as many others have done, that in one respect the very success of Lyell's masterpiece has been the reason why its originality and influence have not been so fully recognised as they deserved to be. Written as the book was before its author had arrived at the age of thirty, no less than eleven editions of the "Principles" were called for in his lifetime. With the most scrupulous care, Lyell, devoting all his time and energies to the task of collecting and sifting all evidence bearing on the subjects of his work, revised and re-revised it; and as in each edition, eliminations, modifications, corrections, and additions were made, the book, while it increased in value as a storehouse of facts, lost much of its freshness, vigour and charm as a piece of connected reasoning.

Darwin undoubtedly realised this when he wrote concerning the "Principles", "the first edition, my old true love, which I never deserted for the later editions." ("M.L." II. page 222.) Huxley once told me that when, in later life, he read the first edition, he was both surprised and delighted, feeling as if it were a new book to him. (I have before me a letter which illustrates this feeling on Huxley's part. He had lamented to me that he did not possess a copy of the first edition of the "Principles", when, shortly afterwards, I picked up a dilapidated copy on a bookstall; this I had bound and sent to my old teacher and colleague. His reply is characteristic:

October 8, 1884.

My Dear Judd,

You could not have made me a more agreeable present than the copy of the first edition of Lyell, which I find on my table. I have never been able to meet with the book, and your copy is, as the old woman said of her Bible, "the best of books in the best of bindings."

Ever yours sincerely,

T.H. Huxley.

(I cannot refrain from relating an incident which very strikingly exemplifies the affection for one another felt by Lyell and Huxley. In his last illness, when confined to his bed, Lyell heard that Huxley was to lecture at the Royal Institution on the "Results of the 'Challenger' expedition": he begged me to attend the lecture and bring him an account of it. Happening to mention this to Huxley, he at once undertook to go to Lyell in my place, and he did so on the morning following his lecture. I shall never forget the look of gratitude on the face of the invalid when he told me, shortly afterwards, how Huxley had sat by his bedside and "repeated the whole lecture to him.")

Darwin's generous nature seems often to have made him experience a fear lest he should do less than justice to his "dear old master," and to the influence that the "Principles of Geology" had in moulding his mind. In 1845 he wrote to Lyell, "I have long wished, not so much for your sake, as for my own feelings of honesty, to acknowledge more plainly than by mere reference, how much I geologically owe you. Those authors, however, who like you, educate people's minds as well as teach them special facts, can never, I should think, have full justice done them except by posterity, for the mind thus insensibly improved can hardly perceive its own upward ascent." ("L.L." I. pages 337-8.) In another letter, to Leonard Horner, he says: "I always feel as if my books came half out of Lyell's brain, and that I never acknowledge this sufficiently." ("M.L." II. page 117.) Darwin's own most favourite book, the "Narrative of the Voyage", was dedicated to Lyell in glowing terms; and in the "Origin of Species" he wrote of "Lyell's grand work on the "Principles of Geology", which the future historian will recognise as having produced a revolution in Natural Science." "What glorious good that work has done" he fervently exclaims on another occasion. ("L.L." I. page 342.)

To the very end of his life, as all who were in the habit of talking with Darwin can testify, this sense of his indebtedness to Lyell remained with him. In his "Autobiography", written in 1876, the year after Lyell's death, he spoke in the warmest terms of the value to him of the "Principles" while on the voyage and of the aid afforded to him by Lyell on his return to England. ("L.L." I. page 62.) But the year before his own death, Darwin felt constrained to return to the subject and to place on record a final appreciation—one as honourable to the writer as it is to his lost friend:

"I saw more of Lyell than of any other man, both before and after my marriage. His mind was characterised, as it appeared to me, by clearness, caution, sound judgment, and a good deal of originality. When I made any remark to him on Geology, he never rested until he saw the whole case clearly, and often made me see it more clearly than I had done before. He would advance all possible objections to my suggestion, and even after these were exhausted would remain long dubious. A second characteristic was his hearty sympathy with the work of other scientific men... His delight in science was ardent, and he felt the keenest interest in the future progress of mankind. He was very kind-hearted... His candour was highly remarkable. He exhibited this by becoming a convert to the Descent theory, though he had gained much fame by opposing Lamarck's views, and this after he had grown old."

"THE SCIENCE OF GEOLOGY IS ENORMOUSLY INDEBTED TO LYELL— MORE SO, AS I BELIEVE, THAN TO ANY OTHER MAN WHO EVER LIVED." ("L.L." I. pages 71-2 (the italics are mine.))

Those who knew Lyell intimately will recognise the truth of the portrait drawn by his dearest friend, and I believe that posterity will endorse Darwin's deliberate verdict concerning the value of his labours.

It was my own good fortune, to be brought into close contact with these two great men during the later years of their life, and I may perhaps be permitted to put on record the impressions made upon me during friendly intercourse with both.

In some respects, there was an extraordinary resemblance in their modes and habits of thought, between Lyell and Darwin; and this likeness was also seen in their modesty, their deference to the opinion of younger men, their enthusiasm for science, their freedom from petty jealousies and their righteous indignation for what was mean and unworthy in others. But yet there was a difference. Both Lyell and Darwin were cautious, but perhaps Lyell carried his caution to the verge of timidity. I think Darwin possessed, and Lyell lacked, what I can only describe by the theological term, "faith—the substance of things hoped for, the evidence of things not seen." Both had been constrained to feel that the immutability of species could not be maintained. Both, too, recognised the fact that it would be useless to proclaim this conviction, unless prepared with a satisfactory alternative to what Huxley called "the Miltonic hypothesis." But Darwin's conviction was so far vital and operative that it sustained him while working unceasingly for twenty-two years in collecting evidence bearing on the question, till at last he was in the position of being able to justify that conviction to others.

And yet Lyell's attitude—and that of Hooker, which was very similar—proved of inestimable service to science, as Darwin often acknowledged. One of the greatest merits of the "Origin of Species" is that so many difficulties and objections are anticipated and fairly met; and this was to a great extent the result of the persistent and very candid—if always friendly—criticism of Lyell and Hooker.

I think the divergence of mental attitude in Lyell and Darwin must be attributed to a difference in temperament, the evidence of which sometimes appears in a very striking manner in their correspondence. Thus in 1838, while they were in the thick of the fight with the Catastrophists of the Geological Society, Lyell wrote characteristically: "I really find, when bringing up my Preliminary Essays in "Principles" to the science of the present day, so far as I know it, that the great outline, and even most of the details, stand so uninjured, and in many cases they are so much strengthened by new discoveries, especially by yours, that we may begin to hope that the great principles there insisted on will stand the test of new discoveries." (Lyell's "Life, Letters and Journals", Vol. II. page 44.) To which the more youthful and impetuous Darwin replies: "BEGIN TO HOPE: why the POSSIBILITY of a doubt has never crossed my mind for many a day. This may be very unphilosophical, but my geological salvation is staked on it... it makes me quite indignant that you should talk of HOPING." ("L.L." I. page 296.)

It was not only Darwin's "geological salvation" that was at stake, when he surrendered himself to his enthusiasm for an idea. To his firm faith in the doctrine of continuity we owe the "Origin of Species"; and while Darwin became the "Paul" of evolution, Lyell long remained the "doubting Thomas."

Many must have felt like H.C. Watson when he wrote: "How could Sir C. Lyell... for thirty years read, write, and think, on the subject of species AND THEIR SUCCESSION, and yet constantly look down the wrong road!" ("L.L." II. page 227.) Huxley attributed this hesitation of Lyell to his "profound antipathy" to the doctrine of the "pithecoid origin of man." ("L.L." II. page 193.) Without denying that this had considerable influence (and those who knew Lyell and his great devotion to his wife and her memory, are aware that he and she felt much stronger convictions concerning such subjects as the immortality of the soul than Darwin was able to confess to) yet I think Darwin had divined the real characteristics of his friend's mind, when he wrote: "He would advance all possible objections... AND EVEN AFTER THESE WERE EXHAUSTED, WOULD REMAIN LONG DUBIOUS."

Very touching indeed was the friendship maintained to the end between these two leaders of thought—free as their intercourse was from any smallest trace of self-seeking or jealousy. When in 1874 I spent some time with Lyell in his Forfarshire home, a communication from Darwin was always an event which made a "red-letter day," as Lyell used to say; and he gave me many indications in his conversation of how strongly he relied upon the opinion of Darwin—more indeed than on the judgment of any other man—this confidence not being confined to questions of science, but extending to those of morals, politics, and religion.

I have heard those who knew Lyell only slightly, speak of his manners as cold and reserved. His complete absorption in his scientific work, coupled with extreme short-sightedness, almost in the end amounting to blindness, may have permitted those having but a casual acquaintance with him to accept such a view. But those privileged to know him intimately recognised the nobleness of his character and can realise the justice and force of Hooker's words when he heard of his death: "My loved, my best friend, for well nigh forty years of my life. The most generous sharer of my own and my family's hopes, joys and sorrows, whose affection for me was truly that of a father and brother combined."

But the strongest of all testimonies to the grandeur of Lyell's character is the lifelong devotion to him of such a man as Darwin. Before the two met, we find Darwin constantly writing of facts and observations that he thinks "will interest Mr Lyell"; and when they came together the mutual esteem rapidly ripened into the warmest affection. Both having the advantage of a moderate independence, permitting of an entire devotion of their lives to scientific research, they had much in common, and the elder man—who had already achieved both scientific and literary distinction—was able to give good advice and friendly help to the younger one. The warmth of their friendship comes out very strikingly in their correspondence. When Darwin first conceived the idea of writing a book on the "species question," soon after his return from the voyage, it was "by following the example of Lyell in Geology" that he hoped to succeed ("L.L." I. page 83.); when in 1844, Darwin had finished his first sketch of the work, and, fearing that his life might not be spared to complete his great undertaking, committed the care of it in a touching letter to his wife, it was his friend Lyell whom he named as her adviser and the possible editor of the book ("L.L." II. pages 17-18.); it was Lyell who, in 1856, induced Darwin to lay the foundations of a treatise ("L.L." I. page 84.) for which the author himself selected the "Principles" as his model; and when the dilemma arose from the receipt of Wallace's essay, it was to Lyell jointly with Hooker that Darwin turned, not in vain, for advice and help.

During the later years of his life, I never heard Darwin allude to his lost friend—and he did so very often—without coupling his name with some term of affection. For a brief period, it is true, Lyell's excessive caution when the "Origin" was published, seemed to try even the patience of Darwin; but when "the master" was at last able to declare himself fully convinced, he was the occasion of more rejoicing on the part of Darwin, than any other convert to his views. The latter was never tired of talking of Lyell's "magnanimity" and asserted that, "To have maintained in the position of a master, one side of a question for thirty years, and then deliberately give it up, is a fact to which I much doubt whether the records of science offer a parallel." ("L.L." II. pages 229-30.)

Of Darwin himself, I can safely affirm that I never knew anyone who had met him, even for the briefest period, who was not charmed by his personality. Who could forget the hearty hand-grip at meeting, the gentle and lingering pressure of the palm at parting, and above all that winning smile which transformed his countenance—so as to make portraits, and even photographs, seem ever afterwards unsatisfying! Looking back, one is indeed tempted to forget the profoundness of the philosopher, in recollection of the loveableness of the man.

XIX. DARWIN'S WORK ON THE MOVEMENTS OF PLANTS. By Francis Darwin,

Honorary Fellow of Christ's College, Cambridge.

My father's interest in plants was of two kinds, which may be roughly distinguished as EVOLUTIONARY and PHYSIOLOGICAL. Thus in his purely evolutionary work, for instance in "The Origin of Species" and in his book on "Variation under Domestication", plants as well as animals served as material for his generalisations. He was largely dependent on the work of others for the facts used in the evolutionary work, and despised himself for belonging to the "blessed gang" of compilers. And he correspondingly rejoiced in the employment of his wonderful power of observation in the physiological problems which occupied so much of his later life. But inasmuch as he felt evolution to be his life's work, he regarded himself as something of an idler in observing climbing plants, insectivorous plants, orchids, etc. In this physiological work he was to a large extent urged on by his passionate desire to understand the machinery of all living things. But though it is true that he worked at physiological problems in the naturalist's spirit of curiosity, yet there was always present to him the bearing of his facts on the problem of evolution. His interests, physiological and evolutionary, were indeed so interwoven that they cannot be sharply separated. Thus his original interest in the fertilisation of flowers was evolutionary. "I was led" ("Life and Letters", I. page 90.), he says, "to attend to the cross-fertilisation of flowers by the aid of insects, from having come to the conclusion in my speculations on the origin of species, that crossing played an important part in keeping specific forms constant." In the same way the value of his experimental work on heterostyled plants crystalised out in his mind into the conclusion that the product of illegitimate unions are equivalent to hybrids—a conclusion of the greatest interest from an evolutionary point of view. And again his work "Cross and Self Fertilisation" may be condensed to a point of view of great importance in reference to the meaning and origin of sexual reproduction. (See Professor Goebel's article in the present volume.)

The whole of his physiological work may be looked at as an illustration of the potency of his theory as an "instrument for the extension of the realm of natural knowledge." (Huxley in Darwin's "Life and Letters." II. page 204.)

His doctrine of natural selection gave, as is well known, an impulse to the investigation of the use of organs—and thus created the great school of what is known in Germany as Biology—a department of science for which no English word exists except the rather vague term Natural History. This was especially the case in floral biology, and it is interesting to see with what hesitation he at first expressed the value of his book on Orchids ("Life and

Letters", III. page 254.), "It will perhaps serve to illustrate how Natural History may be worked under the belief of the modification of species" (1861). And in 1862 he speaks (Loc. cit.) more definitely of the relation of his work to natural selection: "I can show the meaning of some of the apparently meaningless ridges (and) horns; who will now venture to say that this or that structure is useless?" It is the fashion now to minimise the value of this class of work, and we even find it said by a modern writer that to inquire into the ends subserved by organs is not a scientific problem. Those who take this view surely forget that the structure of all living things is, as a whole, adaptive, and that a knowledge of how the present forms come to be what they are includes a knowledge of why they survived. They forget that the SUMMATION of variations on which divergence depends is under the rule of the environment considered as a selective force. They forget that the scientific study of the interdependence of organisms is only possible through a knowledge of the machinery of the units. And that, therefore, the investigation of such widely interesting subjects as extinction and distribution must include a knowledge of function. It is only those who follow this line of work who get to see the importance of minute points of structure and understand as my father did even in 1842, as shown in his sketch of the "Origin" (Now being prepared for publication.), that every grain of sand counts for something in the balance. Much that is confidently stated about the uselessness of different organs would never have been written if the naturalist spirit were commoner nowadays. This spirit is strikingly shown in my father's work on the movements of plants. The circumstance that botanists had not, as a class, realised the interest of the subject accounts for the fact that he was able to gather such a rich harvest of results from such a familiar object as a twining plant. The subject had been investigated by H. von Mohl, Palm, and Dutrochet, but they failed not only to master the problem but (which here concerns us) to give the absorbing interest of Darwin's book to what they discovered.

His work on climbing plants was his first sustained piece of work on the physiology of movement, and he remarks in 1864: "This has been new sort of work for me." ("Life and Letters", III. page 315. He had, however, made a beginning on the movements of Drosera.) He goes on to remark with something of surprise, "I have been pleased to find what a capital guide for observations a full conviction of the change of species is."

It was this point of view that enabled him to develop a broad conception of the power of climbing as an adaptation by means of which plants are enabled to reach the light. Instead of being compelled to construct a stem of sufficient strength to stand alone, they succeed in the struggle by making use of other plants as supports. He showed that the great class of tendril- and root-climbers which do not depend on twining round a pole, like a scarlet-runner, but on attaching themselves as they grow upwards, effect an economy. Thus a Phaseolus has to manufacture a stem three feet in length to reach a height of two feet above the ground, whereas a pea "which had ascended to the same height by the aid of its tendrils, was but little longer than the height reached." ("Climbing Plants" (2nd edition 1875), page 193.)

Thus he was led on to the belief that TWINING is the more ancient form of climbing, and that tendril-climbers have been developed from twiners. In accordance with this view we find LEAF-CLIMBERS, which may be looked on as incipient tendril-bearers, occurring in the same genera with simple twiners. (Loc. cit. page 195.) He called attention to the case of Maurandia semperflorens in which the young flower-stalks revolve spontaneously and are sensitive to a touch, but neither of these qualities is of any perceptible value to the species. This forced him to believe that in other young plants the rudiments of the faculty needed for twining would be found—a prophecy which he made good in his "Power of Movement" many years later.

In "Climbing Plants" he did little more than point out the remarkable fact that the habit of climbing is widely scattered through the vegetable kingdom. Thus climbers are to be found in 35 out of the 59 Phanerogamic Alliances of Lindley, so that "the conclusion is forced on our minds that the capacity of revolving (If a twining plant, e.g. a hop, is observed before it has begun to ascend a pole, it will be noticed that, owing to the curvature of the stem, the tip is not vertical but hangs over in a roughly horizontal position. If such a shoot is watched it will be found that if, for instance, it points to the north at a given hour, it will be found after a short interval pointing north-east, then east, and after about two hours it will once more be looking northward. The curvature of the stem depends on one side growing quicker than the opposite side, and the revolving movement, i.e. circumnutation, depends on the region of quickest growth creeping gradually round the stem from south through west to south again. Other plants, e.g. Phaseolus, revolve in the opposite direction.), on which most climbers depend, is inherent, though undeveloped, in almost every plant in the vegetable kingdom." ("Climbing Plants", page 205.)

In the "Origin" (Edition I. page 427, Edition VI. page 374.) Darwin speaks of the "apparent paradox, that the very same characters are analogical when one class or order is compared with another, but give true affinities when the members of the same class or order are compared one with another." In this way we might perhaps say that the climbing of an ivy and a hop are analogical; the resemblance depending on the adaptive result rather than on community of blood; whereas the relation between a leaf-climber and a true tendril-bearer reveals descent. This particular resemblance was one in which my father took especial delight. He has described an interesting case occurring in the Fumariaceae. ("Climbing Plants", page 195.) "The terminal leaflets of the leaf-climbing Fumaria officinalis are not smaller than the other leaflets; those of the leaf-climbing Adlumia cirrhosa are greatly reduced; those of Corydalis claviculata (a plant which may be indifferently called a leaf-climber or a tendril-bearer) are either reduced to microscopical dimensions or have their blades wholly aborted, so that this plant is actually in a state of transition; and finally in the Dicentra the tendrils are perfectly characterized."

It is a remarkable fact that the quality which, broadly speaking, forms the basis of the climbing habit (namely revolving nutation, otherwise known as circumnutation) subserves two distinct ends. One of these is the finding of a support, and this is common to twiners and tendrils. Here the value ends as far as tendril-climbers are concerned, but in twiners Darwin believed that the act of climbing round a support is a continuation of the revolving movement (circumnutation). If we imagine a man swinging a rope round his head and if we suppose the rope to strike a vertical post, the free end will twine round it. This may serve as a rough model of twining as explained in the "Movements and Habits of Climbing Plants". It is on these points—the nature of revolving nutation and the mechanism of twining—that modern physiologists differ from Darwin. (See the discussion in Pfeffer's "The Physiology of Plants" Eng. Tr. (Oxford, 1906), III. page 34, where the literature is given. Also Jost, "Vorlesungen uber Pflanzenphysiologie", page 562, Jena, 1904.)

Their criticism originated in observations made on a revolving shoot which is removed from the action of gravity by keeping the plant slowly rotating about a horizontal axis by means of the instrument known as a klinostat. Under these conditions circumnutation becomes irregular or ceases altogether. When the same experiment is made with a plant which has twined spirally up a stick, the process of climbing is checked and the last few turns become loosened or actually untwisted. From this it has been argued that Darwin was wrong in his description of circumnutation as an automatic change in the region of quickest growth. When the free end of a revolving shoot points towards the north there is no doubt that the south side has been elongating more than the north; after a time it is plain from the shoot hanging over to the east that the west side of the plant has grown most, and so on. This rhythmic change of the position of the region of greatest growth Darwin ascribes to an unknown internal regulating power. Some modern physiologists, however, attempt to explain the revolving movement as due to a particular form of sensitiveness to gravitation which it is not necessary to discuss in detail in this place. It is sufficient for my purpose to point out that Darwin's explanation of circumnutation is not universally accepted. Personally I believe that circumnutation is automatic—is primarily due to internal stimuli. It is however in some way connected with gravitational sensitiveness, since the movement normally occurs round a vertical line. It is not unnatural that, when the plant has no external stimulus by which the vertical can be recognised, the revolving movement should be upset.

Very much the same may be said of the act of twining, namely that most physiologists refuse to accept Darwin's view (above referred to) that twining is the direct result of circumnutation. Everyone must allow that the two phenomena are in some way connected, since a plant which circumnutates clockwise, i.e. with the sun, twines in the same direction, and vice versa. It must also be granted that geotropism has a bearing on the problem, since all plants twine upwards, and cannot twine along a horizontal support. But how these two factors are combined, and whether any (and if so what) other factors contribute, we cannot say. If we give up Darwin's explanation, we must at the same time say with Pfeffer that "the causes of twining are... unknown." ("The Physiology of Plants", Eng. Tr. (Oxford, 1906), III. page 37.)

Let us leave this difficult question and consider some other points made out in the progress of the work on climbing plants. One result of what he called his "niggling" ("Life and Letters", III. page 312.) work on tendrils was the discovery of the delicacy of their sense of touch, and the rapidity of their movement. Thus in a passion-flower tendril, a bit of platinum wire weighing 1.2 mg. produced curvature ("Climbing Plants", page 171.), as did a loop of cotton weighing 2 mg. Pfeffer ("Untersuchungen a.d. Bot. Inst. z. Tubingen", Bd. I. 1881-85, page 506.), however, subsequently found much greater sensitiveness: thus the tendril of Sicyos angulatus reacted to 0.00025 mg., but this only occurred when the delicate rider of cottonwool fibre was disturbed by the wind. The same author expanded and explained in a most interesting way the meaning of Darwin's observation that tendrils are not stimulated to movement by drops of water resting on them. Pfeffer showed that DIRTY water containing minute particles of clay in suspension acts as a stimulus. He also showed that gelatine acts like pure water; if a smooth glass rod is coated with a 10 per cent solution of gelatine and is then applied to a tendril, no movement occurs in spite of the fact that the gelatine is solid when cold. Pfeffer ("Physiology", Eng. Tr. III. page 52. Pfeffer has pointed out the resemblance between the contact irritability of plants and the human sense of touch. Our skin is not sensitive to uniform pressure such as is produced when the finger is dipped into mercury (Tubingen "Untersuchungen", I. page 504.) generalises the result in the statement that the tendril has a special form of irritability and only reacts to "differences of pressure or variations of pressure in contiguous... regions." Darwin was especially interested in such cases of specialised irritability. For instance in May, 1864, he wrote to Asa Gray ("Life and Letters", III. page 314.) describing the tendrils of Bignonia capreolata, which "abhor a simple stick, do not much relish rough bark, but delight in wool or moss." He received, from Gray, information as to the natural habitat of the species, and finally concluded that the tendrils "are specially adapted to climb trees clothed with lichens, mosses, or other such productions." ("Climbing Plants", page 102.)

Tendrils were not the only instance discovered by Darwin of delicacy of touch in plants. In 1860 he had already begun to observe Sundew (Drosera), and was full of astonishment at its behaviour. He wrote to Sir Joseph Hooker ("Life and Letters", III. page 319.): "I have been working like a madman at Drosera. Here is a fact for you which is certain as you stand where you are, though you won't believe it, that a bit of hair 1/78000 of one grain in weight placed on gland, will cause ONE of the gland-bearing hairs of Drosera to curve inwards." Here again Pfeffer (Pfeffer in "Untersuchungen a. d. Bot. Inst. z. Tubingen", I. page 491.) has, as in so many cases, added important facts to my father's observations. He showed that if the leaf of Drosera is entirely freed from such vibrations as would reach it if observed on an ordinary table, it does not react to small weights, so that in fact it was the vibration of the minute fragment of hair on the gland that produced movement. We may fancifully see an adaptation to the capture of insects—to the dancing of a gnat's foot on the sensitive surface.

Darwin was fond of telling how when he demonstrated the sensitiveness of Drosera to Mr Huxley and (I think) to Sir John Burdon Sanderson, he could perceive (in spite of their courtesy) that they thought the whole thing a delusion. And the story ended with his triumph when Mr Huxley cried out, "It IS moving."

Darwin's work on tendrils has led to some interesting investigations on the mechanisms by which plants perceive stimuli. Thus Pfeffer (Tubingen "Untersuchungen" I. page 524.)

showed that certain epidermic cells occurring in tendrils are probably organs of touch. In these cells the protoplasm burrows as it were into cavities in the thickness of the external cell-walls and thus comes close to the surface, being separated from an object touching the tendril merely by a very thin layer of cell-wall substance. Haberlandt ("Physiologische Pflanzenanatomie", Edition III. Leipzig, 1904. "Sinnesorgane im Pflanzenreich", Leipzig, 1901, and other publications.) has greatly extended our knowledge of vegetable structure in relation to mechanical stimulation. He defines a sense-organ as a contrivance by which the DEFORMATION or forcible change of form in the protoplasm—on which mechanical stimulation depends—is rendered rapid and considerable in amplitude ("Sinnesorgane", page 10). He has shown that in certain papillose and bristle-like contrivances, plants possess such sense-organs; and moreover that these contrivances show a remarkable similarity to corresponding sense-organs in animals.

Haberlandt and Nemec ("Ber. d. Deutschen bot. Gesellschaft", XVIII. 1900. See F. Darwin, Presidential Address to Section K, British Association, 1904.) published independently and simultaneously a theory of the mechanism by which plants are orientated in relation to gravitation. And here again we find an arrangement identical in principle with that by which certain animals recognise the vertical, namely the pressure of free particles on the irritable wall of a cavity. In the higher plants, Nemec and Haberlandt believe that special loose and freely movable starch-grains play the part of the otoliths or statoliths of the crustacea, while the protoplasm lining the cells in which they are contained corresponds to the sensitive membrane lining the otocyst of the animal. What is of special interest in our present connection is that according to this ingenious theory (The original conception was due to Noll ("Heterogene Induction", Leipzig, 1892), but his view differed in essential points from those here given.) the sense of verticality in a plant is a form of contact-irritability. The vertical position is distinguished from the horizontal by the fact that, in the latter case, the loose starch-grains rest on the lateral walls of the cells instead of on the terminal walls as occurs in the normal upright position. It should be added that the statolith theory is still sub judice; personally I cannot doubt that it is in the main a satisfactory explanation of the facts.

With regard to the RAPIDITY of the reaction of tendrils, Darwin records ("Climbing Plants", page 155. Others have observed movement after about 6".) that a Passion-Flower tendril moved distinctly within 25 seconds of stimulation. It was this fact, more than any other, that made him doubt the current explanation, viz. that the movement is due to unequal growth on the two sides of the tendril. The interesting work of Fitting (Pringsheim's "Jahrb." XXXVIII. 1903, page 545.) has shown, however, that the primary cause is not (as Darwin supposed) contraction on the concave, but an astonishingly rapid increase in growth-rate on the convex side.

On the last page of "Climbing Plants" Darwin wrote: "It has often been vaguely asserted that plants are distinguished from animals by not having the power of movement. It should rather be said that plants acquire and display this power only when it is of some advantage to them."

He gradually came to realise the vividness and variety of vegetable life, and that a plant like an animal has capacities of behaving in different ways under different circumstances, in a manner that may be compared to the instinctive movements of animals. This point of view is expressed in well-known passages in the "Power of Movement". ("The Power of Movement in Plants", 1880, pages 571-3.) "It is impossible not to be struck with the resemblance between the... movements of plants and many of the actions performed unconsciously by the lower animals." And again, "It is hardly an exaggeration to say that the tip of the radicle... having the power of directing the movements of the adjoining parts, acts like the brain of one of the lower animals; the brain being seated within the anterior end of the body, receiving impressions from the sense-organs, and directing the several movements."

The conception of a region of perception distinct from a region of movement is perhaps the most fruitful outcome of his work on the movements of plants. But many years before its publication, viz. in 1861, he had made out the wonderful fact that in the Orchid

Catasetum ("Life and Letters", III. page 268.) the projecting organs or antennae are sensitive to a touch, and transmit an influence "for more than one inch INSTANTANEOUSLY," which leads to the explosion or violent ejection of the pollinia. And as we have already seen a similar transmission of a stimulus was discovered by him in Sundew in 1860, so that in 1862 he could write to Hooker ("Life and Letters", III. page 321.): "I cannot avoid the conclusion, that Drosera possesses matter at least in some degree analogous in constitution and function to nervous matter." I propose in what follows to give some account of the observations on the transmission of stimuli given in the "Power of Movement". It is impossible within the space at my command to give anything like a complete account of the matter, and I must necessarily omit all mention of much interesting work. One well-known experiment consisted in putting opaque caps on the tips of seedling grasses (e.g. oat and canary-grass) and then exposing them to light from one side. The difference, in the amount of curvature towards the light, between the blinded and unblinded specimens, was so great that it was concluded that the light-sensitiveness resided exclusively in the tip. The experiment undoubtedly proves that the sensitiveness is much greater in the tip than elsewhere, and that there is a transmission of stimulus from the tip to the region of curvature. But Rothert (Rothert, Cohn's "Beitrage", VII. 1894.) has conclusively proved that the basal part where the curvature occurs is also DIRECTLY sensitive to light. He has shown, however, that in other grasses (Setaria, Panicum) the cotyledon is the only part which is sensitive, while the hypocotyl, where the movement occurs, is not directly sensitive.

It was however the question of the localisation of the gravitational sense in the tip of the seedling root or radicle that aroused most attention, and it was on this question that a controversy arose which has continued to the present day.

The experiment on which Darwin's conclusion was based consisted simply in cutting off the tip, and then comparing the behaviour of roots so treated with that of normal specimens. An uninjured root when placed horizontally regains the vertical by means of a sharp downward curve; not so a decapitated root which continues to grow more or less horizontally. It was argued that this depends on the loss of an organ specialised for the perception of gravity, and residing in the tip of the root; and the experiment (together with certain important variants) was claimed as evidence of the existence of such an organ.

It was at once objected that the amputation of the tip might check curvature by interfering with longitudinal growth, on the distribution of which curvature depends. This objection was met by showing that an injury, e.g. splitting the root longitudinally (See F. Darwin, "Linnean Soc. Journal (Bot)." XIX. 1882, page 218.), which does not remove the tip, but seriously checks growth, does not prevent geotropism. This was of some interest in another and more general way, in showing that curvature and longitudinal growth must be placed in different categories as regards the conditions on which they depend.

Another objection of a much more serious kind was that the amputation of the tip acts as a shock. It was shown by Rothert (See his excellent summary of the subject in "Flora" 1894 (Erganzungsband), page 199.) that the removal of a small part of the cotyledon of Setaria prevents the plant curving towards the light, and here there is no question of removing the sense-organ since the greater part of the sensitive cotyledon is intact. In view of this result it was impossible to rely on the amputations performed on roots as above described.

At this juncture a new and brilliant method originated in Pfeffer's laboratory. (See Pfeffer, "Annals of Botany", VIII. 1894, page 317, and Czapek, Pringsheim's "Jahrb." XXVII. 1895, page 243.) Pfeffer and Czapek showed that it is possible to bend the root of a lupine so that, for instance, the supposed sense-organ at the tip is vertical while the motile region is horizontal. If the motile region is directly sensitive to gravity the root ought to curve downwards, but this did not occur: on the contrary it continued to grow horizontally. This is precisely what should happen if Darwin's theory is the right one: for if the tip is kept vertical, the sense-organ is in its normal position and receives no stimulus from gravitation, and therefore can obviously transmit none to the region of curvature. Unfortunately this method did not convince the botanical world because some of those who repeated Czapek's experiment failed to get his results.

Czapek ("Berichte d. Deutsch. bot. Ges." XV. 1897, page 516, and numerous subsequent papers. English readers should consult Czapek in the "Annals of Botany", XIX. 1905, page 75.) has devised another interesting method which throws light on the problem. He shows that roots, which have been placed in a horizontal position and have therefore been geotropically stimulated, can be distinguished by a chemical test from vertical, i.e. unstimulated roots. The chemical change in the root can be detected before any curvature has occurred and must therefore be a symptom of stimulation, not of movement. It is particularly interesting to find that the change in the root, on which Czapek's test depends, takes place in the tip, i.e. in the region which Darwin held to be the centre for gravitational sensitiveness.

In 1899 I devised a method (F. Darwin, "Annals of Botany", XIII. 1899, page 567.) by which I sought to prove that the cotyledon of Setaria is not only the organ for light-perception, but also for gravitation. If a seedling is supported horizontally by pushing the apical part (cotyledon) into a horizontal tube, the cotyledon will, according to my supposition, be stimulated gravitationally and a stimulus will be transmitted to the basal part of the stem (hypocotyl) causing it to bend. But this curvature merely raises the basal end of the seedling, the sensitive cotyledon remains horizontal, imprisoned in its tube; it will therefore be continually stimulated and will continue to transmit influences to the bending region, which should therefore curl up into a helix or corkscrew-like form,—and this is precisely what occurred.

I have referred to this work principally because the same method was applied to roots by Massart (Massart, "Mem. Couronnes Acad. R. Belg." LXII. 1902.) and myself (F. Darwin, "Linnean Soc. Journ." XXXV. 1902, page 266.) with a similar though less striking result. Although these researches confirmed Darwin's work on roots, much stress cannot be laid on them as there are several objections to them, and they are not easily repeated.

The method which—as far as we can judge at present—seems likely to solve the problem of the root-tip is most ingenious and is due to Piccard. (Pringsheim's "Jahrb." XL. 1904, page 94.)

Andrew Knight's celebrated experiment showed that roots react to centrifugal force precisely as they do to gravity. So that if a bean root is fixed to a wheel revolving rapidly on a horizontal axis, it tends to curve away from the centre in the line of a radius of the wheel. In ordinary demonstrations of Knight's experiment the seed is generally fixed so that the root is at right angles to a radius, and as far as convenient from the centre of rotation. Piccard's experiment is arranged differently. (A seed is depicted below a horizontal dotted line AA, projecting a root upwards.) The root is oblique to the axis of rotation, and the extreme tip projects beyond that axis. Line AA represents the axis of rotation, T is the tip of the root just above the line AA, and B is the region just below line AA in which curvature takes place. If the motile region B is directly sensitive to gravitation (and is the only part which is sensitive) the root will curve (down and away from the vertical) away from the axis of rotation, just as in Knight's experiment. But if the tip T is alone sensitive to gravitation the result will be exactly reversed, the stimulus originating in T and conveyed to B will produce curvature (up towards the vertical). We may think of the line AA as a plane dividing two worlds. In the lower one gravity is of the earthly type and is shown by bodies falling and roots curving downwards: in the upper world bodies fall upwards and roots curve in the same direction. The seedling is in the lower world, but its tip containing the supposed sense-organ is in the strange world where roots curve upwards. By observing whether the root bends up or down we can decide whether the impulse to bend originates in the tip or in the motile region.

Piccard's results showed that both curvatures occurred and he concluded that the sensitive region is not confined to the tip. (Czapek (Pringsheim's "Jahrb." XXXV. 1900, page 362) had previously given reasons for believing that, in the root, there is no sharp line of separation between the regions of perception and movement.)

247

Haberlandt (Pringsheim's "Jahrb." XLV. 1908, page 575.) has recently repeated the experiment with the advantage of better apparatus and more experience in dealing with plants, and has found as Piccard did that both the tip and the curving region are sensitive to gravity, but with the important addition that the sensitiveness of the tip is much greater than that of the motile region. The case is in fact similar to that of the oat and canary-grass. In both instances my father and I were wrong in assuming that the sensitiveness is confined to the tip, yet there is a concentration of irritability in that region and transmission of stimulus is as true for geotropism as it is for heliotropism. Thus after nearly thirty years the controversy of the root-tip has apparently ended somewhat after the fashion of the quarrels at the "Rainbow" in "Silas Marner"—"you're both right and you're both wrong." But the "brain-function" of the root-tip at which eminent people laughed in early days turns out to be an important part of the truth. (By using Piccard's method I have succeeded in showing that the gravitational sensitiveness of the cotyledon of Sorghum is certainly much greater than the sensitiveness of the hypocotyl—if indeed any such sensitiveness exists. See Wiesner's "Festschrift", Vienna, 1908.)

Another observation of Darwin's has given rise to much controversy. ("Power of Movement", page 133.) If a minute piece of card is fixed obliquely to the tip of a root some influence is transmitted to the region of curvature and the root bends away from the side to which the card was attached. It was thought at the time that this proved the root-tip to be sensitive to contact, but this is not necessarily the case. It seems possible that the curvature is a reaction to the injury caused by the alcoholic solution of shellac with which the cards were cemented to the tip. This agrees with the fact given in the "Power of Movement" that injuring the root-tip on one side, by cutting or burning it, induced a similar curvature. On the other hand it was shown that curvature could be produced in roots by cementing cards, not to the naked surface of the root-tip, but to pieces of gold-beaters skin applied to the root; gold-beaters skin being by itself almost without effect. But it must be allowed that, as regards touch, it is not clear how the addition of shellac and card can increase the degree of contact. There is however some evidence that very close contact from a solid body, such as a curved fragment of glass, produces curvature: and this may conceivably be the explanation of the effect of gold-beaters skin covered with shellac. But on the whole it is perhaps safer to classify the shellac experiments with the results of undoubted injury rather than with those of contact.

Another subject on which a good deal of labour was expended is the sleep of leaves, or as Darwin called it their NYCTITROPIC movement. He showed for the first time how widely spread this phenomenon is, and attempted to give an explanation of the use to the plant of the power of sleeping. His theory was that by becoming more or less vertical at night the leaves escape the chilling effect of radiation. Our method of testing this view was to fix some of the leaves of a sleeping plant so that they remained horizontal at night and therefore fully exposed to radiation, while their fellows were partly protected by assuming the nocturnal position. The experiments showed clearly that the horizontal leaves were more injured than the sleeping, i.e. more or less vertical, ones. It may be objected that the danger from cold is very slight in warm countries where sleeping plants abound. But it is quite possible that a lowering of the temperature which produces no visible injury may nevertheless be hurtful by checking the nutritive processes (e.g. translocation of carbohydrates), which go on at night. Stahl ("Bot. Zeitung", 1897, page 81.) however has ingeniously suggested that the exposure of the leaves to radiation is not DIRECTLY hurtful because it lowers the temperature of the leaf, but INDIRECTLY because it leads to the deposition of dew on the leaf-surface. He gives reasons for believing that dew-covered leaves are unable to transpire efficiently, and that the absorption of mineral food-material is correspondingly checked. Stahl's theory is in no way destructive of Darwin's, and it is possible that nyctitropic leaves are adapted to avoid the indirect as well as the direct results of cooling by radiation.

In what has been said I have attempted to give an idea of some of the discoveries brought before the world in the "Power of Movement" (In 1881 Professor Wiesner published his "Das Bewegungsvermogen der Pflanzen", a book devoted to the criticism of

"The Power of Movement in Plants". A letter to Wiesner, published in "Life and Letters", III. page 336, shows Darwin's warm appreciation of his critic's work, and of the spirit in which it is written.) and of the subsequent history of the problems. We must now pass on to a consideration of the central thesis of the book,—the relation of circumnutation to the adaptive curvatures of plants.

Darwin's view is plainly stated on pages 3-4 of the "Power of Movement". Speaking of circumnutation he says, "In this universally present movement we have the basis or groundwork for the acquirement, according to the requirements of the plant, of the most diversified movements." He then points out that curvatures such as those towards the light or towards the centre of the earth can be shown to be exaggerations of circumnutation in the given directions. He finally points out that the difficulty of conceiving how the capacities of bending in definite directions were acquired is diminished by his conception. "We know that there is always movement in progress, and its amplitude, or direction, or both, have only to be modified for the good of the plant in relation with internal or external stimuli."

It may at once be allowed that the view here given has not been accepted by physiologists. The bare fact that circumnutation is a general property of plants (other than climbing species) is not generally rejected. But the botanical world is no nearer to believing in the theory of reaction built on it.

If we compare the movements of plants with those of the lower animals we find a certain resemblance between the two. According to Jennings (H.S. Jennings, "The Behavior of the Lower Animals". Columbia U. Press, N.Y. 1906.) a Paramoecium constantly tends to swerve towards the aboral side of its body owing to certain peculiarities in the set and power of its cilia. But the tendency to swim in a circle, thus produced, is neutralised by the rotation of the creature about its longitudinal axis. Thus the direction of the swerves IN RELATION TO THE PATH of the organism is always changing, with the result that the creature moves in what approximates to a straight line, being however actually a spiral about the general line of progress. This method of motion is strikingly like the circumnutation of a plant, the apex of which also describes a spiral about the general line of growth. A rooted plant obviously cannot rotate on its axis, but the regular series of curvatures of which its growth consists correspond to the aberrations of Paramoecium distributed regularly about its course by means of rotation. (In my address to the Biological Section of the British Association at Cardiff (1891) I have attempted to show the connection between circumnutation and RECTIPETALITY, i.e. the innate capacity of growing in a straight line.) Just as a plant changes its direction of growth by an exaggeration of one of the curvature-elements of which circumnutation consists, so does a Paramoecium change its course by the accentuation of one of the deviations of which its path is built. Jennings has shown that the infusoria, etc., react to stimuli by what is known as the "method of trial." If an organism swims into a region where the temperature is too high or where an injurious substance is present, it changes its course. It then moves forward again, and if it is fortunate enough to escape the influence, it continues to swim in the given direction. If however its change of direction leads it further into the heated or poisonous region it repeats the movement until it emerges from its difficulties. Jennings finds in the movements of the lower organisms an analogue with what is known as pain in conscious organisms. There is certainly this much resemblance that a number of quite different sub-injurious agencies produce in the lower organisms a form of reaction by the help of which they, in a partly fortuitous way, escape from the threatening element in their environment. The higher animals are stimulated in a parallel manner to vague and originally purposeless movements, one of which removes the discomfort under which they suffer, and the organism finally learns to perform the appropriate movement without going through the tentative series of actions.

I am tempted to recognise in circumnutation a similar groundwork of tentative movements out of which the adaptive ones were originally selected by a process rudely representative of learning by experience.

It is, however, simpler to confine ourselves to the assumption that those plants have survived which have acquired through unknown causes the power of reacting in appropriate ways to the external stimuli of light, gravity, etc. It is quite possible to conceive this occurring in plants which have no power of circumnutating—and, as already pointed out, physiologists do as a fact neglect circumnutation as a factor in the evolution of movements. Whatever may be the fate of Darwin's theory of circumnutation there is no doubt that the research he carried out in support of, and by the light of, this hypothesis has had a powerful influence in guiding the modern theories of the behaviour of plants. Pfeffer ("The Physiology of Plants", Eng. Tr. III. page 11.), who more than any one man has impressed on the world a rational view of the reactions of plants, has acknowledged in generous words the great value of Darwin's work in the same direction. The older view was that, for instance, curvature towards the light is the direct mechanical result of the difference of illumination on the lighted and shaded surfaces of the plant. This has been proved to be an incorrect explanation of the fact, and Darwin by his work on the transmission of stimuli has greatly contributed to the current belief that stimuli act indirectly. Thus we now believe that in a root and a stem the mechanism for the perception of gravitation is identical, but the resulting movements are different because the motor-irritabilities are dissimilar in the two cases. We must come back, in fact, to Darwin's comparison of plants to animals. In both there is perceptive machinery by which they are made delicately alive to their environment, in both the existing survivors are those whose internal constitution has enabled them to respond in a beneficial way to the disturbance originating in their sense-organs.

XX. THE BIOLOGY OF FLOWERS. By
K. Goebel, Ph.D.

Professor of Botany in the University of
Munich.

There is scarcely any subject to which Darwin devoted so much time and work as to his researches into the biology of flowers, or, in other words, to the consideration of the question to what extent the structural and physiological characters of flowers are correlated with their function of producing fruits and seeds. We know from his own words what fascination these studies possessed for him. We repeatedly find, for example, in his letters expressions such as this:—"Nothing in my life has ever interested me more than the fertilisation of such plants as Primula and Lythrum, or again Anacamptis or Listera." ("More Letters of Charles Darwin", Vol. II. page 419.)

Expressions of this kind coming from a man whose theories exerted an epoch-making influence, would be unintelligible if his researches into the biology of flowers had been concerned only with records of isolated facts, however interesting these might be. We may at once take it for granted that the investigations were undertaken with the view of following up important problems of general interest, problems which are briefly dealt with in this essay.

Darwin published the results of his researches in several papers and in three larger works, (i) "On the various contrivances by which British and Foreign Orchids are fertilised by insects" (First edition, London, 1862; second edition, 1877; popular edition, 1904.) (ii) "The effects of Cross and Self fertilisation in the vegetable kingdom" (First edition, 1876; second edition, 1878). (iii) "The different forms of Flowers on plants of the same species" (First edition, 1877; second edition, 1880).

Although the influence of his work is considered later, we may here point out that it was almost without a parallel; not only does it include a mass of purely scientific observations, but it awakened interest in very wide circles, as is shown by the fact that we find the results of Darwin's investigations in floral biology universally quoted in school books; they are even willingly accepted by those who, as regards other questions, are opposed to Darwin's views.

The works which we have mentioned are, however, not only of special interest because of the facts they contribute, but because of the MANNER in which the facts are expressed. A superficial reader seeking merely for catch-words will, for instance, probably find the book on cross and self-fertilisation rather dry because of the numerous details which it contains: it is, indeed, not easy to compress into a few words the general conclusions of this volume. But on closer examination, we cannot be sufficiently grateful to the author for the exactness and objectivity with which he enables us to participate in the scheme of his researches. He never tries to persuade us, but only to convince us that his conclusions are based on facts; he always gives prominence to such facts as appear to be in opposition to his opinions,—a feature of his work in accordance with a maxim which he laid down:—"It is a golden rule, which I try to follow, to put every fact which is opposed to one's preconceived opinion in the strongest light." ("More Letters", Vol. II. page 324.)

The result of this method of presentation is that the works mentioned above represent a collection of most valuable documents even for those who feel impelled to draw from the data other conclusions than those of the author. Each investigation is the outcome of a definite question, a "preconceived opinion," which is either supported by the facts or must be abandoned. "How odd it is that anyone should not see that all observation must be for or against some view if it is to be of any service!" (Ibid. Vol. I. page 195.)

The points of view which Darwin had before him were principally the following. In the first place the proof that a large number of the peculiarities in the structure of flowers are not useless, but of the greatest significance in pollination must be of considerable importance for the interpretation of adaptations; "The use of each trifling detail of structure is far from a barren search to those who believe in natural selection." ("Fertilisation of Orchids" (1st edition), page 351; (2nd edition 1904) page 286.) Further, if these structural relations are shown to be useful, they may have been acquired because from the many variations which have occurred along different lines, those have been preserved by natural selection "which are beneficial to the organism under the complex and ever-varying conditions of life." (Ibid. page 351.) But in the case of flowers there is not only the question of adaptation to fertilisation to be considered. Darwin, indeed, soon formed the opinion which he has expressed in the following sentence,—"From my own observations on plants, guided to a certain extent by the experience of the breeders of animals, I became convinced many years ago that it is a general law of nature that flowers are adapted to be crossed, at least occasionally, by pollen from a distinct plant." ("Cross and Self fertilisation" (1st edition), page 6.)

The experience of animal breeders pointed to the conclusion that continual in-breeding is injurious. If this is correct, it raises the question whether the same conclusion holds for plants. As most flowers are hermaphrodite, plants afford much more favourable material than animals for an experimental solution of the question, what results follow from the union of nearly related sexual cells as compared with those obtained by the introduction of new blood. The answer to this question must, moreover, possess the greatest significance for the correct understanding of sexual reproduction in general.

We see, therefore, that the problems which Darwin had before him in his researches into the biology of flowers were of the greatest importance, and at the same time that the point of view from which he attacked the problems was essentially a teleological one.

We may next inquire in what condition he found the biology of flowers at the time of his first researches, which were undertaken about the year 1838. In his autobiography he writes,—"During the summer of 1839, and, I believe, during the previous summer, I was led to attend to the cross-fertilisation of flowers by the aid of insects, from having come to the conclusion in my speculations on the origin of species, that crossing played an important part in keeping specific forms constant." ("The Life and Letters of Charles Darwin", Vol. I. page 90, London, 1888.) In 1841 he became acquainted with Sprengel's work: his researches into the biology of flowers were thus continued for about forty years.

It is obvious that there could only be a biology of flowers after it had been demonstrated that the formation of seeds and fruit in the flower is dependent on pollination and subsequent fertilisation. This proof was supplied at the end of the seventeenth century by R.J. Camerarius (1665-1721). He showed that normally seeds and fruits are developed only when the pollen reaches the stigma. The manner in which this happens was first thoroughly investigated by J.G. Kolreuter (1733-1806 (Kolreuter, "Vorlaufige Nachricht von einigen das Geschlecht der Planzen betreffenden Versuchen und Beobachtungen", Leipzig, 1761; with three supplements, 1763-66. Also, "Mem. de l'acad. St Petersbourg", Vol. XV. 1809.)), the same observer to whom we owe the earliest experiments in hybridisation of real scientific interest. Kolreuter mentioned that pollen may be carried from one flower to another partly by wind and partly by insects. But he held the view, and that was, indeed, the natural assumption, that self-fertilisation usually occurs in a flower, in other words that the pollen of a flower reaches the stigma of the same flower. He demonstrated, however, certain cases in which cross-pollination occurs, that is in which the pollen of another flower of the same species is conveyed to the stigma. He was familiar with the phenomenon, exhibited by numerous flowers, to which Sprengel afterwards applied the term Dichogamy, expressing the fact that the anthers and stigmas of a flower often ripen at different times, a peculiarity which is now recognised as one of the commonest means of ensuring cross-pollination.

With far greater thoroughness and with astonishing power of observation C.K. Sprengel (1750-1816) investigated the conditions of pollination of flowers. Darwin was introduced by that eminent botanist Robert Brown to Sprengel's then but little appreciated work,—"Das entdeckte Geheimniss der Natur im Bau und in der Befruchtung der Blumen" (Berlin, 1793); this is by no means the least service to Botany rendered by Robert Brown.

Sprengel proceeded from a naive teleological point of view. He firmly believed "that the wise Author of nature had not created a single hair without a definite purpose." He succeeded in demonstrating a number of beautiful adaptations in flowers for ensuring pollination; but his work exercised but little influence on his contemporaries and indeed for a long time after his death. It was through Darwin that Sprengel's work first achieved a well deserved though belated fame. Even such botanists as concerned themselves with researches into the biology of flowers appear to have formerly attached much less value to Sprengel's work than it has received since Darwin's time. In illustration of this we may quote C.F. Gartner whose name is rightly held in the highest esteem as that of one of the most eminent hybridologists. In his work "Versuche und Beobachtungen uder die Befruchtungsorgane der vollkommeneren Gewachse und uber die naturliche und kunstliche Befruchtung durch den eigenen Pollen" he also deals with flower-pollination. He recognised the action of the wind, but he believed, in spite of the fact that he both knew and quoted Kolreuter and Sprengel, that while insects assist pollination, they do so only occasionally, and he held that insects are responsible for the conveyance of pollen; thorough investigations would show "that a very small proportion of the plants included in this category require this assistance in their native habitat." (Gartner, "Versucher und Beobachtungen... ", page 335, Stuttgart, 1844.) In the majority of plants self-pollination occurs.

Seeing that even investigators who had worked for several decades at fertilisation-phenomena had not advanced the biology of flowers beyond the initial stage, we cannot be surprised that other botanists followed to even a less extent the lines laid down by Kolreuter and Sprengel. This was in part the result of Sprengel's supernatural teleology and in part due to the fact that his book appeared at a time when other lines of inquiry exerted a dominating influence.

At the hands of Linnaeus systematic botany reached a vigorous development, and at the beginning of the nineteenth century the anatomy and physiology of plants grew from small beginnings to a flourishing branch of science. Those who concerned themselves with flowers endeavoured to investigate their development and structure or the most minute phenomena connected with fertilisation and the formation of the embryo. No room was left for the extension of the biology of flowers on the lines marked out by Kolreuter and Sprengel. Darwin was the first to give new life and a deeper significance to this subject, chiefly because he took as his starting-point the above-mentioned problems, the importance of which is at once admitted by all naturalists.

The further development of floral biology by Darwin is in the first place closely connected with the book on the fertilisation of Orchids. It is noteworthy that the title includes the sentence,—"and on the good effects of intercrossing."

The purpose of the book is clearly stated in the introduction:—"The object of the following work is to show that the contrivances by which Orchids are fertilised, are as varied and almost as perfect as any of the most beautiful adaptations in the animal kingdom; and, secondly, to show that these contrivances have for their main object the fertilisation of each flower by the pollen of another flower." ("Fertilisation of Orchids", page 1.) Orchids constituted a particularly suitable family for such researches. Their flowers exhibit a striking wealth of forms; the question, therefore, whether the great variety in floral structure bears any relation to fertilisation (In the older botanical literature the word fertilisation is usually employed in cases where POLLINATION is really in question: as Darwin used it in this sense it is so used here.) must in this case possess special interest.

Darwin succeeded in showing that in most of the orchids examined self-fertilisation is either an impossibility, or, under natural conditions, occurs only exceptionally. On the other hand these plants present a series of extraordinarily beautiful and remarkable adaptations which ensure the transference of pollen by insects from one flower to another. It is impossible to describe adequately in a few words the wealth of facts contained in the Orchid book. A few examples may, however, be quoted in illustration of the delicacy of the observations and of the perspicuity employed in interpreting the facts.

The majority of orchids differ from other seed plants (with the exception of the Asclepiads) in having no dust-like pollen. The pollen, or more correctly, the pollen-tetrads, remain fastened together as club-shaped pollinia usually borne on a slender pedicel. At the base of the pedicel is a small viscid disc by which the pollinium is attached to the head or proboscis of one of the insects which visit the flower. Darwin demonstrated that in Orchis and other flowers the pedicel of the pollinium, after its removal from the anther, undergoes a curving movement. If the pollinium was originally vertical, after a time it assumed a horizontal position. In the latter position, if the insect visited another flower, the pollinium would exactly hit the sticky stigmatic surface and thus effect fertilisation. The relation between the behaviour of the viscid disc and the secretion of nectar by the flower is especially remarkable. The flowers possess a spur which in some species (e.g. Gymnadenia conopsea, Platanthera bifolia, etc.) contains honey (nectar), which serves as an attractive bait for insects, but in others (e.g. our native species of Orchis) the spur is empty. Darwin held the opinion, confirmed by later investigations, that in the case of flowers without honey the insects must penetrate the wall of the nectarless spurs in order to obtain a nectar-like substance. The glands behave differently in the nectar-bearing and in the nectarless flowers. In the former they are so sticky that they at once adhere to the body of the insect; in the nectarless flowers firm adherence only occurs after the viscid disc has hardened. It is, therefore, adaptively of value that the insects should be detained longer in the nectarless flowers (by having to bore into the spur),—than in flowers in which the nectar is freely exposed. "If this relation, on the one hand, between the viscid matter requiring some little time to set hard, and the nectar being so lodged that moths are delayed in getting it; and, on the other hand, between the viscid matter being at first as viscid as ever it will become, and the nectar lying all ready for rapid suction, be accidental, it is a fortunate accident for the plant. If not accidental, and I cannot believe it to be accidental, what a singular case of adaptation!" ("Fertilisation of Orchids" (1st edition), page 53.)

Among exotic orchids Catasetum is particularly remarkable. One and the same species bears different forms of flowers. The species known as Catasetum tridentatum has pollinia with very large viscid discs; on touching one of the two filaments (antennae) which occur on the gynostemium of the flower the pollinia are shot out to a fairly long distance (as far as 1 metre) and in such manner that they alight on the back of the insect, where they are held. The antennae have, moreover, acquired an importance, from the point of view of the physiology of stimulation, as stimulus-perceiving organs. Darwin had shown that it is only a touch on the antennae that causes the explosion, while contact, blows, wounding, etc. on other places produce no effect. This form of flower proved to be the male. The second form, formerly regarded as a distinct species and named Monachanthus viridis, is shown to be the female flower. The anthers have only rudimentary pollinia and do not open; there are no antennae, but on the other hand numerous seeds are produced. Another type of flower, known as Myanthus barbatus, was regarded by Darwin as a third form: this was afterwards recognised by Rolfe (Rolfe, R.A. "On the sexual forms of Catasetum with special reference to the researches of Darwin and others," "Journ. Linn. Soc." Vol. XXVII. (Botany), 1891, pages 206-225.) as the male flower of another species, Catasetum barbatum Link, an identification in accordance with the discovery made by Cruger in Trinidad that it always remains sterile.

Darwin had noticed that the flowers of Catasetum do not secrete nectar, and he conjectured that in place of it the insects gnaw a tissue in the cavity of the labellum which has a "slightly sweet, pleasant and nutritious taste." This conjecture as well as other conclusions drawn by Darwin from Catasetum have been confirmed by Cruger—assuredly the best proof of the acumen with which the wonderful floral structure of this "most remarkable of the Orchids" was interpreted far from its native habitat.

As is shown by what we have said about Catasetum, other problems in addition to those concerned with fertilisation are dealt with in the Orchid book. This is especially the case in regard to flower morphology. The scope of flower morphology cannot be more clearly and better expressed than by these words: "He will see how curiously a flower may be moulded out of many separate organs—how perfect the cohesion of primordially distinct parts may become,—how organs may be used for purposes widely different from their proper function,—how other organs may be entirely suppressed, or leave mere useless emblems of their former existence." ("Fertilisation of Orchids", page 289.)

In attempting, from this point of view, to refer the floral structure of orchids to their original form, Darwin employed a much more thorough method than that of Robert Brown and others. The result of this was the production of a considerable literature, especially in France, along the lines suggested by Darwin's work. This is the so-called anatomical method, which seeks to draw conclusions as to the morphology of the flower from the course of the vascular bundles in the several parts. (He wrote in one of his letters, "... the destiny of the whole human race is as nothing to the course of vessels of orchids" ("More Letters", Vol. II. page 275.) Although the interpretation of the orchid flower given by Darwin has not proved satisfactory in one particular point—the composition of the labellum—the general results have received universal assent, namely "that all Orchids owe what they have in common to descent from some monocotyledonous plant, which, like so many other plants of the same division, possessed fifteen organs arranged alternately three within three in five whorls." ("Fertilisation of Orchids" (1st edition), page 307.) The alterations which their original form has undergone have persisted so far as they were found to be of use.

We see also that the remarkable adaptations of which we have given some examples are directed towards cross-fertilisation. In only a few of the orchids investigated by Darwin— other similar cases have since been described—was self-fertilisation found to occur regularly or usually. The former is the case in the Bee Ophrys (Ophrys apifera), the mechanism of which greatly surprised Darwin. He once remarked to a friend that one of the things that made him wish to live a few thousand years was his desire to see the extinction of the Bee Ophrys, an end to which he believed its self-fertilising habit was leading. ("Life and Letters", Vol. III. page 276 (footnote).) But, he wrote, "the safest conclusion, as it seems to me, is, that under certain unknown circumstances, and perhaps at very long intervals of time, one individual of the Bee Ophrys is crossed by another." ("Fertilisation of Orchids" page 71.)

If, on the one hand, we remember how much more sure self-fertilisation would be than cross-fertilisation, and, on the other hand, if we call to mind the numerous contrivances for cross-fertilisation, the conclusion is naturally reached that "it is an astonishing fact that self-fertilisation should not have been an habitual occurrence. It apparently demonstrates to us that there must be something injurious in the process. Nature thus tells us, in the most emphatic manner, that she abhors perpetual self-fertilisation... For may we not further infer as probable, in accordance with the belief of the vast majority of the breeders of our domestic productions, that marriage between near relations is likewise in some way injurious, that some unknown great good is derived from the union of individuals which have been kept distinct for many generations?" (Ibid., page 359.)

This view was supported by observations on plants of other families, e.g. Papilionaceae; it could, however, in the absence of experimental proof, be regarded only as a "working hypothesis."

All adaptations to cross-pollination might also be of use simply because they made pollination possible when for any reason self-pollination had become difficult or impossible. Cross-pollination would, therefore, be of use, not as such, but merely as a means of pollination in general; it would to some extent serve as a remedy for a method unsuitable in itself, such as a modification standing in the way of self-pollination, and on the other hand as a means of increasing the chance of pollination in the case of flowers in which self-pollination was possible, but which might, in accidental circumstances, be prevented. It was, therefore, very important to obtain experimental proof of the conclusion to which Darwin was led by the belief of the majority of breeders and by the evidence of the widespread occurrence of cross-pollination and of the remarkable adaptations thereto.

This was supplied by the researches which are described in the two other works named above. The researches on which the conclusions rest had, in part at least, been previously published in separate papers: this is the case as regards the heterostyled plants. The discoveries which Darwin made in the course of his investigations of these plants belong to the most brilliant in biological science.

The case of Primula is now well known. C.K. Sprengel and others were familiar with the remarkable fact that different individuals of the European species of Primula bear differently constructed flowers; some plants possess flowers in which the styles project beyond the stamens attached to the corolla-tube (long-styled form), while in others the stamens are inserted above the stigma which is borne on a short style (short-styled form). It has been shown by Breitenbach that both forms of flower may occur on the same plant, though this happens very rarely. An analogous case is occasionally met with in hybrids, which bear flowers of different colour on the same plant (e.g. Dianthus caryophyllus). Darwin showed that the external differences are correlated with others in the structure of the stigma and in the nature of the pollen. The long-styled flowers have a spherical stigma provided with large stigmatic papillae; the pollen grains are oblong and smaller than those of the short-styled flowers. The number of the seeds produced is smaller and the ovules larger, probably also fewer in number. The short-styled flowers have a smooth compressed stigma and a corolla of somewhat different form; they produce a greater number of seeds.

These different forms of flowers were regarded as merely a case of variation, until Darwin showed "that these heterostyled plants are adapted for reciprocal fertilisation; so that the two or three forms, though all are hermaphrodites, are related to one another almost like the males and females of ordinary unisexual animals." ("Forms of Flowers" (1st edition), page 2.) We have here an example of hermaphrodite flowers which are sexually different. There are essential differences in the manner in which fertilisation occurs. This may be effected in four different ways; there are two legitimate and two illegitimate types of fertilisation. The fertilisation is legitimate if pollen from the long-styled flowers reaches the stigma of the short-styled form or if pollen of the short-styled flowers is brought to the stigma of the long-styled flower, that is the organs of the same length of the two different kinds of flower react on one another. Illegitimate fertilisation is represented by the two kinds of self-fertilisation, also by cross-fertilisation, in which the pollen of the long-styled form reaches the stigma of the same type of flower and, similarly, by cross-pollination in the case of the short-styled flowers.

The applicability of the terms legitimate and illegitimate depends, on the one hand, upon the fact that insects which visit the different forms of flowers pollinate them in the manner suggested; the pollen of the short-styled flowers adhere to that part of the insect's body which touches the stigma of the long-styled flower and vice versa. On the other hand, it is based also on the fact that experiment shows that artificial pollination produces a very different result according as this is legitimate or illegitimate; only the legitimate union ensures complete fertility, the plants thus produced being stronger than those which are produced illegitimately.

If we take 100 as the number of flowers which produce seeds as the result of legitimate fertilisation, we obtain the following numbers from illegitimate fertilisation:

Primula officinalis (P. veris) (Cowslip)... 69 Primula elatior (Oxlip).................. 27 Primula acaulis (P. vulgaris) (Primrose)... 60

Further, the plants produced by the illegitimate method of fertilisation showed, e.g. in P. officinalis, a decrease in fertility in later generations, sterile pollen and in the open a feebler growth. (Under very favourable conditions (in a greenhouse) the fertility of the plants of the fourth generation increases—a point, which in view of various theoretical questions, deserves further investigation.) They behave in fact precisely in the same way as hybrids between species of different genera. This result is important, "for we thus learn that the difficulty in sexually uniting two organic forms and the sterility of their offspring, afford no sure criterion of so-called specific distinctness" ("Forms of Flowers", page 242): the relative or absolute sterility of the illegitimate unions and that of their illegitimate descendants depend exclusively on the nature of the sexual elements and on their inability to combine in a particular manner. This functional difference of sexual cells is characteristic of the behaviour of hybrids as of the illegitimate unions of heterostyled plants. The agreement becomes even closer if we regard the Primula plants bearing different forms of flowers not as belonging to a systematic entity or "species," but as including several elementary species. The legitimately produced plants are thus true hybrids (When Darwin wrote in reference to the different forms of heterostyled plants, "which all belong to the same species as certainly as do the two sexes of the same species" ("Cross and Self fertilisation", page 466), he adopted the term species in a comprehensive sense. The recent researches of Bateson and Gregory ("On the inheritance of Heterostylism in Primula"; "Proc. Roy. Soc." Ser. B, Vol. LXXVI. 1905, page 581) appear to me also to support the view that the results of illegitimate crossing of heterostyled Primulas correspond with those of hybridisation. The fact that legitimate pollen effects fertilisation, even if illegitimate pollen reaches the stigma a short time previously, also points to this conclusion. Self-pollination in the case of the short-styled form, for example, is not excluded. In spite of this, the numerical proportion of the two forms obtained in the open remains approximately the same as when the pollination was exclusively legitimate, presumably because legitimate pollen is prepotent.), with which their behaviour in other respects, as Darwin showed, presents so close an agreement. This view receives support also from the fact that descendants of a flower fertilised illegitimately by pollen from another plant with the same form of flower belong, with few exceptions, to the same type as that of their parents. The two forms of flower, however, behave differently in this respect. Among 162 seedlings of the long-styled illegitimately pollinated plants of Primula officinalis, including five generations, there were 156 long-styled and only six short-styled forms, while as the result of legitimate fertilisation nearly half of the offspring were long-styled and half short-styled. The short-styled illegitimately pollinated form gave five long-styled and nine short-styled; the cause of this difference requires further explanation. The significance of heterostyly, whether or not we now regard it as an arrangement for the normal production of hybrids, is comprehensively expressed by Darwin: "We may feel sure that plants have been rendered heterostyled to ensure cross-fertilisation, for we now know that a cross between the distinct individuals of the same species is highly important for the vigour and fertility of the offspring." ("Forms of Flowers", page 258.) If we remember how important the interpretation of heterostyly has become in all general problems as, for example, those connected with the conditions of the formation of hybrids, a fact which was formerly overlooked, we can appreciate how Darwin was able to say in his autobiography: "I do not think anything in my scientific life has given me so much satisfaction as making out the meaning of the structure of these plants." ("Life and Letters", Vol. I. page 91.)

The remarkable conditions represented in plants with three kinds of flowers, such as Lythrum and Oxalis, agree in essentials with those in Primula. These cannot be considered in detail here; it need only be noted that the investigation of these cases was still more laborious. In order to establish the relative fertility of the different unions in Lythrum salicaria 223 different fertilisations were made, each flower being deprived of its male organs and then dusted with the appropriate pollen.

In the book containing the account of heterostyled plants other species are dealt with which, in addition to flowers opening normally (chasmogamous), also possess flowers which remain closed but are capable of producing fruit. These cleistogamous flowers afford a striking example of habitual self-pollination, and H. von Mohl drew special attention to them as such shortly after the appearance of Darwin's Orchid book. If it were only a question of producing seed in the simplest way, cleistogamous flowers would be the most conveniently constructed. The corolla and frequently other parts of the flower are reduced; the development of the seed may, therefore, be accomplished with a smaller expenditure of building material than in chasmogamous flowers; there is also no loss of pollen, and thus a smaller amount suffices for fertilisation.

Almost all these plants, as Darwin pointed out, have also chasmogamous flowers which render cross-fertilisation possible. His view that cleistogamous flowers are derived from originally chasmogamous flowers has been confirmed by more recent researches. Conditions of nutrition in the broader sense are the factors which determine whether chasmogamous or cleistogamous flowers are produced, assuming, of course, that the plants in question have the power of developing both forms of flower. The former may fail to appear for some time, but are eventually developed under favourable conditions of nourishment. The belief of many authors that there are plants with only cleistogamous flowers cannot therefore be accepted as authoritative without thorough experimental proof, as we are concerned with extra-european plants for which it is often difficult to provide appropriate conditions in cultivation.

Darwin sees in cleistogamous flowers an adaptation to a good supply of seeds with a small expenditure of material, while chasmogamous flowers of the same species are usually cross-fertilised and "their offspring will thus be invigorated, as we may infer from a wide-spread analogy." ("Forms of Flowers" (1st edition), page 341.) Direct proof in support of this has hitherto been supplied in a few cases only; we shall often find that the example set by Darwin in solving such problems as these by laborious experiment has unfortunately been little imitated.

Another chapter of this book treats of the distribution of the sexes in polygamous, dioecious, and gyno-dioecious plants (the last term, now in common use, we owe to Darwin). It contains a number of important facts and discussions and has inspired the experimental researches of Correns and others.

The most important of Darwin's work on floral biology is, however, that on cross and self-fertilisation, chiefly because it states the results of experimental investigations extending over many years. Only such experiments, as we have pointed out, could determine whether cross-fertilisation is in itself beneficial, and self-fertilisation on the other hand injurious; a conclusion which a merely comparative examination of pollination-mechanisms renders in the highest degree probable. Later floral biologists have unfortunately almost entirely confined themselves to observations on floral mechanisms. But there is little more to be gained by this kind of work than an assumption long ago made by C.K. Sprengel that "very many flowers have the sexes separate and probably at least as many hermaphrodite flowers are dichogamous; it would thus appear that Nature was unwilling that any flower should be fertilised by its own pollen."

It was an accidental observation which inspired Darwin's experiments on the effect of cross and self-fertilisation. Plants of Linaria vulgaris were grown in two adjacent beds; in the one were plants produced by cross-fertilisation, that is, from seeds obtained after fertilisation by pollen of another plant of the same species; in the other grew plants produced by self-fertilisation, that is from seed produced as the result of pollination of the same flower. The first were obviously superior to the latter.

Darwin was surprised by this observation, as he had expected a prejudicial influence of self-fertilisation to manifest itself after a series of generations: "I always supposed until lately that no evil effects would be visible until after several generations of self-fertilisation, but now I see that one generation sometimes suffices and the existence of dimorphic plants and all the wonderful contrivances of orchids are quite intelligible to me." ("More Letters", Vol. II. page 373.)

The observations on Linaria and the investigations of the results of legitimate and illegitimate fertilisation in heterostyled plants were apparently the beginning of a long series of experiments. These were concerned with plants of different families and led to results which are of fundamental importance for a true explanation of sexual reproduction.

The experiments were so arranged that plants were shielded from insect-visits by a net. Some flowers were then pollinated with their own pollen, others with pollen from another plant of the same species. The seeds were germinated on moist sand; two seedlings of the same age, one from a cross and the other from a self-fertilised flower, were selected and planted on opposite sides of the same pot. They grew therefore under identical external conditions; it was thus possible to compare their peculiarities such as height, weight, fruiting capacity, etc. In other cases the seedlings were placed near to one another in the open and in this way their capacity of resisting unfavourable external conditions was tested. The experiments were in some cases continued to the tenth generation and the flowers were crossed in different ways. We see, therefore, that this book also represents an enormous amount of most careful and patient original work.

The general result obtained is that plants produced as the result of cross-fertilisation are superior, in the majority of cases, to those produced as the result of self-fertilisation, in height, resistance to external injurious influences, and in seed-production.

Ipomoea purpurea may be quoted as an example. If we express the result of cross-fertilisation by 100, we obtain the following numbers for the fertilised plants.

Generation.	Height.	Number of seeds.	1		100: 76	100: 64	2
100: 79	-	3	100: 68	100: 94	4	100: 86	100: 94	5
100: 75	100: 89	6	100: 72	-	7	100: 81	-	8
100: 85	-	9	100: 79	100: 26 (Number of capsules)	10		100:
54	-						

Taking the average, the ratio as regards growth is 100:77. The considerable superiority of the crossed plants is apparent in the first generation and is not increased in the following generations; but there is some fluctuation about the average ratio. The numbers representing the fertility of crossed and self-fertilised plants are more difficult to compare with accuracy; the superiority of the crossed plants is chiefly explained by the fact that they produce a much larger number of capsules, not because there are on the average more seeds in each capsule. The ratio of the capsules was, e.g. in the third generation, 100:38, that of the seeds in the capsules 100:94. It is also especially noteworthy that in the self-fertilised plants the anthers were smaller and contained a smaller amount of pollen, and in the eighth generation the reduced fertility showed itself in a form which is often found in hybrids, that is the first flowers were sterile. (Complete sterility was not found in any of the plants investigated by Darwin. Others appear to be more sensitive; Cluer found Zea Mais "almost sterile" after three generations of self-fertilisation. (Cf. Fruwirth, "Die Zuchtung der Landwirtschaftlichen Kulturpflanzen", Berlin, 1904, II. page 6.))

The superiority of crossed individuals is not exhibited in the same way in all plants. For example in Eschscholzia californica the crossed seedlings do not exceed the self-fertilised in height and vigour, but the crossing considerably increases the plant's capacity for flower-production, and the seedlings from such a mother-plant are more fertile.

The conception implied by the term crossing requires a closer analysis. As in the majority of plants, a large number of flowers are in bloom at the same time on one and the same plant, it follows that insects visiting the flowers often carry pollen from one flower to another of the same stock. Has this method, which is spoken of as Geitonogamy, the same influence as crossing with pollen from another plant? The results of Darwin's experiments with different plants (Ipomoea purpurea, Digitalis purpurea, Mimulus luteus, Pelargonium, Origanum) were not in complete agreement; but on the whole they pointed to the conclusion that Geitonogamy shows no superiority over self-fertilisation (Autogamy). (Similarly crossing in the case of flowers of Pelargonium zonale, which belong to plants raised from cuttings from the same parent, shows no superiority over self-fertilisation.) Darwin, however, considered it possible that this may sometimes be the case. "The sexual elements in the flowers on the same plant can rarely have been differentiated, though this is possible, as flower-buds are in one sense distinct individuals, sometimes varying and differing from one another in structure or constitution." ("Cross and Self fertilisation" (1st edition), page 444.)

As regards the importance of this question from the point of view of the significance of cross-fertilisation in general, it may be noted that later observers have definitely discovered a difference between the results of autogamy and geitonogamy. Gilley and Fruwirth found that in Brassica Napus, the length and weight of the fruits as also the total weight of the seeds in a single fruit were less in the case of autogamy than in geitonogamy. With Sinapis alba a better crop of seeds was obtained after geitonogamy, and in the Sugar Beet the average weight of a fruit in the case of a self-fertilised plant was 0.009 gr., from geitonogamy 0.012 gr., and on cross-fertilisation 0.013 gr.

On the whole, however, the results of geitonogamy show that the favourable effects of cross-fertilisation do not depend simply on the fact that the pollen of one flower is conveyed to the stigma of another. But the plants which are crossed must in some way be different. If plants of Ipomoea purpurea (and Mimulus luteus) which have been self-fertilised for seven generations and grown under the same conditions of cultivation are crossed together, the plants so crossed would not be superior to the self-fertilised; on the other hand crossing with a fresh stock at once proves very advantageous. The favourable effect of crossing is only apparent, therefore, if the parent plants are grown under different conditions or if they belong to different varieties. "It is really wonderful what an effect pollen from a distinct seedling plant, which has been exposed to different conditions of life, has on the offspring in comparison with pollen from the same flower or from a distinct individual, but which has been long subjected to the same conditions. The subject bears on the very principle of life, which seems almost to require changes in the conditions." ("More Letters", Vol. II. page 406.)

The fertility—measured by the number or weight of the seeds produced by an equal number of plants—noticed under different conditions of fertilisation may be quoted in illustration.

	On crossing with a fresh plants of the fertilisation	On crossing stock	On self-fertilisation same stock
Mimuleus luteus (First and ninth generation)	100	4	3
Eschscholzia californica (second generation)	100	45	40
Dianthus caryophyllus (third and fourth generation)	100	45	33
Petunia violacea	100	54	46

Crossing under very similar conditions shows, therefore, that the difference between the sexual cells is smaller and thus the result of crossing is only slightly superior to that given by self-fertilisation. Is, then, the favourable result of crossing with a foreign stock to be attributed to the fact that this belongs to another systematic entity or to the fact that the plants, though belonging to the same entity were exposed to different conditions? This is a point on which further researches must be taken into account, especially since the analysis of the systematic entities has been much more thorough than formerly. (In the case of garden plants, as Darwin to a large extent claimed, it is not easy to say whether two individuals really belong to the same variety, as they are usually of hybrid origin. In some instances (Petunia, Iberis) the fresh stock employed by Darwin possessed flowers differing in colour from those of the plant crossed with it.) We know that most of Linneaus's species are compound species, frequently consisting of a very large number of smaller or elementary species formerly included under the comprehensive term varieties. Hybridisation has in most cases affected our garden and cultivated plants so that they do not represent pure species but a mixture of species.

But this consideration has no essential bearing on Darwin's point of view, according to which the nature of the sexual cells is influenced by external conditions. Even individuals growing close to one another are only apparently exposed to identical conditions. Their sexual cells may therefore be differently influenced and thus give favourable results on crossing, as "the benefits which so generally follow from a cross between two plants apparently depend on the two differing somewhat in constitution or character." As a matter of fact we are familiar with a large number of cases in which the condition of the reproductive organs is influenced by external conditions. Darwin has himself demonstrated this for self-sterile plants, that is plants in which self-fertilisation produces no result. This self-sterility is affected by climatic conditions: thus in Brazil Eschscholzia californica is absolutely sterile to the pollen of its own flowers; the descendants of Brazilian plants in Darwin's cultures were partially self-fertile in one generation and in a second generation still more so. If one has any doubt in this case whether it is a question of the condition of the style and stigma, which possibly prevents the entrance of the pollen-tube or even its development, rather than that of the actual sexual cells, in other cases there is no doubt that an influence is exerted on the latter.

Janczewski (Janczewski, "Sur les antheres steriles des Groseilliers", "Bull. de l'acad. des sciences de Cracovie", June, 1908.) has recently shown that species of Ribes cultivated under unnatural conditions frequently produce a mixed (i.e. partly useless) or completely sterile pollen, precisely as happens with hybrids. There are, therefore, substantial reasons for the conclusion that conditions of life exert an influence on the sexual cells. "Thus the proposition that the benefit from cross-fertilisation depends on the plants which are crossed having been subjected during previous generations to somewhat different conditions, or to their having varied from some unknown cause as if they had been thus subjected, is securely fortified on all sides." ("Cross and Self fertilisation" (1st edition), page 444.)

We thus obtain an insight into the significance of sexuality. If an occasional and slight alteration in the conditions under which plants and animals live is beneficial (Reasons for this are given by Darwin in "Variation under Domestication" (2nd edition), Vol. II. page 127.), crossing between organisms which have been exposed to different conditions becomes still more advantageous. The entire constitution is in this way influenced from the beginning, at a time when the whole organisation is in a highly plastic state. The total life-energy, so to speak, is increased, a gain which is not produced by asexual reproduction or by the union of sexual cells of plants which have lived under the same or only slightly different conditions. All the wonderful arrangements for cross-fertilisation now appear to be useful adaptations. Darwin was, however, far from giving undue prominence to this point of view, though this has been to some extent done by others. He particularly emphasised the following consideration:—"But we should always keep in mind that two somewhat opposed ends have to be gained; the first and more important one being the production of seeds by any means, and the second, cross-fertilisation." ("Cross and Self fertilisation" (1st edition), page 371.) Just as in some orchids and cleistogamic flowers self-pollination regularly occurs, so it may also occur in other cases. Darwin showed that Pisum sativum and Lathyrus odoratus belong to plants in which self-pollination is regularly effected, and that this accounts for the constancy of certain sorts of these plants, while a variety of form is produced by crossing. Indeed among his culture plants were some which derived no benefit from crossing. Thus in the sixth self-fertilised generation of his Ipomoea cultures the "Hero" made its appearance, a form slightly exceeding its crossed companion in height; this was in the highest degree self-fertile and handed on its characteristics to both children and grandchildren. Similar forms were found in Mimulus luteus and Nicotiana (In Pisum sativum also the crossing of two individuals of the same variety produced no advantage; Darwin attributed this to the fact that the plants had for several generations been self-fertilised and in each generation cultivated under almost the same conditions. Tschermak ("Ueber kunstliche Kreuzung an Pisum sativum") afterwards recorded the same result; but he found on crossing different varieties that usually there was no superiority as regards height over the products of self-fertilisation, while Darwin found a greater height represented by the ratios 100:75 and 100:60.), types which, after self-fertilisation, have an enhanced power of seed-production and of attaining a greater height than the plants of the corresponding generation which are crossed together and self-fertilised and grown under the same conditions. "Some observations made on other plants lead me to suspect that self-fertilisation is in some respects beneficial; although the benefit thus derived is as a rule very small compared with that from a cross with a distinct plant." ("Cross and Self fertilisation", page 350.) We are as ignorant of the reason why plants behave differently when crossed and self-fertilised as we are in regard to the nature of the differentiation of the sexual cells, which determines whether a union of the sexual cells will prove favourable or unfavourable.

It is impossible to discuss the different results of cross-fertilisation; one point must, however, be emphasised, because Darwin attached considerable importance to it. It is inevitable that pollen of different kinds must reach the stigma. It was known that pollen of the same "species" is dominant over the pollen of another species, that, in other words, it is prepotent. Even if the pollen of the same species reaches the stigma rather later than that of another species, the latter does not effect fertilisation.

Darwin showed that the fertilising power of the pollen of another variety or of another individual is greater than that of the plant's own pollen. ("Cross and Self fertilisation", page 391.) This has been demonstrated in the case of Mimulus luteus (for the fixed white-flowering variety) and Iberis umbellata with pollen of another variety, and observations on cultivated plants, such as cabbage, horseradish, etc. gave similar results. It is, however, especially remarkable that pollen of another individual of the same variety may be prepotent over the plant's own pollen. This results from the superiority of plants crossed in this manner over self-fertilised plants. "Scarcely any result from my experiments has surprised me so much as this of the prepotency of pollen from a distinct individual over each plant's own pollen, as proved by the greater constitutional vigour of the crossed seedlings." (Ibid. page 397.) Similarly, in self-fertile plants the flowers of which have not been deprived of the male organs, pollen brought to the stigma by the wind or by insects from another plant effects fertilisation, even if the plant's own pollen has reached the stigma somewhat earlier.

Have the results of his experimental investigations modified the point of view from which Darwin entered on his researches, or not? In the first place the question is, whether or not the opinion expressed in the Orchid book that there is "Something injurious" connected with self-fertilisation, has been confirmed. We can, at all events, affirm that Darwin adhered in essentials to his original position; but self-fertilisation afterwards assumed a greater importance than it formerly possessed. Darwin emphasised the fact that "the difference between the self-fertilised and crossed plants raised by me cannot be attributed to the superiority of the crossed, but to the inferiority of the self-fertilised seedlings, due to the injurious effects of self-fertilisation." (Ibid. page 437.) But he had no doubt that in favourable circumstances self-fertilised plants were able to persist for several generations without crossing. An occasional crossing appears to be useful but not indispensable in all cases; its sporadic occurrence in plants in which self-pollination habitually occurs is not excluded. Self-fertilisation is for the most part relatively and not absolutely injurious and always better than no fertilisation. "Nature abhors perpetual self-fertilisation" (It is incorrect to say, as a writer has lately said, that the aphorism expressed by Darwin in 1859 and 1862, "Nature abhors perpetual self-fertilisation," is not repeated in his later works. The sentence is repeated in "Cross and Self fertilisation" (page 8), with the addition, "If the word perpetual had been omitted, the aphorism would have been false. As it stands, I believe that it is true, though perhaps rather too strongly expressed.") is, however, a pregnant expression of the fact that cross-fertilisation is exceedingly widespread and has been shown in the majority of cases to be beneficial, and that in those plants in which we find self-pollination regularly occurring cross-pollination may occasionally take place.

An attempt has been made to express in brief the main results of Darwin's work on the biology of flowers. We have seen that his object was to elucidate important general questions, particularly the question of the significance of sexual reproduction.

It remains to consider what influence his work has had on botanical science. That this influence has been very considerable, is shown by a glance at the literature on the biology of flowers published since Darwin wrote. Before the book on orchids was published there was nothing but the old and almost forgotten works of Kolreuter and Sprengel with the exception of a few scattered references. Darwin's investigations gave the first stimulus to the development of an extensive literature on floral biology. In Knuth's "Handbuch der Blutenbiologie" ("Handbook of Flower Pollination", Oxford, 1906) as many as 3792 papers on this subject are enumerated as having been published before January 1, 1904. These describe not only the different mechanisms of flowers, but deal also with a series of remarkable adaptations in the pollinating insects. As a fertilising rain quickly calls into existence the most varied assortment of plants on a barren steppe, so activity now reigns in a field which men formerly left deserted. This development of the biology of flowers is of importance not only on theoretical grounds but also from a practical point of view. The rational breeding of plants is possible only if the flower-biology of the plants in question (i.e. the question of the possibility of self-pollination, self-sterility, etc.) is accurately known. And it is also essential for plant-breeders that they should have "the power of fixing each fleeting variety of colour, if they will fertilise the flowers of the desired kind with their own pollen for half-a-dozen generations, and grow the seedlings under the same conditions." ("Cross and Self fertilisation" (1st edition), page 460.)

But the influence of Darwin on floral biology was not confined to the development of this branch of Botany. Darwin's activity in this domain has brought about (as Asa Gray correctly pointed out) the revival of teleology in Botany and Zoology. Attempts were now made to determine, not only in the case of flowers but also in vegetative organs, in what relation the form and function of organs stand to one another and to what extent their morphological characters exhibit adaptation to environment. A branch of Botany, which has since been called Ecology (not a very happy term) has been stimulated to vigorous growth by floral biology.

While the influence of the work on the biology of flowers was extraordinarily great, it could not fail to elicit opinions at variance with Darwin's conclusions. The opposition was based partly on reasons valueless as counterarguments, partly on problems which have still to be solved; to some extent also on that tendency against teleological conceptions which has recently become current. This opposing trend of thought is due to the fact that many biologists are content with teleological explanations, unsupported by proof; it is also closely connected with the fact that many authors estimate the importance of natural selection less highly than Darwin did. We may describe the objections which are based on the widespread occurrence of self-fertilisation and geitonogamy as of little importance. Darwin did not deny the occurrence of self-fertilisation, even for a long series of generations; his law states only that "Nature abhors PERPETUAL self-fertilisation." (It is impossible (as has been attempted) to express Darwin's point of view in a single sentence, such as H. Muller's statement of the "Knight-Darwin law." The conditions of life in organisms are so various and complex that laws, such as are formulated in physics and chemistry, can hardly be conceived.) An exception to this rule would therefore occur only in the case of plants in which the possibility of cross-pollination is excluded. Some of the plants with cleistogamous flowers might afford examples of such cases. We have already seen, however, that such a case has not as yet been shown to occur. Burck believed that he had found an instance in certain tropical plants (Anonaceae, Myrmecodia) of the complete exclusion of cross-fertilisation. The flowers of these plants, in which, however,—in contrast to the cleistogamous flowers—the corolla is well developed, remain closed and fruit is produced.

Loew (E. Loew, "Bemerkungen zu Burck... ", "Biolog. Centralbl." XXVI. (1906).) has shown that cases occur in which cross-fertilisation may be effected even in these "cleistopetalous" flowers: humming birds visit the permanently closed flowers of certain species of Nidularium and transport the pollen. The fact that the formation of hybrids may occur as the result of this shows that pollination may be accomplished.

The existence of plants for which self-pollination is of greater importance than it is for others is by no means contradictory to Darwin's view. Self-fertilisation is, for example, of greater importance for annuals than for perennials as without it seeds might fail to be produced. Even in the case of annual plants with small inconspicuous flowers in which self-fertilisation usually occurs, such as Senecio vulgaris, Capsella bursa-pastoris and Stellaria media, A. Bateson (Anna Bateson, "The effects of cross-fertilisation on inconspicuous flowers", "Annals of Botany", Vol. I. 1888, page 255.) found that cross-fertilisation gave a beneficial result, although only in a slight degree. If the favourable effects of sexual reproduction, according to Darwin's view, are correlated with change of environment, it is quite possible that this is of less importance in plants which die after ripening their seeds ("hapaxanthic") and which in any case constantly change their situation. Objections which are based on the proof of the prevalence of self-fertilisation are not, therefore, pertinent. At first sight another point of view, which has been more recently urged, appears to have more weight.

W. Burck (Burck, "Darwin's Kreuzeungsgesetz... ", "Biol. Centralbl". XXVIII. 1908, page 177.) has expressed the opinion that the beneficial results of cross-fertilisation demonstrated by Darwin concern only hybrid plants. These alone become weaker by self-pollination; while pure species derive no advantage from crossing and no disadvantage from self-fertilisation. It is certain that some of the plants used by Darwin were of hybrid origin. (It is questionable if this was always the case.) This is evident from his statements, which are models of clearness and precision; he says that his Ipomoea plants "were probably the offspring of a cross." ("Cross and Self fertilisation" (1st edition), page 55.) The fixed forms of this plant, such as Hero, which was produced by self-fertilisation, and a form of Mimulus with white flowers spotted with red probably resulted from splitting of the hybrids. It is true that the phenomena observed in self-pollination, e.g. in Ipomoea, agree with those which are often noticed in hybrids; Darwin himself drew attention to this.

Let us next call to mind some of the peculiarities connected with hybridisation. We know that hybrids are often characterized by their large size, rapidity of growth, earlier production of flowers, wealth of flower-production and a longer life; hybrids, if crossed with one of the two parent forms, are usually more fertile than when they are crossed together or with another hybrid. But the characters which hybrids exhibit on self-fertilisation are rather variable. The following instance may be quoted from Gartner: "There are many hybrids which retain the self-fertility of the first generation during the second and later generations, but very often in a less degree; a considerable number, however, become sterile." But the hybrid varieties may be more fertile in the second generation than in the first, and in some hybrids the fertility with their own pollen increases in the second, third, and following generations. (K.F. Gartner, "Versuche uber die Bastarderzeugung", Stuttgart, 1849, page 149.) As yet it is impossible to lay down rules of general application for the self-fertility of hybrids. That the beneficial influence of crossing with a fresh stock rests on the same ground—a union of sexual cells possessing somewhat different characters—as the fact that many hybrids are distinguished by greater luxuriance, wealth of flowers, etc. corresponds entirely with Darwin's conclusions. It seems to me to follow clearly from his investigations that there is no essential difference between cross-fertilisation and hybridisation. The heterostyled plants are normally dependent on a process corresponding to hybridisation. The view that specifically distinct species could at best produce sterile hybrids was always opposed by Darwin. But if the good results of crossing were EXCLUSIVELY dependent on the fact that we are concerned with hybrids, there must then be a demonstration of two distinct things. First, that crossing with a fresh stock belonging to the same systematic entity or to the same hybrid, but cultivated for a considerable time under different conditions, shows no superiority over self-fertilisation, and that in pure species crossing gives no better results than self-pollination. If this were the case, we should be better able to understand why in one plant crossing is advantageous while in others, such as Darwin's Hero and the forms of Mimulus and Nicotiana no advantage is gained; these would then be pure species.

But such a proof has not been supplied; the inference drawn from cleistogamous and cleistopetalous plants is not supported by evidence, and the experiments on geitonogamy and on the advantage of cross-fertilisation in species which are usually self-fertilised are opposed to this view. There are still but few researches on this point; Darwin found that in Ononis minutissima, which produces cleistogamous as well as self-fertile chasmogamous flowers, the crossed and self-fertilised capsules produced seed in the proportion of 100:65 and that the average bore the proportion 100:86. Facts previously mentioned are also applicable to this case. Further, it is certain that the self-sterility exhibited by many plants has nothing to do with hybridisation. Between self-sterility and reduced fertility as the result of self-fertilisation there is probably no fundamental difference.

It is certain that so difficult a problem as that of the significance of sexual reproduction requires much more investigation. Darwin was anything but dogmatic and always ready to alter an opinion when it was not based on definite proof: he wrote, "But the veil of secrecy is as yet far from lifted; nor will it be, until we can say why it is beneficial that the sexual elements should be differentiated to a certain extent, and why, if the differentiation be carried still further, injury follows." He has also shown us the way along which to follow up this problem; it is that of carefully planned and exact experimental research. It may be that eventually many things will be viewed in a different light, but Darwin's investigations will always form the foundation of Floral Biology on which the future may continue to build.

XXI. MENTAL FACTORS IN EVOLUTION. By C. Lloyd Morgan, LL.D., F.R.S.

In developing his conception of organic evolution Charles Darwin was of necessity brought into contact with some of the problems of mental evolution. In "The Origin of Species" he devoted a chapter to "the diversities of instinct and of the other mental faculties in animals of the same class." ("Origin of Species" (6th edition), page 205.) When he passed to the detailed consideration of "The Descent of Man", it was part of his object to show "that there is no fundamental difference between man and the higher mammals in their mental faculties." ("Descent of Man" (2nd edition 1888), Vol. I. page 99; Popular edition page 99.) "If no organic being excepting man," he said, "had possessed any mental power, or if his powers had been of a wholly different nature from those of the lower animals, then we should never have been able to convince ourselves that our high faculties had been gradually developed." (Ibid. page 99.) In his discussion of "The Expression of the Emotions" it was important for his purpose "fully to recognise that actions readily become associated with other actions and with various states of the mind." ("The Expression of the Emotions" (2nd edition), page 32.) His hypothesis of sexual selection is largely dependent upon the exercise of choice on the part of the female and her preference for "not only the more attractive but at the same time the more vigorous and victorious males." ("Descent of Man", Vol. II. page 435.) Mental processes and physiological processes were for Darwin closely correlated; and he accepted the conclusion "that the nervous system not only regulates most of the existing functions of the body, but has indirectly influenced the progressive development of various bodily structures and of certain mental qualities." (Ibid. pages 437, 438.)

Throughout his treatment, mental evolution was for Darwin incidental to and contributory to organic evolution. For specialised research in comparative and genetic psychology, as an independent field of investigation, he had neither the time nor the requisite training. None the less his writings and the spirit of his work have exercised a profound influence on this department of evolutionary thought. And, for those who follow Darwin's lead, mental evolution is still in a measure subservient to organic evolution. Mental processes are the accompaniments or concomitants of the functional activity of specially differentiated parts of the organism. They are in some way dependent on physiological and physical conditions. But though they are not physical in their nature, and though it is difficult or impossible to conceive that they are physical in their origin, they are, for Darwin and his followers, factors in the evolutionary process in its physical or organic aspect. By the physiologist within his special and well-defined universe of discourse they may be properly regarded as epiphenomena; but by the naturalist in his more catholic survey of nature they cannot be so regarded, and were not so regarded by Darwin. Intelligence has contributed to evolution of which it is in a sense a product.

The facts of observation or of inference which Darwin accepted are these: Conscious experience accompanies some of the modes of animal behaviour; it is concomitant with certain physiological processes; these processes are the outcome of development in the individual and evolution in the race; the accompanying mental processes undergo a like development. Into the subtle philosophical questions which arise out of the naive acceptance of such a creed it was not Darwin's province to enter; "I have nothing to do," he said ("Origin of Species" (6th edition), page 205.), "with the origin of the mental powers, any more than I have with that of life itself." He dealt with the natural history of organisms, including not only their structure but their modes of behaviour; with the natural history of the states of consciousness which accompany some of their actions; and with the relation of behaviour to experience. We will endeavour to follow Darwin in his modesty and candour in making no pretence to give ultimate explanations. But we must note one of the implications of this self-denying ordinance of science. Development and evolution imply continuity. For Darwin and his followers the continuity is organic through physical heredity. Apart from speculative hypothesis, legitimate enough in its proper place but here out of court, we know nothing of continuity of mental evolution as such: consciousness appears afresh in each succeeding generation. Hence it is that for those who follow Darwin's lead, mental evolution is and must ever be, within his universe of discourse, subservient to organic evolution. Only in so far as conscious experience, or its neural correlate, effects some changes in organic structure can it influence the course of heredity; and conversely only in so far as changes in organic structure are transmitted through heredity, is mental evolution rendered possible. Such is the logical outcome of Darwin's teaching.

Those who abide by the cardinal results of this teaching are bound to regard all behaviour as the expression of the functional activities of the living tissues of the organism, and all conscious experience as correlated with such activities. For the purposes of scientific treatment, mental processes are one mode of expression of the same changes of which the physiological processes accompanying behaviour are another mode of expression. This is simply accepted as a fact which others may seek to explain. The behaviour itself is the adaptive application of the energies of the organism; it is called forth by some form of presentation or stimulation brought to bear on the organism by the environment. This presentation is always an individual or personal matter. But in order that the organism may be fitted to respond to the presentation of the environment it must have undergone in some way a suitable preparation. According to the theory of evolution this preparation is primarily racial and is transmitted through heredity. Darwin's main thesis was that the method of preparation is predominantly by natural selection. Subordinate to racial preparation, and always dependent thereon, is individual or personal preparation through some kind of acquisition; of which the guidance of behaviour through individually won experience is a typical example. We here introduce the mental factor because the facts seem to justify the inference. Thus there are some modes of behaviour which are wholly and solely dependent upon inherited racial preparation; there are other modes of behaviour which are also dependent, in part at least, on individual preparation. In the former case the behaviour is adaptive on the first occurrence of the appropriate presentation; in the latter case accommodation to circumstances is only reached after a greater or less amount of acquired organic modification of structure, often accompanied (as we assume) in the higher animals by acquired experience. Logically and biologically the two classes of behaviour are clearly distinguishable: but the analysis of complex cases of behaviour where the two factors cooperate, is difficult and requires careful and critical study of life-history.

The foundations of the mental life are laid in the conscious experience that accompanies those modes of behaviour, dependent entirely on racial preparation, which may broadly be described as instinctive. In the eighth chapter of "The Origin of Species" Darwin says ("Origin of Species" (6th edition), page 205.), "I will not attempt any definition of instinct... Every one understands what is meant, when it is said that instinct impels the cuckoo to migrate and to lay her eggs in other birds' nests. An action, which we ourselves require experience to enable us to perform, when performed by an animal, more especially by a very young one, without experience, and when performed by many individuals in the same way, without their knowing for what purpose it is performed, is usually said to be instinctive." And in the summary at the close of the chapter he says ("Origin of Species" (6th edition), page 233.), "I have endeavoured briefly to show that the mental qualities of our domestic animals vary, and that the variations are inherited. Still more briefly I have attempted to show that instincts vary slightly in a state of nature. No one will dispute that instincts are of the highest importance to each animal. Therefore there is no real difficulty, under changing conditions of life, in natural selection accumulating to any extent slight modifications of instinct which are in any way useful. In many cases habit or use and disuse have probably come into play."

Into the details of Darwin's treatment there is neither space nor need to enter. There are some ambiguous passages; but it may be said that for him, as for his followers to-day, instinctive behaviour is wholly the result of racial preparation transmitted through organic heredity. For the performance of the instinctive act no individual preparation under the guidance of personal experience is necessary. It is true that Darwin quotes with approval Huber's saying that "a little dose of judgment or reason often comes into play, even with animals low in the scale of nature." (Ibid. page 205.) But we may fairly interpret his meaning to be that in behaviour, which is commonly called instinctive, some element of intelligent guidance is often combined. If this be conceded the strictly instinctive performance (or part of the performance) is the outcome of heredity and due to the direct transmission of parental or ancestral aptitudes. Hence the instinctive response as such depends entirely on how the nervous mechanism has been built up through heredity; while intelligent behaviour, or the intelligent factor in behaviour, depends also on how the nervous mechanism has been modified and moulded by use during its development and concurrently with the growth of individual experience in the customary situations of daily life. Of course it is essential to the Darwinian thesis that what Sir E. Ray Lankester has termed "educability," not less than instinct, is hereditary. But it is also essential to the understanding of this thesis that the differentiae of the hereditary factors should be clearly grasped.

For Darwin there were two modes of racial preparation, (1) natural selection, and (2) the establishment of individually acquired habit. He showed that instincts are subject to hereditary variation; he saw that instincts are also subject to modification through acquisition in the course of individual life. He believed that not only the variations but also, to some extent, the modifications are inherited. He therefore held that some instincts (the greater number) are due to natural selection but that others (less numerous) are due, or partly due, to the inheritance of acquired habits. The latter involve Lamarckian inheritance, which of late years has been the centre of so much controversy. It is noteworthy however that Darwin laid especial emphasis on the fact that many of the most typical and also the most complex instincts—those of neuter insects—do not admit of such an interpretation. "I am surprised," he says ("Origin of Species" (6th edition), page 233.), "that no one has hitherto advanced this demonstrative case of neuter insects, against the well-known doctrine of inherited habit, as advanced by Lamarck." None the less Darwin admitted this doctrine as supplementary to that which was more distinctively his own—for example in the case of the instincts of domesticated animals. Still, even in such cases, "it may be doubted," he says (Ibid. pages 210, 211.), "whether any one would have thought of training a dog to point, had not some one dog naturally shown a tendency in this line... so that habit and some degree of selection have probably concurred in civilising by inheritance our dogs." But in the interpretation of the instincts of domesticated animals, a more recently suggested hypothesis, that of organic selection (Independently suggested, on somewhat different lines, by Profs. J. Mark Baldwin, Henry F. Osborn and the writer.), may be helpful. According to this hypothesis any intelligent modification of behaviour which is subject to selection is probably coincident in direction with an inherited tendency to behave in this fashion. Hence in such behaviour there are two factors: (1) an incipient variation in the line of such behaviour, and (2) an acquired modification by which the behaviour is carried further along the same line. Under natural selection those organisms in which the two factors cooperate are likely to survive. Under artificial selection they are deliberately chosen out from among the rest.

Organic selection has been termed a compromise between the more strictly Darwinian and the Lamarckian principles of interpretation. But it is not in any sense a compromise. The principle of interpretation of that which is instinctive and hereditary is wholly Darwinian. It is true that some of the facts of observation relied upon by Lamarckians are introduced. For Lamarckians however the modifications which are admittedly factors in survival, are regarded as the parents of inherited variations; for believers in organic selection they are only the foster parents or nurses. It is because organic selection is the direct outcome of and a natural extension of Darwin's cardinal thesis that some reference to it here is justifiable. The matter may be put with the utmost brevity as follows. (1) Variations (V) occur, some of which are in the direction of increased adaptation (+), others in the direction of decreased adaptation (-). (2) Acquired modifications (M) also occur. Some of these are in the direction of increased accommodation to circumstances (+), while others are in the direction of diminished accommodation (-). Four major combinations are

(a) $+ V$ with $+ M$,   (b) $+ V$ with $- M$,   (c) $- V$ with $+ M$,   (d) $- V$ with $- M$.

Of these (d) must inevitably be eliminated while (a) are selected. The predominant survival of (a) entails the survival of the adaptive variations which are inherited. The contributory acquisitions (+M) are not inherited; but they are none the less factors in determining the survival of the coincident variations. It is surely abundantly clear that this is Darwinism and has no tincture of Lamarck's essential principle, the inheritance of acquired characters.

Whether Darwin himself would have accepted this interpretation of some at least of the evidence put forward by Lamarckians is unfortunately a matter of conjecture. The fact remains that in his interpretation of instinct and in allied questions he accepted the inheritance of individually acquired modifications of behaviour and structure.

Darwin was chiefly concerned with instinct from the biological rather than from the psychological point of view. Indeed it must be confessed that, from the latter standpoint, his conception of instinct as a "mental faculty" which "impels" an animal to the performance of certain actions, scarcely affords a satisfactory basis for genetic treatment. To carry out the spirit of Darwin's teaching it is necessary to link more closely biological and psychological evolution. The first step towards this is to interpret the phenomena of instinctive behaviour in terms of stimulation and response. It may be well to take a particular case. Swimming on the part of a duckling is, from the biological point of view, a typical example of instinctive behaviour. Gently lower a recently hatched bird into water: coordinated movements of the limbs follow in rhythmical sequence. The behaviour is new to the individual though it is no doubt closely related to that of walking, which is no less instinctive. There is a group of stimuli afforded by the "presentation" which results from partial immersion: upon this there follows as a complex response an application of the functional activities in swimming; the sequence of adaptive application on the appropriate presentation is determined by racial preparation. We know, it is true, but little of the physiological details of what takes place in the central nervous system; but in broad outline the nature of the organic mechanism and the manner of its functioning may at least be provisionally conjectured in the present state of physiological knowledge. Similarly in the case of the pecking of newly-hatched chicks; there is a visual presentation, there is probably a cooperating group of stimuli from the alimentary tract in need of food, there is an adaptive application of the activities in a definite mode of behaviour. Like data are afforded in a great number of cases of instinctive procedure, sometimes occurring very early in life, not infrequently deferred until the organism is more fully developed, but all of them dependent upon racial preparation. No doubt there is some range of variation in the behaviour, just such variation as the theory of natural selection demands. But there can be no question that the higher animals inherit a bodily organisation and a nervous system, the functional working of which gives rise to those inherited modes of behaviour which are termed instinctive.

It is to be noted that the term "instinctive" is here employed in the adjectival form as a descriptive heading under which may be grouped many and various modes of behaviour due to racial preparation. We speak of these as inherited; but in strictness what is transmitted through heredity is the complex of anatomical and physiological conditions under which, in appropriate circumstances, the organism so behaves. So far the term "instinctive" has a restricted biological connotation in terms of behaviour. But the connecting link between biological evolution and psychological evolution is to be sought,—as Darwin fully realised,—in the phenomena of instinct, broadly considered. The term "instinctive" has also a psychological connotation. What is that connotation?

Let us take the case of the swimming duckling or the pecking chick, and fix our attention on the first instinctive performance. Grant that just as there is, strictly speaking, no inherited behaviour, but only the conditions which render such behaviour under appropriate circumstances possible; so too there is no inherited experience, but only the conditions which render such experience possible; then the cerebral conditions in both cases are the same. The biological behaviour-complex, including the total stimulation and the total response with the intervening or resultant processes in the sensorium, is accompanied by an experience-complex including the initial stimulation-consciousness and resulting response-consciousness. In the experience-complex are comprised data which in psychological analysis are grouped under the headings of cognition, affective tone and conation. But the complex is probably experienced as an unanalysed whole. If then we use the term "instinctive" so as to comprise all congenital modes of behaviour which contribute to experience, we are in a position to grasp the view that the net result in consciousness constitutes what we may term the primary tissue of experience. To the development of this experience each instinctive act contributes. The nature and manner of organisation of this primary tissue of experience are dependent on inherited biological aptitudes; but they are from the outset onwards subject to secondary development dependent on acquired aptitudes. Biological values are supplemented by psychological values in terms of satisfaction or the reverse.

In our study of instinct we have to select some particular phase of animal behaviour and isolate it so far as is possible from the life of which it is a part. But the animal is a going concern, restlessly active in many ways. Many instinctive performances, as Darwin pointed out ("Origin of Species" (6th edition), page 206.), are serial in their nature. But the whole of active life is a serial and coordinated business. The particular instinctive performance is only an episode in a life-history, and every mode of behaviour is more or less closely correlated with other modes. This coordination of behaviour is accompanied by a correlation of the modes of primary experience. We may classify the instinctive modes of behaviour and their accompanying modes of instinctive experience under as many heads as may be convenient for our purposes of interpretation, and label them instincts of self-preservation, of pugnacity, of acquisition, the reproductive instincts, the parental instincts, and so forth. An instinct, in this sense of the term (for example the parental instinct), may be described as a specialised part of the primary tissue of experience differentiated in relation to some definite biological end. Under such an instinct will fall a large number of particular and often well-defined modes of behaviour, each with its own peculiar mode of experience.

It is no doubt exceedingly difficult as a matter of observation and of inference securely based thereon to distinguish what is primary from what is in part due to secondary acquisition—a fact which Darwin fully appreciated. Animals are educable in different degrees; but where they are educable they begin to profit by experience from the first. Only,

268

therefore, on the occasion of the first instinctive act of a given type can the experience gained be weighed as WHOLLY primary; all subsequent performance is liable to be in some degree, sometimes more, sometimes less, modified by the acquired disposition which the initial behaviour engenders. But the early stages of acquisition are always along the lines predetermined by instinctive differentiation. It is the task of comparative psychology to distinguish the primary tissue of experience from its secondary and acquired modifications. We cannot follow up the matter in further detail. It must here suffice to suggest that this conception of instinct as a primary form of experience lends itself better to natural history treatment than Darwin's conception of an impelling force, and that it is in line with the main trend of Darwin's thought.

In a characteristic work,—characteristic in wealth of detail, in closeness and fidelity of observation, in breadth of outlook, in candour and modesty,—Darwin dealt with "The Expression of the Emotions in Man and Animals". Sir Charles Bell in his "Anatomy of Expression" had contended that many of man's facial muscles had been specially created for the sole purpose of being instrumental in the expression of his emotions. Darwin claimed that a natural explanation, consistent with the doctrine of evolution, could in many cases be given and would in other cases be afforded by an extension of the principles he advocated. "No doubt," he said ("Expression of the Emotions", page 13. The passage is here somewhat condensed.), "as long as man and all other animals are viewed as independent creations, an effectual stop is put to our natural desire to investigate as far as possible the causes of Expression. By this doctrine, anything and everything can be equally well explained... With mankind, some expressions... can hardly be understood, except on the belief that man once existed in a much lower and animal-like condition. The community of certain expressions in distinct though allied species... is rendered somewhat more intelligible, if we believe in their descent from a common progenitor. He who admits on general grounds that the structure and habits of all animals have been gradually evolved, will look at the whole subject of Expression in a new and interesting light."

Darwin relied on three principles of explanation. "The first of these principles is, that movements which are serviceable in gratifying some desire, or in relieving some sensation, if often repeated, become so habitual that they are performed, whether or not of any service, whenever the same desire or sensation is felt, even in a very weak degree." (Ibid. page 368.) The modes of expression which fall under this head have become instinctive through the hereditary transmission of acquired habit. "As far as we can judge, only a few expressive movements are learnt by each individual; that is, were consciously and voluntarily performed during the early years of life for some definite object, or in imitation of others, and then became habitual. The far greater number of the movements of expression, and all the more important ones, are innate or inherited; and such cannot be said to depend on the will of the individual. Nevertheless, all those included under our first principle were at first voluntarily performed for a definite object,—namely, to escape some danger, to relieve some distress, or to gratify some desire." (Ibid. pages 373, 374.)

"Our second principle is that of antithesis. The habit of voluntarily performing opposite movements under opposite impulses has become firmly established in us by the practice of our whole lives. Hence, if certain actions have been regularly performed, in accordance with our first principle, under a certain frame of mind, there will be a strong and involuntary tendency to the performance of directly opposite actions, whether or not these are of any use, under the excitement of an opposite frame of mind." ("Expression of the Emotions", page 368.) This principle of antithesis has not been widely accepted. Nor is Darwin's own position easy to grasp.

"Our third principle," he says (Ibid. page 369.), "is the direct action of the excited nervous system on the body, independently of the will, and independently, in large part, of habit. Experience shows that nerve-force is generated and set free whenever the cerebro-spinal system is excited. The direction which this nerve-force follows is necessarily determined by the lines of connection between the nerve-cells, with each other and with various parts of the body."

Lack of space prevents our following up the details of Darwin's treatment of expression. Whether we accept or do not accept his three principles of explanation we must regard his work as a masterpiece of descriptive analysis, packed full of observations possessing lasting value. For a further development of the subject it is essential that the instinctive factors in expression should be more fully distinguished from those which are individually acquired—a difficult task—and that the instinctive factors should be rediscussed in the light of modern doctrines of heredity, with a view to determining whether Lamarckian inheritance, on which Darwin so largely relied, is necessary for an interpretation of the facts.

The whole subject as Darwin realised is very complex. Even the term "expression" has a certain amount of ambiguity. When the emotion is in full flood the animal fights, flees, or faints. Is this full-tide effect to be regarded as expression; or are we to restrict the term to the premonitory or residual effects—the bared canine when the fighting mood is being roused, the ruffled fur when reminiscent representations of the object inducing anger cross the mind? Broadly considered both should be included. The activity of premonitory expression as a means of communication was recognised by Darwin; he might, perhaps, have emphasised it more strongly in dealing with the lower animals. Man so largely relies on a special means of communication, that of language, that he sometimes fails to realise that for animals with their keen powers of perception, and dependent as they are on such means of communication, the more strictly biological means of expression are full of subtle suggestiveness. Many modes of expression, otherwise useless, are signs of behaviour that may be anticipated,—signs which stimulate the appropriate attitude of response. This would not, however, serve to account for the utility of the organic accompaniments—heart-affection, respiratory changes, vaso-motor effects and so forth, together with heightened muscular tone,—on all of which Darwin lays stress ("Expression of the Emotions", pages 65 ff.) under his third principle. The biological value of all this is, however, of great importance, though Darwin was hardly in a position to take it fully into account.

Having regard to the instinctive and hereditary factors of emotional expression we may ask whether Darwin's third principle does not alone suffice as an explanation. Whether we admit or reject Lamarckian inheritance it would appear that all hereditary expression must be due to pre-established connections within the central nervous system and to a transmitted provision for coordinated response under the appropriate stimulation. If this be so, Darwin's first and second principles are subordinate and ancillary to the third, an expression, so far as it is instinctive or hereditary, being "the direct result of the constitution of the nervous system."

Darwin accepted the emotions themselves as hereditary or acquired states of mind and devoted his attention to their expression. But these emotions themselves are genetic products and as such dependent on organic conditions. It remained, therefore, for psychologists who accepted evolution and sought to build on biological foundations to trace the genesis of these modes of animal and human experience. The subject has been independently developed by Professors Lange and James (Cf. William James, "Principles of Psychology", Vol. II. Chap. XXV, London, 1890.); and some modification of their view is regarded by many evolutionists as affording the best explanation of the facts. We must fix our attention on the lower emotions, such as anger or fear, and on their first occurrence in the life of the individual organism. It is a matter of observation that if a group of young birds which have been hatched in an incubator are frightened by an appropriate presentation, auditory or visual, they instinctively respond in special ways. If we speak of this response as the expression, we find that there are many factors. There are certain visible modes of behaviour, crouching at once, scattering and then crouching, remaining motionless, the braced muscles sustaining an attitude of arrest, and so forth. There are also certain visceral or organic effects, such as affections of the heart and respiration. These can be readily observed by taking the young bird in the hand. Other effects cannot be readily observed; vaso-motor changes, affections of the alimentary canal, the skin and so forth. Now the essence of the James-Lange view, as applied to these congenital effects, is that though we are justified in speaking of them as effects of the stimulation, we are not justified, without further evidence, in speaking of them as effects of the emotional state. May it not rather be that the emotion as a primary mode of experience is the concomitant of the net result of the organic situation—the initial presentation, the instinctive mode of behaviour, the visceral disturbances? According to this interpretation the primary tissue of experience of the emotional order, felt as an unanalysed complex, is generated by the stimulation of the sensorium by afferent or incoming physiological impulses from the special senses, from the organs concerned in the responsive behaviour, from the viscera and vaso-motor system.

Some psychologists, however, contend that the emotional experience is generated in the sensorium prior to, and not subsequent to, the behaviour-response and the visceral disturbances. It is a direct and not an indirect outcome of the presentation to the special senses. Be this as it may, there is a growing tendency to bring into the closest possible relation, or even to identify, instinct and emotion in their primary genesis. The central core of all such interpretations is that instinctive behaviour and experience, its emotional accompaniments, and its expression, are but different aspects of the outcome of the same organic occurrences. Such emotions are, therefore, only a distinguishable aspect of the primary tissue of experience and exhibit a like differentiation. Here again a biological foundation is laid for a psychological doctrine of the mental development of the individual.

The intimate relation between emotion as a psychological mode of experience and expression as a group of organic conditions has an important bearing on biological interpretation. The emotion, as the psychological accompaniment of orderly disturbances in the central nervous system profoundly influences behaviour and often renders it more vigorous and more effective. The utility of the emotions in the struggle for existence can, therefore, scarcely be over-estimated. Just as keenness of perception has survival-value; just as it is obviously subject to variation; just as it must be enhanced under natural selection, whether individually acquired increments are inherited or not; and just as its value lies not only in this or that special perceptive act but in its importance for life as a whole; so the vigorous effectiveness of activity has survival-value; it is subject to variation; it must be enhanced under natural selection; and its importance lies not only in particular modes of behaviour but in its value for life as a whole. If emotion and its expression as a congenital endowment are but different aspects of the same biological occurrence; and if this is a powerful supplement to vigour effectiveness and persistency of behaviour, it must on Darwin's principles be subject to natural selection.

If we include under the expression of the emotions not only the premonitory symptoms of the initial phases of the organic and mental state, not only the signs or conditions of half-tide emotion, but the full-tide manifestation of an emotion which dominates the situation, we are naturally led on to the consideration of many of the phenomena which are discussed under the head of sexual selection. The subject is difficult and complex, and it was treated by Darwin with all the strength he could summon to the task. It can only be dealt with here from a special point of view—that which may serve to illustrate the influence of certain mental factors on the course of evolution. From this point of view too much stress can scarcely be laid on the dominance of emotion during the period of courtship and pairing in the more highly organised animals. It is a period of maximum vigour, maximum activity, and, correlated with special modes of behaviour and special organic and visceral accompaniments, a period also of maximum emotional excitement. The combats of males, their dances and aerial evolutions, their elaborate behaviour and display, or the flood of song in birds, are emotional expressions which are at any rate coincident in time with sexual periodicity. From the combat of the males there follows on Darwin's principles the elimination of those which are deficient in bodily vigour, deficient in special structures, offensive or protective, which contribute to success, deficient in the emotional supplement of which persistent and whole-hearted fighting is the expression, and deficient in alertness and skill which are the outcome of the psychological development of the powers of perception. Few biologists question that we have here a mode of selection of much importance, though its influence on psychological evolution often fails to receive its due emphasis. Mr Wallace ("Darwinism", pages 282, 283, London, 1889.) regards it as "a form of natural selection"; "to it," he says, "we must impute the development of the exceptional strength, size, and activity of the male, together with the possession of special offensive and defensive weapons, and of all other characters which arise from the development of these or are correlated with them." So far there is little disagreement among the followers of Darwin—for Mr Wallace, with fine magnanimity, has always preferred to be ranked as such, notwithstanding his right, on which a smaller man would have constantly insisted, to the claim of independent originator of the doctrine of natural selection. So far with regard to sexual selection Darwin and Mr Wallace are agreed; so far and no farther. For Darwin, says Mr Wallace (Ibid. page 283.), "has extended the principle into a totally different field of action, which has none of that character of constancy and of inevitable result that attaches to natural selection, including male rivalry; for by far the larger portion of the phenomena, which he endeavours to explain by the direct action of sexual selection, can only be so explained on the hypothesis that the immediate agency is female choice or preference. It is to this that he imputes the origin of all secondary sexual characters other than weapons of offence and defence... In this extension of sexual selection to include the action of female choice or preference, and in the attempt to give to that choice such wide-reaching effects, I am unable to follow him more than a very little way."

Into the details of Mr Wallace's criticisms it is impossible to enter here. We cannot discuss either the mode of origin of the variations in structure which have rendered secondary sexual characters possible or the modes of selection other than sexual which have rendered them, within narrow limits, specifically constant. Mendelism and mutation theories may have something to say on the subject when these theories have been more fully correlated with the basal principles of selection. It is noteworthy that Mr Wallace says ("Darwinism", pages 283, 284.): "Besides the acquisition of weapons by the male for the purpose of fighting with other males, there are some other sexual characters which may have been produced by natural selection. Such are the various sounds and odours which are peculiar to the male, and which serve as a call to the female or as an indication of his presence. These are evidently a valuable addition to the means of recognition of the two sexes, and are a further indication that the pairing season has arrived; and the production, intensification, and differentiation of these sounds and odours are clearly within the power of natural selection. The same remark will apply to the peculiar calls of birds, and even to the singing of the males." Why the same remark should not apply to their colours and adornments is not obvious. What is obvious is that "means of recognition" and "indication that the pairing season has arrived" are dependent on the perceptive powers of the female who recognises and for whom the indication has meaning. The hypothesis of female preference, stripped of the aesthetic surplusage which is psychologically both unnecessary and unproven, is really only different in degree from that which Mr Wallace admits in principle when he says that it is probable that the female is pleased or excited by the display.

Let us for our present purpose leave on one side and regard as sub judice the question whether the specific details of secondary sexual characters are the outcome of female choice. For us the question is whether certain psychological accompaniments of the pairing situation have influenced the course of evolution and whether these psychological accompaniments are themselves the outcome of evolution. As a matter of observation, specially differentiated modes of behaviour, often very elaborate, frequently requiring highly developed skill, and apparently highly charged with emotional tone, are the precursors of pairing. They are generally confined to the males, whose fierce combats during the period of sexual activity are part of the emotional manifestation. It is inconceivable that they have no biological meaning; and it is difficult to conceive that they have any other biological end than to evoke in the generally more passive female the pairing impulse. They are based on instinctive foundations ingrained in the nervous constitution through natural (or may we not say sexual?) selection in virtue of their profound utility. They are called into play by a specialised presentation such as the sight or the scent of the female at, or a little in advance of, a critical period of the physiological rhythm. There is no necessity that the male should have any knowledge of the end to which his strenuous activity leads up. In presence of the female there is an elaborate application of all the energies of behaviour, just because ages of racial preparation have made him biologically and emotionally what he is—a functionally sexual male that must dance or sing or go through hereditary movements of display, when the appropriate stimulation

272

comes. Of course after the first successful courtship his future behaviour will be in some degree modified by his previous experience. No doubt during his first courtship he is gaining the primary data of a peculiarly rich experience, instinctive and emotional. But the biological foundations of the behaviour of courtship are laid in the hereditary coordinations. It would seem that in some cases, not indeed in all, but perhaps especially in those cases in which secondary sexual behaviour is most highly evolved,—correlative with the ardour of the male is a certain amount of reluctance in the female. The pairing act on her part only takes place after prolonged stimulation, for affording which the behaviour of male courtship is the requisite presentation. The most vigorous, defiant and mettlesome male is preferred just because he alone affords a contributory stimulation adequate to evoke the pairing impulse with its attendant emotional tone.

It is true that this places female preference or choice on a much lower psychological plane than Darwin in some passages seems to contemplate where, for example, he says that the female appreciates the display of the male and places to her credit a taste for the beautiful. But Darwin himself distinctly states ("Descent of Man" (2nd edition), Vol. II. pages 136, 137; (Popular edition), pages 642, 643.) that "it is not probable that she consciously deliberates; but she is most excited or attracted by the most beautiful, or melodious, or gallant males." The view here put forward, which has been developed by Prof. Groos ("The Play of Animals", page 244, London, 1898.), therefore seems to have Darwin's own sanction. The phenomena are not only biological; there are psychological elements as well. One can hardly suppose that the female is unconscious of the male's presence; the final yielding must surely be accompanied by heightened emotional tone. Whether we call it choice or not is merely a matter of definition of terms. The behaviour is in part determined by supplementary psychological values. Prof. Groos regards the coyness of females as "a most efficient means of preventing the too early and too frequent yielding to the sexual impulse." (Ibid. page 283.) Be that as it may, it is, in any case, if we grant the facts, a means through which male sexual behaviour with all its biological and psychological implications, is raised to a level otherwise perhaps unattainable by natural means, while in the female it affords opportunities for the development in the individual and evolution in the race of what we may follow Darwin in calling appreciation, if we empty this word of the aesthetic implications which have gathered round it in the mental life of man.

Regarded from this standpoint sexual selection, broadly considered, has probably been of great importance. The psychological accompaniments of the pairing situation have profoundly influenced the course of biological evolution and are themselves the outcome of that evolution.

Darwin makes only passing reference to those modes of behaviour in animals which go by the name of play. "Nothing," he says ("Descent of Man", Vol. II. page 60; (Popular edition), page 566.), "is more common than for animals to take pleasure in practising whatever instinct they follow at other times for some real good." This is one of the very numerous cases in which a hint of the master has served to stimulate research in his disciples. It was left to Prof. Groos to develop this subject on evolutionary lines and to elaborate in a masterly manner Darwin's suggestion. "The utility of play," he says ("The Play of Animals", page 76.), "is incalculable. This utility consists in the practice and exercise it affords for some of the more important duties of life,"—that is to say, for the performance of activities which will in adult life be essential to survival. He urges (Ibid. page 75.) that "the play of young animals has its origin in the fact that certain very important instincts appear at a time when the animal does not seriously need them." It is, however, questionable whether any instincts appear at a time when they are not needed. And it is questionable whether the instinctive and emotional attitude of the play-fight, to take one example, can be identified with those which accompany fighting in earnest, though no doubt they are closely related and have some common factors. It is probable that play, as preparatory behaviour, differs in biological detail (as it almost certainly does in emotional attributes) from the earnest of after-life and that it has been evolved through differentiation and integration of the primary tissue of experience, as a preparation through which certain essential modes of skill may be acquired—those animals in which the preparatory play-propensity was not inherited in due force and requisite amount being subsequently eliminated in the struggle for existence. In any case there is little question that Prof. Groos is right in basing the play-propensity on instinctive foundations. ("The Play of Animals" page 24.) None the less, as he contends, the essential biological value of play is that it is a means of training the educable nerve-tissue, of developing that part of the brain which is modified by experience and which thus acquires new characters, of elaborating the secondary tissue of experience on the predetermined lines of instinctive differentiation and thus furthering the psychological activities which are included under the comprehensive term "intelligent."

In "The Descent of Man" Darwin dealt at some length with intelligence and the higher mental faculties. ("Descent of Man" (1st edition), Chapters II, III, V; (2nd edition), Chapters III, IV, V.) His object, he says, is to show that there is no fundamental difference between man and the higher mammals in their mental faculties; that these faculties are variable and the variations tend to be inherited; and that under natural selection beneficial variations of all kinds will have been preserved and injurious ones eliminated.

Darwin was too good an observer and too honest a man to minimise the "enormous difference" between the level of mental attainment of civilised man and that reached by any animal. His contention was that the difference, great as it is, is one of degree and not of kind. He realised that, in the development of the mental faculties of man, new factors in evolution have supervened—factors which play but a subordinate and subsidiary part in animal intelligence. Intercommunication by means of language, approbation and blame, and all that arises out of reflective thought, are but foreshadowed in the mental life of animals. Still he contends that these may be explained on the doctrine of evolution. He urges (Ibid. Vol. I. pages 70, 71; (Popular edition), pages 70, 71.)" that man is variable in body and mind; and that the variations are induced, either directly or indirectly, by the same general causes, and obey the same general laws, as with the lower animals." He correlates mental development with the evolution of the brain. (Ibid. page 81.) "As the various mental faculties gradually developed themselves, the brain would almost certainly become larger. No one, I presume, doubts that the large proportion which the size of man's brain bears to his body, compared to the same proportion in the gorilla or orang, is closely connected with his higher mental powers." "With respect to the lower animals," he says ("Descent of Man" (Popular edition), page 82.), "M.E. Lartet ("Comptes Rendus des Sciences", June 1, 1868.), by comparing the crania of tertiary and recent mammals belonging to the same groups, has come to the remarkable conclusion that the brain is generally larger and the convolutions are more complex in the more recent form."

Sir E. Ray Lankester has sought to express in the simplest terms the implications of the increase in size of the cerebrum. "In what," he asks, "does the advantage of a larger cerebral mass consist?" "Man," he replies "is born with fewer ready-made tricks of the nerve-centres—these performances of an inherited nervous mechanism so often called by the ill-defined term 'instincts'—than are the monkeys or any other animal. Correlated with the absence of inherited ready-made mechanism, man has a greater capacity of developing in the course of his individual growth similar nervous mechanisms (similar to but not identical with those of 'instinct') than any other animal... The power of being educated—'educability' as we may term it—is what man possesses in excess as compared with the apes. I think we are justified in forming the hypothesis that it is this 'educability' which is the correlative of the increased size of the cerebrum." There has been natural selection of the more educable animals, for "the character which we describe as 'educability' can be transmitted, it is a congenital character. But the RESULTS of education can NOT be transmitted. In each generation they have to be acquired afresh, and with increased 'educability' they are more readily acquired and a larger variety of them... The fact is that there is no community between the mechanisms of instinct and the mechanisms of intelligence, and that the latter are later in the history of the evolution of the brain than the former and can only develop in proportion as the former become feeble and defective." ("Nature", Vol. LXI. pages 624, 625 (1900).)

In this statement we have a good example of the further development of views which Darwin foreshadowed but did not thoroughly work out. It states the biological case clearly and tersely. Plasticity of behaviour in special accommodation to special circumstances is of survival value; it depends upon acquired characters; it is correlated with increase in size and complexity of the cerebrum; under natural selection therefore the larger and more complex cerebrum as the organ of plastic behaviour has been the outcome of natural selection. We have thus the biological foundations for a further development of genetic psychology.

There are diversities of opinion, as Darwin showed, with regard to the range of instinct in man and the higher animals as contrasted with lower types. Darwin himself said ("Descent of Man", Vol. I. page 100.) that "Man, perhaps, has somewhat fewer instincts than those possessed by the animals which come next to him in the series." On the other hand, Prof. Wm. James says ("Principles of Psychology," Vol. II. page 289.) that man is probably the animal with most instincts. The true position is that man and the higher animals have fewer complete and self-sufficing instincts than those which stand lower in the scale of mental evolution, but that they have an equally large or perhaps larger mass of instinctive raw material which may furnish the stuff to be elaborated by intelligent processes. There is, perhaps, a greater abundance of the primary tissue of experience to be refashioned and integrated by secondary modification; there is probably the same differentiation in relation to the determining biological ends, but there is at the outset less differentiation of the particular and specific modes of behaviour. The specialised instinctive performances and their concomitant experience-complexes are at the outset more indefinite. Only through acquired connections, correlated with experience, do they become definitely organised.

The full working-out of the delicate and subtle relationship of instinct and educability—that is, of the hereditary and the acquired factors in the mental life—is the task which lies before genetic and comparative psychology. They interact throughout the whole of life, and their interactions are very complex. No one can read the chapters of "The Descent of Man" which Darwin devotes to a consideration of the mental characters of man and animals without noticing, on the one hand, how sedulous he is in his search for hereditary foundations, and, on the other hand, how fully he realises the importance of acquired habits of mind. The fact that educability itself has innate tendencies—is in fact a partially differentiated educability—renders the unravelling of the factors of mental progress all the more difficult.

In his comparison of the mental powers of men and animals it was essential that Darwin should lay stress on points of similarity rather than on points of difference. Seeking to establish a doctrine of evolution, with its basal concept of continuity of process and community of character, he was bound to render clear and to emphasise the contention that the difference in mind between man and the higher animals, great as it is, is one of degree and not of kind. To this end Darwin not only recorded a large number of valuable observations of his own, and collected a considerable body of information from reliable sources, he presented the whole subject in a new light and showed that a natural history of mind might be written and that this method of study offered a wide and rich field for investigation. Of course those who regarded the study of mind only as a branch of metaphysics smiled at the philosophical ineptitude of the mere man of science. But the investigation, on natural history lines, has been prosecuted with a large measure of success. Much indeed still remains to be done; for special training is required, and the workers are still few. Promise for the future is however afforded by the fact that investigation is prosecuted on experimental lines and that something like organised methods of research are taking form. There is now but little reliance on casual observations recorded by those who have not undergone the necessary discipline in these methods. There is also some change of emphasis in formulating conclusions. Now that the general evolutionary thesis is fully and freely accepted by those who carry on such researches, more stress is laid on the differentiation of the stages of evolutionary advance than on the fact of their underlying community of nature. The conceptual intelligence which is especially characteristic of the higher mental procedure of man is more firmly distinguished from the perceptual intelligence which he shares with the lower animals—distinguished now as a higher product of evolution, no longer as differing in origin or different in kind. Some progress has been made, on the one hand in rendering an account of intelligent profiting by experience under the guidance of pleasure and pain in the perceptual field, on lines predetermined by instinctive differentiation for biological ends, and on the other hand in elucidating the method of conceptual thought employed, for example, by the investigator himself in interpreting the perceptual experience of the lower animals.

Thus there is a growing tendency to realise more fully that there are two orders of educability—first an educability of the perceptual intelligence based on the biological foundation of instinct, and secondly an educability of the conceptual intelligence which refashions and rearranges the data afforded by previous inheritance and acquisition. It is in relation to this second and higher order of educability that the cerebrum of man shows so large an increase of mass and a yet larger increase of effective surface through its rich convolutions. It is through educability of this order that the human child is brought intellectually and affectively into touch with the ideal constructions by means of which man has endeavoured, with more or less success, to reach an interpretation of nature, and to guide the course of the further evolution of his race—ideal constructions which form part of man's environment.

It formed no part of Darwin's purpose to consider, save in broad outline, the methods, or to discuss in any fulness of detail the results of the process by which a differentiation of the mental faculties of man from those of the lower animals has been brought about—a differentiation the existence of which he again and again acknowledges. His purpose was rather to show that, notwithstanding this differentiation, there is basal community in kind. This must be remembered in considering his treatment of the biological foundations on which man's systems of ethics are built. He definitely stated that he approached the subject "exclusively from the side of natural history." ("Descent of Man", Vol. I. page 149.) His general conclusion is that the moral sense is fundamentally identical with the social instincts, which have been developed for the good of the community; and he suggests that the concept which thus enables us to interpret the biological ground-plan of morals also enables us to frame a rational ideal of the moral end. "As the social instincts," he says (Ibid. page 185.), "both of man and the lower animals have no doubt been developed by nearly the same steps, it would be advisable, if found practicable, to use the same definition in both cases, and to take as the standard of morality, the general good or welfare of the community, rather than the general happiness." But the kind of community for the good of which the social instincts of animals and primitive men were biologically developed may be different from that which is the product of civilisation, as Darwin no doubt realised. Darwin's contention was that conscience is a social instinct and has been evolved because it is useful to the tribe in the struggle for existence against other tribes. On the other hand, J.S. Mill urged that the moral feelings are not innate but acquired, and Bain held the same view, believing that the moral sense is acquired by each individual during his life-time. Darwin, who notes (Ibid. page 150 (footnote).) their opinion with his usual candour, adds that "on the general theory of evolution this is at least extremely improbable. It is impossible to enter into the question here: much turns on the exact connotation of the terms "conscience" and "moral sense," and on the meaning we attach to the statement that the moral sense is fundamentally identical with the social instincts."

Presumably the majority of those who approach the subjects discussed in the third, fourth and fifth chapters of "The Descent of Man" in the full conviction that mental phenomena, not less than organic phenomena, have a natural genesis, would, without hesitation, admit that the intellectual and moral systems of civilised man are ideal constructions, the products of conceptual thought, and that as such they are, in their developed form, acquired. The moral sentiments are the emotional analogues of highly developed concepts. This does not however imply that they are outside the range of natural history treatment. Even though it may be desirable to differentiate the moral conduct of men from the social behaviour of animals (to which some such term as "pre-moral" or "quasi-moral" may be applied), still the fact remains that, as Darwin showed, there is abundant evidence of the occurrence of such social behaviour—social behaviour which, even granted that it is in large part intelligently acquired, and is itself so far a product of educability, is of survival value. It makes for that integration without which no social group could hold together and escape elimination. Furthermore, even if we grant that such behaviour is intelligently acquired, that is to say arises through the modification of hereditary instincts and emotions, the fact remains that only through these instinctive and emotional data is afforded the primary tissue of the experience which is susceptible of such modification.

Darwin sought to show, and succeeded in showing, that for the intellectual and moral life there are instinctive foundations which a biological treatment alone can disclose. It is true that he did not in all cases analytically distinguish the foundations from the superstructure. Even to-day we are scarcely in a position to do so adequately. But his treatment was of great value in giving an impetus to further research. This value indeed can scarcely be overestimated. And when the natural history of the mental operations shall have been written, the cardinal fact will stand forth, that the instinctive and emotional foundations are the outcome of biological evolution and have been ingrained in the race through natural selection. We shall more clearly realise that educability itself is a product of natural selection, though the specific results acquired through cerebral modifications are not transmitted through heredity. It will, perhaps, also be realised that the instinctive foundations of social behaviour are, for us, somewhat out of date and have undergone but little change throughout the progress of civilisation, because natural selection has long since ceased to be the dominant factor in human progress. The history of human progress has been mainly the history of man's higher educability, the products of which he has projected on to his environment. This educability remains on the average what it was a dozen generations ago; but the thought-woven tapestry of his surroundings is refashioned and improved by each succeeding generation. Few men have in greater measure enriched the thought-environment with which it is the aim of education to bring educable human beings into vital contact, than has Charles Darwin. His special field of work was the wide province of biology; but he did much to help us realise that mental factors have contributed to organic evolution and that in man, the highest product of Evolution, they have reached a position of unquestioned supremacy.

XXII. THE INFLUENCE OF THE CONCEPTION OF EVOLUTION ON MODERN PHILOSOPHY. By H. Hoffding.

Professor of Philosophy in the University of Copenhagen.

I.

It is difficult to draw a sharp line between philosophy and natural science. The naturalist who introduces a new principle, or demonstrates a fact which throws a new light on existence, not only renders an important service to philosophy but is himself a philosopher in the broader sense of the word. The aim of philosophy in the stricter sense is to attain points of view from which the fundamental phenomena and the principles of the special sciences can be seen in their relative importance and connection. But philosophy in this stricter sense has always been influenced by philosophy in the broader sense. Greek philosophy came under the influence of logic and mathematics, modern philosophy under the influence of natural science. The name of Charles Darwin stands with those of Galileo,

Newton, and Robert Mayer—names which denote new problems and great alterations in our conception of the universe.

First of all we must lay stress on Darwin's own personality. His deep love of truth, his indefatigable inquiry, his wide horizon, and his steady self-criticism make him a scientific model, even if his results and theories should eventually come to possess mainly an historical interest. In the intellectual domain the primary object is to reach high summits from which wide surveys are possible, to reach them toiling honestly upwards by way of experience, and then not to turn dizzy when a summit is gained. Darwinians have sometimes turned dizzy, but Darwin never. He saw from the first the great importance of his hypothesis, not only because of its solution of the old problem as to the value of the concept of species, not only because of the grand picture of natural evolution which it unrolls, but also because of the life and inspiration its method would impart to the study of comparative anatomy, of instinct and of heredity, and finally because of the influence it would exert on the whole conception of existence. He wrote in his note-book in the year 1837: "My theory would give zest to recent and fossil comparative anatomy; it would lead to the study of instinct, heredity, and mind-heredity, whole (of) metaphysics." ("Life and Letters of Charles Darwin", Vol. I. page 8.)

We can distinguish four main points in which Darwin's investigations possess philosophical importance.

The evolution hypothesis is much older than Darwin; it is, indeed, one of the oldest guessings of human thought. In the eighteenth century it was put forward by Diderot and Lamettrie and suggested by Kant (1786). As we shall see later, it was held also by several philosophers in the first half of the nineteenth century. In his preface to "The Origin of Species", Darwin mentions the naturalists who were his forerunners. But he has set forth the hypothesis of evolution in so energetic and thorough a manner that it perforce attracts the attention of all thoughtful men in a much higher degree than it did before the publication of the "Origin".

And further, the importance of his teaching rests on the fact that he, much more than his predecessors, even than Lamarck, sought a foundation for his hypothesis in definite facts. Modern science began by demanding—with Kepler and Newton—evidence of verae causae; this demand Darwin industriously set himself to satisfy—hence the wealth of material which he collected by his observations and his experiments. He not only revived an old hypothesis, but he saw the necessity of verifying it by facts. Whether the special cause on which he founded the explanation of the origin of species—Natural Selection—is sufficient, is now a subject of discussion. He himself had some doubt in regard to this question, and the criticisms which are directed against his hypothesis hit Darwinism rather than Darwin. In his indefatigable search for empirical evidence he is a model even for his antagonists: he has compelled them to approach the problems of life along other lines than those which were formerly followed.

Whether the special cause to which Darwin appealed is sufficient or not, at least to it is probably due the greater part of the influence which he has exerted on the general trend of thought. "Struggle for existence" and "natural selection" are principles which have been applied, more or less, in every department of thought. Recent research, it is true, has discovered greater empirical discontinuity—leaps, "mutations"—whereas Darwin believed in the importance of small variations slowly accumulated. It has also been shown by the experimental method, which in recent biological work has succeeded Darwin's more historical method, that types once constituted possess great permanence, the fluctuations being restricted within clearly defined boundaries. The problem has become more precise, both as to variation and as to heredity. The inner conditions of life have in both respects shown a greater independence than Darwin had supposed in his theory, though he always admitted that the cause of variation was to him a great enigma, "a most perplexing problem," and that the struggle for life could only occur where variation existed. But, at any rate, it was of the greatest importance that Darwin gave a living impression of the struggle for life which is everywhere going on, and to which even the highest forms of existence must be amenable.

The philosophical importance of these ideas does not stand or fall with the answer to the question, whether natural selection is a sufficient explanation of the origin of species or not: it has an independent, positive value for everyone who will observe life and reality with an unbiassed mind.

In accentuating the struggle for life Darwin stands as a characteristically English thinker: he continues a train of ideas which Hobbes and Malthus had already begun. Moreover in his critical views as to the conception of species he had English forerunners; in the middle ages Occam and Duns Scotus, in the eighteenth century Berkeley and Hume. In his moral philosophy, as we shall see later, he is an adherent of the school which is represented by Hutcheson, Hume and Adam Smith. Because he is no philosopher in the stricter sense of the term, it is of great interest to see that his attitude of mind is that of the great thinkers of his nation.

In considering Darwin's influence on philosophy we will begin with an examination of the attitude of philosophy to the conception of evolution at the time when "The Origin of Species" appeared. We will then examine the effects which the theory of evolution, and especially the idea of the struggle for life, has had, and naturally must have, on the discussion of philosophical problems.

II.

When "The Origin of Species" appeared fifty years ago Romantic speculation, Schelling's and Hegel's philosophy, still reigned on the continent, while in England Positivism, the philosophy of Comte and Stuart Mill, represented the most important trend of thought. German speculation had much to say on evolution, it even pretended to be a philosophy of evolution. But then the word "evolution" was to be taken in an ideal, not in a real, sense. To speculative thought the forms and types of nature formed a system of ideas, within which any form could lead us by continuous transitions to any other. It was a classificatory system which was regarded as a divine world of thought or images, within which metamorphoses could go on—a condition comparable with that in the mind of the poet when one image follows another with imperceptible changes. Goethe's ideas of evolution, as expressed in his "Metamorphosen der Pflanzen und der Thiere", belong to this category; it is, therefore, incorrect to call him a forerunner of Darwin. Schelling and Hegel held the same idea; Hegel expressly rejected the conception of a real evolution in time as coarse and materialistic. "Nature," he says, "is to be considered as a SYSTEM OF STAGES, the one necessarily arising from the other, and being the nearest truth of that from which it proceeds; but not in such a way that the one is NATURALLY generated by the other; on the contrary (their connection lies) in the inner idea which is the ground of nature. The METAMORPHOSIS can be ascribed only to the notion as such, because it alone is evolution... It has been a clumsy idea in the older as well as in the newer philosophy of nature, to regard the transformation and the transition from one natural form and sphere to a higher as an outward and actual production." ("Encyclopaedie der philosophischen Wissenschaften" (4th edition), Berlin, 1845, paragraph 249.)

The only one of the philosophers of Romanticism who believed in a real, historical evolution, a real production of new species, was Oken. ("Lehrbuch der Naturphilosophie", Jena, 1809.) Danish philosophers, such as Treschow (1812) and Sibbern (1846), have also broached the idea of an historical evolution of all living beings from the lowest to the highest. Schopenhauer's philosophy has a more realistic character than that of Schelling's and Hegel's, his diametrical opposites, though he also belongs to the romantic school of thought. His philosophical and psychological views were greatly influenced by French naturalists and philosophers, especially by Cabanis and Lamarck. He praises the "ever memorable Lamarck," because he laid so much stress on the "will to live." But he repudiates as a "wonderful error" the idea that the organs of animals should have reached their present perfection through a development in time, during the course of innumerable generations. It was, he said, a consequence of the low standard of contemporary French philosophy, that Lamarck came to the idea of the construction of living beings in time through succession! ("Ueber den Willen in der Natur" (2nd edition), Frankfurt a. M., 1854, pages 41-43.)

279

The positivistic stream of thought was not more in favour of a real evolution than was the Romantic school. Its aim was to adhere to positive facts: it looked with suspicion on far-reaching speculation. Comte laid great stress on the discontinuity found between the different kingdoms of nature, as well as within each single kingdom. As he regarded as unscientific every attempt to reduce the number of physical forces, so he rejected entirely the hypothesis of Lamarck concerning the evolution of species; the idea of species would in his eyes absolutely lose its importance if a transition from species to species under the influence of conditions of life were admitted. His disciples (Littre, Robin) continued to direct against Darwin the polemics which their master had employed against Lamarck. Stuart Mill, who, in the theory of knowledge, represented the empirical or positivistic movement in philosophy—like his English forerunners from Locke to Hume—founded his theory of knowledge and morals on the experience of the single individual. He sympathised with the theory of the original likeness of all individuals and derived their differences, on which he practically and theoretically laid much stress, from the influence both of experience and education, and, generally, of physical and social causes. He admitted an individual evolution, and, in the human species, an evolution based on social progress; but no physiological evolution of species. He was afraid that the hypothesis of heredity would carry us back to the old theory of "innate" ideas.

Darwin was more empirical than Comte and Mill; experience disclosed to him a deeper continuity than they could find; closer than before the nature and fate of the single individual were shown to be interwoven in the great web binding the life of the species with nature as a whole. And the continuity which so many idealistic philosophers could find only in the world of thought, he showed to be present in the world of reality.

III.

Darwin's energetic renewal of the old idea of evolution had its chief importance in strengthening the conviction of this real continuity in the world, of continuity in the series of form and events. It was a great support for all those who were prepared to base their conception of life on scientific grounds. Together with the recently discovered law of the conservation of energy, it helped to produce the great realistic movement which characterises the last third of the nineteenth century. After the decline of the Romantic movement people wished to have firmer ground under their feet and reality now asserted itself in a more emphatic manner than in the period of Romanticism. It was easy for Hegel to proclaim that "the real" was "the rational," and that "the rational" was "the real": reality itself existed for him only in the interpretation of ideal reason, and if there was anything which could not be merged in the higher unity of thought, then it was only an example of the "impotence of nature to hold to the idea." But now concepts are to be founded on nature and not on any system of categories too confidently deduced a priori. The new devotion to nature had its recompense in itself, because the new points of view made us see that nature could indeed "hold to ideas," though perhaps not to those which we had cogitated beforehand.

A most important question for philosophers to answer was whether the new views were compatible with an idealistic conception of life and existence. Some proclaimed that we have now no need of any philosophy beyond the principles of the conservation of matter and energy and the principle of natural evolution: existence should and could be definitely and completely explained by the laws of material nature. But abler thinkers saw that the thing was not so simple. They were prepared to give the new views their just place and to examine what alterations the old views must undergo in order to be brought into harmony with the new data.

The realistic character of Darwin's theory was shown not only in the idea of natural continuity, but also, and not least, in the idea of the cause whereby organic life advances step by step. This idea—the idea of the struggle for life—implied that nothing could persist, if it had no power to maintain itself under the given conditions. Inner value alone does not decide. Idealism was here put to its hardest trial. In continuous evolution it could perhaps still find an analogy to the inner evolution of ideas in the mind; but in the demand for power

280

in order to struggle with outward conditions Realism seemed to announce itself in its most brutal form. Every form of Idealism had to ask itself seriously how it was going to "struggle for life" with this new Realism.

We will now give a short account of the position which leading thinkers in different countries have taken up in regard to this question.

I. Herbert Spencer was the philosopher whose mind was best prepared by his own previous thinking to admit the theory of Darwin to a place in his conception of the world. His criticism of the arguments which had been put forward against the hypothesis of Lamarck, showed that Spencer, as a young man, was an adherent to the evolution idea. In his "Social Statics" (1850) he applied this idea to human life and moral civilisation. In 1852 he wrote an essay on "The Development Hypothesis", in which he definitely stated his belief that the differentiation of species, like the differentiation within a single organism, was the result of development. In the first edition of his "Psychology" (1855) he took a step which put him in opposition to the older English school (from Locke to Mill): he acknowledged "innate ideas" so far as to admit the tendency of acquired habits to be inherited in the course of generations, so that the nature and functions of the individual are only to be understood through its connection with the life of the species. In 1857, in his essay on "Progress", he propounded the law of differentiation as a general law of evolution, verified by examples from all regions of experience, the evolution of species being only one of these examples. On the effect which the appearance of "The Origin of Species" had on his mind he writes in his "Autobiography": "Up to that time... I held that the sole cause of organic evolution is the inheritance of functionally-produced modifications. The "Origin of Species" made it clear to me that I was wrong, and that the larger part of the facts cannot be due to any such cause... To have the theory of organic evolution justified was of course to get further support for that theory of evolution at large with which... all my conceptions were bound up." (Spencer, "Autobiography", Vol. II. page 50, London, 1904.) Instead of the metaphorical expression "natural selection," Spencer introduced the term "survival of the fittest," which found favour with Darwin as well as with Wallace.

In working out his ideas of evolution, Spencer found that differentiation was not the only form of evolution. In its simplest form evolution is mainly a concentration, previously scattered elements being integrated and losing independent movement. Differentiation is only forthcoming when minor wholes arise within a greater whole. And the highest form of evolution is reached when there is a harmony between concentration and differentiation, a harmony which Spencer calls equilibration and which he defines as a moving equilibrium. At the same time this definition enables him to illustrate the expression "survival of the fittest." "Every living organism exhibits such a moving equilibrium—a balanced set of functions constituting its life; and the overthrow of this balanced set of functions or moving equilibrium is what we call death. Some individuals in a species are so constituted that their moving equilibria are less easily overthrown than those of other individuals; and these are the fittest which survive, or, in Mr Darwin's language, they are the select which nature preserves." (Ibid. page 100.) Not only in the domain of organic life, but in all domains, the summit of evolution is, according to Spencer, characterised by such a harmony—by a moving equilibrium.

Spencer's analysis of the concept of evolution, based on a great variety of examples, has made this concept clearer and more definite than before. It contains the three elements; integration, differentiation and equilibration. It is true that a concept which is to be valid for all domains of experience must have an abstract character, and between the several domains there is, strictly speaking, only a relation of analogy. So there is only analogy between psychical and physical evolution. But this is no serious objection, because general concepts do not express more than analogies between the phenomena which they represent. Spencer takes his leading terms from the material world in defining evolution (in the simplest form) as integration of matter and dissipation of movement; but as he—not always quite consistently (Cf. my letter to him, 1876, now printed in Duncan's "Life and Letters of Herbert Spencer", page 178, London, 1908.)—assumed a correspondence of mind and matter, he could very well give these terms an indirect importance for psychical evolution. Spencer has always, in my opinion with full right, repudiated the ascription of materialism. He is no more a materialist than Spinoza. In his "Principles of Psychology" (paragraph 63) he expressed himself very clearly: "Though it seems easier to translate so-called matter into so-called spirit, than to translate so-called spirit into so-called matter—which latter is indeed wholly impossible—yet no translation can carry us beyond our symbols." These words lead us naturally to a group of thinkers whose starting-point was psychical evolution. But we have still one aspect of Spencer's philosophy to mention.

Spencer founded his "laws of evolution" on an inductive basis, but he was convinced that they could be deduced from the law of the conservation of energy. Such a deduction is, perhaps, possible for the more elementary forms of evolution, integration and differentiation; but it is not possible for the highest form, the equilibration, which is a harmony of integration and differentiation. Spencer can no more deduce the necessity for the eventual appearance of "moving equilibria" of harmonious totalities than Hegel could guarantee the "higher unities" in which all contradictions should be reconciled. In Spencer's hands the theory of evolution acquired a more decidedly optimistic character than in Darwin's; but I shall deal later with the relation of Darwin's hypothesis to the opposition of optimism and pessimism.

II. While the starting-point of Spencer was biological or cosmological, psychical evolution being conceived as in analogy with physical, a group of eminent thinkers—in Germany Wundt, in France Fouillee, in Italy Ardigo—took, each in his own manner, their starting-point in psychical evolution as an original fact and as a type of all evolution, the hypothesis of Darwin coming in as a corroboration and as a special example. They maintain the continuity of evolution; they find this character most prominent in psychical evolution, and this is for them a motive to demand a corresponding continuity in the material, especially in the organic domain.

To Wundt and Fouillee the concept of will is prominent. They see the type of all evolution in the transformation of the life of will from blind impulse to conscious choice; the theories of Lamarck and Darwin are used to support the view that there is in nature a tendency to evolution in steady reciprocity with external conditions. The struggle for life is here only a secondary fact. Its apparent prominence is explained by the circumstance that the influence of external conditions is easily made out, while inner conditions can be verified only through their effects. For Ardigo the evolution of thought was the starting-point and the type: in the evolution of a scientific hypothesis we see a progress from the indefinite (indistinto) to the definite (distinto), and this is a characteristic of all evolution, as Ardigo has pointed out in a series of works. The opposition between indistinto and distinto corresponds to Spencer's opposition between homogeneity and heterogeneity. The hypothesis of the origin of differences of species from more simple forms is a special example of the general law of evolution.

In the views of Wundt and Fouillee we find the fundamental idea of idealism: psychical phenomena as expressions of the innermost nature of existence. They differ from the older Idealism in the great stress which they lay on evolution as a real, historical process which is going on through steady conflict with external conditions. The Romantic dread of reality is broken. It is beyond doubt that Darwin's emphasis on the struggle for life as a necessary condition of evolution has been a very important factor in carrying philosophy back to reality from the heaven of pure ideas. The philosophy of Ardigo, on the other side, appears more as a continuation and deepening of positivism, though the Italian thinker arrived at his point of view independently of French-English positivism. The idea of continuous evolution is here maintained in opposition to Comte's and Mill's philosophy of discontinuity. From Wundt and Fouillee Ardigo differs in conceiving psychical evolution not as an immediate revelation of the innermost nature of existence, but only as a single, though the most accessible example, of evolution.

III. To the French philosophers Boutroux and Bergson, evolution proper is continuous and qualitative, while outer experience and physical science give us fragments only, sporadic processes and mechanical combinations. To Bergson, in his recent work "L'Evolution Creatrice", evolution consists in an elan de vie which to our fragmentary observation and analytic reflexion appears as broken into a manifold of elements and processes. The concept of matter in its scientific form is the result of this breaking asunder, essential for all scientific reflexion. In these conceptions the strongest opposition between inner and outer conditions of evolution is expressed: in the domain of internal conditions spontaneous development of qualitative forms—in the domain of external conditions discontinuity and mechanical combination.

We see, then, that the theory of evolution has influenced philosophy in a variety of forms. It has made idealistic thinkers revise their relation to the real world; it has led positivistic thinkers to find a closer connection between the facts on which they based their views; it has made us all open our eyes for new possibilities to arise through the prima facie inexplicable "spontaneous" variations which are the condition of all evolution. This last point is one of peculiar interest. Deeper than speculative philosophy and mechanical science saw in the days of their triumph, we catch sight of new streams, whose sources and laws we have still to discover. Most sharply does this appear in the theory of mutation, which is only a stronger accentuation of a main point in Darwinism. It is interesting to see that an analogous problem comes into the foreground in physics through the discovery of radioactive phenomena, and in psychology through the assumption of psychical new formations (as held by Boutroux, William James and Bergson). From this side, Darwin's ideas, as well as the analogous ideas in other domains, incite us to renewed examination of our first principles, their rationality and their value. On the other hand, his theory of the struggle for existence challenges us to examine the conditions and discuss the outlook as to the persistence of human life and society and of the values that belong to them. It is not enough to hope (or fear?) the rising of new forms; we have also to investigate the possibility of upholding the forms and ideals which have hitherto been the bases of human life. Darwin has here given his age the most earnest and most impressive lesson. This side of Darwin's theory is of peculiar interest to some special philosophical problems to which I now pass.

IV.

Among philosophical problems the problem of knowledge has in the last century occupied a foremost place. It is natural, then, to ask how Darwin and the hypothesis whose most eminent representative he is, stand to this problem.

Darwin started an hypothesis. But every hypothesis is won by inference from certain presuppositions, and every inference is based on the general principles of human thought. The evolution hypothesis presupposes, then, human thought and its principles. And not only the abstract logical principles are thus presupposed. The evolution hypothesis purports to be not only a formal arrangement of phenomena, but to express also the law of a real process. It supposes, then, that the real data—all that in our knowledge which we do not produce ourselves, but which we in the main simply receive—are subjected to laws which are at least analogous to the logical relations of our thoughts; in other words, it assumes the validity of the principle of causality. If organic species could arise without cause there would be no use in framing hypotheses. Only if we assume the principle of causality, is there a problem to solve.

Though Darwinism has had a great influence on philosophy considered as a striving after a scientific view of the world, yet here is a point of view—the epistemological—where philosophy is not only independent but reaches beyond any result of natural science. Perhaps it will be said: the powers and functions of organic beings only persist (perhaps also only arise) when they correspond sufficiently to the conditions under which the struggle of life is to go on. Human thought itself is, then, a variation (or a mutation) which has been able to persist and to survive. Is not, then, the problem of knowledge solved by the evolution hypothesis? Spencer had given an affirmative answer to this question before the appearance of "The Origin of Species". For the individual, he said, there is an a priori, original, basis (or Anlage) for all mental life; but in the species all powers have developed in reciprocity with external conditions. Knowledge is here considered from the practical point of view, as a weapon in the struggle for life, as an "organon" which has been continuously in use for generations. In recent years the economic or pragmatic epistemology, as developed by Avenarius and Mach in Germany, and by James in America, points in the same direction. Science, it is said, only maintains those principles and presuppositions which are necessary to the simplest and clearest orientation in the world of experience. All assumptions which cannot be applied to experience and to practical work, will successively be eliminated.

In these views a striking and important application is made of the idea of struggle for life to the development of human thought. Thought must, as all other things in the world, struggle for life. But this whole consideration belongs to psychology, not to the theory of knowledge (epistemology), which is concerned only with the validity of knowledge, not with its historical origin. Every hypothesis to explain the origin of knowledge must submit to cross-examination by the theory of knowledge, because it works with the fundamental forms and principles of human thought. We cannot go further back than these forms and principles, which it is the aim of epistemology to ascertain and for which no further reason can be given. (The present writer, many years ago, in his "Psychology" (Copenhagen, 1882; English translation London, 1891), criticised the evolutionistic treatment of the problem of knowledge from the Kantian point of view.)

But there is another side of the problem which is, perhaps, of more importance and which epistemology generally overlooks. If new variations can arise, not only in organic but perhaps also in inorganic nature, new tasks are placed before the human mind. The question is, then, if it has forms in which there is room for the new matter? We are here touching a possibility which the great master of epistemology did not bring to light. Kant supposed confidently that no other matter of knowledge could stream forth from the dark source which he called "the thing-in-itself," than such as could be synthesised in our existing forms of knowledge. He mentions the possibility of other forms than the human, and warns us against the dogmatic assumption that the human conception of existence should be absolutely adequate. But he seems to be quite sure that the thing-in-itself works constantly, and consequently always gives us only what our powers can master. This assumption was a consequence of Kant's rationalistic tendency, but one for which no warrant can be given. Evolutionism and systematism are opposing tendencies which can never be absolutely harmonised one with the other. Evolution may at any time break some form which the system-monger regards as finally established. Darwin himself felt a great difference in looking at variation as an evolutionist and as a systematist. When he was working at his evolution theory, he was very glad to find variations; but they were a hindrance to him when he worked as a systematist, in preparing his work on Cirripedia. He says in a letter: "I had thought the same parts of the same species more resemble (than they do anyhow in Cirripedia) objects cast in the same mould. Systematic work would be easy were it not for this confounded variation, which, however, is pleasant to me as a speculatist, though odious to me as a systematist." ("Life and Letters", Vol. II. page 37.) He could indeed be angry with variations even as an evolutionist; but then only because he could not explain them, not because he could not classify them. "If, as I must think, external conditions produce little DIRECT effect, what the devil determines each particular variation?" (Ibid. page 232.) What Darwin experienced in his particular domain holds good of all knowledge. All knowledge is systematic, in so far as it strives to put phenomena in quite definite relations, one to another. But the systematisation can never be complete. And here Darwin has contributed much to widen the world for us. He has shown us forces and tendencies in nature which make absolute systems impossible, at the same time that they give us new objects and problems. There is still a place for what Lessing called "the unceasing striving after truth," while "absolute truth" (in the sense of a closed system) is unattainable so long as life and experience are going on.

There is here a special remark to be made. As we have seen above, recent research has shown that natural selection or struggle for life is no explanation of variations. Hugo de Vries distinguishes between partial and embryonal variations, or between variations and mutations, only the last-named being heritable, and therefore of importance for the origin of new species. But the existence of variations is not only of interest for the problem of the origin of species; it has also a more general interest. An individual does not lose its importance for knowledge, because its qualities are not heritable. On the contrary, in higher beings at least, individual peculiarities will become more and more independent objects of interest. Knowledge takes account of the biographies not only of species, but also of individuals: it seeks to find the law of development of the single individual. (The new science of Ecology occupies an intermediate position between the biography of species and the biography of individuals. Compare "Congress of Arts and Science", St Louis, Vol. V. 1906 (the Reports of Drude and Robinson) and the work of my colleague E. Warming.) As Leibniz said long ago, individuality consists in the law of the changes of a being. "La loi du changement fait l'individualite de chaque substance." Here is a world which is almost new for science, which till now has mainly occupied itself with general laws and forms. But these are ultimately only means to understand the individual phenomena, in whose nature and history a manifold of laws and forms always cooperate. The importance of this remark will appear in the sequel.

V.

To many people the Darwinian theory of natural selection or struggle for existence seemed to change the whole conception of life, and particularly all the conditions on which the validity of ethical ideas depends. If only that has persistence which can be adapted to a given condition, what will then be the fate of our ideals, of our standards of good and evil? Blind force seems to reign, and the only thing that counts seems to be the most heedless use of power. Darwinism, it was said, has proclaimed brutality. No other difference seems permanent save that between the sound, powerful and happy on the one side, the sick, feeble and unhappy on the other; and every attempt to alleviate this difference seems to lead to general enervation. Some of those who interpreted Darwinism in this manner felt an aesthetic delight in contemplating the heedlessness and energy of the great struggle for existence and anticipated the realisation of a higher human type as the outcome of it: so Nietzsche and his followers. Others recognising the same consequences in Darwinism regarded these as one of the strongest objections against it; so Duhring and Kropotkin (in his earlier works).

This interpretation of Darwinism was frequent in the interval between the two main works of Darwin—"The Origin of Species" and "The Descent of Man". But even during this interval it was evident to an attentive reader that Darwin himself did not found his standard of good and evil on the features of the life of nature he had emphasised so strongly. He did not justify the ways along which nature reached its ends; he only pointed them out. The "real" was not to him, as to Hegel, one with the "rational." Darwin has, indeed, by his whole conception of nature, rendered a great service to ethics in making the difference between the life of nature and the ethical life appear in so strong a light. The ethical problem could now be stated in a sharper form than before. But this was not the first time that the idea of the struggle for life was put in relation to the ethical problem. In the seventeenth century Thomas Hobbes gave the first impulse to the whole modern discussion of ethical principles in his theory of bellum omnium contra omnes. Men, he taught, are in the state of nature enemies one of another, and they live either in fright or in the glory of power. But it was not the opinion of Hobbes that this made ethics impossible. On the contrary, he found a standard for virtue and vice in the fact that some qualities and actions have a tendency to bring us out of the state of war and to secure peace, while other qualities have a contrary tendency. In the eighteenth century even Immanuel Kant's ideal ethics had—so far as can be seen—a similar origin. Shortly before the foundation of his definitive ethics, Kant wrote his "Idee zu einer allgemeinen Weltgeschichte" (1784), where—in a way which reminds us of Hobbes, and is prophetic of Darwin—he describes the forward-driving power of struggle in the human world. It is here as with the struggle of the trees for light and air, through which they compete with one another in height. Anxiety about war can only be allayed by an ordinance which gives everyone his full liberty under acknowledgment of the equal liberty of others. And such ordinance and acknowledgment are also attributes of the content of the moral law, as Kant proclaimed it in the year after the publication of his essay (1785) (Cf. my "History of Modern Philosophy" (English translation London, 1900), I. pages 76-79.) Kant really came to his ethics by the way of evolution, though he afterwards disavowed it. Similarly the same line of thought may be traced in Hegel though it has been disguised in the form of speculative dialectics. ("Herrschaft und Knechtschaft", "Phanomenologie des Geistes", IV. A., Leiden, 1907.) And in Schopenhauer's theory of the blind will to live and its abrogation by the ethical feeling, which is founded on universal sympathy, we have a more individualistic form of the same idea.

It was, then, not entirely a foreign point of view which Darwin introduced into ethical thought, even if we take no account of the poetical character of the word "struggle" and of the more direct adaptation, through the use and non-use of power, which Darwin also emphasised. In "The Descent of Man" he has devoted a special chapter ("The Descent of Man", Vol. I. Ch. iii.) to a discussion of the origin of the ethical consciousness. The characteristic expression of this consciousness he found, just as Kant did, in the idea of "ought"; it was the origin of this new idea which should be explained. His hypothesis was that the ethical "ought" has its origin in the social and parental instincts, which, as well as other instincts (e.g. the instinct of self-preservation), lie deeper than pleasure and pain. In many species, not least in the human species, these instincts are fostered by natural selection; and when the powers of memory and comparison are developed, so that single acts can be valued according to the claims of the deep social instinct, then consciousness of duty and remorse are possible. Blind instinct has developed to conscious ethical will.

As already stated, Darwin, as a moral philosopher belongs to the school that was founded by Shaftesbury, and was afterwards represented by Hutcheson, Hume, Adam Smith, Comte and Spencer. His merit is, first, that he has given this tendency of thought a biological foundation, and that he has stamped on it a doughty character in showing that ethical ideas and sentiments, rightly conceived, are forces which are at work in the struggle for life.

There are still many questions to solve. Not only does the ethical development within the human species contain features still unexplained (The works of Westermarck and Hobhouse throw new light on many of these features.); but we are confronted by the great problem whether after all a genetic historical theory can be of decisive importance here. To every consequent ethical consciousness there is a standard of value, a primordial value which determines the single ethical judgments as their last presupposition, and the "rightness" of this basis, the "value" of this value can as little be discussed as the "rationality" of our logical principles. There is here revealed a possibility of ethical scepticism which evolutionistic ethics (as well as intuitive or rationalistic ethics) has overlooked. No demonstration can show that the results of the ethical development are definitive and universal. We meet here again with the important opposition of systematisation and evolution. There will, I think, always be an open question here, though comparative ethics, of which we have so far only the first attempts, can do much to throw light on it.

It would carry us too far to discuss all the philosophical works on ethics, which have been influenced directly or indirectly by evolutionism. I may, however, here refer to the book of C.M. Williams, "A Review of the Systems of Ethics founded on the Theory of Evolution" (New York and London, 1893.), in which, besides Darwin, the following authors are reviewed: Wallace, Haeckel, Spencer, Fiske, Rolph, Barratt, Stephen, Carneri, Hoffding, Gizycki, Alexander, Ree. As works which criticise evolutionistic ethics from an intuitive point of view and in an instructive way, may be cited: Guyau "La morale anglaise contemporaine" (Paris, 1879.), and Sorley, "Ethics of Naturalism". I will only mention some interesting contributions to ethical discussion which can be found in Darwinism besides the idea of struggle for life.

The attention which Darwin has directed to variations has opened our eyes to the differences in human nature as well as in nature generally. There is here a fact of great importance for ethical thought, no matter from what ultimate premiss it starts. Only from a very abstract point of view can different individuals be treated in the same manner. The most eminent ethical thinkers, men such as Jeremy Bentham and Immanuel Kant, who discussed ethical questions from very opposite standpoints, agreed in regarding all men as equal in respect of ethical endowment. In regard to Bentham, Leslie Stephen remarks: "He is determined to be thoroughly empirical, to take men as he found them. But his utilitarianism supposed that men's views of happiness and utility were uniform and clear, and that all that was wanted was to show them the means by which their ends could be reached." ("English literature and society in the eighteenth century", London, 1904, page 187.) And Kant supposed that every man would find the "categorical imperative" in his consciousness, when he came to sober reflexion, and that all would have the same qualifications to follow it. But if continual variations, great or small, are going on in human nature, it is the duty of ethics to make allowance for them, both in making claims, and in valuing what is done. A new set of ethical problems have their origin here. (Cf. my paper, "The law of relativity in Ethics," "International Journal of Ethics", Vol. I. 1891, pages 37-62.) It is an interesting fact that Stuart Mill's book "On Liberty" appeared in the same year as "The Origin of Species". Though Mill agreed with Bentham about the original equality of all men's endowments, he regarded individual differences as a necessary result of physical and social influences, and he claimed that free play shall be allowed to differences of character so far as is possible without injury to other men. It is a condition of individual and social progress that a man's mode of action should be determined by his own character and not by tradition and custom, nor by abstract rules. This view was to be corroborated by the theory of Darwin.

But here we have reached a point of view from which the criticism, which in recent years has often been directed against Darwin—that small variations are of no importance in the struggle for life—is of no weight. From an ethical standpoint, and particularly from the ethical standpoint of Darwin himself, it is a duty to foster individual differences that can be valuable, even though they can neither be of service for physical preservation nor be physically inherited. The distinction between variation and mutation is here without importance. It is quite natural that biologists should be particularly interested in such variations as can be inherited and produce new species. But in the human world there is not only a physical, but also a mental and social heredity. When an ideal human character has taken form, then there is shaped a type, which through imitation and influence can become an important factor in subsequent development, even if it cannot form a species in the biological sense of the word. Spiritually strong men often succumb in the physical struggle for life; but they can nevertheless be victorious through the typical influence they exert, perhaps on very distant generations, if the remembrance of them is kept alive, be it in legendary or in historical form. Their very failure can show that a type has taken form which is maintained at all risks, a standard of life which is adhered to in spite of the strongest opposition. The question "to be or not to be" can be put from very different levels of being: it has too often been considered a consequence of Darwinism that this question is only to be put from the lowest level. When a stage is reached, where ideal (ethical, intellectual, aesthetic) interests are concerned, the struggle for life is a struggle for the preservation of this stage. The giving up of a higher standard of life is a sort of death; for there is not only a physical, there is also a spiritual, death.

VI.

The Socratic character of Darwin's mind appears in his wariness in drawing the last consequences of his doctrine, in contrast both with the audacious theories of so many of his followers and with the consequences which his antagonists were busy in drawing. Though he, as we have seen, saw from the beginning that his hypothesis would occasion "a whole of metaphysics," he was himself very reserved as to the ultimate questions, and his answers to such questions were extorted from him.

As to the question of optimism and pessimism, Darwin held that though pain and suffering were very often the ways by which animals were led to pursue that course of action which is most beneficial to the species, yet pleasurable feelings were the most habitual guides. "We see this in the pleasure from exertion, even occasionally from great exertion of the body or mind, in the pleasure of our daily meals, and especially in the pleasure derived from sociability, and from loving our families." But there was to him so much suffering in the world that it was a strong argument against the existence of an intelligent First Cause. ("Life and Letters" Vol. I. page 310.)

It seems to me that Darwin was not so clear on another question, that of the relation between improvement and adaptation. He wrote to Lyell: "When you contrast natural selection and 'improvement,' you seem always to overlook... that every step in the natural selection of each species implies improvement in that species IN RELATION TO ITS CONDITION OF LIFE... Improvement implies, I suppose, EACH FORM OBTAINING MANY PARTS OR ORGANS, all excellently adapted for their functions." "All this," he adds, "seems to me quite compatible with certain forms fitted for simple conditions, remaining unaltered, or being degraded." (Ibid. Vol. II. page 177.) But the great question is, if the conditions of life will in the long run favour "improvement" in the sense of differentiation (or harmony of differentiation and integration). Many beings are best adapted to their conditions of life if they have few organs and few necessities. Pessimism would not only be the consequence, if suffering outweighed happiness, but also if the most elementary forms of happiness were predominant, or if there were a tendency to reduce the standard of life to the simplest possible, the contentment of inertia or stable equilibrium. There are animals which are very highly differentiated and active in their young state, but later lose their complex organisation and concentrate themselves on the one function of nutrition. In the human world analogies to this sort of adaptation are not wanting. Young "idealists" very often end as old "Philistines." Adaptation and progress are not the same.

Another question of great importance in respect to human evolution is, whether there will be always a possibility for the existence of an impulse to progress, an impulse to make great claims on life, to be active and to alter the conditions of life instead of adapting to them in a passive manner. Many people do not develop because they have too few necessities, and because they have no power to imagine other conditions of life than those under which they live. In his remarks on "the pleasure from exertion" Darwin has a point of contact with the practical idealism of former times—with the ideas of Lessing and Goethe, of Condorcet and Fichte. The continual striving which was the condition of salvation to Faust's soul, is also the condition of salvation to mankind. There is a holy fire which we ought to keep burning, if adaptation is really to be improvement. If, as I have tried to show in my "Philosophy of Religion", the innermost core of all religion is faith in the persistence of value in the world, and if the highest values express themselves in the cry "Excelsior!" then the capital point is, that this cry should always be heard and followed. We have here a corollary of the theory of evolution in its application to human life.

Darwin declared himself an agnostic, not only because he could not harmonise the large amount of suffering in the world with the idea of a God as its first cause, but also because he "was aware that if we admit a first cause, the mind still craves to know whence it came and how it arose." ("Life and Letters", Vol. I. page 306.) He saw, as Kant had seen before him and expressed in his "Kritik der Urtheilskraft", that we cannot accept either of the only two possibilities which we are able to conceive: chance (or brute force) and design. Neither mechanism nor teleology can give an absolute answer to ultimate questions. The universe, and especially the organic life in it, can neither be explained as a mere combination of absolute elements nor as the effect of a constructing thought. Darwin concluded, as Kant, and before him Spinoza, that the oppositions and distinctions which our experience presents, cannot safely be regarded as valid for existence in itself. And, with Kant and Fichte, he found his stronghold in the conviction that man has something to do, even if he cannot solve all enigmas. "The safest conclusion seems to me that the whole subject is beyond the scope of man's intellect; but man can do his duty." (Ibid. page 307.)

Is this the last word of human thought? Does not the possibility, that man can do his duty, suppose that the conditions of life allow of continuous ethical striving, so that there is a certain harmony between cosmic order and human ideals? Darwin himself has shown how the consciousness of duty can arise as a natural result of evolution. Moreover there are lines of evolution which have their end in ethical idealism, in a kingdom of values, which must struggle for life as all things in the world must do, but a kingdom which has its firm foundation in reality.

XXIII. DARWINISM AND SOCIOLOGY. By C. Bougle.

Professor of Social Philosophy in the University of Toulouse and Deputy-Professor at the Sorbonne, Paris.

How has our conception of social phenomena, and of their history, been affected by Darwin's conception of Nature and the laws of its transformations? To what extent and in what particular respects have the discoveries and hypotheses of the author of "The Origin of Species" aided the efforts of those who have sought to construct a science of society?

To such a question it is certainly not easy to give any brief or precise answer. We find traces of Darwinism almost everywhere. Sociological systems differing widely from each other have laid claim to its authority; while, on the other hand, its influence has often made itself felt only in combination with other influences. The Darwinian thread is worked into a hundred patterns along with other threads.

To deal with the problem, we must, it seems, first of all distinguish the more general conclusions in regard to the evolution of living beings, which are the outcome of Darwinism, from the particular explanations it offers of the ways and means by which that evolution is effected. That is to say, we must, as far as possible, estimate separately the influence of Darwin as an evolutionist and Darwin as a selectionist.

The nineteenth century, said Cournot, has witnessed a mighty effort to "reintegrer l'homme dans la nature." From divers quarters there has been a methodical reaction against the persistent dualism of the Cartesian tradition, which was itself the unconscious heir of the Christian tradition. Even the philosophy of the eighteenth century, materialistic as were for the most part the tendencies of its leaders, seemed to revere man as a being apart, concerning whom laws might be formulated a priori. To bring him down from his pedestal there was needed the marked predominance of positive researches wherein no account was taken of the "pride of man." There can be no doubt that Darwin has done much to familiarise us with this attitude. Take for instance the first part of "The Descent of Man": it is an accumulation of typical facts, all tending to diminish the distance between us and our brothers, the lower animals. One might say that the naturalist had here taken as his motto, "Whosoever shall exalt himself shall be abased; and he that shall humble himself shall be exalted." Homologous structures, the survival in man of certain organs of animals, the rudiments in the animal of certain human faculties, a multitude of facts of this sort, led Darwin to the conclusion that there is no ground for supposing that the "king of the universe" is exempt from universal laws. Thus belief in the imperium in imperio has been, as it were, whittled away by the progress of the naturalistic spirit, itself continually strengthened by the conquests of the natural sciences. The tendency may, indeed, drag the social sciences into overstrained analogies, such, for instance, as the assimilation of societies to organisms. But it will, at least, have had the merit of helping sociology to shake off the pre-conception that the groups formed by men are artificial, and that history is completely at the mercy of chance. Some years before the appearance of "The Origin of Species", Auguste Comte had pointed out the importance, as regards the unification of positive knowledge, of the conviction that the social world, the last refuge of spiritualism, is itself subject to determininism. It cannot be doubted that the movement of thought which Darwin's discoveries promoted contributed to the spread of this conviction, by breaking down the traditional barrier which cut man off from Nature.

But Nature, according to modern naturalists, is no immutable thing: it is rather perpetual movement, continual progression. Their discoveries batter a breach directly into the Aristotelian notion of species; they refuse to see in the animal world a collection of immutable types, distinct from all eternity, and corresponding, as Cuvier said, to so many particular thoughts of the Creator. Darwin especially congratulated himself upon having been able to deal this doctrine the coup de grace: immutability is, he says, his chief enemy; and he is concerned to show—therein following up Lyell's work—that everything in the organic world, as in the inorganic, is explained by insensible but incessant transformations. "Nature makes no leaps"—"Nature knows no gaps": these two dicta form, as it were, the two landmarks between which Darwin's idea of transformation is worked out. That is to say, the development of Darwinism is calculated to further the application of the philosophy of Becoming to the study of human institutions.

The progress of the natural sciences thus brings unexpected reinforcements to the revolution which the progress of historical discipline had begun. The first attempt to constitute an actual science of social phenomena—that, namely, of the economists—had resulted in laws which were called natural, and which were believed to be eternal and universal, valid for all times and all places. But this perpetuality, brother, as Knies said, of the immutability of the old zoology, did not long hold out against the ever swelling tide of the historical movement. Knowledge of the transformations that had taken place in language, of the early phases of the family, of religion, of property, had all favoured the revival of the Heraclitean view: panta rei. As to the categories of political economy, it was soon to be recognised, as by Lassalle, that they too are only historical. The philosophy of history, moreover, gave expression under various forms to the same tendency. Hegel declares that "all that is real is rational," but at the same time he shows that all that is real is ephemeral, and that for history there is nothing fixed beneath the sun. It is this sense of universal evolution that Darwin came with fresh authority to enlarge. It was in the name of biological facts themselves that he taught us to see only slow metamorphoses in the history of institutions, and to be always on the outlook for survivals side by side with rudimentary forms. Anyone who reads "Primitive Culture", by Tylor,—a writer closely connected with Darwin—will be able to estimate the services which these cardinal ideas were to render to the social sciences when the age of comparative research had succeeded to that of a priori construction.

Let us note, moreover, that the philosophy of Becoming in passing through the Darwinian biology became, as it were, filtered: it got rid of those traces of finalism, which, under different forms, it had preserved through all the systems of German Romanticism. Even in Herbert Spencer, it has been plausibly argued, one can detect something of that sort of mystic confidence in forces spontaneously directing life, which forms the very essence of those systems. But Darwin's observations were precisely calculated to render such an hypothesis futile. At first people may have failed to see this; and we call to mind the ponderous sarcasms of Flourens when he objected to the theory of Natural Selection that it attributed to nature a power of free choice. "Nature endowed with will! That was the final error of last century; but the nineteenth no longer deals in personifications." (P. Flourens, "Examen du Livre de M. Darwin sur l'Origine des Especes", page 53, Paris, 1864. See also Huxley, "Criticisms on the 'Origin of Species'", "Collected Essays", Vol. II, page 102, London, 1902.) In fact Darwin himself put his readers on their guard against the metaphors he was obliged to use. The processes by which he explains the survival of the fittest are far from affording any indication of the design of some transcendent breeder. Nor, if we look closely, do they even imply immanent effort in the animal; the sorting out can be brought about mechanically, simply by the action of the environment. In this connection Huxley could with good reason maintain that Darwin's originality consisted in showing how harmonies which hitherto had been taken to imply the agency of intelligence and will could be explained without any such intervention. So, when later on, objective sociology declares that, even when social phenomena are in question, all finalist preconceptions must be distrusted if a science is to be constituted, it is to Darwin that its thanks are due; he had long been clearing paths for it which lay well away from the old familiar road trodden by so many theories of evolution.

This anti-finalist doctrine, when fully worked out, was, moreover, calculated to aid in the needful dissociation of two notions: that of evolution and that of progress. In application to society these had long been confounded; and, as a consequence, the general idea seemed to be that only one type of evolution was here possible. Do we not detect such a view in Comte's sociology, and perhaps even in Herbert Spencer's? Whoever, indeed, assumes an end for evolution is naturally inclined to think that only one road leads to that end. But those whose minds the Darwinian theory has enlightened are aware that the transformations of living beings depend primarily upon their conditions, and that it is these conditions which are the agents of selection from among individual variations. Hence, it immediately follows that transformations are not necessarily improvements. Here, Darwin's thought hesitated. Logically his theory proves, as Ray Lankester pointed out, that the struggle for existence may have as its outcome degeneration as well as amelioration: evolution may be regressive as well as progressive. Then, too—and this is especially to be borne in mind—each species takes its good where it finds it, seeks its own path and survives as best it can. Apply this notion to society and you arrive at the theory of multilinear evolution. Divergencies will no longer surprise you. You will be forewarned not to apply to all civilisations the same measure of progress, and you will recognise that types of evolution may differ just as social species themselves differ. Have we not here one of the conceptions which mark off sociology proper from the old philosophy of history?

But if we are to estimate the influence of Darwinism upon sociological conceptions, we must not dwell only upon the way in which Darwin impressed the general notion of evolution upon the minds of thinkers. We must go into details. We must consider the influence of the particular theories by which he explained the mechanism of this evolution. The name of the author of "The Origin of Species" has been especially attached, as everyone knows, to the doctrines of "natural selection" and of "struggle for existence," completed by the notion of "individual variation." These doctrines were turned to account by very different schools of social philosophy. Pessimistic and optimistic, aristocratic and democratic, individualistic and socialistic systems were to war with each other for years by casting scraps of Darwinism at each other's heads.

It was the spectacle of human contrivance that suggested to Darwin his conception of natural selection. It was in studying the methods of pigeon breeders that he divined the processes by which nature, in the absence of design, obtains analogous results in the differentiation of types. As soon as the importance of artificial selection in the transformation of species of animals was understood, reflection naturally turned to the human species, and the question arose, How far do men observe, in connection with themselves, those laws of which they make practical application in the case of animals? Here we come upon one of the ideas which guided the researches of Galton, Darwin's cousin. The author of "Inquiries into Human Faculty and its Development" ("Inquiries into Human Faculty", pages 1, 2, 3 sq., London, 1883.), has often expressed his surprise that, considering all the precautions taken, for example, in the breeding of horses, none whatever are taken in the breeding of the human species. It seems to be forgotten that the species suffers when the "fittest" are not able to perpetuate their type. Ritchie, in his "Darwinism and Politics" ("Darwinism and Politics" pages 9, 22, London, 1889.) reminds us of Darwin's remark that the institution of the peerage might be defended on the ground that peers, owing to the prestige they enjoy, are enabled to select as wives "the most beautiful and charming women out of the lower ranks." ("Life and Letters of Charles Darwin", II. page 385.) But, says Galton, it is as often as not "heiresses" that they pick out, and birth statistics seem to show that these are either less robust or less fecund than others. The truth is that considerations continue to preside over marriage which are entirely foreign to the improvement of type, much as this is a condition of general progress. Hence the importance of completing Odin's and De Candolle's statistics which are designed to show how characters are incorporated in organisms, how they are transmitted, how lost, and according to what law eugenic elements depart from the mean or return to it.

But thinkers do not always content themselves with undertaking merely the minute researches which the idea of Selection suggests. They are eager to defend this or that thesis. In the name of this idea certain social anthropologists have recast the conception of the process of civilisation, and have affirmed that Social Selection generally works against the trend of Natural Selection. Vacher de Lapouge—following up an observation by Broca on the point—enumerates the various institutions, or customs, such as the celibacy of priests and military conscription, which cause elimination or sterilisation of the bearers of certain superior qualities, intellectual or physical. In a more general way he attacks the democratic movement, a movement, as P. Bourget says, which is "anti-physical" and contrary to the natural laws of progress; though it has been inspired "by the dreams of that most visionary of all centuries, the eighteenth." (V. de Lapouge, "Les Selections sociales", page 259, Paris, 1896.) The "Equality" which levels down and mixes (justly condemned, he holds, by the Comte de Gobineau), prevents the aristocracy of the blond dolichocephales from holding the position and playing the part which, in the interests of all, should belong to them. Otto Ammon, in his "Natural Selection in Man", and in "The Social Order and its Natural Bases" ("Die naturliche Auslese beim Menschen", Jena, 1893; "Die Gesellschaftsordnung und ihre naturlichen Grundlagen". "Entwurf einer Sozialanthropologie", Jena, 1896.), defended analogous doctrines in Germany; setting the curve representing frequency of talent over against that of income, he attempted to show that all democratic measures which aim at promoting the rise in the social scale of the talented are useless, if not dangerous; that they only increase the panmixia, to the great detriment of the species and of society.

Among the aristocratic theories which Darwinism has thus inspired we must reckon that of Nietzsche. It is well known that in order to complete his philosophy he added biological studies to his philological; and more than once in his remarks upon the "Wille zur Macht" he definitely alludes to Darwin; though it must be confessed that it is generally in order to proclaim the in sufficiency of the processes by which Darwin seeks to explain the genesis of species. Nevertheless, Nietzsche's mind is completely possessed by an ideal of Selection. He, too, has a horror of panmixia. The naturalists' conception of "the fittest" is joined by him to that of the "hero" of romance to furnish a basis for his doctrine of the Superman. Let us hasten to add, moreover, that at the very moment when support was being sought in the theory of Selection for the various forms of the aristocratic doctrine, those same forms were being battered down on another side by means of that very theory. Attention was drawn to the fact that by virtue of the laws which Darwin himself had discovered isolation leads to etiolation. There is a risk that the privilege which withdraws the privileged elements of Society from competition will cause them to degenerate. In fact, Jacoby in his "Studies in Selection, in connexion with Heredity in Man", ("Etudes sur la Selection dans ses rapports avec l'heredite chez l'homme", Paris, page 481, 1881.), concludes that "sterility, mental debility, premature death and, finally, the extinction of the stock were not specially and exclusively the fate of sovereign dynasties; all privileged classes, all families in exclusively elevated positions share the fate of reigning families, although in a minor degree and in direct proportion to the loftiness of their social standing. From the mass of human beings spring individuals, families, races, which tend to raise themselves above the common level; painfully they climb the rugged heights, attain the summits of power, of wealth, of intelligence, of talent, and then, no sooner are they there than they topple down and disappear in gulfs of mental and physical degeneracy." The demographical researches of Hansen ("Die drei Bevolkerungsstufen", Munich, 1889.) (following up and completing Dumont's) tended, indeed, to show that urban as well as feudal aristocracies, burgher classes as well as noble castes, were liable to become effete. Hence it might well be concluded that the democratic movement, operating as it does to break down class barriers, was promoting instead of impeding human selection.

So we see that, according to the point of view, very different conclusions have been drawn from the application of the Darwinian idea of Selection to human society. Darwin's other central idea, closely bound up with this, that, namely, of the "struggle for existence" also has been diversely utilised. But discussion has chiefly centered upon its signification. And while some endeavour to extend its application to everything, we find others trying to limit its range. The conception of a "struggle for existence" has in the present day been taken up into the social sciences from natural science, and adopted. But originally it descended from social science to natural. Darwin's law is, as he himself said, only Malthus' law generalised and extended to the animal world: a growing disproportion between the supply of food and the number of the living is the fatal order whence arises the necessity of universal struggle, a struggle which, to the great advantage of the species, allows only the best equipped individuals to survive. Nature is regarded by Huxley as an immense arena where all living beings are gladiators. ("Evolution and Ethics", page 200; "Collected Essays", Vol. IX, London, 1894.)

Such a generalisation was well adapted to feed the stream of pessimistic thought; and it furnished to the apologists of war, in particular, new arguments, weighted with all the authority which in these days attaches to scientific deliverances. If people no longer say, as Bonald did, and Moltke after him, that war is a providential fact, they yet lay stress on the point that it is a natural fact. To the peace party Dragomirov's objection is urged that its attempts are contrary to the fundamental laws of nature, and that no sea wall can hold against breakers that come with such gathered force.

But in yet another quarter Darwinism was represented as opposed to philanthropic intervention. The defenders of the orthodox political economy found in it support for their tenets. Since in the organic world universal struggle is the condition of progress, it seemed obvious that free competition must be allowed to reign unchecked in the economic world. Attempts to curb it were in the highest degree imprudent. The spirit of Liberalism here seemed in conformity with the trend of nature: in this respect, at least, contemporary naturalism, offspring of the discoveries of the nineteenth century, brought reinforcements to the individualist doctrine, begotten of the speculations of the eighteenth: but only, it appeared, to turn mankind away for ever from humanitarian dreams. Would those whom such conclusions repelled be content to oppose to nature's imperatives only the protests of the heart? There were some who declared, like Brunetiere, that the laws in question, valid though they might be for the animal kingdom, were not applicable to the human. And so a return was made to the classic dualism. This indeed seems to be the line that Huxley took, when, for instance, he opposed to the cosmic process an ethical process which was its reverse.

But the number of thinkers whom this antithesis does not satisfy grows daily. Although the pessimism which claims authorisation from Darwin's doctrines is repugnant to them, they still are unable to accept the dualism which leaves a gulf between man and nature. And their endeavour is to link the two by showing that while Darwin's laws obtain in both kingdoms, the conditions of their application are not the same: their forms, and, consequently, their results, vary with the varying mediums in which the struggle of living beings takes place, with the means these beings have at disposal, with the ends even which they propose to themselves.

Here we have the explanation of the fact that among determined opponents of war partisans of the "struggle for existence" can be found: there are disciples of Darwin in the peace party. Novicow, for example, admits the "combat universel" of which Le Dantec ("Les Luttes entre Societies humaines et leurs phases successives", Paris, 1893,) speaks; but he remarks that at different stages of evolution, at different stages of life the same weapons are not necessarily employed. Struggles of brute force, armed hand to hand conflicts, may have been a necessity in the early phases of human societies. Nowadays, although competition may remain inevitable and indispensable, it can assume milder forms. Economic rivalries, struggles between intellectual influences, suffice to stimulate progress: the processes which these admit are, in the actual state of civilisation, the only ones which attain their end without waste, the only ones logical. From one end to the other of the ladder of life, struggle is the order of the day; but more and more as the higher rungs are reached, it takes on characters which are proportionately more "humane."

Reflections of this kind permit the introduction into the economic order of limitations to the doctrine of "laisser faire, laisser passer." This appeals, it is said, to the example of nature where creatures, left to themselves, struggle without truce and without mercy; but the fact is forgotten that upon industrial battlefields the conditions are different. The competitors here are not left simply to their natural energies: they are variously handicapped. A rich store of artificial resources exists in which some participate and others do not. The sides then are unequal; and as a consequence the result of the struggle is falsified. "In the animal world," said De Laveleye ("Le socialisme contemporain", page 384 (6th edition), Paris, 1891.), criticising Spencer, "the fate of each creature is determined by its individual qualities; whereas in civilised societies a man may obtain the highest position and the most beautiful wife because he is rich and well-born, although he may be ugly, idle or improvident; and then it is he who will perpetuate the species. The wealthy man, ill constituted, incapable, sickly, enjoys his riches and establishes his stock under the protection of the laws." Haycraft in England and Jentsch in Germany have strongly emphasised these "anomalies," which nevertheless are the rule. That is to say that even from a Darwinian point of view all social reforms can readily be justified which aim at diminishing, as Wallace said, inequalities at the start.

But we can go further still. Whence comes the idea that all measures inspired by the sentiment of solidarity are contrary to Nature's trend? Observe her carefully, and she will not give lessons only in individualism. Side by side with the struggle for existence do we not find in operation what Lanessan calls "association for existence." Long ago, Espinas had drawn attention to "societies of animals," temporary or permanent, and to the kind of morality that arose in them. Since then, naturalists have often insisted upon the importance of various forms of symbiosis. Kropotkin in "Mutual Aid" has chosen to enumerate many examples of altruism furnished by animals to mankind. Geddes and Thomson went so far as to maintain that "Each of the greater steps of progress is in fact associated with an increased measure of subordination of individual competition to reproductive or social ends, and of interspecific competition to co-operative association." (Geddes and Thomson, "The Evolution of Sex", page 311, London, 1889.) Experience shows, according to Geddes, that the types which are fittest to surmount great obstacles are not so much those who engage in the fiercest competitive struggle for existence, as those who contrive to temper it. From all these observations there resulted, along with a limitation of Darwinian pessimism, some encouragement for the aspirations of the collectivists.

And Darwin himself would, doubtless, have subscribed to these rectifications. He never insisted, like his rival, Wallace, upon the necessity of the solitary struggle of creatures in a state of nature, each for himself and against all. On the contrary, in "The Descent of Man", he pointed out the serviceableness of the social instincts, and corroborated Bagehot's statements when the latter, applying laws of physics to politics, showed the great advantage societies derived from intercourse and communion. Again, the theory of sexual evolution which makes the evolution of types depend increasingly upon preferences, judgments, mental factors, surely offers something to qualify what seems hard and brutal in the theory of natural selection.

But, as often happens with disciples, the Darwinians had out-Darwined Darwin. The extravagancies of social Darwinism provoked a useful reaction; and thus people were led to seek, even in the animal kingdom, for facts of solidarity which would serve to justify humane effort.

On quite another line, however, an attempt has been made to connect socialist tendencies with Darwinian principles. Marx and Darwin have been confronted; and writers have undertaken to show that the work of the German philosopher fell readily into line with that of the English naturalist and was a development of it. Such has been the endeavour of Ferri in Italy and of Woltmann in Germany, not to mention others. The founders of "scientific socialism" had, moreover, themselves thought of this reconciliation. They make more than one allusion to Darwin in works which appeared after 1859. And sometimes they use his theory to define by contrast their own ideal. They remark that the capitalist system, by giving free course to individual competition, ends indeed in a bellum omnium contra omnes; and they make it clear that Darwinism, thus understood, is as repugnant to them as to Duhring.

But it is at the scientific and not at the moral point of view that they place themselves when they connect their economic history with Darwin's work. Thanks to this unifying hypothesis, they claim to have constructed—as Marx does in his preface to "Das Kapital"— a veritable natural history of social evolution. Engels speaks in praise of his friend Marx as having discovered the true mainspring of history hidden under the veil of idealism and sentimentalism, and as having proclaimed in the primum vivere the inevitableness of the struggle for existence. Marx himself, in "Das Kapital", indicated another analogy when he dwelt upon the importance of a general technology for the explanation of this psychology:— a history of tools which would be to social organs what Darwinism is to the organs of animal species. And the very importance they attach to tools, to apparatus, to machines, abundantly proves that neither Marx nor Engels were likely to forget the special characters which mark off the human world from the animal. The former always remains to a great extent an artificial world. Inventions change the face of its institutions. New modes of production revolutionise not only modes of government, but modes even of collective thought. Therefore it is that the evolution of society is controlled by laws special to it, of which the spectacle of nature offers no suggestion.

If, however, even in this special sphere, it can still be urged that the evolution of the material conditions of society is in accord with Darwin's theory, it is because the influence of the methods of production is itself to be explained by the incessant strife of the various classes with each other. So that in the end Marx, like Darwin, finds the source of all progress in struggle. Both are grandsons of Heraclitus:—polemos pater panton. It sometimes happens, in these days, that the doctrine of revolutionary socialism is contrasted as rude and healthy with what may seem to be the enervating tendency of "solidarist" philanthropy: the apologists of the doctrine then pride themselves above all upon their faithfulness to Darwinian principles.

So far we have been mainly concerned to show the use that social philosophies have made of the Darwinian laws for practical purposes: in order to orientate society towards their ideals each school tries to show that the authority of natural science is on its side. But even in the most objective of theories, those which systematically make abstraction of all political tendencies in order to study the social reality in itself, traces of Darwinism are readily to be found.

Let us take for example Durkheim's theory of Division of Labour ("De la Division du Travail social", Paris, 1893.) The conclusions he derives from it are that whenever professional specialisation causes multiplication of distinct branches of activity, we get organic solidarity—implying differences—substituted for mechanical solidarity, based upon likenesses. The umbilical cord, as Marx said, which connects the individual consciousness with the collective consciousness is cut. The personality becomes more and more emancipated. But on what does this phenomenon, so big with consequences, itself depend? The author goes to social morphology for the answer: it is, he says, the growing density of population which brings with it this increasing differentiation of activities. But, again, why? Because the greater density, in thrusting men up against each other, augments the intensity of their competition for the means of existence; and for the problems which society thus has to face differentiation of functions presents itself as the gentlest solution.

Here one sees that the writer borrows directly from Darwin. Competition is at its maximum between similars, Darwin had declared; different species, not laying claim to the same food, could more easily coexist. Here lay the explanation of the fact that upon the same oak hundreds of different insects might be found. Other things being equal, the same applies to society. He who finds some unadopted speciality possesses a means of his own for getting a living. It is by this division of their manifold tasks that men contrive not to crush each other. Here we obviously have a Darwinian law serving as intermediary in the explanation of that progress of division of labour which itself explains so much in the social evolution.

And we might take another example, at the other end of the series of sociological systems. G. Tarde is a sociologist with the most pronounced anti-naturalistic views. He has attempted to show that all application of the laws of natural science to society is misleading. In his "Opposition Universelle" he has directly combatted all forms of sociological Darwinism. According to him the idea that the evolution of society can be traced on the same plan as the evolution of species is chimerical. Social evolution is at the mercy of all kinds of inventions, which by virtue of the laws of imitation modify, through individual to individual, through neighbourhood to neighbourhood, the general state of those beliefs and desires which are the only "quantities" whose variation matters to the sociologist. But, it may be rejoined, that however psychical the forces may be, they are none the less subject to Darwinian laws. They compete with each other; they struggle for the mastery of minds. Between types of ideas, as between organic forms, selection operates. And though it may be that these types are ushered into the arena by unexpected discoveries, we yet recognise in the psychological accidents, which Tarde places at the base of everything, near relatives of those small accidental variations upon which Darwin builds. Thus, accepting Tarde's own representations, it is quite possible to express in Darwinian terms, with the necessary transpositions, one of the most idealistic sociologies that have ever been constructed.

These few examples suffice. They enable us to estimate the extent of the field of influence of Darwinism. It affects sociology not only through the agency of its advocates but through that of its opponents. The questionings to which it has given rise have proved no less fruitful than the solutions it has suggested. In short, few doctrines, in the history of social philosophy, will have produced on their passage a finer outcrop of ideas.

XXIV. THE INFLUENCE OF DARWIN UPON RELIGIOUS THOUGHT. By P.N. Waggett, M.A., S.S.J.E.

I.

The object of this paper is first to point out certain elements of the Darwinian influence upon Religious thought, and then to show reason for the conclusion that it has been, from a Christian point of view, satisfactory. I shall not proceed further to urge that the Christian apologetic in relation to biology has been successful. A variety of opinions may be held on this question, without disturbing the conclusion that the movements of readjustment have been beneficial to those who remain Christians, and this by making them more Christian and not only more liberal. The theologians may sometimes have retreated, but there has been an advance of theology. I know that this account incurs the charge of optimism. It is not the worst that could be made. The influence has been limited in personal range, unequal, even divergent, in operation, and accompanied by the appearance of waste and mischievous products. The estimate which follows requires for due balance a full development of many qualifying considerations. For this I lack space, but I must at least distinguish my view from the popular one that our difficulties about religion and natural science have come to an end.

Concerning the older questions about origins—the origin of the world, of species, of man, of reason, conscience, religion—a large measure of understanding has been reached by some thoughtful men. But meanwhile new questions have arisen, questions about conduct, regarding both the reality of morals and the rule of right action for individuals and societies. And these problems, still far from solution, may also be traced to the influence of Darwin. For they arise from the renewed attention to heredity, brought about by the search for the causes of variation, without which the study of the selection of variations has no sufficient basis.

Even the existing understanding about origins is very far from universal. On these points there were always thoughtful men who denied the necessity of conflict, and there are still thoughtful men who deny the possibility of a truce.

It must further be remembered that the earlier discussion now, as I hope to show, producing favourable results, created also for a time grave damage, not only in the disturbance of faith and the loss of men—a loss not repaired by a change in the currents of debate—but in what I believe to be a still more serious respect. I mean the introduction of a habit of facile and untested hypothesis in religious as in other departments of thought.

Darwin is not responsible for this, but he is in part the cause of it. Great ideas are dangerous guests in narrow minds; and thus it has happened that Darwin—the most patient of scientific workers, in whom hypothesis waited upon research, or if it provisionally outstepped it did so only with the most scrupulously careful acknowledgment—has led smaller and less conscientious men in natural science, in history, and in theology to an over-eager confidence in probable conjecture and a loose grip upon the facts of experience. It is not too much to say that in many quarters the age of materialism was the least matter-of-fact age conceivable, and the age of science the age which showed least of the patient temper of inquiry.

I have indicated, as shortly as I could, some losses and dangers which in a balanced account of Darwin's influence would be discussed at length.

One other loss must be mentioned. It is a defect in our thought which, in some quarters, has by itself almost cancelled all the advantages secured. I mean the exaggerated emphasis on uniformity or continuity; the unwillingness to rest any part of faith or of our practical expectation upon anything that from any point of view can be called exceptional. The high degree of success reached by naturalists in tracing, or reasonably conjecturing, the small beginnings of great differences, has led the inconsiderate to believe that anything may in time become anything else.

It is true that this exaggeration of the belief in uniformity has produced in turn its own perilous reaction. From refusing to believe whatever can be called exceptional, some have come to believe whatever can be called wonderful.

But, on the whole, the discontinuous or highly various character of experience received for many years too little deliberate attention. The conception of uniformity which is a necessity of scientific description has been taken for the substance of history. We have accepted a postulate of scientific method as if it were a conclusion of scientific demonstration. In the name of a generalisation which, however just on the lines of a particular method, is the prize of a difficult exploit of reflexion, we have discarded the direct impressions of experience; or, perhaps it is more true to say, we have used for the criticism of alleged experiences a doctrine of uniformity which is only valid in the region of abstract science. For every science depends for its advance upon limitation of attention, upon the selection out of the whole content of consciousness of that part or aspect which is measurable by the method of the science. Accordingly there is a science of life which rightly displays the unity underlying all its manifestations. But there is another view of life, equally valid, and practically sometimes more important, which recognises the immediate and lasting effect of crisis, difference, and revolution. Our ardour for the demonstration of uniformity of process and of minute continuous change needs to be balanced by a recognition of the catastrophic element in experience, and also by a recognition of the exceptional significance for us of events which may be perfectly regular from an impersonal point of view.

An exorbitant jealousy of miracle, revelation, and ultimate moral distinctions has been imported from evolutionary science into religious thought. And it has been a damaging influence, because it has taken men's attention from facts, and fixed them upon theories.

II.

With this acknowledgment of important drawbacks, requiring many words for their proper description, I proceed to indicate certain results of Darwin's doctrine which I believe to be in the long run wholly beneficial to Christian thought. These are:

The encouragement in theology of that evolutionary method of observation and study, which has shaped all modern research:

The recoil of Christian apologetics towards the ground of religious experience, a recoil produced by the pressure of scientific criticism upon other supports of faith:

The restatement, or the recovery of ancient forms of statement, of the doctrines of Creation and of divine Design in Nature, consequent upon the discussion of evolution and of natural selection as its guiding factor.

(1) The first of these is quite possibly the most important of all. It was well defined in a notable paper read by Dr Gore, now Bishop of Birmingham, to the Church Congress at Shrewsbury in 1896. We have learnt a new caution both in ascribing and in denying significance to items of evidence, in utterance or in event. There has been, as in art, a study of values, which secures perspective and solidity in our representation of facts. On the one hand, a given utterance or event cannot be drawn into evidence as if all items were of equal consequence, like sovereigns in a bag. The question whence and whither must be asked, and the particular thing measured as part of a series. Thus measured it is not less truly important, but it may be important in a lower degree. On the other hand, and for exactly the same reason, nothing that is real is unimportant. The "failures" are not mere mistakes. We see them, in St Augustine's words, as "scholar's faults which men praise in hope of fruit."

We cannot safely trace the origin of the evolutionistic method to the influence of natural science. The view is tenable that theology led the way. Probably this is a case of alternate and reciprocal debt. Quite certainly the evolutionist method in theology, in Christian history, and in the estimate of scripture, has received vast reinforcement from biology, in which evolution has been the ever present and ever victorious conception.

(2) The second effect named is the new willingness of Christian thinkers to take definite account of religious experience. This is related to Darwin through the general pressure upon religious faith of scientific criticism. The great advance of our knowledge of organisms has been an important element in the general advance of science. It has acted, by the varied requirements of the theory of organisms, upon all other branches of natural inquiry, and it held for a long time that leading place in public attention which is now occupied by speculative physics. Consequently it contributed largely to our present estimation of science as the supreme judge in all matters of inquiry (F.R. Tennant: "The Being of God in the light of Physical Science", in "Essays on some theological questions of the day". London, 1905.), to the supposed destruction of mystery and the disparagement of metaphysic which marked the last age, as well as to the just recommendation of scientific method in branches of learning where the direct acquisitions of natural science had no place.

Besides this, the new application of the idea of law and mechanical regularity to the organic world seemed to rob faith of a kind of refuge. The romantics had, as Berthelot ("Evolutionisme et Platonisme", pages 45, 46, 47. Paris, 1908.) shows, appealed to life to redress the judgments drawn from mechanism. Now, in Spencer, evolution gave us a vitalist mechanic or mechanical vitalism, and the appeal seemed cut off. We may return to this point later when we consider evolution; at present I only endeavour to indicate that general pressure of scientific criticism which drove men of faith to seek the grounds of reassurance in a science of their own; in a method of experiment, of observation, of hypothesis checked by known facts. It is impossible for me to do more than glance across the threshold of this subject. But it is necessary to say that the method is in an elementary stage of revival. The imposing success that belongs to natural science is absent: we fall short of the unchallengeable unanimity of the Biologists on fundamentals. The experimental method with its sure repetitions cannot be applied to our subject-matter. But we have something like the observational method of palaeontology and geographical distribution; and in biology there are still men who think that the large examination of varieties by way of geography and the search of strata is as truly scientific, uses as genuinely the logical method of difference, and is as fruitful in sure conclusions as the quasi-chemical analysis of Mendelian laboratory work, of which last I desire to express my humble admiration. Religion also has its observational work in the larger and possibly more arduous manner.

But the scientific work in religion makes its way through difficulties and dangers. We are far from having found the formula of its combination with the historical elements of our apologetic. It is exposed, therefore, to a damaging fire not only from unspiritualist psychology and pathology but also from the side of scholastic dogma. It is hard to admit on equal terms a partner to the old undivided rule of books and learning. With Charles Lamb, we cry in some distress, "must knowledge come to me, if it come at all, by some awkward experiment of intuition, and no longer by this familiar process of reading?" ("Essays of Elia", "New Year's Eve", page 41; Ainger's edition. London, 1899.) and we are answered that the old process has an imperishable value, only we have not yet made clear its connection with other contributions. And all the work is young, liable to be drawn into unprofitable excursions, side-tracked by self-deceit and pretence; and it fatally attracts, like the older mysticism, the curiosity and the expository powers of those least in sympathy with it, ready writers who, with all the air of extended research, have been content with narrow grounds for induction. There is a danger, besides, which accompanies even the most genuine work of this science and must be provided against by all its serious students. I mean the danger of unbalanced introspection both for individuals and for societies; of a preoccupation comparable to our modern social preoccupation with bodily health; of reflection upon mental states not accompanied by exercise and growth of the mental powers; the danger of contemplating will and neglecting work, of analysing conviction and not criticising evidence.

Still, in spite of dangers and mistakes, the work remains full of hopeful indications, and, in the best examples (Such an example is given in Baron F. von Hugel's recently finished book, the result of thirty years' research: "The Mystical Element of Religion, as studied in Saint Catherine of Genoa and her Friends". London, 1908.), it is truly scientific in its determination to know the very truth, to tell what we think, not what we think we ought to think. (G. Tyrrell, in "Mediaevalism", has a chapter which is full of the important MORAL element in a scientific attitude. "The only infallible guardian of truth is the spirit of truthfulness." "Mediaevalism" page 182, London, 1908.), truly scientific in its employment of hypothesis and verification, and in growing conviction of the reality of its subject-matter through the repeated victories of a mastery which advances, like science, in the Baconian road of obedience. It is reasonable to hope that progress in this respect will be more rapid and sure when religious study enlists more men affected by scientific desire and endowed with scientific capacity.

The class of investigating minds is a small one, possibly even smaller than that of reflecting minds. Very few persons at any period are able to find out anything whatever. There are few observers, few discoverers, few who even wish to discover truth. In how many societies the problems of philology which face every person who speaks English are left unattempted! And if the inquiring or the successfully inquiring class of minds is small, much smaller, of course, is the class of those possessing the scientific aptitude in an eminent degree. During the last age this most distinguished class was to a very great extent absorbed in the study of phenomena, a study which had fallen into arrears. For we stood possessed, in rudiment, of means of observation, means for travelling and acquisition, qualifying men for a larger knowledge than had yet been attempted. These were now to be directed with new accuracy and ardour upon the fabric and behaviour of the world of sense. Our debt to the great masters in physical science who overtook and almost out-stripped the task cannot be measured; and, under the honourable leadership of Ruskin, we may all well do penance if we have failed "in the respect due to their great powers of thought, or in the admiration due to the far scope of their discovery." ("Queen of the Air", Preface, page vii. London, 1906.) With what miraculous mental energy and divine good fortune—as Romans said of their soldiers—did our men of curiosity face the apparently impenetrable mysteries of nature! And how natural it was that immense accessions of knowledge, unrelated to the spiritual facts of life, should discredit Christian faith, by the apparent superiority of the new work to the feeble and unprogressive knowledge of Christian believers! The day is coming when men of this mental character and rank, of this curiosity, this energy and this good fortune in investigation, will be employed in opening mysteries of a spiritual nature. They will silence with masterful witness the over-confident denials of naturalism. They will be in danger of the widespread recognition which thirty years ago accompanied every utterance of Huxley, Tyndall, Spencer. They will contribute, in spite of adulation, to the advance of sober religious and moral science.

And this result will be due to Darwin, first because by raising the dignity of natural science, he encouraged the development of the scientific mind; secondly because he gave to religious students the example of patient and ardent investigation; and thirdly because by the pressure of naturalistic criticism the religious have been driven to ascertain the causes of their own convictions, a work in which they were not without the sympathy of men of science. (The scientific rank of its writer justifies the insertion of the following letter from the late Sir John Burdon-Sanderson to me. In the lecture referred to I had described the methods of Professor Moseley in teaching Biology as affording a suggestion of the scientific treatment of religion.)

Oxford, April 30, 1902.

Dear Sir,

I feel that I must express to you my thanks for the discourse which I had the pleasure of listening to yesterday afternoon.

I do not mean to say that I was able to follow all that you said as to the identity of Method in the two fields of Science and Religion, but I recognise that the "mysticism" of which you spoke gives us the only way by which the two fields can be brought into relation.

Among much that was memorable, nothing interested me more than what you said of Moseley.

No one, I am sure, knew better than you the value of his teaching and in what that value consisted.

Yours faithfully

J. Burdon-Sanderson. 31-2.)

In leaving the subject of scientific religious inquiry, I will only add that I do not believe it receives any important help—and certainly it suffers incidentally much damaging interruption—from the study of abnormal manifestations or abnormal conditions of personality.

(3) Both of the above effects seem to me of high, perhaps the very highest, importance to faith and to thought. But, under the third head, I name two which are more directly traceable to the personal work of Darwin, and more definitely characteristic of the age in which his influence was paramount: viz. the influence of the two conceptions of evolution and natural selection upon the doctrine of creation and of design respectively.

It is impossible here, though it is necessary for a complete sketch of the matter, to distinguish the different elements and channels of this Darwinian influence; in Darwin's own writings, in the vigorous polemic of Huxley, and strangely enough, but very actually for popular thought, in the teaching of the definitely anti-Darwinian evolutionist Spencer.

Under the head of the directly and purely Darwinian elements I should class as preeminent the work of Wallace and of Bates; for no two sets of facts have done more to fix in ordinary intelligent minds a belief in organic evolution and in natural selection as its guiding factor than the facts of geographical distribution and of protective colour and mimicry. The facts of geology were difficult to grasp and the public and theologians heard more often of the imperfection than of the extent of the geological record. The witness of embryology, depending to a great extent upon microscopic work, was and is beyond the appreciation of persons occupied in fields of work other than biology.

III.

From the influence in religion of scientific modes of thought we pass to the influence of particular biological conceptions. The former effect comes by way of analogy, example, encouragement and challenge; inspiring or provoking kindred or similar modes of thought in the field of theology; the latter by a collision of opinions upon matters of fact or conjecture which seem to concern both science and religion.

In the case of Darwinism the story of this collision is familiar, and falls under the heads of evolution and natural selection, the doctrine of descent with modification, and the doctrine of its guidance or determination by the struggle for existence between related varieties. These doctrines, though associated and interdependent, and in popular thought not only combined but confused, must be considered separately. It is true that the ancient doctrine of Evolution, in spite of the ingenuity and ardour of Lamarck, remained a dream tantalising the intellectual ambition of naturalists, until the day when Darwin made it conceivable by suggesting the machinery of its guidance. And, further, the idea of natural selection has so effectively opened the door of research and stimulated observation in a score of principal directions that, even if the Darwinian explanation became one day much less convincing than, in spite of recent criticism, it now is, yet its passing, supposing it to pass, would leave the doctrine of Evolution immeasurably and permanently strengthened. For in the interests of the theory of selection, "Fur Darwin," as Muller wrote, facts have been collected which remain in any case evidence of the reality of descent with modification.

But still, though thus united in the modern history of convictions, though united and confused in the collision of biological and traditional opinion, yet evolution and natural selection must be separated in theological no less than in biological estimation. Evolution seemed inconsistent with Creation; natural selection with Providence and Divine design.

Discussion was maintained about these points for many years and with much dark heat. It ranged over many particular topics and engaged minds different in tone, in quality, and in accomplishment. There was at most times a degree of misconception. Some naturalists attributed to theologians in general a poverty of thought which belonged really to men of a particular temper or training. The "timid theism" discerned in Darwin by so cautious a theologian as Liddon (H.P. Liddon, "The Recovery of S. Thomas"; a sermon preached in St Paul's, London, on April 23rd, 1882 (the Sunday after Darwin's death).) was supposed by many biologists to be the necessary foundation of an honest Christianity. It was really more characteristic of devout NATURALISTS like Philip Henry Gosse, than of religious believers as such. (Dr Pusey ("Unscience not Science adverse to Faith" 1878) writes: "The questions as to 'species,' of what variations the animal world is capable, whether the species be more or fewer, whether accidental variations may become hereditary... and the like, naturally fall under the province of science. In all these questions Mr Darwin's careful observations gained for him a deserved approbation and confidence.") The study of theologians more considerable and even more typically conservative than Liddon does not confirm the description of religious intolerance given in good faith, but in serious ignorance, by a disputant so acute, so observant and so candid as Huxley. Something hid from each other's knowledge the devoted pilgrims in two great ways of thought. The truth may be, that naturalists took their view of what creation was from Christian men of science who naturally looked in their own special studies for the supports and illustrations of their religious belief. Of almost every laborious student it may be said "Hic ab arte sua non recessit." And both the believing and the denying naturalists, confining habitual attention to a part of experience, are apt to affirm and deny with trenchant vigour and something of a narrow clearness "Qui respiciunt ad pauca, de facili pronunciant." (Aristotle, in Bacon, quoted by Newman in his "Idea of a University", page 78. London, 1873.)

Newman says of some secular teachers that "they persuade the world of what is false by urging upon it what is true." Of some early opponents of Darwin it might be said by a candid friend that, in all sincerity of devotion to truth, they tried to persuade the world of what is true by urging upon it what is false. If naturalists took their version of orthodoxy from amateurs in theology, some conservative Christians, instead of learning what evolution meant to its regular exponents, took their view of it from celebrated persons, not of the front rank in theology or in thought, but eager to take account of public movements and able to arrest public attention.

Cleverness and eloquence on both sides certainly had their share in producing the very great and general disturbance of men's minds in the early days of Darwinian teaching. But by far the greater part of that disturbance was due to the practical novelty and the profound importance of the teaching itself, and to the fact that the controversy about evolution quickly became much more public than any controversy of equal seriousness had been for many generations.

We must not think lightly of that great disturbance because it has, in some real sense, done its work, and because it is impossible in days of more coolness and light, to recover a full sense of its very real difficulties.

Those who would know them better should add to the calm records of Darwin ("Life and Letters" and "More Letters of Charles Darwin".) and to the story of Huxley's impassioned championship, all that they can learn of George Romanes. ("Life and Letters", London, 1896. "Thoughts on Religion", London, 1895. "Candid Examination of Theism", London, 1878.) For his life was absorbed in this very struggle and reproduced its stages. It began in a certain assured simplicity of biblical interpretation; it went on, through the glories and adventures of a paladin in Darwin's train, to the darkness and dismay of a man who saw all his most cherished beliefs rendered, as he thought, incredible. ("Never in the history of man has so terrific a calamity befallen the race as that which all who look may now (viz. in consequence of the scientific victory of Darwin) behold advancing as a deluge, black with destruction, resistless in might, uprooting our most cherished hopes, engulphing our most precious creed, and burying our highest life in mindless destruction."—"A Candid Examination of Theism", page 51.) He lived to find the freer faith for which process and purpose are not irreconcilable, but necessary to one another. His development, scientific, intellectual and moral, was itself of high significance; and its record is of unique value to our own generation, so near the age of that doubt and yet so far from it; certainly still much in need of the caution and courage by which past endurance prepares men for new emergencies. We have little enough reason to be sure that in the discussions awaiting us we shall do as well as our predecessors in theirs. Remembering their endurance of mental pain, their ardour in mental labour, the heroic temper and the high sincerity of controversialists on either side, we may well speak of our fathers in such words of modesty and self-judgment as Drayton used when he sang the victors of Agincourt. The progress of biblical study, in the departments of Introduction and Exegesis, resulting in the recovery of a point of view anciently tolerated if not prevalent, has altered some of the conditions of that discussion. In the years near 1858, the witness of Scripture was adduced both by Christian advocates and their critics as if unmistakeably irreconcilable with Evolution.

Huxley ("Science and Christian Tradition". London, 1904.) found the path of the blameless naturalist everywhere blocked by "Moses": the believer in revelation was generally held to be forced to a choice between revealed cosmogony and the scientific account of origins. It is not clear how far the change in Biblical interpretation is due to natural science, and how far to the vital movements of theological study which have been quite independent of the controversy about species. It belongs to a general renewal of Christian movement, the recovery of a heritage. "Special Creation"—really a biological rather than a theological conception,—seems in its rigid form to have been a recent element even in English biblical orthodoxy.

The Middle Ages had no suspicion that religious faith forbad inquiry into the natural origination of the different forms of life. Bartholomaeus Anglicus, an English Franciscan of the thirteenth century, was a mutationist in his way, as Aristotle, "the Philosopher" of the Christian Schoolmen, had been in his. So late as the seventeenth century, as we learn not only from early proceedings of the Royal Society, but from a writer so homely and so regularly pious as Walton, the variation of species and "spontaneous" generations had no theological bearing, except as instances of that various wonder of the world which in devout minds is food for devotion.

It was in the eighteenth century that the harder statement took shape. Something in the preciseness of that age, its exaltation of law, its cold passion for a stable and measured universe, its cold denial, its cold affirmation of the power of God, a God of ice, is the occasion of that rigidity of religious thought about the living world which Darwin by accident challenged, or rather by one of those movements of genius which, Goethe ("No productiveness of the highest kind... is in the power of anyone."—"Conversations of Goethe with Eckermann and Soret". London, 1850.) declares, are "elevated above all earthly control."

If religious thought in the eighteenth century was aimed at a fixed and nearly finite world of spirit, it followed in all these respects the secular and critical lead. ("La philosophie reformatrice du XVIIIe siecle" (Berthelot, "Evolutionisme et Platonisme", Paris, 1908, page 45.) ramenait la nature et la societe a des mecanismes que la pensee reflechie peut concevoir et recomposer." In fact, religion in a mechanical age is condemned if it takes any but a mechanical tone. Butler's thought was too moving, too vital, too evolutionary, for the sceptics of his time. In a rationalist, encyclopaedic period, religion also must give hard outline to its facts, it must be able to display its secret to any sensible man in the language used by all sensible men. Milton's prophetic genius furnished the eighteenth century, out of the depth of the passionate age before it, with the theological tone it was to need. In spite of the austere magnificence of his devotion, he gives to smaller souls a dangerous lead. The rigidity of Scripture exegesis belonged to this stately but imperfectly sensitive mode of thought. It passed away with the influence of the older rationalists whose precise denials matched the precise and limited affirmations of the static orthodoxy.

I shall, then, leave the specially biblical aspect of the debate—interesting as it is and even useful, as in Huxley's correspondence with the Duke of Argyll and others in 1892 ("Times", 1892, passim.)—in order to consider without complication the permanent elements of Christian thought brought into question by the teaching of evolution.

Such permanent elements are the doctrine of God as Creator of the universe, and the doctrine of man as spiritual and unique. Upon both the doctrine of evolution seemed to fall with crushing force.

With regard to Man I leave out, acknowledging a grave omission, the doctrine of the Fall and of Sin. And I do so because these have not yet, as I believe, been adequately treated: here the fruitful reaction to the stimulus of evolution is yet to come. The doctrine of sin, indeed, falls principally within the scope of that discussion which has followed or displaced the Darwinian; and without it the Fall cannot be usefully considered. For the question about the Fall is a question not merely of origins, but of the interpretation of moral facts whose moral reality must first be established.

I confine myself therefore to Creation and the dignity of man.

The meaning of evolution, in the most general terms, is that the differentiation of forms is not essentially separate from their behaviour and use; that if these are within the scope of study, that is also; that the world has taken the form we see by movements not unlike those we now see in progress; that what may be called proximate origins are continuous in the way of force and matter, continuous in the way of life, with actual occurrences and actual characteristics. All this has no revolutionary bearing upon the question of ultimate origins. The whole is a statement about process. It says nothing to metaphysicians about cause. It simply brings within the scope of observation or conjecture that series of changes which has given their special characters to the different parts of the world we see. In particular, evolutionary science aspires to the discovery of the process or order of the appearance of life itself: if it were to achieve its aim it could say nothing of the cause of this or indeed of the most familiar occurrences. We should have become spectators or convinced historians of an event which, in respect of its cause and ultimate meaning, would be still impenetrable.

With regard to the origin of species, supposing life already established, biological science has the well founded hopes and the measure of success with which we are all familiar. All this has, it would seem, little chance of collision with a consistent theism, a doctrine which has its own difficulties unconnected with any particular view of order or process. But when it was stated that species had arisen by processes through which new species were still being made, evolutionism came into collision with a statement, traditionally religious, that species were formed and fixed once for all and long ago.

What is the theological import of such a statement when it is regarded as essential to belief in God? Simply that God's activity, with respect to the formation of living creatures, ceased at some point in past time.

"God rested" is made the touchstone of orthodoxy. And when, under the pressure of the evidences, we found ourselves obliged to acknowledge and assert the present and persistent power of God, in the maintenance and in the continued formation of "types," what happened was the abolition of a time-limit. We were forced only to a bolder claim, to a theistic language less halting, more consistent, more thorough in its own line, as well as better qualified to assimilate and modify such schemes as Von Hartmann's philosophy of the unconscious—a philosophy, by the way, quite intolerant of a merely mechanical evolution. (See Von Hartmann's "Wahrheit und Irrthum in Darwinismus". Berlin, 1875.)

Here was not the retrenchment of an extravagant assertion, but the expansion of one which was faltering and inadequate. The traditional statement did not need paring down so as to pass the meshes of a new and exacting criticism. It was itself a net meant to surround and enclose experience; and we must increase its size and close its mesh to hold newly disclosed facts of life. The world, which had seemed a fixed picture or model, gained first perspective and then solidity and movement. We had a glimpse of organic HISTORY; and Christian thought became more living and more assured as it met the larger view of life.

However unsatisfactory the new attitude might be to our critics, to Christians the reform was positive. What was discarded was a limitation, a negation. The movement was essentially conservative, even actually reconstructive. For the language disused was a language inconsistent with the definitions of orthodoxy; it set bounds to the infinite, and by implication withdrew from the creative rule all such processes as could be brought within the descriptions of research. It ascribed fixity and finality to that "creature" in which an apostle taught us to recognise the birth-struggles of an unexhausted progress. It tended to banish mystery from the world we see, and to confine it to a remote first age.

In the reformed, the restored, language of religion, Creation became again not a link in a rational series to complete a circle of the sciences, but the mysterious and permanent relation between the infinite and the finite, between the moving changes we know in part, and the Power, after the fashion of that observation, unknown, which is itself "unmoved all motion's source." (Hymn of the Church— Rerum Deus tenax vigor, Immotus in te permanens.)

With regard to man it is hardly necessary, even were it possible, to illustrate the application of this bolder faith. When the record of his high extraction fell under dispute, we were driven to a contemplation of the whole of his life, rather than of a part and that part out of sight. We remembered again, out of Aristotle, that the result of a process interprets its beginnings. We were obliged to read the title of such dignity as we may claim, in results and still more in aspirations.

Some men still measure the value of great present facts in life—reason and virtue and sacrifice—by what a self-disparaged reason can collect of the meaner rudiments of these noble gifts. Mr Balfour has admirably displayed the discrepancy, in this view, between the alleged origin and the alleged authority of reason. Such an argument ought to be used not to discredit the confident reason, but to illuminate and dignify its dark beginnings, and to show that at every step in the long course of growth a Power was at work which is not included in any term or in all the terms of the series.

I submit that the more men know of actual Christian teaching, its fidelity to the past, and its sincerity in face of discovery, the more certainly they will judge that the stimulus of the doctrine of evolution has produced in the long run vigour as well as flexibility in the doctrine of Creation and of man.

I pass from Evolution in general to Natural Selection.

The character in religious language which I have for short called mechanical was not absent in the argument from design as stated before Darwin. It seemed to have reference to a world conceived as fixed. It pointed, not to the plastic capacity and energy of living matter, but to the fixed adaptation of this and that organ to an unchanging place or function.

Mr Hobhouse has given us the valuable phrase "a niche of organic opportunity." Such a phrase would have borne a different sense in non-evolutionary thought. In that thought, the opportunity was an opportunity for the Creative Power, and Design appeared in the preparation of the organism to fit the niche. The idea of the niche and its occupant growing together from simpler to more complex mutual adjustment was unwelcome to this teleology. If the adaptation was traced to the influence, through competition, of the environment, the old teleology lost an illustration and a proof. For the cogency of the proof in every instance depended upon the absence of explanation. Where the process of adaptation was discerned, the evidence of Purpose or Design was weak. It was strong only when the natural antecedents were not discovered, strongest when they could be declared undiscoverable.

Paley's favourite word is "Contrivance"; and for him contrivance is most certain where production is most obscure. He points out the physiological advantage of the valvulae conniventes to man, and the advantage for teleology of the fact that they cannot have been formed by "action and pressure." What is not due to pressure may be attributed to design, and when a "mechanical" process more subtle than pressure was suggested, the case for design was so far weakened. The cumulative proof from the multitude of instances began to disappear when, in selection, a natural sequence was suggested in which all the adaptations might be reached by the motive power of life, and especially when, as in Darwin's teaching, there was full recognition of the reactions of life to the stimulus of circumstance. "The organism fits the niche," said the teleologist, "because the Creator formed it so as to fit." "The organism fits the niche," said the naturalist, "because unless it fitted it could not exist." "It was fitted to survive," said the theologian. "It survives because it fits," said the selectionist. The two forms of statement are not incompatible; but the new statement, by provision of an ideally universal explanation of process, was hostile to a doctrine of purpose which relied upon evidences always exceptional however numerous. Science persistently presses on to find the universal machinery of adaptation in this planet; and whether this be found in selection, or in direct-effect, or in vital reactions resulting in large changes, or in a combination of these and other factors, it must always be opposed to the conception of a Divine Power here and there but not everywhere active.

For science, the Divine must be constant, operative everywhere and in every quality and power, in environment and in organism, in stimulus and in reaction, in variation and in struggle, in hereditary equilibrium, and in "the unstable state of species"; equally present on both sides of every strain, in all pressures and in all resistances, in short in the general wonder of life and the world. And this is exactly what the Divine Power must be for religious faith.

The point I wish once more to make is that the necessary readjustment of teleology, so as to make it depend upon the contemplation of the whole instead of a part, is advantageous quite as much to theology as to science. For the older view failed in courage. Here again our theism was not sufficiently theistic.

Where results seemed inevitable, it dared not claim them as God-given. In the argument from Design it spoke not of God in the sense of theology, but of a Contriver, immensely, not infinitely wise and good, working within a world, the scene, rather than the ever dependent outcome, of His Wisdom; working in such emergencies and opportunities as occurred, by forces not altogether within His control, towards an end beyond Himself. It gave us, instead of the awful reverence due to the Cause of all substance and form, all love and wisdom, a dangerously detached appreciation of an ingenuity and benevolence meritorious in aim and often surprisingly successful in contrivance.

The old teleology was more useful to science than to religion, and the design-naturalists ought to be gratefully remembered by Biologists. Their search for evidences led them to an eager study of adaptations and of minute forms, a study such as we have now an incentive to in the theory of Natural Selection. One hardly meets with the same ardour in microscopical research until we come to modern workers. But the argument from Design was never of great importance to faith. Still, to rid it of this character was worth all the stress and anxiety of the gallant old war. If Darwin had done nothing else for us, we are to-day deeply in his debt for this. The world is not less venerable to us now, not less eloquent of the causing mind, rather much more eloquent and sacred. But our wonder is not that "the underjaw of the swine works under the ground" or in any or all of those particular adaptations which Paley collected with so much skill, but that a purpose transcending, though resembling, our own purposes, is everywhere manifest; that what we live in is a whole, mutually sustaining, eventful and beautiful, where the "dead" forces feed the energies of life, and life sustains a stranger existence, able in some real measure to contemplate the whole, of which, mechanically considered, it is a minor product and a rare ingredient. Here, again, the change was altogether positive. It was not the escape of a vessel in a storm with loss of spars and rigging, not a shortening of sail to save the masts and make a port of refuge. It was rather the emergence from narrow channels to an open sea. We had propelled the great ship, finding purchase here and there for slow and uncertain movement. Now, in deep water, we spread large canvas to a favouring breeze.

The scattered traces of design might be forgotten or obliterated. But the broad impression of Order became plainer when seen at due distance and in sufficient range of effect, and the evidence of love and wisdom in the universe could be trusted more securely for the loss of the particular calculation of their machinery.

Many other topics of faith are affected by modern biology. In some of these we have learnt at present only a wise caution, a wise uncertainty. We stand before the newly unfolded spectacle of suffering, silenced; with faith not scientifically reassured but still holding fast certain other clues of conviction. In many important topics we are at a loss. But in others, and among them those I have mentioned, we have passed beyond this negative state and find faith positively strengthened and more fully expressed.

We have gained also a language and a habit of thought more fit for the great and dark problems that remain, less liable to damaging conflicts, equipped for more rapid assimilation of knowledge. And by this change biology itself is a gainer. For, relieved of fruitless encounters with popular religion, it may advance with surer aim along the path of really scientific life-study which was reopened for modern men by the publication of "The Origin of Species".

Charles Darwin regretted that, in following science, he had not done "more direct good" ("Life and Letters", Vol. III. page 359.) to his fellow-creatures. He has, in fact, rendered substantial service to interests bound up with the daily conduct and hopes of common men; for his work has led to improvements in the preaching of the Christian faith.

XXV. THE INFLUENCE OF DARWINISM ON THE STUDY OF RELIGIONS. By Jane Ellen Harrison.

Hon. D.Litt. (Durham), Hon. LL.D. (Aberdeen), Staff Lecturer and sometime Fellow of Newnham College, Cambridge.

Corresponding member of the German Archaeological Institute.

The title of my paper might well have been "the creation by Darwinism of the scientific study of Religions," but that I feared to mar my tribute to a great name by any shadow of exaggeration. Before the publication of "The Origin of Species" and "The Descent of Man", even in the eighteenth century, isolated thinkers, notably Hume and Herder, had conjectured that the orthodox beliefs of their own day were developments from the cruder superstitions of the past. These were however only particular speculations of individual sceptics. Religion was not yet generally regarded as a proper subject for scientific study, with facts to be collected and theories to be deduced. A Congress of Religions such as that recently held at Oxford would have savoured of impiety.

In the brief space allotted me I can attempt only two things; first, and very briefly, I shall try to indicate the normal attitude towards religion in the early part of the last century; second, and in more detail, I shall try to make clear what is the outlook of advanced thinkers to-day. (To be accurate I ought to add "in Europe." I advisedly omit from consideration the whole immense field of Oriental mysticism, because it has remained practically untouched by the influence of Darwinism.) From this second inquiry it will, I hope, be abundantly manifest that it is the doctrine of evolution that has made this outlook possible and even necessary.

The ultimate and unchallenged presupposition of the old view was that religion was a DOCTRINE, a body of supposed truths. It was in fact what we should now call Theology, and what the ancients called Mythology. Ritual was scarcely considered at all, and, when considered, it was held to be a form in which beliefs, already defined and fixed as dogma, found a natural mode of expression. This, it will be later shown, is a profound error or rather a most misleading half-truth. Creeds, doctrines, theology and the like are only a part, and at first the least important part, of religion.

Further, and the fact is important, this DOGMA, thus supposed to be the essential content of the "true" religion, was a teleological scheme complete and unalterable, which had been revealed to man once and for all by a highly anthropomorphic God, whose existence was assumed. The duty of man towards this revelation was to accept its doctrines and obey its precepts. The notion that this revelation had grown bit by bit out of man's consciousness and that his business was to better it would have seemed rank blasphemy. Religion, so conceived, left no place for development. "The Truth" might be learnt, but never critically examined; being thus avowedly complete and final, it was doomed to stagnation.

The details of this supposed revelation seem almost too naive for enumeration. As Hume observed, "popular theology has a positive appetite for absurdity." It is sufficient to recall that "revelation" included such items as the Creation (It is interesting to note that the very word "Creator" has nowadays almost passed into the region of mythology. Instead we have "L'Evolution Creatrice".) of the world out of nothing in six days; the making of Eve from one of Adam's ribs; the Temptation by a talking snake; the confusion of tongues at the tower of Babel; the doctrine of Original Sin; a scheme of salvation which demanded the Virgin Birth, Vicarious Atonement, and the Resurrection of the material body. The scheme was unfolded in an infallible Book, or, for one section of Christians, guarded by the tradition of an infallible Church, and on the acceptance or refusal of this scheme depended an eternity of weal or woe. There is not one of these doctrines that has not now been recast, softened down, mysticised, allegorised into something more conformable with modern thinking. It is hard for the present generation, unless their breeding has been singularly archaic, to realise that these amazing doctrines were literally held and believed to constitute the very essence of religion; to doubt them was a moral delinquency.

It had not, however, escaped the notice of travellers and missionaries that savages carried on some sort of practices that seemed to be religious, and believed in some sort of spirits or demons. Hence, beyond the confines illuminated by revealed truth, a vague region was assigned to NATURAL Religion. The original revelation had been kept intact only by one chosen people, the Jews, by them to be handed on to Christianity. Outside the borders of this Goshen the world had sunk into the darkness of Egypt. Where analogies between savage cults and the Christian religions were observed, they were explained as degradations; the heathen had somehow wilfully "lost the light." Our business was not to study but, exclusively, to convert them, to root out superstition and carry the torch of revelation to "Souls in heathen darkness lying." To us nowadays it is a commonplace of anthropological research that we must seek for the beginnings of religion in the religions of primitive peoples, but in the last century the orthodox mind was convinced that it possessed a complete and luminous ready-made revelation; the study of what was held to be a mere degradation seemed idle and superfluous.

But, it may be asked, if, to the orthodox, revealed religion was sacrosanct and savage religion a thing beneath consideration, why did not the sceptics show a more liberal spirit, and pursue to their logical issue the conjectures they had individually hazarded? The reason is simple and significant. The sceptics too had not worked free from the presupposition that the essence of religion is dogma. Their intellectualism, expressive of the whole eighteenth century, was probably in England strengthened by the Protestant doctrine of an infallible Book. Hume undoubtedly confused religion with dogmatic theology. The attention of orthodox and sceptics alike was focussed on the truth or falsity of certain propositions. Only a few minds of rare quality were able dimly to conceive that religion might be a necessary step in the evolution of human thought.

It is not a little interesting to note that Darwin, who was leader and intellectual king of his generation, was also in this matter to some extent its child. His attitude towards religion is stated clearly, in Chapter VIII. of the "Life and Letters". (Vol. I. page 304. For Darwin's religious views see also "Descent of Man", 1871, Vol. I. page 65; 2nd edition. Vol. I. page 142.) On board the "Beagle" he was simply orthodox and was laughed at by several of the officers for quoting the Bible as an unanswerable authority on some point of morality. By 1839 he had come to see that the Old Testament was no more to be trusted than the sacred books of the Hindoos. Next went the belief in miracles, and next Paley's "argument from design" broke down before the law of natural selection; the suffering so manifest in nature is seen to be compatible rather with Natural Selection than with the goodness and omnipotence of God. Darwin felt to the full all the ignorance that lay hidden under specious phrases like "the plan of creation" and "Unity of design." Finally, he tells us "the mystery of the beginning of all things is insoluble by us; and I for one must be content to remain an Agnostic."

The word Agnostic is significant not only of the humility of the man himself but also of the attitude of his age. Religion, it is clear, is still conceived as something to be KNOWN, a matter of true or false OPINION. Orthodox religion was to Darwin a series of erroneous hypotheses to be bit by bit discarded when shown to be untenable. The ACTS of religion which may result from such convictions, i.e. devotion in all its forms, prayer, praise, sacraments, are left unmentioned. It is clear that they are not, as now to us, sociological survivals of great interest and importance, but rather matters too private, too personal, for discussion.

Huxley, writing in the "Contemporary Review" (1871.), says, "In a dozen years "The Origin of Species" has worked as complete a revolution in biological science as the "Principia" did in astronomy." It has done so because, in the words of Helmholtz, it contained "an essentially new creative thought," that of the continuity of life, the absence of breaks. In the two most conservative subjects, Religion and Classics, this creative ferment was slow indeed to work. Darwin himself felt strongly "that a man should not publish on a subject to which he has not given special and continuous thought," and hence wrote little on religion and with manifest reluctance, though, as already seen, in answer to pertinacious inquiry he gave an outline of his own views. But none the less he foresaw that his doctrine must have, for the history of man's mental evolution, issues wider than those with which he was prepared personally to deal. He writes, in "The Origin of Species" (6th edition, page 428.), "In the future I see open fields for far more important researches. Psychology will be securely based on the foundation already well laid by Mr Herbert Spencer, that of the necessary acquirement of each mental power and capacity by gradation."

Nowhere, it is true, does Darwin definitely say that he regarded religion as a set of phenomena, the development of which may be studied from the psychological standpoint.

Rather we infer from his PIETY—in the beautiful Roman sense—towards tradition and association, that religion was to him in some way sacrosanct. But it is delightful to see how his heart went out towards the new method in religious study which he had himself, if half-unconsciously, inaugurated. Writing in 1871 to Dr Tylor, on the publication of his "Primitive Culture", he says ("Life and Letters", Vol. III. page 151.), "It is wonderful how you trace animism from the lower races up the religious belief of the highest races. It will make me for the future look at religion—a belief in the soul, etc.—from a new point of view."

Psychology was henceforth to be based on "the necessary acquirement of each mental capacity by gradation." With these memorable words the door closes on the old and opens on the new horizon. The mental focus henceforth is not on the maintaining or refuting of an orthodoxy but on the genesis and evolution of a capacity, not on perfection but on process. Continuous evolution leaves no gap for revelation sudden and complete. We have henceforth to ask, not when was religion revealed or what was the revelation, but how did religious phenomena arise and develop. For an answer to this we turn with new and reverent eyes to study "the heathen in his blindness" and the child "born in sin." We still indeed send out missionaries to convert the heathen, but here at least in Cambridge before they start they attend lectures on anthropology and comparative religion. The "decadence" theory is dead and should be buried.

The study of primitive religions then has been made possible and even inevitable by the theory of Evolution. We have now to ask what new facts and theories have resulted from that study. This brings us to our second point, the advanced outlook on religion to-day.

The view I am about to state is no mere personal opinion of my own. To my present standpoint I have been led by the investigations of such masters as Drs Wundt, Lehmann, Preuss, Bergson, Beck and in our own country Drs Tylor and Frazer. (I can only name here the books that have specially influenced my own views. They are W. Wundt, "Volkerpsychologie", Leipzig, 1900, P. Beck, "Die Nachahmung", Leipzig, 1904, and "Erkenntnisstheorie des primitiven Denkens" in "Zeitschrift f. Philos. und Philos. Kritik", 1903, page 172, and 1904, page 9. Henri Bergson, "L'Evolution Creatrice" and "Matiere et Memoire", 1908, K. Th. Preuss, various articles published in the "Globus" (see page 507, note 1), and in the "Archiv. f. Religionswissenschaft", and for the subject of magic, MM. Hubert et Mauss, "Theorie generale de la Magie", in "L'Annee Sociologique", VII.)

Religion always contains two factors. First, a theoretical factor, what a man THINKS about the unseen—his theology, or, if we prefer so to call it, his mythology. Second, what he DOES in relation to this unseen—his ritual. These factors rarely if ever occur in complete separation; they are blended in very varying proportions. Religion we have seen was in the last century regarded mainly in its theoretical aspect as a doctrine. Greek religion for example meant to most educated persons Greek mythology. Yet even a cursory examination shows that neither Greek nor Roman had any creed or dogma, any hard and fast formulation of belief. In the Greek Mysteries (See my "Prolegomena to the Study of Greek Religion", page 155, Cambridge, 1903.) only we find what we should call a Confiteor; and this is not a confession of faith, but an avowal of rites performed. When the religion of primitive peoples came to be examined it was speedily seen that though vague beliefs necessarily abound, definite creeds are practically non-existent. Ritual is dominant and imperative.

This predominance and priority of ritual over definite creed was first forced upon our notice by the study of savages, but it promptly and happily joined hands with modern psychology. Popular belief says, I think, therefore I act; modern scientific psychology says, I act (or rather, REact to outside stimulus), and so I come to think. Thus there is set going a recurrent series: act and thought become in their turn stimuli to fresh acts and thoughts. In examining religion as envisaged to-day it would therefore be more correct to begin with the practice of religion, i.e. ritual, and then pass to its theory, theology or mythology. But it will be more convenient to adopt the reverse method. The theoretical content of religion is to those of us who are Protestants far more familiar and we shall thus proceed from the known to the comparatively unknown.

I shall avoid all attempt at rigid definition. The problem before the modern investigator is, not to determine the essence and definition of religion but to inquire how religious phenomena, religious ideas and practices arose. Now the theoretical content of religion, the domain of theology or mythology, is broadly familiar to all. It is the world of the unseen, the supersensuous; it is the world of what we call the soul and the supposed objects of the soul's perception, sprites, demons, ghosts and gods. How did this world grow up?

We turn to our savages. Intelligent missionaries of bygone days used to ply savages with questions such as these: Had they any belief in God? Did they believe in the immortality of the soul? Taking their own clear-cut conceptions, discriminated by a developed terminology, these missionaries tried to translate them into languages that had neither the words nor the thoughts, only a vague, inchoate, tangled substratum, out of which these thoughts and words later differentiated themselves. Let us examine this substratum.

Nowadays we popularly distinguish between objective and subjective; and further, we regard the two worlds as in some sense opposed. To the objective world we commonly attribute some reality independent of consciousness, while we think of the subjective as dependent for its existence on the mind. The objective world consists of perceptible things, or of the ultimate constituents to which matter is reduced by physical speculation. The subjective world is the world of beliefs, hallucinations, dreams, abstract ideas, imaginations and the like. Psychology of course knows that the objective and subjective worlds are interdependent, inextricably intertwined, but for practical purposes the distinction is convenient.

But primitive man has not yet drawn the distinction between objective and subjective. Nay, more, it is foreign to almost the whole of ancient philosophy. Plato's Ideas (I owe this psychological analysis of the elements of the primitive supersensuous world mainly to Dr Beck, "Erkenntnisstheorie des primitiven Denkens", see page 498, note 1.), his Goodness, Truth, Beauty, his class-names, horse, table, are it is true dematerialised as far as possible, but they have outside existence, apart from the mind of the thinker, they have in some shadowy way spatial extension. Yet ancient philosophies and primitive man alike needed and possessed for practical purposes a distinction which served as well as our subjective and objective. To the primitive savage all his thoughts, every object of which he was conscious, whether by perception or conception, had reality, that is, it had existence outside himself, but it might have reality of various kinds or different degrees.

It is not hard to see how this would happen. A man's senses may mislead him. He sees the reflection of a bird in a pond. To his eyes it is a real bird. He touches it, HE PUTS IT TO THE TOUCH, and to his touch it is not a bird at all. It is real then, but surely not quite so real as a bird that you can touch. Again, he sees smoke. It is real to his eyes. He tries to grasp it, it vanishes. The wind touches him, but he cannot see it, which makes him feel uncanny. The most real thing is that which affects most senses and especially what affects the sense of touch. Apparently touch is the deepest down, most primitive, of senses. The rest are specialisations and complications. Primitive man has no formal rubric "optical delusion," but he learns practically to distinguish between things that affect only one sense and things that affect two or more—if he did not he would not survive. But both classes of things are real to him. Percipi est esse.

So far, primitive man has made a real observation; there are things that appeal to one sense only. But very soon creeps in confusion fraught with disaster. He passes naturally enough, being economical of any mental effort, from what he really sees but cannot feel to what he thinks he sees, and gives to it the same secondary reality. He has dreams, visions, hallucinations, nightmares. He dreams that an enemy is beating him, and he wakes rubbing his head. Then further he remembers things; that is, for him, he sees them. A great chief died the other day and they buried him, but he sees him still in his mind, sees him in his war-paint, splendid, victorious. So the image of the past goes together with his dreams and visions to the making of this other less real, but still real world, his other-world of the supersensuous, the supernatural, a world, the outside existence of which, independent of himself, he never questions.

And, naturally enough, the future joins the past in this supersensuous world. He can hope, he can imagine, he can prophesy. And again the images of his hope are real; he sees them with that mind's eye which as yet he has not distinguished from his bodily eye. And so the supersensuous world grows and grows big with the invisible present, and big also with the past and the future, crowded with the ghosts of the dead and shadowed with oracles and portents. It is this supersensuous, supernatural world which is the eternity, the other-world, of primitive religion, not an endlessness of time, but a state removed from full sensuous reality, a world in which anything and everything may happen, a world peopled by demonic ancestors and liable to a splendid vagueness, to a "once upon a time-ness" denied to the present. It not unfrequently happens that people who know that the world nowadays obeys fixed laws have no difficulty in believing that six thousand years ago man was made direct from a lump of clay, and woman was made from one of man's superfluous ribs.

The fashioning of the supersensuous world comes out very clearly in primitive man's views about the soul and life after death. Herbert Spencer noted long ago the influence of dreams in forming a belief in immortality, but being very rational himself, he extended to primitive man a quite alien quality of rationality. Herbert Spencer argued that when a savage has a dream he seeks to account for it, and in so doing invents a spirit world. The mistake here lies in the "seeks to account for it." (Primitive man, as Dr Beck observes, is not impelled by an Erkenntnisstrieb. Dr Beck says he has counted upwards of 30 of these mythological Triebe (tendencies) with which primitive man has been endowed.) Man is at first too busy LIVING to have any time for disinterested THINKING. He dreams a dream and it is real for him. He does not seek to account for it any more than for his hands and feet. He cannot distinguish between a CONception and a PERception, that is all. He remembers his ancestors or they appear to him in a dream; therefore they are alive still, but only as a rule to about the third generation. Then he remembers them no more and they cease to be.

Next as regards his own soul. He feels something within him, his life-power, his will to live, his power to act, his personality—whatever we like to call it. He cannot touch this thing that is himself, but it is real. His friend too is alive and one day he is dead; he cannot move, he cannot act. Well, something has gone that was his friend's self. He has stopped breathing. Was it his breath? or he is bleeding; is it his blood? This life-power IS something; does it live in his heart or his lungs or his midriff? He did not see it go; perhaps it is like wind, an anima, a Geist, a ghost. But again it comes back in a dream, only looking shadowy; it is not the man's life, it is a thin copy of the man; it is an "image" (eidolon). It is like that shifting distorted thing that dogs the living man's footsteps in the sunshine; it is a "shade" (skia). (The two conceptions of the soul, as a life-essence, inseparable from the body, and as a separable phantom seem to occur in most primitive systems. They are distinct conceptions but are inextricably blended in savage thought. The two notions Korperseele and Psyche have been very fully discussed in Wundt's "Volkerpsychologie" II. pages 1-142, Leipzig, 1900.)

Ghosts and sprites, ancestor worship, the soul, oracles, prophecy; all these elements of the primitive supersensuous world we willingly admit to be the proper material of religion; but other elements are more surprising; such are class-names, abstract ideas, numbers, geometrical figures. We do not nowadays think of these as of religious content, but to primitive men they were all part of the furniture of his supernatural world.

With respect to class-names, Dr Tylor ("Primitive Culture", Vol. II. page 245 (4th edition), 1903.) has shown how instructive are the first attempts of the savage to get at the idea of a class. Things in which similarity is observed, things indeed which can be related at all are to the savage KINDRED. A species is a family or a number of individuals with a common god to look after them. Such for example is the Finn doctrine of the haltia. Every object has its haltia, but the haltiat were not tied to the individual, they interested themselves in every member of the species. Each stone had its haltia, but that haltia was interested in other stones; the individuals disappeared, the haltia remained.

Nor was it only class-names that belonged to the supersensuous world. A man's own proper-name is a sort of spiritual essence of him, a kind of soul to be carefully concealed. By pronouncing a name you bring the thing itself into being. When Elohim would create Day "he called out to the Light 'Day,' and to the Darkness he called out 'Night'"; the great magician pronounced the magic Names and the Things came into being. "In the beginning was the Word" is literally true, and this reflects the fact that our CONCEPTUAL world comes into being by the mental process of naming. (For a full discussion of this point see Beck, "Nachahmung" page 41, "Die Sprache".) In old times people went further; they thought that by naming events they could bring them to be, and custom even to-day keeps up the inveterate magical habit of wishing people "Good Morning" and a "Happy Christmas."

Number, too, is part of the supersensuous world that is thoroughly religious. We can see and touch seven apples, but seven itself, that wonderful thing that shifts from object to object, giving it its SEVENness, that living thing, for it begets itself anew in multiplication— surely seven is a fit denizen of the upper-world. Originally all numbers dwelt there, and a certain supersensuous sanctity still clings to seven and three. We still say "Holy, Holy, Holy," and in some mystic way feel the holier.

The soul and the supersensuous world get thinner and thinner, rarer and more rarified, but they always trail behind them clouds of smoke and vapour from the world of sense and space whence they have come. It is difficult for us even nowadays to use the word "soul" without lapsing into a sensuous mythology. The Cartesians' sharp distinction between res extensa non cogitans and res cogitans non extansa is remote.

So far then man, through the processes of his thinking, has provided himself with a supersensuous world, the world of sense-delusion, of smoke and cloud, of dream and phantom, of imagination, of name and number and image. The natural course would now seem to be that this supersensuous world should develop into the religious world as we know it, that out of a vague animism with ghosts of ancestors, demons, and the like, there should develop in due order momentary gods (Augenblicks-Gotter), tribal gods, polytheism, and finally a pure monotheism.

This course of development is usually assumed, but it is not I think quite what really happens. The supersensuous world as we have got it so far is too theoretic to be complete material of religion. It is indeed only one factor, or rather it is as it were a lifeless body that waits for a living spirit to possess and inform it. Had the theoretic factor remained uninformed it would eventually have separated off into its constituent elements of error and truth, the error dying down as a belated metaphysic, the truth developing into a correct and scientific psychology of the subjective. But man has ritual as well as mythology; that is, he feels and acts as well as thinks; nay more he probably feels and acts long before he definitely thinks. This contradicts all our preconceived notions of theology. Man, we imagine, believes in a god or gods and then worships. The real order seems to be that, in a sense presently to be explained, he worships, he feels and acts, and out of his feeling and action, projected into his confused thinking, he develops a god. We pass therefore to our second factor in religion:—ritual.

The word "ritual" brings to our modern minds the notion of a church with a priesthood and organised services. Instinctively we think of a congregation meeting to confess sins, to receive absolution, to pray, to praise, to listen to sermons, and possibly to partake of sacraments. Were we to examine these fully developed phenomena we should hardly get further in the analysis of our religious conceptions than the notion of a highly anthropomorphic god approached by purely human methods of personal entreaty and adulation.

Further, when we first come to the study of primitive religions we expect a priori to find the same elements, though in a ruder form. We expect to see "The heathen in his blindness bow down to wood and stone," but the facts that actually confront us are startlingly dissimilar. Bowing down to wood and stone is an occupation that exists mainly in the minds of hymn-writers. The real savage is more actively engaged. Instead of asking a god to do what he wants done, he does it or tries to do it himself; instead of prayers he utters spells. In a word he is busy practising magic, and above all he is strenuously engaged in dancing magical dances. When the savage wants rain or wind or sunshine, he does not go to church; he summons his tribe and they dance a rain-dance or wind-dance or sun-dance. When a savage goes to war we must not picture his wife on her knees at home praying for the absent; instead we must picture her dancing the whole night long; not for mere joy of heart or to pass the weary hours; she is dancing his war-dance to bring him victory.

Magic is nowadays condemned alike by science and by religion; it is both useless and impious. It is obsolete, and only practised by malign sorcerers in obscure holes and corners. Undoubtedly magic is neither religion nor science, but in all probability it is the spiritual protoplasm from which religion and science ultimately differentiated. As such the doctrine of evolution bids us scan it closely. Magic may be malign and private; nowadays it is apt to be both. But in early days magic was as much for good as for evil; it was publicly practised for the common weal.

The gist of magic comes out most clearly in magical dances. We think of dancing as a light form of recreation, practised by the young from sheer joie de vivre and unsuitable for the mature. But among the Tarahumares (Carl Lumholtz, "Unknown Mexico", page 330, London, 1903.) in Mexico the word for dancing, nolavoa, means "to work." Old men will reproach young men saying "Why do you not go to work?" meaning why do you not dance instead of only looking on. The chief religious sin of which the Tarahumare is conscious is that he has not danced enough and not made enough tesvino, his cereal intoxicant.

Dancing then is to the savage WORKING, DOING, and the dance is in its origin an imitation or perhaps rather an intensification of processes of work. (Karl Bucher, "Arbeit und Rhythmus", Leipzig (3rd edition), 1902, passim.) Repetition, regular and frequent, constitutes rhythm and rhythm heightens the sense of will power in action. Rhythmical action may even, as seen in the dances of Dervishes, produce a condition of ecstasy. Ecstasy among primitive peoples is a condition much valued; it is often, though not always, enhanced by the use of intoxicants. Psychologically the savage starts from the sense of his own will power, he stimulates it by every means at his command. Feeling his will strongly and knowing nothing of natural law he recognises no limits to his own power; he feels himself a magician, a god; he does not pray, he WILLS. Moreover he wills collectively (The subject of collective hallucination as an element in magic has been fully worked out by MM. Hubert and Mauss. "Theorie generale de la Magie", In "L'Annee Sociologique", 1902—3, page 140.), reinforced by the will and action of his whole tribe. Truly of him it may be said "La vie deborde l'intelligence, l'intelligence c'est un retrecissement." (Henri Bergson, "L'Evolution Creatrice", page 50.)

The magical extension and heightening of personality come out very clearly in what are rather unfortunately known as MIMETIC dances. Animal dances occur very frequently among primitive peoples. The dancers dress up as birds, beasts, or fishes, and reproduce the characteristic movements and habits of the animals impersonated. (So characteristic is this impersonation in magical dancing that among the Mexicans the word for magic, navali, means "disguise." K. Th. Preuss, "Archiv f. Religionswissenschaft", 1906, page 97.) A very common animal dance is the frog-dance. When it rains the frogs croak. If you desire rain you dress up like a frog and croak and jump. We think of such a performance as a conscious imitation. The man, we think, is more or less LIKE a frog. That is not how primitive man thinks; indeed, he scarcely thinks at all; what HE wants done the frog can do by croaking and jumping, so he croaks and jumps and, for all he can, BECOMES a frog. "L'intelligence animale JOUE sans doute les representations plutot qu'elle ne les pense." (Bergson, "L'Evolution Creatrice", page 205.)

We shall best understand this primitive state of mind if we study the child "born in sin." If a child is "playing at lions" he does not IMITATE a lion, i.e. he does not consciously try to be a thing more or less like a lion, he BECOMES one. His reaction, his terror, is the same as if the real lion were there. It is this childlike power of utter impersonation, of BEING the thing we act or even see acted, this extension and intensification of our own personality that lives deep down in all of us and is the very seat and secret of our joy in the drama.

A child's mind is indeed throughout the best clue to the understanding of savage magic. A young and vital child knows no limit to his own will, and it is the only reality to him. It is not that he wants at the outset to fight other wills, but that they simply do not exist for him. Like the artist he goes forth to the work of creation, gloriously alone. His attitude towards other recalcitrant wills is "they simply must." Let even a grown man be intoxicated, be in love, or subject to an intense excitement, the limitations of personality again fall away. Like the omnipotent child he is again a god, and to him all things are possible. Only when he is old and weary does he cease to command fate.

The Iroquois (Hewitt, "American Anthropologist", IV. I. page 32, 1902, N.S.) of North America have a word, orenda, the meaning of which is easier to describe than to define, but it seems to express the very soul of magic. This orenda is your power to do things, your force, sometimes almost your personality. A man who hunts well has much and good orenda; the shy bird who escapes his snares has a fine orenda. The orenda of the rabbit controls the snow and fixes the depth to which it will fall. When a storm is brewing the magician is said to be making its orenda. When you yourself are in a rage, great is your orenda. The notes of birds are utterances of their orenda. When the maize is ripening, the Iroquois know it is the sun's heat that ripens it, but they know more; it is the cigala makes the sun to shine and he does it by chirping, by uttering his orenda. This orenda is sometimes very like the Greek thumos, your bodily life, your vigour, your passion, your power, the virtue that is in you to feel and do. This notion of orenda, a sort of pan-vitalism, is more fluid than animism, and probably precedes it. It is the projection of man's inner experience, vague and unanalysed, into the outer world.

The mana of the Melanesians (Codrington, "The Melanesians", pages 118, 119, 192, Oxford, 1891.) is somewhat more specialised—all men do not possess mana—but substantially it is the same idea. Mana is not only a force, it is also an action, a quality, a state, at once a substantive, an adjective, and a verb. It is very closely neighboured by the idea of sanctity. Things that have mana are tabu. Like orenda it manifests itself in noises, but specially mysterious ones, it is mana that is rustling in the trees. Mana is highly contagious, it can pass from a holy stone to a man or even to his shadow if it cross the stone. "All Melanesian religion," Dr Codrington says, "consists in getting mana for oneself or getting it used for one's benefit." (Codrington, "The Melanesians", page 120, Oxford, 1891.)

Specially instructive is a word in use among the Omaka (See Prof. Haddon, "Magic and Fetishism", page 60, London, 1906. Dr Vierkandt ("Globus", July, 1907, page 41) thinks that "Fernzauber" is a later development from Nahzauber.), wazhin-dhedhe, "directive energy, to send." This word means roughly what we should call telepathy, sending out your thought or will-power to influence another and affect his action. Here we seem to get light on what has always been a puzzle, the belief in magic exercised at a distance. For the savage will, distance is practically non-existent, his intense desire feels itself as non-spatial. (This notion of mana, orenda, wazhin-dhedhe and the like lives on among civilised peoples in such words as the Vedic brahman in the neuter, familiar to us in its masculine form Brahman. The neuter, brahman, means magic power of a rite, a rite itself, formula, charm, also first principle, essence of the universe. It is own cousin to the Greek dunamis and phusis. See MM. Hubert et Mauss, "Theorie generale de la Magie", page 117, in "L'Annee Sociologique", VII.)

Through the examination of primitive ritual we have at last got at one tangible, substantial factor in religion, a real live experience, the sense, that is, of will, desire, power actually experienced in person by the individual, and by him projected, extended into the rest of the world.

At this stage it may fairly be asked, though the question cannot with any certainty be answered, "at what point in the evolution of man does this religious experience come in?"

So long as an organism reacts immediately to outside stimulus, with a certainty and conformity that is almost chemical, there is, it would seem, no place, no possibility for magical experience. But when the germ appears of an intellect that can foresee an end not immediately realised, or rather when a desire arises that we feel and recognise as not satisfied, then comes in the sense of will and the impulse magically to intensify that will. The animal it would seem is preserved by instinct from drawing into his horizon things which do not immediately subserve the conservation of his species. But the moment man's life-power began to make on the outside world demands not immediately and inevitably realised in action (I owe this observation to Dr K. Th. Preuss. He writes ("Archiv f. Relig." 1906, page 98), "Die Betonung des Willens in den Zauberakten ist der richtige Kern. In der Tat muss der Mensch den Willen haben, sich selbst und seiner Umgebung besondere Fahigkeiten zuzuschreiben, und den Willen hat er, sobald sein Verstand ihn befahigt, EINE UBER DEN INSTINKT HINAUSGEHEN DER FURSORGE fur sich zu zeigen. SO LANGE IHN DER INSTINKT ALLEIN LEITET, KONNEN ZAUBERHANDLUNGEN NICHT ENSTEHEN." For more detailed analysis of the origin of magic, see Dr Preuss "Ursprung der Religion und Kunst", "Globus", LXXXVI. and LXXXVII.), then a door was opened to magic, and in the train of magic followed errors innumerable, but also religion, philosophy, science and art.

The world of mana, orenda, brahman is a world of feeling, desiring, willing, acting. What element of thinking there may be in it is not yet differentiated out. But we have already seen that a supersensuous world of thought grew up very early in answer to other needs, a world of sense-illusions, shadows, dreams, souls, ghosts, ancestors, names, numbers, images, a world only wanting as it were the impulse of mana to live as a religion. Which of the two worlds, the world of thinking or the world of doing, developed first it is probably idle to inquire. (If external stimuli leave on organisms a trace or record such as is known as an Engram, this physical basis of memory and hence of thought is almost coincident with reaction of the most elementary kind. See Mr Francis Darwin's Presidential Address to the British Association, Dublin, 1908, page 8, and again Bergson places memory at the very root of conscious existence, see "L'Evolution Creatrice", page 18, "le fond meme de notre existence consciente est memoire, c'est a dire prolongation du passee dans le present," and again "la duree mord dans le temps et y laisse l'enpreint de son dent," and again, "l'Evolution implique une continuation reelle du passee par le present.")

It is more important to ask, Why do these two worlds join? Because, it would seem, mana, the egomaniac or megalomaniac element, cannot get satisfied with real things, and therefore goes eagerly out to a false world, the supersensuous other-world whose growth we have sketched. This junction of the two is fact, not fancy. Among all primitive peoples dead men, ghosts, spirits of all kinds, become the chosen vehicle of mana. Even to this day it is sometimes urged that religion, i.e. belief in the immortality of the soul, is true "because it satisfies the deepest craving of human nature." The two worlds, of mana and magic on the one hand, of ghosts and other-world on the other, combine so easily because they have the same laws, or rather the same comparative absence of law. As in the world of dreams and ghosts, so in the world of mana, space and time offer no obstacles; with magic all things are possible. In the one world what you imagine is real; in the other what you desire is ipso facto accomplished. Both worlds are egocentric, megalomaniac, filled to the full with unbridled human will and desire.

We are all of us born in sin, in that sin which is to science "the seventh and deadliest," anthropomorphism, we are egocentric, ego-projective. Hence necessarily we make our gods in our own image. Anthropomorphism is often spoken of in books on religion and mythology as if it were a last climax, a splendid final achievement in religious thought. First, we are told, we have the lifeless object as god (fetichism), then the plant or animal (phytomorphism, theriomorphism), and last God is incarnate in the human form divine. This way of putting things is misleading. Anthropomorphism lies at the very beginning of our consciousness. Man's first achievement in thought is to realise that there is anything at all not himself, any object to his subject. When he has achieved however dimly this distinction, still for long, for very long he can only think of those other things in terms of himself; plants and animals are people with ways of their own, stronger or weaker than himself but to all intents and purposes human.

Again the child helps us to understand our own primitive selves. To children animals are always people. You promise to take a child for a drive. The child comes up beaming with a furry bear in her arms. You say the bear cannot go. The child bursts into tears. You think it is because the child cannot endure to be separated from a toy. It is no such thing. It is the intolerable hurt done to the bear's human heart—a hurt not to be healed by any proffer of buns. He wanted to go, but he was a shy, proud bear, and he would not say so.

The relation of magic to religion has been much disputed. According to one school religion develops out of magic, according to another, though they ultimately blend, they are at the outset diametrically opposed, magic being a sort of rudimentary and mistaken science (This view held by Dr Frazer is fully set forth in his "Golden Bough" (2nd edition), pages 73-79, London, 1900. It is criticised by Mr R.R. Marett in "From Spell to Prayer", "Folk-Lore" XI. 1900, page 132, also very fully by MM. Hubert and Mauss, "Theorie generale de la Magie", in "L'Annee Sociologique", VII. page 1, with Mr Marett's view and with that of MM. Hubert and Mauss I am in substantial agreement.), religion having to do from the outset with spirits.

But, setting controversy aside, at the present stage of our inquiry their relation becomes, I think, fairly clear. Magic is, if my view (This view as explained above is, I believe, my own most serious contribution to the subject. In thinking it out I was much helped by Prof. Gilbert Murray.) be correct, the active element which informs a supersensuous world fashioned to meet other needs. This blend of theory and practice it is convenient to call religion. In practice the transition from magic to religion, from Spell to Prayer, has always been found easy. So long as mana remains impersonal you order it about; when it is personified and bulks to the shape of an overgrown man, you drop the imperative and cringe before it. "My will be done" is magic, "Thy Will be done" is the last word in religion. The moral discipline involved in the second is momentous, the intellectual advance not striking.

I have spoken of magical ritual as though it were the informing life-spirit without which religion was left as an empty shell. Yet the word ritual does not, as normally used, convey to our minds this notion of intense vitalism. Rather we associate ritual with something cut and dried, a matter of prescribed form and monotonous repetition. The association is correct; ritual tends to become less and less informed by the life-impulse, more and more externalised. Dr Beck ("Die Nachahmung und ihre Bedeutung fur Psychologie und Volkerkunde", Leipzig, 1904.) in his brilliant monograph on "Imitation" has laid stress on the almost boundless influence of the imitation of one man by another in the evolution of civilisation. Imitation is one of the chief spurs to action. Imitation begets custom, custom begets sanctity. At first all custom is sacred. To the savage it is as much a religious duty to tattoo himself as to sacrifice to his gods. But certain customs naturally survive, because they are really useful; they actually have good effects, and so need no social sanction. Others are really useless; but man is too conservative and imitative to abandon them. These become ritual. Custom is cautious, but la vie est aleatoire. (Bergson, op. cit. page 143.)

Dr Beck's remarks on ritual are I think profoundly true and suggestive, but with this reservation—they are true of ritual only when uninformed by personal experience. The very elements in ritual on which Dr Beck lays such stress, imitation, repetition, uniformity and social collectivity, have been found by the experience of all time to have a twofold influence—they inhibit the intellect, they stimulate and suggest emotion, ecstasy, trance. The Church of Rome knows what she is about when she prescribes the telling of the rosary. Mystery-cults and sacraments, the lineal descendants of magic, all contain rites charged with suggestion, with symbols, with gestures, with half-understood formularies, with all the apparatus of appeal to emotion and will—the more unintelligible they are the better they serve their purpose of inhibiting thought. Thus ritual deadens the intellect and stimulates will, desire, emotion. "Les operations magiques... sont le resultat d'une science et d'une habitude qui exaltent la volonte humaine au-dessus de ses limites habituelles." (Eliphas Levi, "Dogme et Rituel de la haute Magie", II. page 32, Paris, 1861, and "A defence of Magic", by Evelyn Underhill, "Fortnightly Review", 1907.) It is this personal EXPERIENCE, this exaltation, this sense of immediate, non-intellectual revelation, of mystical oneness with all things, that again and again rehabilitates a ritual otherwise moribund.

To resume. The outcome of our examination of ORIGINES seems to be that religious phenomena result from two delusive processes—a delusion of the non-critical intellect, a delusion of the over-confident will. Is religion then entirely a delusion? I think not. (I am deeply conscious that what I say here is a merely personal opinion or sentiment, unsupported and perhaps unsupportable by reason, and very possibly quite worthless, but for fear of misunderstanding I prefer to state it.) Every dogma religion has hitherto produced is probably false, but for all that the religious or mystical spirit may be the only way of apprehending some things and these of enormous importance. It may also be that the contents of this mystical apprehension cannot be put into language without being falsified and misstated, that they have rather to be felt and lived than uttered and intellectually analysed, and thus do not properly fall under the category of true or false, in the sense in which these words are applied to propositions; yet they may be something for which "true" is our nearest existing word and are often, if not necessary at least highly advantageous to life. That is why man through a series of more or less grossly anthropomorphic mythologies and theologies with their concomitant rituals tries to restate them. Meantime we need not despair. Serious psychology is yet young and has only just joined hands with physiology. Religious students are still hampered by mediaevalisms such as Body and Soul, and by the perhaps scarcely less mythological segregations of Intellect, Emotion, Will. But new facts (See the "Proceedings" of the Society for Psychical Research, London, passim, and especially Vols. VII.-XV. For a valuable collection of the phenomena of mysticism, see William James, "Varieties of Religious Experience", Edinburgh, 1901-2.) are accumulating, facts about the formation and flux of personality, and the relations between the conscious and the sub-conscious. Any moment some great imagination may leap out into the dark, touch the secret places of life, lay bare the cardinal mystery of the marriage of the spatial with the non-spatial. It is, I venture to think, towards the apprehension of such mysteries, not by reason only, but by man's whole personality, that the religious spirit in the course of its evolution through ancient magic and modern mysticism is ever blindly yet persistently moving.

Be this as it may, it is by thinking of religion in the light of evolution, not as a revelation given, not as a realite faite but as a process, and it is so only, I think, that we attain to a spirit of real patience and tolerance. We have ourselves perhaps learnt laboriously something of the working of natural law, something of the limitations of our human will, and we have therefore renounced the practice of magic. Yet we are bidden by those in high places to pray "Sanctify this water to the mystical washing away of sin." Mystical in this connection spells magical, and we have no place for a god-magician: the prayer is to us unmeaning, irreverent. Or again, after much toil we have ceased, or hope we have ceased, to think anthropomorphically. Yet we are invited to offer formal thanks to God for a meal of flesh whose sanctity is the last survival of that sacrifice of bulls and goats he has renounced. Such a ritual confuses our intellect and fails to stir our emotion. But to others this ritual, magical or anthropomorphic as it is, is charged with emotional impulse, and others, a still larger number, think that they act by reason when really they are hypnotised by suggestion and tradition; their fathers did this or that and at all costs they must do it. It was good that primitive man in his youth should bear the yoke of conservative custom; from each man's neck that yoke will fall, when and because he has outgrown it. Science teaches us to await that moment with her own inward and abiding patience. Such a patience, such a gentleness we may well seek to practise in the spirit and in the memory of Darwin.

XXVI. EVOLUTION AND THE
SCIENCE OF LANGUAGE. By P. Giles,
M.A., LL.D. (Aberdeen),

Reader in Comparative Philology in the
University of Cambridge.

In no study has the historical method had a more salutary influence than in the Science of Language. Even the earliest records show that the meaning of the names of persons, places, and common objects was then, as it has always been since, a matter of interest to mankind. And in every age the common man has regarded himself as competent without special training to explain by inspection (if one may use a mathematical phrase) the meaning of any words that attracted his attention. Out of this amateur etymologising has sprung a great amount of false history, a kind of historical mythology invented to explain familiar names. A single example will illustrate the tendency. According to the local legend the ancestor of the Earl of Erroll—a husbandman who stayed the flight of his countrymen in the battle of Luncarty and won the victory over the Danes by the help of the yoke of his oxen—exhausted with the fray uttered the exclamation "Hoch heigh!" The grateful king about to ennoble the victorious ploughman at once replied:

"Hoch heigh! said ye   And Hay shall ye be."

The Norman origin of the name Hay is well-known, and the battle of Luncarty long preceded the appearance of Normans in Scotland, but the legend nevertheless persists.

Though the earliest European treatise on philological questions which is now extant—the "Cratylus" of Plato,—as might be expected from its authorship, contains some acute thinking and some shrewd guesses, yet the work as a whole is infantine in its handling of language, and it has been doubted whether Plato was more than half serious in some of the suggestions which he puts forward. (For an account of the "Cratylus" with references to other literature see Sandys' "History of Classical Scholarship", I. page 92 ff., Cambridge, 1903.) In the hands of the Romans things were worse even than they had been in the hands of Plato and his Greek successors. The lack of success on the part of Varro and later Roman writers may have been partly due to the fact that, from the etymological point of view, Latin is a much more difficult language than Greek; it is by no means so closely connected with Greek as the ancients imagined, and they had no knowledge of the Celtic languages from which, on some sides at least, much greater light on the history of the Latin language might have been obtained. Roman civilisation was a late development compared with Greek, and its records dating earlier than 300 B.C.—a period when the best of Greek literature was already in existence—are very few and scanty. Varro it is true was much more of an antiquary than Plato, but his extant works seem to show that he was rather a "dungeon of learning" than an original thinker.

A scientific knowledge of language can be obtained only by comparison of different languages of the same family and the contrasting of their characteristics with those of another family or other families. It never occurred to the Greeks that any foreign language was worthy of serious study. Herodotus and other travellers and antiquaries indeed picked up individual words from various languages, either as being necessary in communication with the inhabitants of the countries where they sojourned, or because of some point which interested them personally. Plato and others noticed the similarity of some Phrygian words to Greek, but no systematic comparison seems ever to have been instituted.

In the Middle Ages the treatment of language was in a sense more historical. The Middle Ages started with the hypothesis, derived from the book of Genesis, that in the early world all men were of one language and of one speech. Though on the same authority they believed that the plain of Shinar has seen that confusion of tongues whence sprang all the languages upon earth, they seem to have considered that the words of each separate language were nevertheless derived from this original tongue. And as Hebrew was the language of the

Chosen People, it was naturally assumed that this original tongue was Hebrew. Hence we find Dante declaring in his treatise on the Vulgar Tongue (Dante "de Vulgari Eloquio", I. 4.) that the first word man uttered in Paradise must have been "El," the Hebrew name of his Maker, while as a result of the fall of Adam, the first utterance of every child now born into this world of sin and misery is "heu," Alas! After the splendidly engraved bronze plates containing, as we now know, ritual regulations for certain cults, were discovered in 1444 at the town of Gubbio, in Umbria, they were declared, by some authorities, to be written in excellent Hebrew. The study of them has been the fascination and the despair of many a philologist. Thanks to the devoted labours of numerous scholars, mainly in the last sixty years, the general drift of these inscriptions is now known. They are the only important records of the ancient Umbrian language, which was related closely to that of the Samnites and, though not so closely, to that of the Romans on the other side of the Apennines. Yet less than twenty years ago a book was published in Germany, which boasts itself the home of Comparative Philology, wherein the German origin of the Umbrian language was no less solemnly demonstrated than had been its Celtic origin by Sir William Betham in 1842.

It is good that the study of language should be historical, but the first requisite is that the history should be sound. How little had been learnt of the true history of language a century ago may be seen from a little book by Stephen Weston first published in 1802 and several times reprinted, where accidental assonance is considered sufficient to establish connection. Is there not a word "bad" in English and a word "bad" in Persian which mean the same thing? Clearly therefore Persian and English must be connected. The conclusion is true, but it is drawn from erroneous premises. As stated, this identity has no more value than the similar assonance between the English "cover" and the Hebrew "kophar", where the history of "cover" as coming through French from a Latin "co-operire" was even in 1802 well-known to many. To this day, in spite of recent elaborate attempts (Most recently in H. Moller's "Semitisch und Indogermanisch", Erster Teil, Kopenhagen, 1907.) to establish connection between the Indo-Germanic and the Semitic families of languages, there is no satisfactory evidence of such relation between these families. This is not to deny the possibility of such a connection at a very early period; it is merely to say that through the lapse of long ages all trustworthy record of such relationship, if it ever existed, has been, so far as present knowledge extends, obliterated.

But while Stephen Weston was publishing, with much public approval, his collection of amusing similarities between languages—similarities which proved nothing—the key to the historical study of at least one family of languages had already been found by a learned Englishman in a distant land. In 1783 Sir William Jones had been sent out as a judge in the supreme court of judicature in Bengal. While still a young man at Oxford he was noted as a linguist; his reputation as a Persian scholar had preceded him to the East. In the intervals of his professional duties he made a careful study of the language which was held sacred by the natives of the country in which he was living. He was mainly instrumental in establishing a society for the investigation of language and related subjects. He was himself the first president of the society, and in the "third anniversary discourse" delivered on February 2, 1786, he made the following observations: "The Sanscrit language, whatever be its antiquity, is of a wonderful structure; more perfect than the GREEK, more copious than the LATIN, and more exquisitely refined than either, yet bearing to both of them a stronger affinity, both in the roots of verbs and in the forms of grammar, than could possibly have been produced by accident; so strong indeed, that no philologer could examine them all three, without believing them to have sprung from some common source, which, perhaps, no longer exists: there is a similar reason, though not quite so forcible, for supposing that both the Gothick and the Celtick, though blended with a very different idiom, had the same origin with the Sanscrit; and the old Persian might be added to the same family, if this was the place for discussing any question concerning the antiquities of Persia." ("Asiatic Researches", I. page 422, "Works of Sir W. Jones", I. page 26, London, 1799.)

No such epoch-making discovery was probably ever announced with less flourish of trumpets. Though Sir William Jones lived for eight years more and delivered other anniversary discourses, he added nothing of importance to this utterance. He had neither the time nor the health that was needed for the prosecution of so arduous an undertaking.

But the good seed did not fall upon stony ground. The news was speedily conveyed to Europe. By a happy chance, the sudden renewal of war between France and England in 1803 gave Friedrich Schlegel the opportunity of learning Sanscrit from Alexander Hamilton, an Englishman who, like many others, was confined in Paris during the long struggle with Napoleon. The influence of Schlegel was not altogether for good in the history of this research, but he was inspiring. Not upon him but upon Franz Bopp, a struggling German student who spent some time in Paris and London a dozen years later, fell the mantle of Sir William Jones. In Bopp's Comparative Grammar of the Indo-Germanic languages which appeared in 1833, three-quarters of a century ago, the foundations of Comparative Philology were laid. Since that day the literature of the subject has grown till it is almost, if not altogether, beyond the power of any single man to cope with it. But long as the discourse may be, it is but the elaboration of the text that Sir William Jones supplied.

With the publication of Bopp's Comparative Grammar the historical study of language was put upon a stable footing. Needless to say much remained to be done, much still remains to be done. More than once there has been danger of the study following erroneous paths. Its terminology and its point of view have in some degree changed. But nothing can shake the truth of the statement that the Indo-Germanic languages constitute in themselves a family sprung from the same source, marked by the same characteristics, and differentiated from all other languages by formation, by vocabulary, and by syntax. The historical method was applied to language long before it reached biology. Nearly a quarter of a century before Charles Darwin was born, Sir William Jones had made the first suggestion of a comparative study of languages. Bopp's Comparative Grammar began to be published nine years before the first draft of Darwin's treatise on the Origin of Species was put on paper in 1842.

It is not therefore on the history of Comparative Philology in general that the ideas of Darwin have had most influence. Unfortunately, as Jowett has said in the introduction to his translation of Plato's "Republic", most men live in a corner. The specialisation of knowledge has many advantages, but it has also disadvantages, none worse perhaps than that it tends to narrow the specialist's horizon and to make it more difficult for one worker to follow the advances that are being made by workers in other departments. No longer is it possible as in earlier days for an intellectual prophet to survey from a Pisgah height all the Promised Land. And the case of linguistic research has been specially hard. This study has, if the metaphor may be allowed, a very extended frontier. On one side it touches the domain of literature, on other sides it is conterminous with history, with ethnology and anthropology, with physiology in so far as language is the production of the brain and tissues of a living being, with physics in questions of pitch and stress accent, with mental science in so far as the principles of similarity, contrast, and contiguity affect the forms and the meanings of words through association of ideas. The territory of linguistic study is immense, and it has much to supply which might be useful to the neighbours who border on that territory. But they have not regarded her even with that interest which is called benevolent because it is not actively maleficent. As Horne Tooke remarked a century ago, Locke had found a whole philosophy in language. What have the philosophers done for language since? The disciples of Kant and of Wilhelm von Humboldt supplied her plentifully with the sour grapes of metaphysics; otherwise her neighbours have left her severely alone save for an occasional "Ausflug," on which it was clear they had sadly lost their bearings. Some articles in Psychological Journals, Wundt's great work on "Volkerpsychologie" (Erster Band: "Die Sprache", Leipzig, 1900. New edition, 1904. This work has been fertile in producing both opponents and supporters. Delbruck, "Grundfragen der Sprachforschung", Strassburg, 1901, with a rejoinder by Wundt, "Sprachgeschichte" and "Sprachpsychologie", Leipzig, 1901; L. Sutterlin, "Das Wesen der Sprachgebilde", Heidelberg, 1902; von Rozwadowski, "Wortbildung und Wortbedeutung", Heidelberg, 1904; O. Dittrich, "Grundzuge der Sprachpsychologie", Halle, 1904, Ch. A. Sechehaye, "Programme et methode de la linguistique theorique", Paris, 1908.), and Mauthner's brilliantly written "Beitrage zu einer Kritik der Sprache" (In three parts: (i) "Sprache und Psychologie, (ii) "Zur Sprachwissenschaft", both Stuttgart 1901, (iii) "Zur

Grammatic und Logik" (with index to all three volumes), Stuttgart and Berlin, 1902.) give some reason to hope that, on one side at least, the future may be better than the past.

Where Charles Darwin's special studies came in contact with the Science of Language was over the problem of the origin and development of language. It is curious to observe that, where so many fields of linguistic research have still to be reclaimed—many as yet can hardly be said to be mapped out,—the least accessible field of all—that of the Origin of Language—has never wanted assiduous tillers. Unfortunately it is a field beyond most others where it may be said that

"Wilding oats and luckless darnel grow."

If Comparative Philology is to work to purpose here, it must be on results derived from careful study of individual languages and groups of languages. But as yet the group which Sir William Jones first mapped out and which Bopp organised is the only one where much has been achieved. Investigation of the Semitic group, in some respects of no less moment in the history of civilisation and religion, where perhaps the labour of comparison is not so difficult, as the languages differ less among themselves, has for some reason strangely lagged behind. Some years ago in the "American Journal of Philology" Paul Haupt pointed out that if advance was to be made, it must be made according to the principles which had guided the investigation of the Indo-Germanic languages to success, and at last a Comparative Grammar of an elaborate kind is in progress also for the Semitic languages. (Brockelmann, "Vergleichende Grammatik der semitischen Sprachen", Berlin, 1907 ff. Brockelmann and Zimmern had earlier produced two small hand-books. The only large work was William Wright's "Lectures on the Comparative Grammar of the Semitic Languages", Cambridge, 1890.) For the great group which includes Finnish, Hungarian, Turkish and many languages of northern Asia, a beginning, but only a beginning has been made. It may be presumed from the great discoveries which are in progress in Turkestan that presently much more will be achieved in this field. But for a certain utterance to be given by Comparative Philology on the question of the origin of language it is necessary that not merely for these languages but also for those in other quarters of the globe, the facts should be collected, sifted and tabulated. England rules an empire which contains a greater variety of languages by far than were ever held under one sway before. The Government of India is engaged in producing, under the editorship of Dr Grierson, a linguistic survey of India, a remarkable undertaking and, so far as it has gone, a remarkable achievement. Is it too much to ask that, with the support of the self-governing colonies, a similar survey should be undertaken for the whole of the British Empire?

Notwithstanding the great number of books that have been written on the origin of language in the last three and twenty centuries, the results of the investigation which can be described as certain are very meagre. The question originally raised was whether language came into being thesei or phusei, by convention or by nature. The first alternative, in its baldest form at least, has passed from out the field of controversy. No one now claims that names were given to living things or objects or activities by formal agreement among the members of an early community, or that the first father of mankind passed in review every living thing and gave it its name. Even if the record of Adam's action were to be taken literally there would still remain the question, whence had he this power? Did he develop it himself or was it a miraculous gift with which he was endowed at his creation? If the latter, then as Wundt says ("Volkerpsychologie", I. 2, page 585.), "the miracle of language is subsumed in the miracle of creation." If Adam developed language of himself, we are carried over to the alternative origin of phusei. On this hypothesis we must assume that the natural growth which modern theories of development regard as the painful progress of multitudinous generations was contracted into the experience of a single individual.

But even if the origin of language is admitted to be NATURAL there may still be much variety of signification attached to the word: NATURE, like most words which are used by philosophers, has accumulated many meanings, and as research into the natural world proceeds, is accumulating more.

Forty years ago an animated controversy raged among the supporters of the theories which were named for short the bow-wow, the pooh-pooh and the ding-dong theories of the origin of language. The third, which was the least tenacious of life, was made known to the English-speaking world by the late Professor Max Muller who, however, when questioned, repudiated it as his own belief. ("Science of Thought", London, 1887, page 211.) It was taken by him from Heyse's lectures on language which were published posthumously by Steinthal. Put shortly the theory is that "everything which is struck, rings. Each substance has its peculiar ring. We can tell the more or less perfect structure of metals by their vibrations, by the answer which they give. Gold rings differently from tin, wood rings differently from stone; and different sounds are produced according to the nature of each percussion. It may be the same with man, the most highly organised of nature's work." (Max Muller as above, translating from Heyse.) Max Muller's repudiation of this theory was, however, not very whole-hearted for he proceeds later in the same argument: "Heyse's theory, which I neither adopted nor rejected, but which, as will be seen, is by no means incompatible with that which for many years has been gaining on me, and which of late has been so clearly formulated by Professor Noire, has been assailed with ridicule and torn to pieces, often by persons who did not even suspect how much truth was hidden behind its paradoxical appearance. We are still very far from being able to identify roots with nervous vibrations, but if it should appear hereafter that sensuous vibrations supply at least the raw material of roots, it is quite possible that the theory, proposed by Oken and Heyse, will retain its place in the history of the various attempts at solving the problem of the origin of language, when other theories, which in our own days were received with popular applause, will be completely forgotten." ("Science of Thought", page 212.)

Like a good deal else that has been written on the origin of language, this statement perhaps is not likely to be altogether clear to the plain man, who may feel that even the "raw material of roots" is some distance removed from nervous vibrations, though obviously without the existence of afferent and efferent nerves articulate speech would be impossible. But Heyse's theory undoubtedly was that every thought or idea which occurred to the mind of man for the first time had its own special phonetic expression, and that this responsive faculty, when its object was thus fulfilled, became extinct. Apart from the philosophical question whether the mind acts without external stimulus, into which it is not necessary to enter here, it is clear that this theory can neither be proved nor disproved, because it postulates that this faculty existed only when language first began, and later altogether disappeared. As we have already seen, it is impossible for us to know what happened at the first beginnings of language, because we have no information from any period even approximately so remote; nor are we likely to attain it. Even in their earliest stages the great families of language which possess a history extending over many centuries—the Indo-Germanic and the Semitic—have very little in common. With the exception of Chinese, the languages which are apparently of a simpler or more primitive formation have either a history which, compared with that of the families mentioned, is very short, or, as in the case of the vast majority, have no history beyond the time extending only over a few years or, at most, a few centuries when they have been observed by competent scholars of European origin. But, if we may judge by the history of geology and other studies, it is well to be cautious in assuming for the first stages of development forces which do not operate in the later, unless we have direct evidence of their existence.

It is unnecessary here to enter into a prolonged discussion of the other views christened by Max Muller, not without energetic protest from their supporters, the bow-wow and pooh-pooh theories of language. Suffice it to say that the former recognises as a source of language the imitation of the sounds made by animals, the fall of bodies into water or on to solid substances and the like, while the latter, also called the interjectional theory, looks to the natural ejaculations produced by particular forms of effort for the first beginnings of speech. It would be futile to deny that some words in most languages come from imitation, and that others, probably fewer in number, can be traced to ejaculations. But if either of these sources alone or both in combination gave rise to primitive speech, it clearly must have been a simple form of language and very limited in amount. There is no reason to think that it was otherwise. Presumably in its earliest stages language only indicated the most elementary ideas, demands for food or the gratification of other appetites, indications of danger, useful animals and plants. Some of these, such as animals or indications of danger, could often be easily represented by imitative sounds: the need for food and the like could be indicated by gesture and natural cries. Both sources are verae causae; to them Noire, supported by Max Muller, has added another which has sometimes been called the Yo-heave-ho theory. Noire contends that the real crux in the early stages of language is for primitive man to make other primitive men understand what he means. The vocal signs which commend themselves to one may not have occurred to another, and may therefore be unintelligible. It may be admitted that this difficulty exists, but it is not insuperable. The old story of the European in China who, sitting down to a meal and being doubtful what the meat in the dish might be, addressed an interrogative Quack-quack? to the waiter and was promptly answered by Bow-wow, illustrates a simple situation where mutual understanding was easy. But obviously many situations would be more complex than this, and to grapple with them Noire has introduced his theory of communal action. "It was common effort directed to a common object, it was the most primitive (uralteste) labour of our ancestors, from which sprang language and the life of reason." (Noire "Der Ursprung der Sprache", page 331, Mainz, 1877.) As illustrations of such common effort he cites battle cries, the rescue of a ship running on shore (a situation not likely to occur very early in the history of man), and others. Like Max Muller he holds that language is the utterance and the organ of thought for mankind, the one characteristic which separates man from the brute. "In common action the word was first produced; for long it was inseparably connected with action; through long-continued connection it gradually became the firm, intelligible symbol of action, and then in its development indicated also things of the external world in so far as the action affected them and finally the sound began to enter into a connexion with them also." (Op. cit. page 339.) In so far as this theory recognises language as a social institution it is undoubtedly correct. Darwin some years before Noire had pointed to the same social origin of language in the fourth chapter of his work on "The Expression of the Emotions in Man and Animals". "Naturalists have remarked, I believe with truth, that social animals, from habitually using their vocal organs as a means of intercommunication, use them on other occasions much more freely than other animals... The principle, also, of association, which is so widely extended in its power, has likewise played its part. Hence it allows that the voice, from having been employed as a serviceable aid under certain conditions, inducing pleasure, pain, rage, etc., is commonly used whenever the same sensations or emotions are excited, under quite different conditions, or in a lesser degree." ("The Expression of the Emotions", page 84 (Popular Edition, 1904).

Darwin's own views on language which are set forth most fully in "The Descent of Man" (page 131 ff. (Popular Edition, 1906).) are characterised by great modesty and caution. He did not profess to be a philologist and the facts are naturally taken from the best known works of the day (1871). In the notes added to the second edition he remarks on Max Muller's denial of thought without words, "what a strange definition must here be given to the word thought!" (Op. cit. page 135, footnote 63.) He naturally finds the origin of language in "the imitation and modification of various natural sounds, the voices of other animals, and man's own instinctive cries aided by signs and gestures (op. cit. page 132.)... As the voice was used more and more, the vocal organs would have been strengthened and perfected through the principle of the inherited effects of use; and this would have reacted on the power of speech." (Op. cit. page 133.) On man's own instinctive cries, he has more to say in "The Expression of the Emotions". (Page 93 (Popular Edition, 1904) and elsewhere.) These remarks have been utilised by Prof. Jespersen of Copenhagen in propounding an ingenious theory of his own to the effect that speech develops out of singing. ("Progress in Language", page 361, London, 1894.)

For many years and in many books Max Muller argued against Darwin's views on evolution on the one ground that thought is impossible without speech; consequently as speech is confined to the human race, there is a gulf which cannot be bridged between man and all other creatures. (Some interesting comments on the theory will be found in a lecture

323

on "Thought and Language" in Samuel Butler's "Essays on Life, Art and Science", London, 1908.) On the title-page of his "Science of Thought" he put the two sentences "No Reason without Language: No Language without Reason." It may be readily admitted that the second dictum is true, that no language properly so-called can exist without reason. Various birds can learn to repeat words or sentences used by their masters or mistresses. In most cases probably the birds do not attach their proper meaning to the words they have learnt; they repeat them in season and out of season, sometimes apparently for their own amusement, generally in the expectation, raised by past experience, of being rewarded for their proficiency. But even here it is difficult to prove a universal negative, and most possessors of such pets would repudiate indignantly the statement that the bird did not understand what was said to it, and would also contend that in many cases the words which it used were employed in their ordinary meaning. The first dictum seems to be inconsistent with fact. The case of deaf mutes, such as Laura Bridgeman, who became well educated, or the still more extraordinary case of Helen Keller, deaf, dumb, and blind, who in spite of these disadvantages has learnt not only to reason but to reason better than the average of persons possessed of all their senses, goes to show that language and reason are not necessarily always in combination. Reason is but the conscious adaptation of means to ends, and so defined is a faculty which cannot be denied to many of the lower animals. In these days when so many books on Animal Intelligence are issued from the press, it seems unnecessary to labour the point. Yet none of these animals, except by parrot-imitation, makes use of speech, because man alone possesses in a sufficient degree of development the centres of nervous energy which are required for the working of articulation in speech. On this subject much investigation was carried on during the last years of Darwin's life and much more in the period since his death. As early as 1861 Broca, following up observations made by earlier French writers, located the centre of articulate speech in the third left frontal convolution of the brain. In 1876 he more definitely fixed the organ of speech in "the posterior two-fifths of the third frontal convolution" (Macnamara, "Human Speech", page 197, London, 1908.), both sides and not merely the left being concerned in speech production. Owing however to the greater use by most human beings of the right side of the body, the left side of the brain, which is the motor centre for the right side of the body, is more highly developed than its right side, which moves the left side of the body. The investigations of Professors Ferrier, Sherrington and Grunbaum have still more precisely defined the relations between brain areas and certain groups of muscles. One form of aphasia is the result of injury to or disease in the third frontal convolution because the motor centre is no longer equal to the task of setting the necessary muscles in motion. In the brain of idiots who are unable to speak, the centre for speech is not developed. (Op. cit. page 226.) In the anthropoid apes the brain is similarly defective, though it has been demonstrated by Professors Cunningham and Marchand "that there is a tendency, especially in the gorilla's brain, for the third frontal convolution to assume the human form... But if they possessed a centre for speech, those parts of the hemispheres of their brains which form the mechanism by which intelligence is elaborated are so ill-developed, as compared with the rest of their bodies, that we can not conceive, even with more perfect frontal convolutions, that these animals could formulate ideas expressible in intelligent speech." (Op. cit. page 223.)

While Max Muller's theory is Shelley's

"He gave man speech, and speech created thought, Which is the measure of the universe" ("Prometheus Unbound" II. 4.),

it seems more probable that the development was just the opposite—that the development of new activities originated new thoughts which required new symbols to express them, symbols which may at first have been, even to a greater extent than with some of the lower races at present, sign language as much as articulation. When once the faculty of articulation was developed, which, though we cannot trace the process, was probably a very gradual growth, there is no reason to suppose that words developed in any other way then they do at present. An erroneous notion of the development of language has become widely spread through the adoption of the metaphorical term "roots" for the irreducible elements of human speech. Men never talked in roots; they talked in words. Many words of kindred meaning have a part in common, and a root is nothing but that common part stripped of all additions. In some cases it is obvious that one word is derived from another by the addition of a fresh element; in other cases it is impossible to say which of two kindred words is the more primitive. A root is merely a convenient term for an abstraction. The simplest word may be called a root, but it is nevertheless a word. How are new words added to a language in the present day? Some communities, like the Germans, prefer to construct new words for new ideas out of the old material existing in the language; others, like the English, prefer to go to the ancient languages of Greece and Rome for terms to express new ideas. The same chemical element is described in the two languages as sour stuff (Sauerstoff) and as oxygen. Both terms mean the same thing etymologically as well as in fact. On behalf of the German method, it may be contended that the new idea is more closely attached to already existing ideas, by being expressed in elements of the language which are intelligible even to the meanest capacity. For the English practice it may be argued that, if we coin a new word which means one thing, and one thing only, the idea which it expresses is more clearly defined than if it were expressed in popularly intelligible elements like "sour stuff." If the etymological value of words were always present in the minds of their users, "oxygen" would undoubtedly have an advantage over "sour stuff" as a technical term. But the tendency in language is to put two words of this kind which express but one idea under a single accent, and when this has taken place, no one but the student of language any longer observes what the elements really mean. When the ordinary man talks of a "blackbird" it is certainly not present to his consciousness that he is talking of a black bird, unless for some reason conversation has been dwelling upon the colour rather than other characteristics of the species.

But, it may be said, words like "oxygen" are introduced by learned men, and do not represent the action of the man in the street, who, after all, is the author of most additions to the stock of human language. We may go back therefore some four centuries to a period, when scientific study was only in its infancy, and see what process was followed. With the discovery of America new products never seen before reached Europe, and these required names. Three of the most characteristic were tobacco, the potato, and the turkey. How did these come to be so named? The first people to import these products into Europe were naturally the Spanish discoverers. The first of these words—tobacco—appears in forms which differ only slightly in the languages of all civilised countries: Spanish tabaco, Italian tabacco, French tabac, Dutch and German tabak, Swedish tobak, etc. The word in the native dialect of Hayti is said to have been tabaco, but to have meant not the plant (According to William Barclay, "Nepenthes, or the Virtue of Tobacco", Edinburgh, 1614, "the countrey which God hath honoured and blessed with this happie and holy herbe doth call it in their native language 'Petum'.") but the pipe in which it was smoked. It thus illustrates a frequent feature of borrowing—that the word is not borrowed in its proper signification, but in some sense closely allied thereto, which a foreigner, understanding the language with difficulty, might readily mistake for the real meaning. Thus the Hindu practice of burning a wife upon the funeral pyre of her husband is called in English "suttee", this word being in fact but the phonetic spelling of the Sanskrit "sati", "a virtuous woman," and passing into its English meaning because formerly the practice of self-immolation by a wife was regarded as the highest virtue.

The name of the potato exhibits greater variety. The English name was borrowed from the Spanish "patata", which was itself borrowed from a native word for the "yam" in the dialect of Hayti. The potato appeared early in Italy, for the mariners of Genoa actively followed the footsteps of their countryman Columbus in exploring America. In Italian generally the form "patata" has survived. The tubers, however, also suggested a resemblance to truffles, so that the Italian word "tartufolo", a diminutive of the Italian modification of the Latin "terrae tuber" was applied to them. In the language of the Rhaetian Alps this word appears as "tartufel". From there it seems to have passed into Germany where potatoes were not cultivated extensively till the eighteenth century, and "tartufel" has in later times through some popular etymology been metamorphosed into "Kartoffel". In France the shape of the tubers suggested the name of earth-apple (pomme de terre), a name also adopted in Dutch (aard-appel), while dialectically in German a form "Grumbire" appears, which is a corruption of "Grund-birne", "ground pear". (Kluge "Etymologisches Worterbuch der deutschen Sprache" (Strassburg), s.v. "Kartoffel".) Here half the languages have adopted the original American word for an allied plant, while others have adopted a name originating in some more or less fanciful resemblance discovered in the tubers; the Germans alone in Western Europe, failing to see any meaning in their borrowed name, have modified it almost beyond recognition. To this English supplies an exact parallel in "parsnep" which, though representing the Latin "pastinaca" through the Old French "pastenaque", was first assimilated in the last syllable to the "nep" of "turnep" ("pasneppe" in Elizabethan English), and later had an "r" introduced into the first syllable, apparently on the analogy of "parsley".

The turkey on the other hand seems never to be found with its original American name. In England, as the name implies, the turkey cock was regarded as having come from the land of the Turks. The bird no doubt spread over Europe from the Italian seaports. The mistake, therefore, was not unnatural, seeing that these towns conducted a great trade with the Levant, while the fact that America when first discovered was identified with India helped to increase the confusion. Thus in French the "coq d'Inde" was abbreviated to "d'Inde" much as "turkey cock" was to "turkey"; the next stage was to identify "dinde" as a feminine word and create a new "dindon" on the analogy of "chapon" as the masculine. In Italian the name "gallo d'India" besides survives, while in German the name "Truthahn" seems to be derived onomatopoetically from the bird's cry, though a dialectic "Calecutischer Hahn" specifies erroneously an origin for the bird from the Indian Calicut. In the Spanish "pavo", on the other hand, there is a curious confusion with the peacock. Thus in these names for objects of common knowledge, the introduction of which into Europe can be dated with tolerable definiteness, we see evinced the methods by which in remoter ages objects were named. The words were borrowed from the community whence came the new object, or the real or fancied resemblance to some known object gave the name, or again popular etymology might convert the unknown term into something that at least approached in sound a well-known word.

"The Origin of Species" had not long been published when the parallelism of development in natural species and in languages struck investigators. At the time, one of the foremost German philologists was August Schleicher, Professor at Jena. He was himself keenly interested in the natural sciences, and amongst his colleagues was Ernst Haeckel, the protagonist in Germany of the Darwinian theory. How the new ideas struck Schleicher may be seen from the following sentences by his colleague Haeckel. "Speech is a physiological function of the human organism, and has been developed simultaneously with its organs, the larynx and tongue, and with the functions of the brain. Hence it will be quite natural to find in the evolution and classification of languages the same features as in the evolution and classification of organic species. The various groups of languages that are distinguished in philology as primitive, fundamental, parent, and daughter languages, dialects, etc., correspond entirely in their development to the different categories which we classify in zoology and botany as stems, classes, orders, families, genera, species and varieties. The relation of these groups, partly coordinate and partly subordinate, in the general scheme is just the same in both cases; and the evolution follows the same lines in both." (Haeckel, "The Evolution of Man", page 485, London, 1905. This represents Schleicher's own words: Was die Naturforscher als Gattung bezeichnen wurden, heisst bei den Glottikern Sprachstamm, auch Sprachsippe; naher verwandte Gattungen bezeichnen sie wohl auch als Sprachfamilien einer Sippe oder eines Sprachstammes... Die Arten einer Gattung nennen wir Sprachen eines Stammes; die Unterarten einer Art sind bei uns die Dialekte oder Mundarten einer Sprache; den Varietaten und Spielarten entsprechen die Untermundarten oder Nebenmundarten und endlich den einzelnen Individuen die Sprechweise der einzelnen die Sprachen redenden Menschen. "Die Darwinische Theorie und die Sprachwissenschaft", Weimar, 1863, page 12 f. Darwin makes a more cautious statement about the classification of languages in "The Origin of Species", page 578, (Popular Edition, 1900).) These views were set forth in an open letter addressed to Haeckel in 1863 by Schleicher entitled, "The Darwinian theory and the science of language". Unfortunately Schleicher's views went a good deal farther than is indicated in the extract given above. He appended to the pamphlet a genealogical tree of the Indo-Germanic languages which, though to a large extent confirmed by later research, by the dichotomy of each branch into two other branches, led the unwary reader to suppose their phylogeny (to use Professor Haeckel's term) was more regular than our evidence warrants.

Without qualification Schleicher declared languages to be "natural organisms which originated unconditioned by the human will, developed according to definite laws, grow old and die; they also are characterised by that series of phenomena which we designate by the term 'Life.' Consequently Glottic, the science of language, is a natural science; its method is in general the same as that of the other natural sciences." ("Die Darwinische Theorie", page 6 f.) In accordance with this view he declared (op. cit. page 23.) that the root in language might be compared with the simple cell in physiology, the linguistic simple cell or root being as yet not differentiated into special organs for the function of noun, verb, etc.

In this probably all recent philologists admit that Schleicher went too far. One of the most fertile theories in the modern science of language originated with him, and was further developed by his pupil, August Leskien ("Die Declination im Slavisch-litanischen und Germanischen", Leipzig, 1876; Osthoff and Brugmann, "Morphologische Untersuchungen", I. (Introduction), 1878. The general principles of this school were formulated (1880) in a fuller form in H. Paul's "Prinzipien der Sprachgeschichte", Halle (3rd edition, 1898). Paul and Wundt (in his "Volkerpsychologie") deal largely with the same matter, but begin their investigations from different points of view, Paul being a philologist with leanings to philosophy and Wundt a philosopher interested in language.), and by Leskien's colleagues and friends, Brugmann and Osthoff. This was the principle that phonetic laws have no exceptions. Under the influence of this generalisation much greater precision in etymology was insisted upon, and a new and remarkably active period in the study of language began. Stated broadly in the fashion given above the principle is not true. A more accurate statement would be that an original sound is represented in a given dialect at a given time and in a given environment only in one way; provided that the development of the original sound into its representation in the given dialect has not been influenced by the working of analogy.

It is this proviso that is most important for the characterisation of the science of language. As I have said elsewhere, it is at this point that this science parts company with the natural sciences. "If the chemist compounds two pure simple elements, there can be but one result, and no power of the chemist can prevent it. But the minds of men do act upon the sounds which they produce. The result is that, when this happens, the phonetic law which would have acted in the case is stopped, and this particular form enters on the same course of development as other forms to which it does not belong." (P. Giles, "Short Manual of Comparative Philology", 2nd edition, page 57, London, 1901.)

Schleicher was wrong in defining a language to be an organism in the sense in which a living being is an organism. Regarded physiologically, language is a function or potentiality of certain human organs; regarded from the point of view of the community it is of the nature of an institution. (This view of language is worked out at some length by Prof. W.D. Whitney in an article in the "Contemporary Review" for 1875, page 713 ff. This article forms part of a controversy with Max Muller, which is partly concerned with Darwin's views on language. He criticises Schleicher's views severely in his "Oriental and Linguistic Studies", page 298 ff., New York, 1873. In this volume will be found criticisms of various other views mentioned in this essay.) More than most influences it conduces to the binding together of the elements that form a state. That geographical or other causes may effectively counteract the influence of identity of language is obvious. One need only read the history of ancient Greece, or observe the existing political separation of Germany and Austria, of Great Britain and the United States of America. But however analogous to an organism, language is not an organism. In a less degree Schleicher, by defining languages as such, committed the same mistake which Bluntschli made regarding the State, and which led him to declare that the State is by nature masculine and the Church feminine. (Bluntschli, "Theory of the State", page 24, Second English Edition, Oxford, 1892.) The views of Schleicher were to some extent injurious to the proper methods of linguistic study. But this misfortune was much more than fully compensated by the inspiration which his ideas, collected and modified by his disciples, had upon the science. In spite of the difference which the psychological element represented by analogy makes between the science of language and the natural sciences, we are entitled to say of it as Schleicher said of Darwin's theory of the origin of species, "it depends upon observation, and is essentially an attempt at a history of development."

Other questions there are in connection with language and evolution which require investigation—the survival of one amongst several competing words (e.g. why German keeps only as a high poetic word "ross", which is identical in origin with the English work-a-day "horse", and replaces it by "pferd", whose congener the English "palfrey" is almost confined to poetry and romance), the persistence of evolution till it becomes revolution in languages like English or Persian which have practically ceased to be inflectional languages, and many other problems. Into these Darwin did not enter, and they require a fuller investigation than is possible within the limits of the present paper.

XXVII. DARWINISM AND HISTORY.
By J.B. Bury, Litt.D., LL.D.

Regius Professor of Modern History in the University of Cambridge.

1. Evolution, and the principles associated with the Darwinian theory, could not fail to exert a considerable influence on the studies connected with the history of civilised man. The speculations which are known as "philosophy of history," as well as the sciences of anthropology, ethnography, and sociology (sciences which though they stand on their own feet are for the historian auxiliary), have been deeply affected by these principles. Historiographers, indeed, have with few exceptions made little attempt to apply them; but the growth of historical study in the nineteenth century has been determined and characterised by the same general principle which has underlain the simultaneous developments of the study of nature, namely the GENETIC idea. The "historical" conception of nature, which has produced the history of the solar system, the story of the earth, the genealogies of telluric organisms, and has revolutionised natural science, belongs to the same order of thought as the conception of human history as a continuous, genetic, causal process—a conception which has revolutionised historical research and made it scientific. Before proceeding to consider the application of evolutionary principles, it will be pertinent to notice the rise of this new view.

2. With the Greeks and Romans history had been either a descriptive record or had been written in practical interests. The most eminent of the ancient historians were pragmatical; that is, they regarded history as an instructress in statesmanship, or in the art of war, or in morals. Their records reached back such a short way, their experience was so brief, that they never attained to the conception of continuous process, or realised the significance of time; and they never viewed the history of human societies as a phenomenon to be investigated for its own sake. In the middle ages there was still less chance of the emergence of the ideas of progress and development. Such notions were excluded by the fundamental doctrines of the dominant religion which bounded and bound men's minds. As the course of history was held to be determined from hour to hour by the arbitrary will of an extra-cosmic person, there could be no self-contained causal development, only a dispensation imposed from

without. And as it was believed that the world was within no great distance from the end of this dispensation, there was no motive to take much interest in understanding the temporal, which was to be only temporary.

The intellectual movements of the fifteenth and sixteenth centuries prepared the way for a new conception, but it did not emerge immediately. The historians of the Renaissance period simply reverted to the ancient pragmatical view. For Machiavelli, exactly as for Thucydides and Polybius, the use of studying history was instruction in the art of politics. The Renaissance itself was the appearance of a new culture, different from anything that had gone before; but at the time men were not conscious of this; they saw clearly that the traditions of classical antiquity had been lost for a long period, and they were seeking to revive them, but otherwise they did not perceive that the world had moved, and that their own spirit, culture, and conditions were entirely unlike those of the thirteenth century. It was hardly till the seventeenth century that the presence of a new age, as different from the middle ages as from the ages of Greece and Rome, was fully realised. It was then that the triple division of ancient, medieval, and modern was first applied to the history of western civilisation. Whatever objections may be urged against this division, which has now become almost a category of thought, it marks a most significant advance in man's view of his own past. He has become conscious of the immense changes in civilisation which have come about slowly in the course of time, and history confronts him with a new aspect. He has to explain how those changes have been produced, how the transformations were effected. The appearance of this problem was almost simultaneous with the rise of rationalism, and the great historians and thinkers of the eighteenth century, such as Montesquieu, Voltaire, Gibbon, attempted to explain the movement of civilisation by purely natural causes. These brilliant writers prepared the way for the genetic history of the following century. But in the spirit of the Aufklarung, that eighteenth-century Enlightenment to which they belonged, they were concerned to judge all phenomena before the tribunal of reason; and the apotheosis of "reason" tended to foster a certain superior a priori attitude, which was not favourable to objective treatment and was incompatible with a "historical sense." Moreover the traditions of pragmatical historiography had by no means disappeared.

3. In the first quarter of the nineteenth century the meaning of genetic history was fully realised. "Genetic" perhaps is as good a word as can be found for the conception which in this century was applied to so many branches of knowledge in the spheres both of nature and of mind. It does not commit us to the doctrine proper of evolution, nor yet to any teleological hypothesis such as is implied in "progress." For history it meant that the present condition of the human race is simply and strictly the result of a causal series (or set of causal series)—a continuous succession of changes, where each state arises causally out of the preceding; and that the business of historians is to trace this genetic process, to explain each change, and ultimately to grasp the complete development of the life of humanity. Three influential writers, who appeared at this stage and helped to initiate a new period of research, may specially be mentioned. Ranke in 1824 definitely repudiated the pragmatical view which ascribes to history the duties of an instructress, and with no less decision renounced the function, assumed by the historians of the Aufklarung, to judge the past; it was his business, he said, merely to show how things really happened. Niebuhr was already working in the same spirit and did more than any other writer to establish the principle that historical transactions must be related to the ideas and conditions of their age. Savigny about the same time founded the "historical school" of law. He sought to show that law was not the creation of an enlightened will, but grew out of custom and was developed by a series of adaptations and rejections, thus applying the conception of evolution. He helped to diffuse the notion that all the institutions of a society or a notion are as closely interconnected as the parts of a living organism.

4. The conception of the history of man as a causal development meant the elevation of historical inquiry to the dignity of a science. Just as the study of bees cannot become scientific so long as the student's interest in them is only to procure honey or to derive moral lessons from the labours of "the little busy bee," so the history of human societies cannot become the object of pure scientific investigation so long as man estimates its value in

pragmatical scales. Nor can it become a science until it is conceived as lying entirely within a sphere in which the law of cause and effect has unreserved and unrestricted dominion. On the other hand, once history is envisaged as a causal process, which contains within itself the explanation of the development of man from his primitive state to the point which he has reached, such a process necessarily becomes the object of scientific investigation and the interest in it is scientific curiosity.

At the same time, the instruments were sharpened and refined. Here Wolf, a philologist with historical instinct, was a pioneer. His "Prolegomena" to Homer (1795) announced new modes of attack. Historical investigation was soon transformed by the elaboration of new methods.

5. "Progress" involves a judgment of value, which is not involved in the conception of history as a genetic process. It is also an idea distinct from that of evolution. Nevertheless it is closely related to the ideas which revolutionised history at the beginning of the last century; it swam into men's ken simultaneously; and it helped effectively to establish the notion of history as a continuous process and to emphasise the significance of time. Passing over earlier anticipations, I may point to a "Discours" of Turgot (1750), where history is presented as a process in which "the total mass of the human race" "marches continually though sometimes slowly to an ever increasing perfection." That is a clear statement of the conception which Turgot's friend Condorcet elaborated in the famous work, published in 1795, "Esquisse d'un tableau historique des progres de l'esprit humain". This work first treated with explicit fulness the idea to which a leading role was to fall in the ideology of the nineteenth century. Condorcet's book reflects the triumphs of the Tiers etat, whose growing importance had also inspired Turgot; it was the political changes in the eighteenth century which led to the doctrine, emphatically formulated by Condorcet, that the masses are the most important element in the historical process. I dwell on this because, though Condorcet had no idea of evolution, the pre-dominant importance of the masses was the assumption which made it possible to apply evolutional principles to history. And it enabled Condorcet himself to maintain that the history of civilisation, a progress still far from being complete, was a development conditioned by general laws.

6. The assimilation of society to an organism, which was a governing notion in the school of Savigny, and the conception of progress, combined to produce the idea of an organic development, in which the historian has to determine the central principle or leading character. This is illustrated by the apotheosis of democracy in Tocqueville's "Democratie en Amerique", where the theory is maintained that "the gradual and progressive development of equality is at once the past and the future of the history of men." The same two principles are combined in the doctrine of Spencer (who held that society is an organism, though he also contemplated its being what he calls a "super-organic aggregate") (A society presents suggestive analogies with an organism, but it certainly is not an organism, and sociologists who draw inferences from the assumption of its organic nature must fall into error. A vital organism and a society are radically distinguished by the fact that the individual components of the former, namely the cells, are morphologically as well as functionally differentiated, whereas the individuals which compose a society are morphologically homogeneous and only functionally differentiated. The resemblances and the differences are worked out in E. de Majewski's striking book "La Science de la Civilisation", Paris, 1908.), that social evolution is a progressive change from militarism to industrialism.

7. the idea of development assumed another form in the speculations of German idealism. Hegel conceived the successive periods of history as corresponding to the ascending phases or ideas in the self-evolution of his Absolute Being. His "Lectures on the Philosophy of History" were published in 1837 after his death. His philosophy had a considerable effect, direct and indirect, on the treatment of history by historians, and although he was superficial and unscientific himself in dealing with historical phenomena, he contributed much towards making the idea of historical development familiar. Ranke was influenced, if not by Hegel himself, at least by the Idealistic philosophies of which Hegel's was the greatest. He was inclined to conceive the stages in the process of history as marked

by incarnations, as it were, of ideas, and sometimes speaks as if the ideas were independent forces, with hands and feet. But while Hegel determined his ideas by a priori logic, Ranke obtained his by induction—by a strict investigation of the phenomena; so that he was scientific in his method and work, and was influenced by Hegelian prepossessions only in the kind of significance which he was disposed to ascribe to his results. It is to be noted that the theory of Hegel implied a judgment of value; the movement was a progress towards perfection.

8. In France, Comte approached the subject from a different side, and exercised, outside Germany, a far wider influence than Hegel. The 4th volume of his "Cours de philosophie positive", which appeared in 1839, created sociology and treated history as a part of this new science, namely as "social dynamics." Comte sought the key for unfolding historical development, in what he called the social-psychological point of view, and he worked out the two ideas which had been enunciated by Condorcet: that the historian's attention should be directed not, as hitherto, principally to eminent individuals, but to the collective behaviour of the masses, as being the most important element in the process; and that, as in nature, so in history, there are general laws, necessary and constant, which condition the development. The two points are intimately connected, for it is only when the masses are moved into the foreground that regularity, uniformity, and law can be conceived as applicable. To determine the social-psychological laws which have controlled the development is, according to Comte, the task of sociologists and historians.

9. The hypothesis of general laws operative in history was carried further in a book which appeared in England twenty years later and exercised an influence in Europe far beyond its intrinsic merit, Buckle's "History of Civilisation in England" (1857-61). Buckle owed much to Comte, and followed him, or rather outdid him, in regarding intellect as the most important factor conditioning the upward development of man, so that progress, according to him, consisted in the victory of the intellectual over the moral laws.

10. The tendency of Comte and Buckle to assimilate history to the sciences of nature by reducing it to general "laws," derived stimulus and plausibility from the vista offered by the study of statistics, in which the Belgian Quetelet, whose book "Sur l'homme" appeared in 1835, discerned endless possibilities. The astonishing uniformities which statistical inquiry disclosed led to the belief that it was only a question of collecting a sufficient amount of statistical material, to enable us to predict how a given social group will act in a particular case. Bourdeau, a disciple of this school, looks forward to the time when historical science will become entirely quantitative. The actions of prominent individuals, which are generally considered to have altered or determined the course of things, are obviously not amenable to statistical computation or explicable by general laws. Thinkers like Buckle sought to minimise their importance or explain them away.

11. These indications may suffice to show that the new efforts to interpret history which marked the first half of the nineteenth century were governed by conceptions closely related to those which were current in the field of natural science and which resulted in the doctrine of evolution. The genetic principle, progressive development, general laws, the significance of time, the conception of society as an organic aggregate, the metaphysical theory of history as the self-evolution of spirit,—all these ideas show that historical inquiry had been advancing independently on somewhat parallel lines to the sciences of nature. It was necessary to bring this out in order to appreciate the influence of Darwinism.

12. In the course of the dozen years which elapsed between the appearances of "The Origin of Species" (observe that the first volume of Buckle's work was published just two years before) and of "The Descent of Man" (1871), the hypothesis of Lamarck that man is the co-descendant with other species of some lower extinct form was admitted to have been raised to the rank of an established fact by most thinkers whose brains were not working under the constraint of theological authority.

One important effect of the discovery of this fact (I am not speaking now of the Darwinian explanation) was to assign to history a definite place in the coordinated whole of knowledge, and relate it more closely to other sciences. It had indeed a defined logical place in systems such as Hegel's and Comte's; but Darwinism certified its standing convincingly and without more ado. The prevailing doctrine that man was created ex abrupto had placed history in an isolated position, disconnected with the sciences of nature. Anthropology, which deals with the animal anthropos, now comes into line with zoology, and brings it into relation with history. (It is to be observed that history is not only different in scope but) not coextensive with anthropology IN TIME. For it deals only with the development of man in societies, whereas anthropology includes in its definition the proto-anthropic period when anthropos was still non-social, whether he lived in herds like the chimpanzee, or alone like the male ourang-outang. (It has been well shown by Majewski that congregations—herds, flocks, packs, etc.—of animals are not SOCIETIES; the characteristic of a society is differentiation of function. Bee hives, ant hills, may be called quasi-societies; but in their case the classes which perform distinct functions are morphologically different.) Man's condition at the present day is the result of a series of transformations, going back to the most primitive phase of society, which is the ideal (unattainable) beginning of history. But that beginning had emerged without any breach of continuity from a development which carries us back to a quadrimane ancestor, still further back (according to Darwin's conjecture) to a marine animal of the ascidian type, and then through remoter periods to the lowest form of organism. It is essential in this theory that though links have been lost there was no break in the gradual development; and this conception of a continuous progress in the evolution of life, resulting in the appearance of uncivilised Anthropos, helped to reinforce, and increase a belief in, the conception of the history of civilised Anthropos as itself also a continuous progressive development.

13. Thus the diffusion of the Darwinian theory of the origin of man, by emphasising the idea of continuity and breaking down the barriers between the human and animal kingdoms, has had an important effect in establishing the position of history among the sciences which deal with telluric development. The perspective of history is merged in a larger perspective of development. As one of the objects of biology is to find the exact steps in the genealogy of man from the lowest organic form, so the scope of history is to determine the stages in the unique causal series from the most rudimentary to the present state of human civilisation.

It is to be observed that the interest in historical research implied by this conception need not be that of Comte. In the Positive Philosophy history is part of sociology; the interest in it is to discover the sociological laws. In the view of which I have just spoken, history is permitted to be an end in itself; the reconstruction of the genetic process is an independent interest. For the purpose of the reconstruction, sociology, as well as physical geography, biology, psychology, is necessary; the sociologist and the historian play into each other's hands; but the object of the former is to establish generalisations; the aim of the latter is to trace in detail a singular causal sequence.

14. The success of the evolutional theory helped to discredit the assumption or at least the invocation of transcendent causes. Philosophically of course it is compatible with theism, but historians have for the most part desisted from invoking the naive conception of a "god in history" to explain historical movements. A historian may be a theist; but, so far as his work is concerned, this particular belief is otiose. Otherwise indeed (as was remarked above) history could not be a science; for with a deus ex machina who can be brought on the stage to solve difficulties scientific treatment is a farce. The transcendent element had appeared in a more subtle form through the influence of German philosophy. I noticed how Ranke is prone to refer to ideas as if they were transcendent existences manifesting themselves in the successive movements of history. It is intelligible to speak of certain ideas as controlling, in a given period,—for instance, the idea of nationality; but from the scientific point of view, such ideas have no existence outside the minds of individuals and are purely psychical forces; and a historical "idea," if it does not exist in this form, is merely a way of expressing a synthesis of the historian himself.

15. From the more general influence of Darwinism on the place of history in the system of human knowledge, we may turn to the influence of the principles and methods by which

Darwin explained development. It had been recognised even by ancient writers (such as Aristotle and Polybius) that physical circumstances (geography, climate) were factors conditioning the character and history of a race or society. In the sixteenth century Bodin emphasised these factors, and many subsequent writers took them into account. The investigations of Darwin, which brought them into the foreground, naturally promoted attempts to discover in them the chief key to the growth of civilisation. Comte had expressly denounced the notion that the biological methods of Lamarck could be applied to social man. Buckle had taken account of natural influences, but had relegated them to a secondary plane, compared with psychological factors. But the Darwinian theory made it tempting to explain the development of civilisation in terms of "adaptation to environment," "struggle for existence," "natural selection," "survival of the fittest," etc. (Recently O. Seeck has applied these principles to the decline of Graeco-Roman civilisation in his "Untergang der antiken Welt", 2 volumes, Berlin, 1895, 1901.)

The operation of these principles cannot be denied. Man is still an animal, subject to zoological as well as mechanical laws. The dark influence of heredity continues to be effective; and psychical development had begun in lower organic forms,—perhaps with life itself. The organic and the social struggles for existence are manifestations of the same principle. Environment and climatic influence must be called in to explain not only the differentiation of the great racial sections of humanity, but also the varieties within these sub-species and, it may be, the assimilation of distinct varieties. Ritter's "Anthropogeography" has opened a useful line of research. But on the other hand, it is urged that, in explaining the course of history, these principles do not take us very far, and that it is chiefly for the primitive ultra-prehistoric period that they can account for human development. It may be said that, so far as concerns the actions and movements of men which are the subject of recorded history, physical environment has ceased to act mechanically, and in order to affect their actions must affect their wills first; and that this psychical character of the causal relations substantially alters the problem. The development of human societies, it may be argued, derives a completely new character from the dominance of the conscious psychical element, creating as it does new conditions (inventions, social institutions, etc.) which limit and counteract the operation of natural selection, and control and modify the influence of physical environment. Most thinkers agree now that the chief clews to the growth of civilisation must be sought in the psychological sphere. Imitation, for instance, is a principle which is probably more significant for the explanation of human development than natural selection. Darwin himself was conscious that his principles had only a very restricted application in this sphere, as is evident from his cautious and tentative remarks in the 5th chapter of his "Descent of Man". He applied natural selection to the growth of the intellectual faculties and of the fundamental social instincts, and also to the differentiation of the great races or "sub-species" (Caucasian, African, etc.) which differ in anthropological character. (Darwinian formulae may be suggestive by way of analogy. For instance, it is characteristic of social advance that a multitude of inventions, schemes and plans are framed which are never carried out, similar to, or designed for the same end as, an invention or plan which is actually adopted because it has chanced to suit better the particular conditions of the hour (just as the works accomplished by an individual statesman, artist or savant are usually only a residue of the numerous projects conceived by his brain). This process in which so much abortive production occurs is analogous to elimination by natural selection.)

16. But if it is admitted that the governing factors which concern the student of social development are of the psychical order, the preliminary success of natural science in explaining organic evolution by general principles encouraged sociologists to hope that social evolution could be explained on general principles also. The idea of Condorcet, Buckle, and others, that history could be assimilated to the natural sciences was powerfully reinforced, and the notion that the actual historical process, and every social movement involved in it, can be accounted for by sociological generalisations, so-called "laws," is still entertained by many, in one form or another. Dissentients from this view do not deny that the generalisations at which the sociologist arrives by the comparative method, by the analysis of

social factors, and by psychological deduction may be an aid to the historian; but they deny that such uniformities are laws or contain an explanation of the phenomena. They can point to the element of chance coincidence. This element must have played a part in the events of organic evolution, but it has probably in a larger measure helped to determine events in social evolution. The collision of two unconnected sequences may be fraught with great results. The sudden death of a leader or a marriage without issue, to take simple cases, has again and again led to permanent political consequences. More emphasis is laid on the decisive actions of individuals, which cannot be reduced under generalisations and which deflect the course of events. If the significance of the individual will had been exaggerated to the neglect of the collective activity of the social aggregate before Condorcet, his doctrine tended to eliminate as unimportant the roles of prominent men, and by means of this elimination it was possible to found sociology. But it may be urged that it is patent on the face of history that its course has constantly been shaped and modified by the wills of individuals (We can ignore here the metaphysical question of freewill and determinism. For the character of the individual's brain depends in any case on ante-natal accidents and coincidences, and so it may be said that the role of individuals ultimately depends on chance,—the accidental coincidence of independent sequences.), which are by no means always the expression of the collective will; and that the appearance of such personalities at the given moments is not a necessary outcome of the conditions and cannot be deduced. Nor is there any proof that, if such and such an individual had not been born, some one else would have arisen to do what he did. In some cases there is no reason to think that what happened need ever have come to pass. In other cases, it seems evident that the actual change was inevitable, but in default of the man who initiated and guided it, it might have been postponed, and, postponed or not, might have borne a different cachet. I may illustrate by an instance which has just come under my notice. Modern painting was founded by Giotto, and the Italian expedition of Charles VIII, near the close of the sixteenth century, introduced into France the fashion of imitating Italian painters. But for Giotto and Charles VIII, French painting might have been very different. It may be said that "if Giotto had not appeared, some other great initiator would have played a role analogous to his, and that without Charles VIII there would have been the commerce with Italy, which in the long run would have sufficed to place France in relation with Italian artists. But the equivalent of Giotto might have been deferred for a century and probably would have been different; and commercial relations would have required ages to produce the rayonnement imitatif of Italian art in France, which the expedition of the royal adventurer provoked in a few years." (I have taken this example from G. Tarde's "La logique sociale" 2 (page 403), Paris, 1904, where it is used for quite a different purpose.) Instances furnished by political history are simply endless. Can we conjecture how events would have moved if the son of Philip of Macedon had been an incompetent? The aggressive action of Prussia which astonished Europe in 1740 determined the subsequent history of Germany; but that action was anything but inevitable; it depended entirely on the personality of Frederick the Great.

Hence it may be argued that the action of individual wills is a determining and disturbing factor, too significant and effective to allow history to be grasped by sociological formulae. The types and general forms of development which the sociologist attempts to disengage can only assist the historian in understanding the actual course of events. It is in the special domains of economic history and Culturgeschichte which have come to the front in modern times that generalisation is most fruitful, but even in these it may be contended that it furnishes only partial explanations.

17. The truth is that Darwinism itself offers the best illustration of the insufficiency of general laws to account for historical development. The part played by coincidence, and the part played by individuals—limited by, and related to, general social conditions—render it impossible to deduce the course of the past history of man or to predict the future. But it is just the same with organic development. Darwin (or any other zoologist) could not deduce the actual course of evolution from general principles. Given an organism and its environment, he could not show that it must evolve into a more complex organism of a definite pre-determined type; knowing what it has evolved into, he could attempt to discover

and assign the determining causes. General principles do not account for a particular sequence; they embody necessary conditions; but there is a chapter of accidents too. It is the same in the case of history.

18. Among the evolutional attempts to subsume the course of history under general syntheses, perhaps the most important is that of Lamprecht, whose "kulturhistorische Methode," which he has deduced from and applied to German history, exhibits the (indirect) influence of the Comtist school. It is based upon psychology, which, in his view, holds among the sciences of mind (Geisteswissenschaften) the same place (that of a Grundwissenschaft) which mechanics holds among the sciences of nature. History, by the same comparison, corresponds to biology, and, according to him, it can only become scientific if it is reduced to general concepts (Begriffe). Historical movements and events are of a psychical character, and Lamprecht conceives a given phase of civilisation as "a collective psychical condition (seelischer Gesamtzustand)" controlling the period, "a diapason which penetrates all psychical phenomena and thereby all historical events of the time." ("Die kulturhistorische Methode", Berlin, 1900, page 26.) He has worked out a series of such phases, "ages of changing psychical diapason," in his "Deutsche Geschichte" with the aim of showing that all the feelings and actions of each age can be explained by the diapason; and has attempted to prove that these diapasons are exhibited in other social developments, and are consequently not singular but typical. He maintains further that these ages succeed each other in a definite order; the principle being that the collective psychical development begins with the homogeneity of all the individual members of a society and, through heightened psychical activity, advances in the form of a continually increasing differentiation of the individuals (this is akin to the Spencerian formula). This process, evolving psychical freedom from psychical constraint, exhibits a series of psychical phenomena which define successive periods of civilisation. The process depends on two simple principles, that no idea can disappear without leaving behind it an effect or influence, and that all psychical life, whether in a person or a society, means change, the acquisition of new mental contents. It follows that the new have to come to terms with the old, and this leads to a synthesis which determines the character of a new age. Hence the ages of civilisation are defined as the "highest concepts for subsuming without exception all psychical phenomena of the development of human societies, that is, of all historical events." (Ibid. pages 28, 29.) Lamprecht deduces the idea of a special historical science, which might be called "historical ethnology," dealing with the ages of civilisation, and bearing the same relation to (descriptive or narrative) history as ethnology to ethnography. Such a science obviously corresponds to Comte's social dynamics, and the comparative method, on which Comte laid so much emphasis, is the principal instrument of Lamprecht.

19. I have dwelt on the fundamental ideas of Lamprecht, because they are not yet widely known in England, and because his system is the ablest product of the sociological school of historians. It carries the more weight as its author himself is a historical specialist, and his historical syntheses deserve the most careful consideration. But there is much in the process of development which on such assumptions is not explained, especially the initiative of individuals. Historical development does not proceed in a right line, without the choice of diverging. Again and again, several roads are open to it, of which it chooses one—why? On Lamprecht's method, we may be able to assign the conditions which limit the psychical activity of men at a particular stage of evolution, but within those limits the individual has so many options, such a wide room for moving, that the definition of those conditions, the "psychical diapasons," is only part of the explanation of the particular development. The heel of Achilles in all historical speculations of this class has been the role of the individual.

The increasing prominence of economic history has tended to encourage the view that history can be explained in terms of general concepts or types. Marx and his school based their theory of human development on the conditions of production, by which, according to them, all social movements and historical changes are entirely controlled. The leading part which economic factors play in Lamprecht's system is significant, illustrating the fact that economic changes admit most readily this kind of treatment, because they have been less subject to direction or interference by individual pioneers.

Perhaps it may be thought that the conception of SOCIAL ENVIRONMENT (essentially psychical), on which Lamprecht's "psychical diapasons" depend, is the most valuable and fertile conception that the historian owes to the suggestion of the science of biology—the conception of all particular historical actions and movements as (1) related to and conditioned by the social environment, and (2) gradually bringing about a transformation of that environment. But no given transformation can be proved to be necessary (pre-determined). And types of development do not represent laws; their meaning and value lie in the help they may give to the historian, in investigating a certain period of civilisation, to enable him to discover the interrelations among the diverse features which it presents. They are, as some one has said, an instrument of heuretic method.

20. The men engaged in special historical researches—which have been pursued unremittingly for a century past, according to scientific methods of investigating evidence (initiated by Wolf, Niebuhr, Ranke)—have for the most part worked on the assumptions of genetic history or at least followed in the footsteps of those who fully grasped the genetic point of view. But their aim has been to collect and sift evidence, and determine particular facts; comparatively few have given serious thought to the lines of research and the speculations which have been considered in this paper. They have been reasonably shy of compromising their work by applying theories which are still much debated and immature. But historiography cannot permanently evade the questions raised by these theories. One may venture to say that no historical change or transformation will be fully understood until it is explained how social environment acted on the individual components of the society (both immediately and by heredity), and how the individuals reacted upon their environment. The problem is psychical, but it is analogous to the main problem of the biologist.

XXVIII. THE GENESIS OF DOUBLE STARS. By Sir George Darwin, K.C.B., F.R.S.

Plumian Professor of Astronomy and Experimental Philosophy in the University of Cambridge.

In ordinary speech a system of any sort is said to be stable when it cannot be upset easily, but the meaning attached to the word is usually somewhat vague. It is hardly surprising that this should be the case, when it is only within the last thirty years, and principally through the investigations of M. Poincaré, that the conception of stability has, even for physicists, assumed a definiteness and clearness in which it was previously lacking. The laws which govern stability hold good in regions of the greatest diversity; they apply to the motion of planets round the sun, to the internal arrangement of those minute corpuscles of which each chemical atom is constructed, and to the forms of celestial bodies. In the present essay I shall attempt to consider the laws of stability as relating to the last case, and shall discuss the succession of shapes which may be assumed by celestial bodies in the course of their evolution. I believe further that homologous conceptions are applicable in the consideration of the transmutations of the various forms of animal and of vegetable life and in other regions of thought. Even if some of my readers should think that what I shall say on this head is fanciful, yet at least the exposition will serve to illustrate the meaning to be attached to the laws of stability in the physical universe.

I propose, therefore, to begin this essay by a sketch of the principles of stability as they are now formulated by physicists.

I.

If a slight impulse be imparted to a system in equilibrium one of two consequences must ensue; either small oscillations of the system will be started, or the disturbance will increase without limit and the arrangement of the system will be completely changed. Thus a stick may be in equilibrium either when it hangs from a peg or when it is balanced on its point. If in the first case the stick is touched it will swing to and fro, but in the second case it will topple over. The first position is a stable one, the second is unstable. But this case is too simple to illustrate all that is implied by stability, and we must consider cases of stable and of unstable motion. Imagine a satellite and its planet, and consider each of them to be of indefinitely small size, in fact particles; then the satellite revolves round its planet in an ellipse. A small disturbance imparted to the satellite will only change the ellipse to a small amount, and so the motion is said to be stable. If, on the other hand, the disturbance were to make the satellite depart from its initial elliptic orbit in ever widening circuits, the motion would be unstable. This case affords an example of stable motion, but I have adduced it principally with the object of illustrating another point not immediately connected with stability, but important to a proper comprehension of the theory of stability.

The motion of a satellite about its planet is one of revolution or rotation. When the satellite moves in an ellipse of any given degree of eccentricity, there is a certain amount of rotation in the system, technically called rotational momentum, and it is always the same at every part of the orbit. (Moment of momentum or rotational momentum is measured by the momentum of the satellite multiplied by the perpendicular from the planet on to the direction of the path of the satellite at any instant.)

Now if we consider all the possible elliptic orbits of a satellite about its planet which have the same amount of "rotational momentum," we find that the major axis of the ellipse described will be different according to the amount of flattening (or the eccentricity) of the ellipse described. A figure titled "A 'family' of elliptic orbits with constant rotational momentum" (Fig. 1) illustrates for a given planet and satellite all such orbits with constant rotational momentum, and with all the major axes in the same direction. It will be observed that there is a continuous transformation from one orbit to the next, and that the whole forms a consecutive group, called by mathematicians "a family" of orbits. In this case the rotational momentum is constant and the position of any orbit in the family is determined by the length of the major axis of the ellipse; the classification is according to the major axis, but it might have been made according to anything else which would cause the orbit to be exactly determinate.

I shall come later to the classification of all possible forms of ideal liquid stars, which have the same amount of rotational momentum, and the classification will then be made according to their densities, but the idea of orderly arrangement in a "family" is just the same.

We thus arrive at the conception of a definite type of motion, with a constant amount of rotational momentum, and a classification of all members of the family, formed by all possible motions of that type, according to the value of some measurable quantity (this will hereafter be density) which determines the motion exactly. In the particular case of the elliptic motion used for illustration the motion was stable, but other cases of motion might be adduced in which the motion would be unstable, and it would be found that classification in a family and specification by some measurable quantity would be equally applicable.

A complex mechanical system may be capable of motion in several distinct modes or types, and the motions corresponding to each such type may be arranged as before in families. For the sake of simplicity I will suppose that only two types are possible, so that there will only be two families; and the rotational momentum is to be constant. The two types of motion will have certain features in common which we denote in a sort of shorthand by the letter A. Similarly the two types may be described as A + a and A + b, so

337

that a and b denote the specific differences which discriminate the families from one another. Now following in imagination the family of the type A + a, let us begin with the case where the specific difference a is well marked. As we cast our eyes along the series forming the family, we find the difference a becoming less conspicuous. It gradually dwindles until it disappears; beyond this point it either becomes reversed, or else the type has ceased to be a possible one. In our shorthand we have started with A + a, and have watched the characteristic a dwindling to zero. When it vanishes we have reached a type which may be specified as A; beyond this point the type would be A - a or would be impossible.

Following the A + b type in the same way, b is at first well marked, it dwindles to zero, and finally may become negative. Hence in shorthand this second family may be described as A + b,... A,... A - b.

In each family there is one single member which is indistinguishable from a member of the other family; it is called by Poincare a form of bifurcation. It is this conception of a form of bifurcation which forms the important consideration in problems dealing with the forms of liquid or gaseous bodies in rotation.

But to return to the general question,—thus far the stability of these families has not been considered, and it is the stability which renders this way of looking at the matter so valuable. It may be proved that if before the point of bifurcation the type A + a was stable, then A + b must have been unstable. Further as a and b each diminish A + a becomes less pronouncedly stable, and A + b less unstable. On reaching the point of bifurcation A + a has just ceased to be stable, or what amounts to the same thing is just becoming unstable, and the converse is true of the A + b family. After passing the point of bifurcation A + a has become definitely unstable and A + b has become stable. Hence the point of bifurcation is also a point of "exchange of stabilities between the two types." (In order not to complicate unnecessarily this explanation of a general principle I have not stated fully all the cases that may occur. Thus: firstly, after bifurcation A + a may be an impossible type and A + a will then stop at this point; or secondly, A + b may have been an impossible type before bifurcation, and will only begin to be a real one after it; or thirdly, both A + a and A + b may be impossible after the point of bifurcation, in which case they coalesce and disappear. This last case shows that types arise and disappear in pairs, and that on appearance or before disappearance one must be stable and the other unstable.)

In nature it is of course only the stable types of motion which can persist for more than a short time. Thus the task of the physical evolutionist is to determine the forms of bifurcation, at which he must, as it were, change carriages in the evolutionary journey so as always to follow the stable route. He must besides be able to indicate some natural process which shall correspond in effect to the ideal arrangement of the several types of motion in families with gradually changing specific differences. Although, as we shall see hereafter, it may frequently or even generally be impossible to specify with exactness the forms of bifurcation in the process of evolution, yet the conception is one of fundamental importance.

The ideas involved in this sketch are no doubt somewhat recondite, but I hope to render them clearer to the non-mathematical reader by homologous considerations in other fields of thought (I considered this subject in my Presidential address to the British Association in 1905, "Report of the 75th Meeting of the British Assoc." (S. Africa, 1905), London, 1906, page 3. Some reviewers treated my speculations as fanciful, but as I believe that this was due generally to misapprehension, and as I hold that homologous considerations as to stability and instability are really applicable to evolution of all sorts, I have thought it well to return to the subject in the present paper.), and I shall pass on thence to illustrations which will teach us something of the evolution of stellar systems.

States or governments are organised schemes of action amongst groups of men, and they belong to various types to which generic names, such as autocracy, aristocracy or democracy, are somewhat loosely applied. A definite type of government corresponds to one of our types of motion, and while retaining its type it undergoes a slow change as the civilisation and character of the people change, and as the relationship of the nation to other nations changes. In the language used before, the government belongs to a family, and as time advances we proceed through the successive members of the family. A government possesses a certain degree of stability—hardly measurable in numbers however—to resist disintegrating influences such as may arise from wars, famines, and internal dissensions. This stability gradually rises to a maximum and gradually declines. The degree of stability at any epoch will depend on the fitness of some leading feature of the government to suit the slowly altering circumstances, and that feature corresponds to the characteristic denoted by a in the physical problem. A time at length arrives when the stability vanishes, and the slightest shock will overturn the government. At this stage we have reached the crisis of a point of bifurcation, and there will then be some circumstance, apparently quite insignificant and almost unnoticed, which is such as to prevent the occurrence of anarchy. This circumstance or condition is what we typified as b. Insignificant although it may seem, it has started the government on a new career of stability by imparting to it a new type. It grows in importance, the form of government becomes obviously different, and its stability increases. Then in its turn this newly acquired stability declines, and we pass on to a new crisis or revolution. There is thus a series of "points of bifurcation" in history at which the continuity of political history is maintained by means of changes in the type of government. These ideas seem, to me at least, to give a true account of the history of states, and I contend that it is no mere fanciful analogy but a true homology, when in both realms of thought—the physical and the political—we perceive the existence of forms of bifurcation and of exchanges of stability.

Further than this, I would ask whether the same train of ideas does not also apply to the evolution of animals? A species is well adapted to its environment when the individual can withstand the shocks of famine or the attacks and competition of other animals; it then possesses a high degree of stability. Most of the casual variations of individuals are indifferent, for they do not tell much either for or against success in life; they are small oscillations which leave the type unchanged. As circumstances change, the stability of the species may gradually dwindle through the insufficiency of some definite quality, on which in earlier times no such insistent demands were made. The individual animals will then tend to fail in the struggle for life, the numbers will dwindle and extinction may ensue. But it may be that some new variation, at first of insignificant importance, may just serve to turn the scale. A new type may be formed in which the variation in question is preserved and augmented; its stability may increase and in time a new species may be produced.

At the risk of condemnation as a wanderer beyond my province into the region of biological evolution, I would say that this view accords with what I understand to be the views of some naturalists, who recognise the existence of critical periods in biological history at which extinction occurs or which form the starting-point for the formation of new species. Ought we not then to expect that long periods will elapse during which a type of animal will remain almost constant, followed by other periods, enormously long no doubt as measured in the life of man, of acute struggle for existence when the type will change more rapidly? This at least is the view suggested by the theory of stability in the physical universe. (I make no claim to extensive reading on this subject, but refer the reader for example to a paper by Professor A.A.W. Hubrecht on "De Vries's theory of Mutations", "Popular Science Monthly", July 1904, especially to page 213.)

And now I propose to apply these ideas of stability to the theory of stellar evolution, and finally to illustrate them by certain recent observations of a very remarkable character.

Stars and planets are formed of materials which yield to the enormous forces called into play by gravity and rotation. This is obviously true if they are gaseous or fluid, and even solid matter becomes plastic under sufficiently great stresses. Nothing approaching a complete study of the equilibrium of a heterogeneous star has yet been found possible, and we are driven to consider only bodies of simpler construction. I shall begin therefore by explaining what is known about the shapes which may be assumed by a mass of incompressible liquid of uniform density under the influences of gravity and of rotation. Such a liquid mass may be regarded as an ideal star, which resembles a real star in the fact that it is formed of gravitating and rotating matter, and because its shape results from the forces to which it is subject. It is unlike a star in that it possesses the attributes of incompressibility and of uniform density. The difference between the real and the ideal is doubtless great, yet the similarity is great enough to allow us to extend many of the conclusions as to ideal liquid stars to the conditions which must hold good in reality. Thus with the object of obtaining some insight into actuality, it is justifiable to discuss an avowedly ideal problem at some length.

The attraction of gravity alone tends to make a mass of liquid assume the shape of a sphere, and the effects of rotation, summarised under the name of centrifugal force, are such that the liquid seeks to spread itself outwards from the axis of rotation. It is a singular fact that it is unnecessary to take any account of the size of the mass of liquid under consideration, because the shape assumed is exactly the same whether the mass be small or large, and this renders the statement of results much easier than would otherwise be the case.

A mass of liquid at rest will obviously assume the shape of a sphere, under the influence of gravitation, and it is a stable form, because any oscillation of the liquid which might be started would gradually die away under the influence of friction, however small. If now we impart to the whole mass of liquid a small speed of rotation about some axis, which may be called the polar axis, in such a way that there are no internal currents and so that it spins in the same way as if it were solid, the shape will become slightly flattened like an orange. Although the earth and the other planets are not homogeneous they behave in the same way, and are flattened at the poles and protuberant at the equator. This shape may therefore conveniently be described as planetary.

If the planetary body be slightly deformed the forces of restitution are slightly less than they were for the sphere; the shape is stable but somewhat less so than the sphere. We have then a planetary spheroid, rotating slowly, slightly flattened at the poles, with a high degree of stability, and possessing a certain amount of rotational momentum. Let us suppose this ideal liquid star to be somewhere in stellar space far removed from all other bodies; then it is subject to no external forces, and any change which ensues must come from inside. Now the amount of rotational momentum existing in a system in motion can neither be created nor destroyed by any internal causes, and therefore, whatever happens, the amount of rotational momentum possessed by the star must remain absolutely constant.

A real star radiates heat, and as it cools it shrinks. Let us suppose then that our ideal star also radiates and shrinks, but let the process proceed so slowly that any internal currents generated in the liquid by the cooling are annulled so quickly by fluid friction as to be insignificant; further let the liquid always remain at any instant incompressible and homogeneous. All that we are concerned with is that, as time passes, the liquid star shrinks, rotates in one piece as if it were solid, and remains incompressible and homogeneous. The condition is of course artificial, but it represents the actual processes of nature as well as may be, consistently with the postulated incompressibility and homogeneity. (Mathematicians are accustomed to regard the density as constant and the rotational momentum as increasing. But the way of looking at the matter, which I have adopted, is easier of comprehension, and it comes to the same in the end.)

The shrinkage of a constant mass of matter involves an increase of its density, and we have therefore to trace the changes which supervene as the star shrinks, and as the liquid of which it is composed increases in density. The shrinkage will, in ordinary parlance, bring the weights nearer to the axis of rotation. Hence in order to keep up the rotational momentum, which as we have seen must remain constant, the mass must rotate quicker. The greater speed of rotation augments the importance of centrifugal force compared with that of gravity, and as the flattening of the planetary spheroid was due to centrifugal force, that flattening is increased; in other words the ellipticity of the planetary spheroid increases.

As the shrinkage and corresponding increase of density proceed, the planetary spheroid becomes more and more elliptic, and the succession of forms constitutes a family classified

340

according to the density of the liquid. The specific mark of this family is the flattening or ellipticity.

Now consider the stability of the system, we have seen that the spheroid with a slow rotation, which forms our starting-point, was slightly less stable than the sphere, and as we proceed through the family of ever flatter ellipsoids the stability continues to diminish. At length when it has assumed the shape shown in a figure titled "Planetary spheroid just becoming unstable" (Fig. 2.) where the equatorial and polar axes are proportional to the numbers 1000 and 583, the stability has just disappeared. According to the general principle explained above this is a form of bifurcation, and corresponds to the form denoted A. The specific difference a of this family must be regarded as the excess of the ellipticity of this figure above that of all the earlier ones, beginning with the slightly flattened planetary spheroid. Accordingly the specific difference a of the family has gradually diminished from the beginning and vanishes at this stage.

According to Poincare's principle the vanishing of the stability serves us with notice that we have reached a figure of bifurcation, and it becomes necessary to inquire what is the nature of the specific difference of the new family of figures which must be coalescent with the old one at this stage. This difference is found to reside in the fact that the equator, which in the planetary family has hitherto been circular in section, tends to become elliptic. Hitherto the rotational momentum has been kept up to its constant value partly by greater speed of rotation and partly by a symmetrical bulging of the equator. But now while the speed of rotation still increases (The mathematician familiar with Jacobi's ellipsoid will find that this is correct, although in the usual mode of exposition, alluded to above in a footnote, the speed diminishes.), the equator tends to bulge outwards at two diametrically opposite points and to be flattened midway between these protuberances. The specific difference in the new family, denoted in the general sketch by b, is this ellipticity of the equator. If we had traced the planetary figures with circular equators beyond this stage A, we should have found them to have become unstable, and the stability has been shunted off along the A + b family of forms with elliptic equators.

This new series of figures, generally named after the great mathematician Jacobi, is at first only just stable, but as the density increases the stability increases, reaches a maximum and then declines. As this goes on the equator of these Jacobian figures becomes more and more elliptic, so that the shape is considerably elongated in a direction at right angles to the axis of rotation.

At length when the longest axis of the three has become about three times as long as the shortest (The three axes of the ellipsoid are then proportional to 1000, 432, 343.), the stability of this family of figures vanishes, and we have reached a new form of bifurcation and must look for a new type of figure along which the stable development will presumably extend. Two sections of this critical Jacobian figure, which is a figure of bifurcation, are shown by the dotted lines in a figure titled "The 'pear-shaped figure' and the Jocobian figure from which it is derived" (Fig. 3.) comprising two figures, one above the other: the upper figure is the equatorial section at right angles to the axis of rotation, the lower figure is a section through the axis.

Now Poincare has proved that the new type of figure is to be derived from the figure of bifurcation by causing one of the ends to be prolonged into a snout and by bluntening the other end. The snout forms a sort of stalk, and between the stalk and the axis of rotation the surface is somewhat flattened. These are the characteristics of a pear, and the figure has therefore been called the "pear-shaped figure of equilibrium." The firm line shows this new type of figure, whilst, as already explained, the dotted line shows the form of bifurcation from which it is derived. The specific mark of this new family is the protrusion of the stalk together with the other corresponding smaller differences. If we denote this difference by c, while A + b denotes the Jacobian figure of bifurcation from which it is derived, the new family may be called A + b + c, and c is zero initially. According to my calculations this series of figures is stable (M. Liapounoff contends that for constant density the new series of figures, which M. Poincare discovered, has less rotational momentum than that of the figure

341

of bifurcation. If he is correct, the figure of bifurcation is a limit of stable figures, and none can exist with stability for greater rotational momentum. My own work seems to indicate that the opposite is true, and, notwithstanding M. Liapounoff's deservedly great authority, I venture to state the conclusions in accordance with my own work.), but I do not know at what stage of its development it becomes unstable.

Professor Jeans has solved a problem which is of interest as throwing light on the future development of the pear-shaped figure, although it is of a still more ideal character than the one which has been discussed. He imagines an INFINITELY long circular cylinder of liquid to be in rotation about its central axis. The existence is virtually postulated of a demon who is always occupied in keeping the axis of the cylinder straight, so that Jeans has only to concern himself with the stability of the form of the section of the cylinder, which as I have said is a circle with the axis of rotation at the centre. He then supposes the liquid forming the cylinder to shrink in diameter, just as we have done, and finds that the speed of rotation must increase so as to keep up the constancy of the rotational momentum. The circularity of section is at first stable, but as the shrinkage proceeds the stability diminishes and at length vanishes. This stage in the process is a form of bifurcation, and the stability passes over to a new series consisting of cylinders which are elliptic in section. The circular cylinders are exactly analogous with our planetary spheroids, and the elliptic ones with the Jacobian ellipsoids.

With further shrinkage the elliptic cylinders become unstable, a new form of bifurcation is reached, and the stability passes over to a series of cylinders whose section is pear-shaped. Thus far the analogy is complete between our problem and Jeans's, and in consequence of the greater simplicity of the conditions, he is able to carry his investigation further. He finds that the stalk end of the pear-like section continues to protrude more and more, and the flattening between it and the axis of rotation becomes a constriction. Finally the neck breaks and a satellite cylinder is born. Jeans's figure for an advanced stage of development is shown in a figure titled "Section of a rotating cylinder of liquid" (Fig. 4.), but his calculations do not enable him actually to draw the state of affairs after the rupture of the neck.

There are certain difficulties in admitting the exact parallelism between this problem and ours, and thus the final development of our pear-shaped figure and the end of its stability in a form of bifurcation remain hidden from our view, but the successive changes as far as they have been definitely traced are very suggestive in the study of stellar evolution.

Attempts have been made to attack this problem from the other end. If we begin with a liquid satellite revolving about a liquid planet and proceed backwards in time, we must make the two masses expand so that their density will be diminished. Various figures have been drawn exhibiting the shapes of two masses until their surfaces approach close to one another and even until they just coalesce, but the discussion of their stability is not easy. At present it would seem to be impossible to reach coalescence by any series of stable transformations, and if this is so Professor Jeans's investigation has ceased to be truly analogous to our problem at some undetermined stage. However this may be this line of research throws an instructive light on what we may expect to find in the evolution of real stellar systems.

In the second part of this paper I shall point out the bearing which this investigation of the evolution of an ideal liquid star may have on the genesis of double stars.

II.

There are in the heavens many stars which shine with a variable brilliancy. Amongst these there is a class which exhibits special peculiarities; the members of this class are generally known as Algol Variables, because the variability of the star Beta Persei or Algol was the first of such cases to attract the attention of astronomers, and because it is perhaps still the most remarkable of the whole class. But the circumstances which led to this discovery were so extraordinary that it seems worth while to pause a moment before entering on the subject.

John Goodricke, a deaf-mute, was born in 1764; he was grandson and heir of Sir John Goodricke of Ribston Hall, Yorkshire. In November 1782, he noted that the brilliancy of

Algol waxed and waned (It is said that Georg Palitzch, a farmer of Prohlis near Dresden, had about 1758 already noted the variability of Algol with the naked eye. "Journ. Brit. Astron. Assoc." Vol. XV. (1904-5), page 203.), and devoted himself to observing it on every fine night from the 28th December 1782 to the 12th May 1783. He communicated his observations to the Royal Society, and suggested that the variation in brilliancy was due to periodic eclipses by a dark companion star, a theory now universally accepted as correct. The Royal Society recognised the importance of the discovery by awarding to Goodricke, then only 19 years of age, their highest honour, the Copley medal. His later observations of Beta Lyrae and of Delta Cephei were almost as remarkable as those of Algol, but unfortunately a career of such extraordinary promise was cut short by death, only a fortnight after his election to the Royal Society. ("Dict. of National Biography"; article Goodricke (John). The article is by Miss Agnes Clerke. It is strange that she did not then seem to be aware that he was a deaf-mute, but she notes the fact in her "Problems of Astrophysics", page 337, London, 1903.)

It was not until 1889 that Goodricke's theory was verified, when it was proved by Vogel that the star was moving in an orbit, and in such a manner that it was only possible to explain the rise and fall in the luminosity by the partial eclipse of a bright star by a dark companion.

The whole mass of the system of Algol is found to be half as great again as that of our sun, yet the two bodies complete their orbit in the short period of 2d 20h 48m 55s. The light remains constant during each period, except for 9h 20m when it exhibits a considerable fall in brightness (Clerke, "Problems of Astrophysics" page 302 and chapter XVIII.); the curve which represents the variation in the light is shown in a figure titled "The light-curve and system of Beta Lyrae" (Fig. 7.).

The spectroscope has enabled astronomers to prove that many stars, although apparently single, really consist of two stars circling around one another (If a source of light is approaching with a great velocity the waves of light are crowded together, and conversely they are spaced out when the source is receding. Thus motion in the line of sight virtually produces an infinitesimal change of colour. The position of certain dark lines in the spectrum affords an exceedingly accurate measurement of colour. Thus displacements of these spectral lines enables us to measure the velocity of the source of light towards or away from the observer.); they are known as spectroscopic binaries. Campbell of the Lick Observatory believes that about one star in six is a binary ("Astrophysical Journ." Vol. XIII. page 89, 1901. See also A. Roberts, "Nature", Sept. 12, 1901, page 468.); thus there must be many thousand such stars within the reach of our spectroscopes.

The orientation of the planes of the orbits of binary stars appears to be quite arbitrary, and in general the star does not vary in brightness. Amongst all such orbits there must be some whose planes pass nearly through the sun, and in these cases the eclipse of one of the stars by the other becomes inevitable, and in each circuit there will occur two eclipses of unequal intensities.

It is easy to see that in the majority of such cases the two components must move very close to one another.

The coincidence between the spectroscopic and the photometric evidence permits us to feel complete confidence in the theory of eclipses. When then we find a star with a light-curve of perfect regularity and with a characteristics of that of Algol, we are justified in extending the theory of eclipses to it, although it may be too faint to permit of adequate spectroscopic examination. This extension of the theory secures a considerable multiplication of the examples available for observation, and some 30 have already been discovered.

Dr Alexander Roberts, of Lovedale in Cape Colony, truly remarks that the study of Algol variables "brings us to the very threshold of the question of stellar evolution." ("Proc. Roy. Soc. Edinburgh", XXIV. Part II. (1902), page 73.) It is on this account that I propose to explain in some detail the conclusion to which he and some other observers have been led.

Although these variable stars are mere points of light, it has been proved by means of the spectroscope that the law of gravitation holds good in the remotest regions of stellar space, and further it seems now to have become possible even to examine the shapes of stars by indirect methods, and thus to begin the study of their evolution. The chain of reasoning which I shall explain must of necessity be open to criticism, yet the explanation of the facts by the theory is so perfect that it is not easy to resist the conviction that we are travelling along the path of truth.

The brightness of a star is specified by what is called its "magnitude." The average brightness of all the stars which can just be seen with the naked eye defines the sixth magnitude. A star which only gives two-fifths as much light is said to be of the seventh magnitude; while one which gives 2 1/2 times as much light is of the fifth magnitude, and successive multiplications or divisions by 2 1/2 define the lower or higher magnitudes. Negative magnitudes have clearly to be contemplated; thus Sirius is of magnitude minus 1.4, and the sun is of magnitude minus 26.

The definition of magnitude is also extended to fractions; for example, the lights given by two candles which are placed at 100 feet and 100 feet 6 inches from the observer differ in brightness by one-hundredth of a magnitude.

A great deal of thought has been devoted to the measurement of the brightness of stars, but I will only describe one of the methods used, that of the great astronomer Argelander. In the neighbourhood of the star under observation some half dozen standard stars are selected of known invariable magnitudes, some being brighter and some fainter than the star to be measured; so that these stars afford a visible scale of brightness. Suppose we number them in order of increasing brightness from 1 to 6; then the observer estimates that on a given night his star falls between stars 2 and 3, on the next night, say between 3 and 4, and then again perhaps it may return to between 2 and 3, and so forth. With practice he learns to evaluate the brightness down to small fractions of a magnitude, even a hundredth part of a magnitude is not quite negligible.

For example, in observing the star RR Centauri five stars were in general used for comparison by Dr Roberts, and in course of three months he secured thereby 300 complete observations. When the period of the cycle had been ascertained exactly, these 300 values were reduced to mean values which appertained to certain mean places in the cycle, and a mean light-curve was obtained in this way. Figures titled "Light curve of RR Centauri" (Fig. 5) and "The light-curve and system of Beta Lyrae" (Fig. 7) show examples of light curves.

I shall now follow out the results of the observation of RR Centauri not only because it affords the easiest way of explaining these investigations, but also because it is one of the stars which furnishes the most striking results in connection with the object of this essay. (See "Monthly notices R.A.S." Vol. 63, 1903, page 527.) This star has a mean magnitude of about 7 1/2, and it is therefore invisible to the naked eye. Its period of variability is 14h 32m 10s.76, the last refinement of precision being of course only attained in the final stages of reduction. Twenty-nine mean values of the magnitude were determined, and they were nearly equally spaced over the whole cycle of changes. The black dots in Fig. 5 exhibit the mean values determined by Dr Roberts. The last three dots on the extreme right are merely the same as the first three on the extreme left, and are repeated to show how the next cycle would begin. The smooth dotted curve will be explained hereafter, but, by reference to the scale of magnitudes on the margins of the figure, it may be used to note that the dots might be brought into a perfectly smooth curve by shifting some few of the dots by about a hundredth of a magnitude.

This light-curve presents those characteristics which are due to successive eclipses, but the exact form of the curve must depend on the nature of the two mutually eclipsing stars. If we are to interpret the curve with all possible completeness, it is necessary to make certain assumptions as to the stars. It is assumed then that the stars are equally bright all over their disks, and secondly that they are not surrounded by an extensive absorptive atmosphere. This last appears to me to be the most dangerous assumption involved in the whole theory.

Making these assumptions, however, it is found that if each of the eclipsing stars were spherical it would not be possible to generate such a curve with the closest accuracy. The two stars are certainly close together, and it is obvious that in such a case the tidal forces exercised by each on the other must be such as to elongate the figure of each towards the other. Accordingly it is reasonable to adopt the hypothesis that the system consists of a pair of elongated ellipsoids, with their longest axes pointed towards one another. No supposition is adopted a priori as to the ratio of the two masses, or as to their relative size or brightness, and the orbit may have any degree of eccentricity. These last are all to be determined from the nature of the light-curve.

In the case of RR Centauri, however, Dr Roberts finds the conditions are best satisfied by supposing the orbit to be circular, and the sizes and masses of the components to be equal, while their luminosities are to one another in the ratio of 4 to 3. As to their shapes he finds them to be so much elongated that they overlap, as exhibited in his figure titled "The shape of the star RR Centauri" (Fig. 6.). The dotted curve shows a form of equilibrium of rotating liquid as computed by me some years before, and it was added for the sake of comparison.

On turning back to Fig. 5 the reader will see in the smooth dotted curve the light variation which would be exhibited by such a binary system as this. The curve is the result of computation and it is impossible not to be struck by the closeness of the coincidence with the series of black dots which denote the observations.

It is virtually certain that RR Centauri is a case of an eclipsing binary system, and that the two stars are close together. It is not of course proved that the figures of the stars are ellipsoids, but gravitation must deform them into a pair of elongated bodies, and, on the assumptions that they are not enveloped in an absorptive atmosphere and that they are ellipsoidal, their shapes must be as shown in the figure.

This light-curve gives an excellent illustration of what we have reason to believe to be a stage in the evolution of stars, when a single star is proceeding to separate into a binary one.

As the star is faint, there is as yet no direct spectroscopic evidence of orbital motion. Let us turn therefore to the case of another star, namely V Puppis, in which such evidence does already exist. I give an account of it, because it presents a peculiarly interesting confirmation of the correctness of the theory.

In 1895 Pickering announced in the "Harvard Circular" No. 14 that the spectroscopic observations at Arequipa proved V Puppis to be a double star with a period of 3d 2h 46m. Now when Roberts discussed its light-curve he found that the period was 1d 10h 54m 27s, and on account of this serious discrepancy he effected the reduction only on the simple assumption that the two stars were spherical, and thus obtained a fairly good representation of the light-curve. It appeared that the orbit was circular and that the two spheres were not quite in contact. Obviously if the stars had been assumed to be ellipsoids they would have been found to overlap, as was the case for RR Centauri. ("Astrophysical Journ." Vol. XIII. (1901), page 177.) The matter rested thus for some months until the spectroscopic evidence was re-examined by Miss Cannon on behalf of Professor Pickering, and we find in the notes on page 177 of Vol. XXVIII. of the "Annals of the Harvard Observatory" the following: "A.G.C. 10534. This star, which is the Algol variable V Puppis, has been found to be a spectroscopic binary. The period 1d.454 (i.e. 1d 10h 54m) satisfies the observations of the changes in light, and of the varying separation of the lines of the spectrum. The spectrum has been examined on 61 plates, on 23 of which the lines are double." Thus we have valuable evidence in confirmation of the correctness of the conclusions drawn from the light-curve. In the circumstances, however, I have not thought it worth while to reproduce Dr Roberts's provisional figure.

I now turn to the conclusions drawn a few years previously by another observer, where we shall find the component stars not quite in contact. This is the star Beta Lyrae which was observed by Goodricke, Argelander, Belopolsky, Schur, Markwick and by many others. The spectroscopic method has been successfully applied in this case, and the component stars are proved to move in an orbit about one another. In 1897, Mr. G.W. Myers applied the theory of eclipses to the light-curve, on the hypothesis that the stars are elongated ellipsoids, and he obtained the interesting results exhibited in Fig. 7. ("Astrophysical Journ." Vol. VII. (1898), page 1.)

The period of Beta Lyrae is relatively long, being 12d 21h 47m, the orbit is sensibly eccentric, and the two spheroids are not so much elongated as was the case with RR Centauri. The mass of the system is enormous, one of the two stars being 10 times and the other 21 times as heavy as our sun.

Further illustrations of this subject might be given, but enough has been said to explain the nature of the conclusions which have been drawn from this class of observation.

In my account of these remarkable systems the consideration of one very important conclusion has been purposely deferred. Since the light-curve is explicable by eclipses, it follows that the sizes of the two stars are determinable relatively to the distance between them. The period of their orbital motion is known, being identical with the complete period of the variability of their light, and an easy application of Kepler's law of periodic times enables us to compute the sum of the masses of the two stars divided by the cube of the distance between their centres. Now the sizes of the bodies being known, the mean density of the whole system may be calculated. In every case that density has been found to be much less than the sun's, and indeed the average of a number of mean densities which have been determined only amounts to one-eighth of that of the sun. In some cases the density is extremely small, and in no case is it quite so great as half the solar density.

It would be absurd to suppose that these stars can be uniform in density throughout, and from all that is known of celestial bodies it is probable that they are gaseous in their external parts with great condensation towards their centres. This conclusion is confirmed by arguments drawn from the theory of rotating masses of liquid. (See J.H. Jeans, "On the density of Algol variables", "Astrophysical Journal." Vol. XXII. (1905), page 97.)

Although, as already explained, a good deal is known about the shapes and the stability of figures consisting of homogeneous incompressible liquid in rotation, yet comparatively little has hitherto been discovered about the equilibrium of rotating gaseous stars. The figures calculated for homogeneous liquid can obviously only be taken to afford a general indication of the kind of figure which we might expect to find in the stellar universe. Thus the dotted curve in Fig. 5, which exhibits one of the figures which I calculated, has some interest when placed alongside the figures of the stars in RR Centauri, as computed from the observations, but it must not be accepted as the calculated form of such a system. I have moreover proved more recently that such a figure of homogeneous liquid is unstable. Notwithstanding this instability it does not necessarily follow that the analogous figure for compressible fluid is also unstable, as will be pointed out more fully hereafter.

Professor Jeans has discussed in a paper of great ability the difficult problems offered by the conditions of equilibrium and of stability of a spherical nebula. ("Phil. Trans. R.S." Vol. CXCIX. A (1902), page 1. See also A. Roberts, "S. African Assoc. Adv. Sci." Vol. I. (1903), page 6.) In a later paper ("Astrophysical Journ." Vol. XXII. (1905), page 97.), in contrasting the conditions which must govern the fission of a star into two parts when the star is gaseous and compressible with the corresponding conditions in the case of incompressible liquid, he points out that for a gaseous star (the agency which effects the separation will no longer be rotation alone; gravitation also will tend towards separation... From numerical results obtained in the various papers of my own,... I have been led to the conclusion that a gravitational instability of the kind described must be regarded as the primary agent at work in the actual evolution of the universe, Laplace's rotation playing only the secondary part of separating the primary and satellite after the birth of the satellite.)

It is desirable to add a word in explanation of the expression "gravitational instability" in this passage. It means that when the concentration of a gaseous nebula (without rotation) has proceeded to a certain stage, the arrangement in spherical layers of equal density becomes unstable, and a form of bifurcation has been reached. For further concentration concentric spherical layers become unstable, and the new stable form involves a concentration about two centres. The first sign of this change is that the spherical layers cease to be quite concentric and then the layers of equal density begin to assume a somewhat pear-shaped form analogous to that which we found to occur under rotation for an incompressible liquid. Accordingly it appears that while a sphere of liquid is stable a sphere of gas may become unstable. Thus the conditions of stability are different in these two simple cases, and it is likely that while certain forms of rotating liquid are unstable the analogous forms for gas may be stable. This furnishes a reason why it is worth while to consider the unstable forms of rotating liquid.

There can I think be little doubt but that Jeans is right in looking to gravitational instability as the primary cause of fission, but when we consider that a binary system, with a mass larger than the sun's, is found to rotate in a few hours, there seems reason to look to rotation as a contributory cause scarcely less important than the primary one.

With the present extent of our knowledge it is only possible to reconstruct the processes of the evolution of stars by means of inferences drawn from several sources. We have first to rely on the general principles of stability, according to which we are to look for a series of families of forms, each terminating in an unstable form, which itself becomes the starting-point of the next family of stable forms. Secondly we have as a guide the analogy of the successive changes in the evolution of ideal liquid stars; and thirdly we already possess some slender knowledge as to the equilibrium of gaseous stars.

From these data it is possible to build up in outline the probable history of binary stars. Originally the star must have been single, it must have been widely diffused, and must have been endowed with a slow rotation. In this condition the strata of equal density must have been of the planetary form. As it cooled and contracted the symmetry round the axis of rotation must have become unstable, through the effects of gravitation, assisted perhaps by the increasing speed of rotation. (I learn from Professor Jeans that he now (December 1908) believes that he can prove that some small amount of rotation is necessary to induce instability in the symmetrical arrangement.) The strata of equal density must then become somewhat pear-shaped, and afterwards like an hour-glass, with the constriction more pronounced in the internal than in the external strata. The constrictions of the successive strata then begin to rupture from the inside progressively outwards, and when at length all are ruptured we have the twin stars portrayed by Roberts and by others.

As we have seen, the study of the forms of equilibrium of rotating liquid is almost complete, and Jeans has made a good beginning in the investigation of the equilibrium of gaseous stars, but much more remains to be discovered. The field for the mathematician is a wide one, and in proportion as the very arduous exploration of that field is attained so will our knowledge of the processes of cosmical evolution increase.

From the point of view of observation, improved methods in the use of the spectroscope and increase of accuracy in photometry will certainly lead to a great increase in our knowledge within the next few years. Probably the observational advance will be more rapid than that of theory, for we know how extraordinary has been the success attained within the last few years, and the theory is one of extreme difficulty; but the two ought to proceed together hand in hand. Human life is too short to permit us to watch the leisurely procedure of cosmical evolution, but the celestial museum contains so many exhibits that it may become possible, by the aid of theory, to piece together bit by bit the processes through which stars pass in the course of their evolution.

In the sketch which I have endeavoured to give of this fascinating subject, I have led my reader to the very confines of our present knowledge. It is not much more than a quarter of a century since this class of observation has claimed the close attention of astronomers; something considerable has been discovered already and there seems scarcely a discernible limit to what will be known in this field a century from now. Some of the results which I have set forth may then be shown to be false, but it seems profoundly improbable that we are being led astray by a Will-of-the-Wisp.

347

XXIX. THE EVOLUTION OF MATTER. By W.C.D. Whetham, M.A., F.R.S.

Trinity College, Cambridge.

The idea of evolution in the organic world, made intelligible by the work of Charles Darwin, has little in common with the recent conception of change in certain types of matter. The discovery that a process of disintegration may take place in some at least of the chemical atoms, previously believed to be indestructible and unalterable, has modified our view of the physical universe, even as Darwin's scheme of the mode of evolution changed the trend of thought concerning the organic world. Both conceptions have in common the idea of change throughout extended realms of space and time, and, therefore, it is perhaps not unfitting that some account of the most recent physical discoveries should be included in the present volume.

The earliest conception of the evolution of matter is found in the speculation of the Greeks. Leucippus and Democritus imagined unchanging eternal atoms, Heracleitus held that all things were in a continual state of flux—Panta rei.

But no one in the Ancient World—no one till quite modern times—could appreciate the strength of the position which the theory of the evolution of matter must carry before it wins the day. Vague speculation, even by the acute minds of philosophers, is of little use in physical science before experimental facts are available. The true problems at issue cannot even be formulated, much less solved, till the humble task of the observer and experimenter has given us a knowledge of the phenomena to be explained.

It was only through the atomic theory, at first apparently diametrically opposed to it, that the conception of evolution in the physical world was to gain an established place. For a century the atomic theory, when put into a modern form by Dalton, led farther and farther away from the idea of change in matter. The chemical elements seemed quite unalterable, and the atoms, of which each element in modern view is composed, bore to Clerk Maxwell, writing about 1870, "the stamp of manufactured articles" exactly similar in kind, unchanging, eternal.

Nevertheless throughout these years, on the whole so unfavourable to its existence, there persisted the idea of a common origin of the distinct kinds of matter known to chemists. Indeed, this idea of unity in substance in nature seems to accord with some innate desire or intimate structure of the human mind. As Mr Arthur Balfour well puts it, "There is no a priori reason that I know of for expecting that the material world should be a modification of a single medium, rather than a composite structure built out of sixty or seventy elementary substances, eternal and eternally different. Why then should we feel content with the first hypothesis and not with the second? Yet so it is. Men of science have always been restive under the multiplication of entities. They have eagerly watched for any sign that the different chemical elements own a common origin, and are all compounded out of some primordial substance. Nor, for my part, do I think that such instincts should be ignored... that they exist is certain; that they modify the indifferent impartiality of pure empiricism can hardly be denied." ("Report of the 74th Meeting of the British Association" (Presidential Address, Cambridge 1904), page 9, London, 1905.)

When Dalton's atomic theory had been in existence some half century, it was noted that certain numerical relations held good between the atomic weights of elements chemically similar to one another. Thus the weight (88) of an atom of strontium compared with that of hydrogen as unity, is about the mean of those of calcium (40) and barium (137). Such relations, in this and other chemical groups, were illustrated by Beguyer de Chancourtois in 1862 by the construction of a spiral diagram in which the atomic weights are placed in order round a cylinder and elements chemically similar are found to fall on vertical lines.

Newlands seems to have been the first to see the significance of such a diagram. In his "law of octaves," formulated in 1864, he advanced the hypothesis that, if arranged in order of rising atomic weight, the elements fell into groups, so that each eighth element was chemically similar. Stated thus, the law was too definite; no room was left for newly-discovered elements, and some dissimilar elements were perforce grouped together.

But in 1869 Mendeleeff developed Newland's hypothesis in a form that attracted at once general attention. Placing the elements in order of rising atomic weight, but leaving a gap where necessary to bring similar elements into vertical columns, he obtained a periodic table with natural vacancies to be filled as new elements were discovered, and with a certain amount of flexibility at the ends of the horizontal lines. From the position of the vacancies, the general chemical and physical properties of undiscovered elements could be predicted, and the success of such predictions gave a striking proof of the usefulness of Mendeleeff's generalisation.

When the chemical and physical properties of the elements were known to be periodic functions of their atomic weights, the idea of a common origin and common substance became much more credible. Differences in atomic weight and differences in properties alike might reasonably be explained by the differences in the amount of the primordial substance present in the various atoms; an atom of oxygen being supposed to be composed of sixteen times as much stuff as the atom of hydrogen, but to be made of the same ultimate material. Speculations about the mode of origin of the elements now began to appear, and put on a certain air of reality. Of these speculations perhaps the most detailed was that of Crookes, who imagined an initial chaos of a primordial medium he named protyle, and a process of periodic change in which the chemical elements successively were precipitated.

From another side too, suggestions were put forward by Sir Norman Lockyer and others that the differences in spectra observed in different classes of stars, and produced by different conditions in the laboratory, were to be explained by changes in the structure of the vibrating atoms.

The next step in advance gave a theoretical basis for the idea of a common structure of matter, and was taken in an unexpected direction. Clerk Maxwell's electromagnetic theory of light, accepted in England, was driven home to continental minds by the confirmatory experiments of Hertz, who in 1888 detected and measured the electromagnetic waves that Maxwell had described twenty years earlier. But, if light be an electromagnetic phenomenon, the light waves radiated by hot bodies must take their origin in the vibrations of electric systems. Hence within the atoms must exist electric charges capable of vibration. On these lines Lorentz and Larmor have developed an electronic theory of matter, which is imagined in its essence to be a conglomerate of electric charges, with electro-magnetic inertia to explain mechanical inertia. (Larmor, "Aether and Matter", Cambridge, 1900.) The movement of electric charges would be affected by a magnetic field, and hence the discovery by Zeeman that the spectral lines of sodium were doubled by a strong magnetic force gave confirmatory evidence to the theory of electrons.

Then came J.J. Thomson's great discovery of minute particles, much smaller than any chemical atom, forming a common constituent of many different kinds of matter. (Thomson, "Conduction of Electricity through Gases" (2nd edition), Cambridge, 1906.) If an electric discharge be passed between metallic terminals through a glass vessel containing air at very low pressure, it is found that rectilinear rays, known as cathode rays, proceed from the surface of the cathode or negative terminal. Where these rays strike solid objects, they give rise to the Rontgen rays now so well known; but it is with the cathode rays themselves that we are concerned. When they strike an insulated conductor, they impart to it a negative charge, and Thomson found that they were deflected from their path both by magnetic and electric forces in the direction in which negatively electrified particles would be deflected.

Cathode rays then were accepted as flights of negatively charged particles, moving with high velocities. The electric and magnetic deflections give two independent measurements which may be made on a cathode ray, and both the deflections involve theoretically three unknown quantities, the mass of the particles, their electric charge and their velocity. There is strong cumulative evidence that all such particles possess the same charge, which is identical with that associated with a univalent atom in electrolytic liquids. The number of unknown quantities was thus reduced to two—the mass and the velocity. The measurement of the magnetic and electric deflections gave two independent relations between the unknowns, which could therefore be determined. The velocities of the cathode ray particles were found to vary round a value about one-tenth that of light, but the mass was found always to be the same within the limits of error, whatever the nature of the terminals, of the residual gas in the vessel, and of the conditions of the experiment. The mass of a cathode ray particle, or corpuscle, as Thomson, adopting Newton's name, called it, is about the eight-hundredth part of the mass of a hydrogen atom.

These corpuscles, found in so many different kinds of substance, are inevitably regarded as a common constituent of matter. They are associated each with a unit of negative electricity. Now electricity in motion possesses electromagnetic energy, and produces effects like those of mechanical inertia. In other words, an electric charge possesses mass, and there is evidence to show that the effective mass of a corpuscle increases as its velocity approaches that of light in the way it would do if all its mass were electromagnetic. We are led therefore to regard the corpuscle from one aspect as a disembodied charge of electricity, and to identify it with the electron of Lorentz and Larmor.

Thus, on this theory, matter and electricity are identified; and a great simplification of our conception of the physical structure of Nature is reached. Moreover, from our present point of view, a common basis for matter suggests or implies a common origin, and a process of development possibly intelligible to our minds. The idea of the evolution of matter becomes much more probable.

The question of the nature and physical meaning of a corpuscle or electron remains for consideration. On the hypothesis of a universal luminiferous aether, Larmor has suggested a centre of aethereal strain "a place where the continuity of the medium has been broken and cemented together again (to use a crude but effective image) without accurately fitting the parts, so that there is a residual strain all round the place." (Larmor, loc. cit.) Thus he explains in quasi-mechanical terms the properties of an electron. But whether we remain content for the time with our identification of matter and electricity, or attempt to express both of them in terms of hypothetical aether, we have made a great step in advance on the view that matter is made up of chemical atoms fundamentally distinct and eternally isolated.

Such was the position when the phenomena of radio-activity threw a new light on the problem, and, for the first time in the history of science, gave definite experimental evidence of the transmutation of matter from one chemical element to another.

In 1896 H. Becquerel discovered that compounds of the metal uranium continually emitted rays capable of penetrating opaque screens and affecting photographic plates. Like cathode and Rontgen rays, the rays from uranium make the air through which they pass a conductor of electricity, and this property gives the most convenient method of detecting the rays and of measuring their intensity. An electroscope may be made of a strip of gold-leaf attached to an insulated brass plate and confined in a brass vessel with glass windows. When the gold-leaf is electrified, it is repelled from the similarly electrified brass plate, and the angle at which it stands out measures the electrification. Such a system, if well insulated, holds its charge for hours, the leakage of electricity through the air being very slow. But, if radio-active radiation reach the air within, the gold-leaf falls, and the rate of its fall, as watched through a microscope with a scale in the eye-piece, measures the intensity of the radiation. With some form of this simple instrument, or with the more complicated quadrant electrometer, most radio-active measurements have been made.

It was soon discovered that the activity of uranium compounds was proportional to the amount of uranium present in them. Thus radio-activity is an atomic property dependent on the amount of an element and independent of its state of chemical combination.

In a search for radio-activity in different minerals, M. and Mme Curie found a greater effect in pitch-blende than its contents of uranium warranted, and, led by the radio-active property alone, they succeeded, by a long series of chemical separations, in isolating compounds of a new and intensely radio-active substance which they named radium.

Radium resembles barium in its chemical properties, and is precipitated with barium in the ordinary course of chemical analysis. It is separated by a prolonged course of successive crystallisation, the chloride of radium being less soluble than that of barium, and therefore sooner separated from an evaporating solution. When isolated, radium chloride has a composition, which, on the assumption that one atom of metal combines with two of chlorine as in barium chloride, indicates that the relative weight of the atom of radium is about 225. As thus prepared, radium is a well-marked chemical element, forming a series of compounds analogous to those of barium and showing a characteristic line spectrum. But, unlike most other chemical elements, it is intensely radio-active, and produces effects some two million times greater than those of uranium.

In 1899 E. Rutherford, then of Montreal, discovered that the radiation from uranium, thorium and radium was complex. (Rutherford, "Radio-activity" (2nd edition), Cambridge, 1905.) Three types of rays were soon distinguished. The first, named by Rutherford alpha-rays, are absorbed by thin metal foil or a few centimetres of air. When examined by measurements of the deflections caused by magnetic and electric fields, the alpha-rays are found to behave as would positively electrified particles of the magnitude of helium atoms possessing a double ionic charge and travelling with a velocity about one-tenth that of light. The second or beta type of radiation is much more penetrating. It will pass through a considerable thickness of metallic foil, or many centimetres of air, and still affect photographic plates or discharge electroscopes. Magnetic and electric forces deflect beta-rays much more than alpha-rays, indicating that, although the speed is greater, approaching in some cases within five per cent. that of light, the mass is very much less. The beta-rays must be streams of particles, identical with those of cathode rays, possessing the minute mass of J.J. Thomson's corpuscle, some eight-hundredth part of that of a hydrogen atom. A third or gamma type of radiation was also detected. More penetrating even than beta-rays, the gamma-rays have never been deflected by any magnetic or electric force yet applied. Like Rontgen rays, it is probable that gamma-rays are wave-pulses in the luminiferous aether, though the possibility of explaining them as flights of non-electrified particles is before the minds of some physicists.

Still another kind of radiation has been discovered more recently by Thomson, who has found that in high vacua, rays become apparent which are absorbed at once by air at any ordinary pressure.

The emission of all these different types of radiation involves a continual drain of energy from the radio-active body. When M. and Mme Curie had prepared as much as a gramme of radium chloride, the energy of the radiation became apparent as an evolution of heat. The radium salt itself, and the case containing it, absorbed the major part of the radiation, and were thus maintained at a temperature measurably higher than that of the surroundings. The rate of thermal evolution was such that it appeared that one gramme of pure radium must emit about 100 gramme-calories of heat in an hour. This observation, naturally as it follows from the phenomena previously discovered, first called attention to the question of the source of the energy which maintains indefinitely and without apparent diminution the wonderful stream of radiation proceeding from a radio-active substance. In the solution of this problem lies the point of the present essay.

In order to appreciate the evidence which bears on the question we must now describe two other series of phenomena.

351

It is a remarkable fact that the intensity of the radiation from a radio-active body is independent of the external conditions of temperature, pressure, etc. which modify so profoundly almost all other physical and chemical processes. Exposure to the extreme cold of liquid air, or to the great heat of a furnace, leaves the radio-activity of a substance unchanged, apparent exceptions to this statement having been traced to secondary causes.

Then, it is found that radio-activity is always accompanied by some chemical change; a new substance always appears as the parent substance emits these radiations. Thus by chemical reactions it is possible to separate from uranium and thorium minute quantities of radio-active materials to which the names of uranium-X and thorium-X have been given. These bodies behave differently from their parents uranium and thorium, and show all the signs of distinct chemical individuality. They are strongly radio-active, while, after the separation, the parents uranium and thorium are found to have lost some of their radio-activity. If the X-substances be kept, their radio-activity decays, while that of the uranium or thorium from which they were obtained gradually rises to the initial value it had before the separation. At any moment, the sum of the radio-activity is constant, the activity lost by the product being equal to that gained by the parent substance. These phenomena are explained if we suppose that the X-product is slowly produced in the substance of the parent, and decays at a constant rate. Uranium, as usually seen, contains a certain amount of uranium-X, and its radio-activity consists of two parts—that of the uranium itself, and that of the X product. When the latter is separated by means of its chemical reactions, its radio-activity is separated also, and the rates of decay and recovery may be examined.

Radium and thorium, but not uranium, give rise to radio-active gases which have been called emanations. Rutherford has shown that their radio-activity, like that of the X products, suffers decay, while the walls of the vessel in which the emanation is confined, become themselves radio-active. If washed with certain acids, however, the walls lose their activity, which is transferred to the acid, and can be deposited by evaporation from it on to a solid surface. Here again it is clear that the emanation gives rise to a radio-active substance which clings to the walls of the vessel, and is soluble in certain liquids, but not in others.

We shall return to this point, and trace farther the history of the radio-active matter. At present we wish to emphasise the fact that, as in other cases, the radio-activity of the emanation is accompanied by the appearance of a new kind of substance with distinct chemical properties.

We are now in a position to consider as a whole the evidence on the question of the source of radio-active energy.

(1) Radio-activity is accompanied by the appearance of new chemical substances. The energy liberated is therefore probably due to the associated chemical change. (2) The activity of a series of compounds is found to accompany the presence of a radio-active element, the activity of each compound depends only on the contents of the element, and is independent of the nature of its combination. Thus radio-activity is a property of the element, and is not affected by its state of isolation or chemical combination. (3) The radio-activity of a simple transient product decays in a geometrical progression, the loss per second being proportional to the mass of substance still left at the moment, and independent of its state of concentration or dilution. This type of reaction is well known in chemistry to mark a mono-molecular change, where each molecule is dissociated or altered in structure independently. If two or more molecules were concerned simultaneously, the rate of reaction would depend on the nearness of the molecules to each other, that is, to the concentration of the material. (4) The amount of energy liberated by the change of a given mass of material far transcends the amount set free by any known ordinary chemical action. The activity of radium decays so slowly that it would not sink to half its initial value in less than some two thousand years, and yet one gramme of radium emits about 100 calories of heat during each hour of its existence.

The energy of radio-activity is due to chemical change, but clearly to no chemical change hitherto familiar to science. It is an atomic property, characteristic of a given element, and the atoms undergo the change individually, not by means of interaction among each other.

The conclusion is irresistible that we are dealing with a fundamental change in the structure of the individual atoms, which, one by one, are dissociating into simpler parts. We are watching the disintegration of the "atoms" of the chemist, hitherto believed indestructible and eternal, and measuring the liberation of some of the long-suspected store of internal atomic energy. We have stumbled on the transmutation dreamed by the alchemist, and discovered the process of a veritable evolution of matter.

The transmutation theory of radio-activity was formulated by Rutherford (Rutherford, "Radio-activity" (2nd edition), Cambridge, 1905, page 307.) and Soddy in 1903. By its light, all recent work on the subject has been guided; it has stood the supreme test of a hypothesis, and shown power to suggest new investigations and to co-ordinate and explain them, when carried out. We have summarised the evidence which led to the conception of the theory; we have now to consider the progress which has been made in tracing the successive disintegration of radio-active atoms.

Soon after the statement of the transmutation theory, a striking verification of one of its consequences appeared. The measurement of the magnetic and electric deflection of the alpha-rays suggested to Rutherford the idea that the stream of projectiles of which they consisted was a flight of helium atoms. Ramsay and Soddy, confining a minute bubble of radium emanation in a fine glass tube, were able to watch the development of the helium spectrum as, day by day, the emanation decayed. By isolating a very narrow pencil of alpha-rays, and watching through a microscope their impact on a fluorescent screen, Rutherford has lately counted the individual alpha-projectiles, and confirmed his original conclusion that their mass corresponded to that of helium atoms and their charge to double that on a univalent atom. ("Proc. Roy. Soc." A, page 141, 1908.) Still more recently, he has collected the alpha-particles shot through an extremely thin wall of glass, and demonstrated by direct spectroscopic evidence the presence of helium. ("Phil. Mag." February 1909.)

But the most thorough investigation of a radio-active pedigree is found in Rutherford's classical researches on the successive disintegration products of radium, in order to follow the evidence on which his results are founded, we must describe more fully the process of decay of the activity of a simple radio-active substance. The decay of activity of the body known as uranium-X is shown in a falling curve (Fig. 1.). It will be seen that, in each successive 22 days, the activity falls to half the value it possessed at the beginning.

This change in a geometrical progression is characteristic of simple radio-active processes, and can be expressed mathematically by a simple exponential formula.

As we have said above, solid bodies exposed to the emanations of radium or thorium become coated with a radio-active deposit. The rate of decay of this activity depends on the time of exposure to the emanation, and does not always show the usual simple type of curve. Thus the activity of a rod exposed to radium emanation for 1 minute decays in accordance with a curve (Fig. 2) which represents the activity as measured by the alpha-rays. If the electroscope be screened from the alpha-rays, it is found that the activity of the rod in beta-an gamma-rays increases for some 35 minutes and then diminishes (Fig. 3.).

These complicated relations have been explained satisfactorily and completely by Rutherford on the hypothesis of successive changes of the radio-active matter into one new body after another. (Rutherford, "Radio-activity" (2nd edition), Cambridge, 1905, page 379.) The experimental curve represents the resultant activity of all the matter present at a given moment, and the process of disentangling the component effects consists in finding a number of curves, which express the rise and fall of activity of each kind of matter as it is produced and decays, and, fitted together, give the curve of the experiments.

Other methods of investigation also are open. They have enabled Rutherford to complete the life-history of radium and its products, and to clear up doubtful points left by the analysis of the curves. By the removal of the emanation, the activity of radium itself has been shown to consist solely of alpha-rays. This removal can be effected by passing air through the solution of a radium salt. The emanation comes away, and the activity of the deposit which it leaves behind decays rapidly to a small fraction of its initial value. Again, some of the active deposits of the emanation are more volatile than others, and can be separated from them by the agency of heat.

From such evidence Rutherford has traced a long series of disintegration products of radium, all but the first of which exist in much too minute quantities to be detected otherwise than by their radio-activities. Moreover, two of these products are not themselves appreciably radio-active, though they are born from radio-active parents, and give rise to a series of radio-active descendants. Their presence is inferred from such evidence as the rise of beta and gamma radio-activity in the solid newly deposited by the emanation; this rise measuring the growth of the first radio-active offspring of one of the non-active bodies. Some of the radium products give out alpha-rays only, one beta- and gamma-rays, while one yields all three types of radiation. The pedigree of the radium family may be expressed in the following table, the time noted in the second column being the time required for a given quantity to be half transformed into its next derivative.

Time of half   Radio-   Properties   decay   activity
Radium   About 2600 years   alpha rays   Element chemically analogous
to barium.   Emanation   3.8 days   alpha rays   Chemically inert gas;
condenses at -150 deg C.   Radium-A   3 minutes   alpha rays   Behaves as a solid
deposited on   surfaces; concentrated on a
negative electrode.   Radium-B   21 minutes   no rays   Soluble in strong acids;
volatile at a white heat; more   volatile than A or C.
Radium-C   28 minutes   alpha, beta,   Soluble in strong acids; less
gamma rays   volatile than B.   Radium-D   about 40 years   no rays   Soluble in
strong acids; volatile   below 1000 deg C.   Radium-E
6 days   beta, gamma   Non-volatile at 1000 deg C.   rays
Radium-F   143 days   alpha rays   Volatile at 1000 deg C.
Deposited from solution on a   bismuth plate.

Of these products, A, B, and C constitute that part of the active deposit of the emanation which suffers rapid decay and nearly disappears in a few hours. Radium-D, continually producing its short-lived descendants E and F, remains for years on surfaces once exposed to the emanation, and makes delicate radio-active researches impossible in laboratories which have been contaminated by an escape of radium emanation.

A somewhat similar pedigree has been made out in the case of thorium. Here thorium-X is interposed between thorium and its short-lived emanation, which decays to half its initial quantity in 54 seconds. Two active deposits, thorium A and B, arise successively from the emanation. In uranium, we have the one obvious derivative uranium-X, and the question remains whether this one descent can be connected with any other individual or family. Uranium is long-lived, and emits only alpha-rays. Uranium-X decays to half value in 22 days, giving out beta- and gamma-rays. Since our evidence goes to show that radio-activity is generally accompanied by the production of new elements, it is natural to search for the substance of uranium-X in other forms, and perhaps under other names, rather than to surrender immediately our belief in the conservation of matter.

With this idea in mind we see at once the significance of the constitution of uranium minerals. Formed in the remote antiquity of past geological ages, these minerals must become store-houses of all the products of uranium except those which may have escaped as gases or possibly liquids. Even gases may be expected to some extent to be retained by occlusion. Among the contents of uranium minerals, then, we may look for the descendants of the parent uranium. If the descendants are permanent or more long-lived than uranium, they will accumulate continually. If they are short-lived, they will accumulate at a steady rate till enough is formed for the quantity disintegrating to be equal to the quantity developed. A state of mobile equilibrium will then be reached, and the amount of the product will remain constant. This constant amount of substance will depend only on the amount of uranium which is its source, and, for different minerals, if all the product is retained, the quantity of the product will be proportional to the quantity of uranium. In a series of analyses of

354

uranium minerals, therefore, we ought to be able to pick out its more short-lived descendants by seeking for instances of such proportionality.

Now radium itself is a constituent of uranium minerals, and two series of experiments by R.J. Strutt and B.B. Boltwood have shown that the content of radium, as measured by the radio-activity of the emanation, is directly proportional to the content of uranium. (Strutt, "Proc. Roy. Soc." A, February 1905; Boltwood, "Phil. Mag." April, 1905.) In Boltwood's investigation, some twenty minerals, with amounts of uranium varying from that in a specimen of uraninite with 74.65 per cent., to that in a monazite with 0.30 per cent., gave a ratio of uranium to radium, constant within about one part in ten.

The conclusion is irresistible that radium is a descendant of uranium, though whether uranium is its parent or a more remote ancestor requires further investigation by the radio-active genealogist. On the hypothesis of direct parentage, it is easy to calculate that the amount of radium produced in a month by a kilogramme of a uranium salt would be enough to be detected easily by the radio-activity of its emanation. The investigation has been attempted by several observers, and the results, especially those of a careful experiment of Boltwood, show that from purified uranium salts the growth of radium, if appreciable at all, is much less than would be found if the radium was the first product of change of the uranium. It is necessary, therefore, to look for one or more intermediate substances.

While working in 1899 with the uranium residues used by M. and Mme Curie for the preparation of radium, Debierne discovered and partially separated another radio-active element which he called actinium. It gives rise to an intermediate product actinium-X, which yields an emanation with the short half-life of 3.9 seconds. The emanation deposits two successive disintegration products actinium-A and actinium-B.

Evidence gradually accumulated that the amounts of actinium in radio-active minerals were, roughly at any rate, proportional to the amounts of uranium. This result pointed to a lineal connection between them, and led Boltwood to undertake a direct attack on the problem. Separating a quantity of actinium from a kilogramme of ore, Boltwood observed a growth of 8.5 x (10 to the power -9) gramme of radium in 193 days, agreeing with that indicated by theory within the limits of experimental error. ("American Journal of Science", December, 1906.) We may therefore insert provisionally actinium and its series of derivatives between uranium and radium in the radio-active pedigree.

Turning to the other end of the radium series we are led to ask what becomes of radium-F when in turn it disintegrates? What is the final non-active product of the series of changes we have traced from uranium through actinium and radium?

One such product has been indicated above. The alpha-ray particles appear to possess the mass of helium atoms, and the growth of helium has been detected by its spectrum in a tube of radium emanation. Moreover, helium is found occluded in most if not all radio-active minerals in amount which approaches, but never exceeds, the quantity suggested by theory. We may safely regard such helium as formed by the accumulation of alpha-ray particles given out by successive radio-active changes.

In considering the nature of the residue left after the expulsion of the five alpha-particles, and the consequent passage of radium to radium-F we are faced by the fact that lead is a general constituent of uranium minerals. Five alpha-particles, each of atomic weight 4, taken from the atomic weight (about 225) of radium gives 205—a number agreeing fairly well with the 207 of lead. Since lead is more permanent than uranium, it must steadily accumulate, no radio-active equilibrium will be reached, and the amount of lead will depend on the age of the mineral as well as on the quantity of uranium present in it. In primary minerals from the same locality, Boltwood has shown that the contents of lead are proportional to the amounts of uranium, while, accepting this theory, the age of minerals with a given content of uranium may be calculated from the amount of lead they contain. The results vary from 400 to 2000 million years. ("American Journal of Science", October, 1905, and February, 1907.)

We can now exhibit in tabular form the amazing pedigree of radio-active change shown by this one family of elements. An immediate descent is indicated by >, while one which may either be immediate or involve an intermediate step is shown by.... No place is found in this pedigree for thorium and its derivatives. They seem to form a separate and independent radio-active family.

	Atomic	Weight	Time	of	half		Radio-Activity
decay	Uranium	238.5			alpha	Uranium-X	?
22 days		beta, gamma	...	Actinium	?	?	no rays
Actinium-X	?	10.2 days	alpha (beta, gamma)			Actinium	
Emanation	?	3.9 seconds	alpha	Actinium-A	?		35.7
minutes	no rays	Actinium-B	?		2.15 minutes		alpha, beta,
gamma	...	Radium	225		about 2600 years	alpha	Radium
Emanation	?	3.8 days	alpha	Radium-A	?		3
minutes	alpha	Radium-B	?		21 minutes		no rays
Radium-C	?	28 minutes	alpha, beta, gamma			Radium-D	
?	about 40 years	no rays	Radium-E	?		6 days	beta
(gamma)	Radium-F	?		143 days	alpha	...	Lead
207	?	no rays					

As soon as the transmutation theory of radio-activity was accepted, it became natural to speculate about the intimate structure of the radio-active atoms, and the mode in which they broke up with the liberation of some of their store of internal energy. How could we imagine an atomic structure which would persist unchanged for long periods of time, and yet eventually spontaneously explode, as here an atom and there an atom reached a condition of instability?

The atomic theory of corpuscles or electrons fortunately was ready to be applied to this new problem. Of the resulting speculations the most detailed and suggestive is that of J.J. Thomson. ("Phil. Mag." March, 1904.) Thomson regards the atom as composed of a number of mutually repelling negative corpuscles or electrons held together by some central attractive force which he represents by supposing them immersed in a uniform sphere of positive electricity. Under the action of the two forces, the electrons space themselves in symmetrical patterns, which depend on the number of electrons. Three place themselves at the corner of an equilateral triangle, four at those of a square, and five form a pentagon. With six, however, the single ring becomes unstable, one corpuscle moves to the middle and five lie round it. But if we imagine the system rapidly to rotate, the centrifugal force would enable the six corpuscles to remain in a single ring. Thus internal kinetic energy would maintain a configuration which would become unstable as the energy drained away. Now in a system of electrons, electromagnetic radiation would result in a loss of energy, and at one point of instability we might well have a sudden spontaneous redistribution of the constituents, taking place with an explosive violence, and accompanied by the ejection of a corpuscle as a beta-ray, or of a large fragment of the atom as an alpha-ray.

The discovery of the new property of radio-activity in a small number of chemical elements led physicists to ask whether the property might not be found in other elements, though in a much less striking form. Are ordinary materials slightly radio-active? Does the feeble electric conductivity always observed in the air contained within the walls of an electroscope depend on ionizing radiations from the material of the walls themselves? The question is very difficult, owing to the wide distribution of slight traces of radium. Contact with radium emanation results in a deposit of the fatal radium-D, which in 40 years is but half removed. Is the "natural" leak of a brass electroscope due to an intrinsic radio-activity of brass, or to traces of a radio-active impurity on its surface? Long and laborious researches have succeeded in establishing the existence of slight intrinsic radio-activity in a few metals such as potassium, and have left the wider problem still unsolved.

It should be noted, however, that, even if ordinary elements are not radio-active, they may still be undergoing spontaneous disintegration. The detection of ray-less changes by Rutherford, when those changes are interposed between two radio-active transformations which can be followed, show that spontaneous transmutation is possible without measureable radio-activity. And, indeed, any theory of disintegration, such as Thomson's corpuscular hypothesis, would suggest that atomic rearrangements are of much more general occurrence than would be apparent to one who could observe them only by the effect of the projectiles, which, in special cases, owing to some peculiarity of atomic configuration, happened to be shot out with the enormous velocity needed to ionize the surrounding gas. No evidence for such ray-less changes in ordinary elements is yet known, perhaps none may ever be obtained; but the possibility should not be forgotten.

In the strict sense of the word, the process of atomic disintegration revealed to us by the new science of radio-activity can hardly be called evolution. In each case radio-active change involves the breaking up of a heavier, more complex atom into lighter and simpler fragments. Are we to regard this process as characteristic of the tendencies in accord with which the universe has reached its present state, and is passing to its unknown future? Or have we chanced upon an eddy in a backwater, opposed to the main stream of advance? In the chaos from which the present universe developed, was matter composed of large highly complex atoms, which have formed the simpler elements by radio-active or ray-less disintegration? Or did the primaeval substance consist of isolated electrons, which have slowly come together to form the elements, and yet have left here and there an anomaly such as that illustrated by the unstable family of uranium and radium, or by some such course are returning to their state of primaeval simplicity?

www.ingramcontent.com/pod-product-compliance
Lightning Source LLC
Chambersburg PA
CBHW070850180526
45168CB00005B/1755